DYNAMICS OF HUMAN REPRODUCTION

FOUNDATIONS OF HUMAN BEHAVIOR
An Aldine de Gruyter Series of Texts and Monographs

SERIES EDITOR
Sarah Blaffer Hrdy
University of California, Davis

Richard D. Alexander, **The Biology of Moral Systems**

Laura L. Betzig, **Despotism and Differential Reproduction: A Darwinian View of History**

Russell L. Ciochon and John G. Heagle (Eds.), **Primate Evolution and Human Origins**

Martin Daly and Margo Wilson, **Homicide**

Irenäus Eibl-Fibesfeldt, **Human Ethology**

Richard J. Gelles and Jane B. Lancaster (Eds.), **Child Abuse and Neglect: Biosocial Dimensions**

Kathleen R. Gibson and Anne C. Petersen (Eds.), **Brain Maturation and Cognitive Development: Comparative and Cross-Cultural Perspectives**

Barry S. Hewlett (Ed.), **Father-Child Relations: Cultural and Biosocial Contexts**

Warren G. Kinzey (Ed.), **New World Primates: Ecology, Evolution and Behavior**

Frederick F. Grine (Ed.), **Evolutionary History of the "Robust" Australopithecines**

Jane B. Lancaster, Jeanne Altmann, Alice S. Rossi, and Lonnie R. Sherrod (Eds.), **Parenting Across the Life Span: Biosocial Dimensions**

Jane B. Lancaster and Beatrix A. Hamburg (Eds.), **School Age Pregnancy and Parenthood: Biosocial Dimensions**

Jonathan Marks, **Human Biodiversity: Genes, Race, and History**

Richard B. Potts, **Early Hominid Activities at Olduvai**

Eric Alden Smith, **Inujjuamiut Foraging Strategies**

Eric Alden Smith and Bruce Winterhalder (Eds.), **Evolutionary Ecology and Human Behavior**

Patricia Stuart-Macadam and Katherine Dettwyler, **Breastfeeding: A Biocultural Perspective**

Patricia Stuart-Macadam and Susan Kent (Eds.), **Diet, Demography, and Disease: Changing Perspectives on Anemia**

Wenda R. Trevathan, **Human Birth: An Evolutionary Perspective**

James W. Wood, **Dynamics of Human Reproduction: Biology, Biometry, Demography**

DYNAMICS OF HUMAN REPRODUCTION
Biology, Biometry, Demography

James W. Wood

with an Appendix by Kenneth L. Campbell and James W. Wood

ALDINE DE GRUYTER
New York

About the Author

James W. Wood is Professor, Department of Anthropology, and Senior Scientist, Population Research Institute, at the Pennsylvania State University. He was previously affiliated with the Universities of Michigan and Wisconsin. Dr. Wood's research on anthropological demography and population biology has been widely published in books and professional journals.

ALDINE DE GRUYTER
A division of Walter de Gruyter, Inc.
200 Saw Mill River Road
Hawthorne, New York 10532

This publication is printed on acid-free paper

Library of Congress Cataloging-in-Publication Data
Wood, James W., 1949–
   Dynamics of human reproduction : biology, biometry, demography /
James W. Wood.
      p.   cm. — (Foundations of human behavior)
   Includes bibliographical references and index.
   ISBN 0–202–01179–8. –– ISBN 0–202–01180–1 (pbk.)
   1. Birth intervals.   2. Fertility, Human.   I. Title.   II. Series.
HB902.W66   1994
304.6'32—dc20   93-36613
                                                                            CIP

Manufactured in the United States of America

10 9 8 7 6 5 4 3 2 1

To Pat
For Good Reason

Contents

PART III
BEYOND THE PROXIMATE DETERMINANTS

Preface

I was ever of opinion that the honest man who married and brought up a large family did more service than he who continued single and only talked of population.

Oliver Goldsmith, *The Vicar of Wakefield* (1766)

Humans are sometimes viewed as if poised between two worlds, the biological and the cultural. In reality this dualism is false. Culture is an integral part of human biology and can no more be separated from its physiological base than software can be run without hardware. By the same token, it would be absurd to think of biology as nothing more than a residual part of human experience, left over by partitioning out the effects of culture. The two are inseparable.

Nowhere should this be more obvious than in the study of human reproduction. And yet that area of research has developed largely as two separate disciplines, one a branch of the social sciences (demography) and the other purely biological and clinical (reproductive physiology). Until recently, there has been virtually no overlap or communication between the two fields. This disjunction has created shortcomings on both sides. The demographic approach, with its emphasis on aggregate phenomena and social process, has failed to appreciate the possibility that individuals and indeed populations may differ with respect to the physiological determinants of fertility. All individuals, in this view, share essentially the same set of biological "susceptibilities" (with due allowance for age and sex), so that all the *important* variation in human fertility is attributable solely to behavioral or social differences. This view, I believe, is one that most reproductive biologists—at least those with any field experience—would reject. The physiological approach, on the other hand, is empirically rich but theoretically impoverished, and it is especially poor at making precise numerical predictions about the timing of births—which, after all, is what the whole area of study is ultimately about.

In recent years, a third approach to the investigation of human reproduction has emerged, one I have chosen to call the "biometric" approach. Biometry is the application of modern statistical methods to biological processes occurring in living organisms (Snedecor, 1954).[1] The biometric approach to human reproduction has its roots in the pioneering work of Raymond Pearl and, more recently, Louis Henry and Mindel Sheps. The biometric approach attempts to link the individual and aggregate levels of analysis, and it encompasses all the major determinants of fertility, both behavioral and biological. As such, it holds forth the promise of important new advances in our understanding of human reproduction. To date, however, this approach has been adopted almost exclusively by researchers with social science backgrounds. Not unnaturally, those researchers have been reluctant to make direct use of biological data, preferring instead to draw biological inferences indirectly from the sort of survey, census, and vital registration data familiar to social scientists. This has given the approach a parochial cast that has lessened the likelihood of meaningful dialogue with biologists. As a consequence, the biometric approach has not yet had the impact on reproductive biology that it deserves. Indeed, reproductive biologists and clinicians scarcely know it exists.

This book is an attempt to weave together the physiological, demographic, and biometric approaches to human fertility in a way that will encourage future interdisciplinary research. The book is not intended as a basic textbook in reproductive biology, much less in demography or biometry; the coverage of each of these disciplines is far too selective. Rather, the book aims at answering a single question: why does fertility, the number of live births, vary from couple to couple within any particular population, and from population to population across the human species as a whole? To answer this question will require more than just a catalog of known facts about human reproduction; it will require the development of a coherent theoretical and analytical framework within which to make sense of those facts. The particular framework I have chosen for this book is structured in terms of birth interval components and the timing of reproductive events—the *pace of childbearing*, as it is often called. The task at hand is to learn how the proximate determinants of fertility (in the sense used by Kingsley Davis and Judith Blake, and later by John Bongaarts and others) affect the pace of childbearing and thereby influence the total number of offspring produced over the course of an individual's reproductive life.

Throughout the book, emphasis will be placed on so-called *natural-fertility populations*—loosely, populations in which couples do not deliberately attempt to limit their reproduction using effective methods of contraception or induced abortion in order to achieve a desired family size. In view of the profound importance of deliberate birth control in our own society, this emphasis may seem contrived. Following Louis Henry, however, I will argue that the interaction of biology and behavior in determining the timing of reproduction is far easier to study in natural-fertility settings, since underlying physiological processes are not obscured by family planning practices. Moreover, the study of natural-fertility populations has enormous relevance for understanding the impact of contraception and induced abortion in contemporary societies. Such populations provide a yardstick against which to gauge what our own fertility might be in the absence of artificial controls.

The concept of natural fertility is also of great importance from an evolutionary and comparative point of view. Most of human evolution has been spent under a regime of natural fertility, and our nearest primate relatives still live under such a regime (at least outside the laboratory and zoo). It makes little sense to seek evolutionary continuities in populations whose reproductive patterns are completely dominated by contraception and abortion.

One feature of this book that may disappoint some readers is that the material is presented in a comparatively decontextualized way—that is, with little attention to the social and environmental context within which human reproduction takes place. This "failure" reflects a conscious decision on my part. Apart from not wishing to lengthen an already long book, my decision was prompted by three considerations. First, research on contextual factors, while very active at present, is also recent, and it seems premature to attempt to summarize it. Second, at the current stage of research I deem it of special importance to present a consensus view (insofar as one exists) of *how the basic reproductive system operates in time*. Third, it is my belief that the best way to proceed in studying contextual factors is to make use of an analytical framework couched in terms of the basic features of the reproductive process. In other words, to understand more remote influences on fertility, we must investigate their effects on the proximate determinants. None of this is to say that contextual factors are uninteresting or unimportant. On the contrary, a primary motive for studying the proximate determinants is to use the knowledge gained from that study to further our understanding of contextual factors. But we need to understand the proximate determinants first.

This book is divided into three major sections. In the first three chapters, the concept of natural fertility is examined in detail, empirical patterns of variation in natural fertility are discussed, and most important, a framework is presented for analyzing the proximate behavioral and biological determinants of natural fertility. In the second section, consisting of Chapters 4–11, the major proximate determinants are investigated in detail. Throughout that section, each determinant is considered more or less in isolation from all the others, admittedly a somewhat artificial way to view the reproductive process. To correct this view, the final chapter discusses the joint effects of the proximate determinants. The purpose of that chapter is integrative: it attempts to show how the various determinants act jointly and in interaction to produce a particular, observed level of reproductive output. (The last chapter will also have some preliminary things to say about a few of the contextual factors that I find especially interesting.) The analytical framework developed at the beginning of the book will be applied in consistent fashion throughout so that, by the end of the book, the reader should have a clear sense of how variation in natural fertility is caused by the same set of factors operating differentially in differing populations.

This book is intended to be sufficiently comprehensive and self-contained for use either as a textbook for graduate students or as a reference work for researchers in demography, reproductive biology, anthropology, and related fields. Because special attention is given to the quantitative effects of the proximate determinants on fertility differentials, a certain amount of mathematical modeling and statistical analysis is unavoidable. However, the mathematical treatment has been kept as informal, intuitive, and heuristic as possible, and it should be accessible to most readers. (I will not be offended to learn that some readers choose to skim the more mathematical sections.) Those who seek a more formal treatment should consult *Mathematical Models of Conception and Birth* by Mindel Sheps and Jane Menken (University of Chicago Press, Chicago, 1973). Readers who wish more substantive detail on modern reproductive biology can turn to the magnificent two-volume compilation on *The Physiology of Reproduction* (second edition), edited by Ernst Knobil and his associates (Raven Press, New York, 1994), or *Reproductive Endocrinology*, edited by Samuel Yen and Robert Jaffe (W. B. Saunders, Philadelphia, 1991); those who want to learn more about the methodology of reproductive biology should consult *An Introduction to Radioimmunoassay and Related Techniques* by T. Chard (Elsevier Science, Amsterdam, 1987).

An unusual feature of this book is its heavy reliance on graphic materials to convey basic information. The book contains more than 200 figures, each an integral part of the presentation. Although the figures provide an extremely efficient way to summarize large amounts of information, they represent a substantial cost to the publisher. The current number of figures, which is about 70 percent of what was originally planned, represents an unequal compromise between more (the theoretically desirable) and fewer (the financially feasible). We were able to come down in favor of *more* figures by "ganging" them at the end of each chapter, thus reducing production costs considerably. Although this strategy may make the book less convenient for the reader—for which I apologize—it has allowed us to retain most of the figures while keeping the price of the book affordable. For the use of these figures, I want to express my gratitude to the many authors, volume editors, and copyright holders who granted me permission to redraw and otherwise adapt their materials to my own purpose here. In particular, I want to thank Dr. P.M. Wassarman of the Roche Institute of Molecular Biology, whose name will be credited several times over in the middle chapters; and among publishers, the American Association for the Advancement of Science, publisher of *Science;* Blackie & Son, Ltd.; Rockefeller University Press, publisher of the *Journal of Cell Biology;* W.B. Saunders Co.; and *Scientific American.*

Most authors of book-length manuscripts incur many intellectual debts. In a work such as this, which attempts to integrate material from so many different fields, the author is especially dependent upon informed readers, and writing this book has made me painfully aware of my own limitations. I wish to express my profound gratitude to Gillian Bentley, Kenneth Campbell, Nancy Gonlin, Darryl Holman, Patricia Johnson, Cheryl Stroud, and Maxine Weinstein, each of whom read either the entire manuscript or those chapters most closely related to his or her own specialty. To all of them I am grateful, not only for their improvements on my prose, but also for saving me from some horrendous gaffes. And to Gillian Bentley and Kenneth Campbell special thanks are due, to the former for making substantive contributions to the material on male reproductive senescence in Chapter 11, and to the latter for co-authoring the Appendix. Since this book originated from materials I prepared for a graduate course on the Biometry of Human Reproduction, I also owe a debt to the many students who have deepened and extended my own thinking over the years. Although the students involved are too numerous for an exhaustive list, I would like to single out Darryl Holman, Bettina Shell, Beverly Strassman, and David Tracer for their special contributions. That they have gone on to launch their own professional careers in closely related fields of study, I find deeply gratifying. I am indebted to two successive editors at Aldine de Gruyter, Trev Leger and Richard Koffler, the first of whom saw prospects for the book when it was little more than a mass of notes, and the second of whom nurtured, cajoled, and blustered until that prospect became a finished manuscript. June-el Piper did a magnificent job of copy-editing the manuscript for Aldine, clarifying numerous ambiguities and making sure my penchant for dangling participles did not become a public spectacle. Arlene Perazzini, Managing Editor at Aldine de Gruyter, has been unfailingly helpful throughout the production phase of this project. I am also indebted to Brian Baker, Lara Glembock, and Loukas Kalisperis of the Stuckeman Cad Lab in Penn State's

Department of Architecture for transforming my crude drawings into publishable figures. And I want to give particular thanks to Kenneth Campbell, Patricia Johnson, and Maxine Weinstein, whose continuing collaboration with me over the years has truly made this book possible. Ken, Max, and I all owe a special debt of gratitude to the Andrew W. Mellon Foundation, which, by funding a postdoctoral research program at the University of Michigan on the biological and demographic determinants of human reproduction, brought us all together. Any errors in this book should therefore be blamed on the late Andrew W. Mellon. Carolyn Makinson, who directed the Mellon program, deserves the thanks of everyone who now toils in this field for fostering it at an early, critical stage of its history when many funding agencies found it difficult to see past traditional disciplinary boundaries. Finally, I would like to thank my wife and children for their forbearance during the long struggle to write this book, but I don't have a wife or any children. My apologies to Oliver Goldsmith.

NOTE

1. The terms "biometry" and "biometrics" also have more restricted technical meanings in the fields of quantitative genetics and actuarial sciences. The general sense in which I use them here dates back at least to Galton (1889), who was one of the first persons to use these terms.

PART I

Natural Fertility

1

Introduction

This book is about the basic mechanisms underlying observed patterns of human reproduction. A single question is addressed throughout: Why do individuals (or couples) have one child, two children, eight children, or whatever, over the course of their lives? In other words, this book asks how the levels of fertility observed in human populations are achieved. *Fertility* is defined by demographers as the production of a live birth, that is, a child born alive (Pressat, 1985). As such, it is to be distinguished from *fecundity*, which is defined as the biological capacity to reproduce. Fertility has the advantage of being marked by a more or less unambiguous, countable event, a birth. This fact has made it the focus of study by demographers for many decades. Fecundity, in contrast, is a theoretical potential and therefore inherently more difficult to measure. Clearly, however, fecundity is crucial to fertility: fecundity interacts with various behavioral processes—at the very least, the frequency and timing of sexual intercourse, and in some cases, deliberate family planning—to determine the level of fertility. Unfortunately, the inherent "fuzziness" of the concept of fecundity has made it inaccessible to empirical analysis. One aim of this book is to decompose fecundity into its major physiological components, which are observable, at least in principle.

The primary focus of the book is fertility *variation* and its causes. Variation will be considered along three dimensions: within individuals over the reproductive life course, among individuals within the same population, and among different human populations. Chapter 2 presents some of the basic empirical findings about the extent of such variation;

3

the rest of the book is devoted to accounting for those findings. Chapter 3 argues that the explanatory approach most appropriate for this task should satisfy two requirements: (1) it should be framed in terms of a small set of basic and universal mechanisms known in demography as the *proximate determinants of fertility* (Bongaarts, 1978) and (2) it should focus upon the *timing* of reproductive events across the individual life course. Temporal aspects of reproduction are of fundamental significance from several points of view. From a purely biological perspective, individual fitness is the summation of reproductive events over the life course as a whole, discounted for any prereproductive deaths among the resulting offspring. Thus, to understand how the human reproductive process evolved, we need to know how it operates in real time. Second, for the demographer, the most powerful models for investigating the effects of fertility on the behavior of populations, so-called stable population models, are themselves dynamic (Coale, 1972; Keyfitz, 1985). For investigators interested primarily in informed policy formulation, the timing of births forms the fundamental framework within which couples make decisions about future reproduction and family planning.

Paradoxically, the reproductive patterns characteristic of contemporary industrialized societies, about which we know a great deal, tell us remarkably little about how the reproductive system functions in time. This is true for a simple reason: the pace of childbearing in such societies is overwhelmingly dominated by modern contraception and induced abortion. For illustration, we can look to the contemporary United States. The average number of live births experienced by U.S. women during their reproductive lives has declined in recent years to about 1.9 (Mosher and Pratt, 1990b). Since this number is below the absolute minimum (2+) needed to replace each couple and thus maintain total population size, it cannot be a long-term characteristic of the human species as a whole. In fact, it is a quite recent achievement reflecting widespread and highly efficient use of modern contraceptives and induced abortion. And the trend toward lower fertility is expected to continue: approximately 80 percent of women in the United States today expect to have three or fewer children during their lifetimes, and almost 10 percent plan to have no children at all (Pratt et al., 1984).

Not only has the total number of offspring declined among U.S. women, there has been a dramatic shift in the timing of reproduction across the female reproductive span.[1] Increasingly, reproduction is being concentrated within a narrow "window," reflecting an increase in the age at which childbearing begins and a simultaneous decrease in the age at which childbearing ends. Between 1970 and 1982, the percentage of first births occurring in women 25 or older nearly doubled, increasing from 19 percent to 36 percent (Baldwin and Nord, 1984). Consistent with that change, the mean age at first birth for U.S. women born in 1940 was 24.7 years; for women born in 1956, it is projected to be between 27.1 and 27.5 (Evans, 1986). The reproductive window has been closing from the other end as well: births to women 35 and above declined from 11 percent of all births in 1950 to 5 percent in 1980 (Baldwin and Nord, 1984). This change reflects the fact that U.S. women have become remarkably adept at terminating reproduction. At present, approximately 23 percent of all married women of childbearing age in the United States have been surgically sterilized, most of them through tubal ligation (Mosher and Pratt, 1990a).

The pattern of reproduction now typical of the United States and other industrialized nations stands in marked contrast to that observed in much of the rural developing world, especially where effective family planning programs have not been implemented. For example, rural women in the Indian state of Punjab, studied by Harvard epidemiologists from 1953 to 1960, could expect to have six or seven children over their reproductive spans, if they were fortunate enough to survive to menopause (Wyon and Gordon, 1971). The average age at which these women experienced their first births was approximately 17 years; thereafter they could expect subsequent births to follow at roughly 32-month intervals up to about age 40 or so. Fewer than 2 percent of Punjabi women attained age 45 without having had at least one child (Wyon and Gordon, 1971).

As subsequent chapters will show, not all women in all preindustrial societies have had reproductive patterns identical to those observed in the Punjab study. Nonetheless, rural Punjabi women are far more typical of the patterns that must have prevailed over vast stretches of human history and prehistory than are contemporary U.S. women. We stand to learn more about how the human reproductive system is "designed" to function in time by studying Punjabi women—and women in other preindustrial societies—than by studying modern American women. Consequently, this book will have little more to say about reproduction in populations where modern methods of birth control are widely available, except when data from such populations shed light upon the reproductive process as it operates in preindustrial settings.

The focus on preindustrial societies can be justified on several grounds. Generalizations about the level of fertility in such societies have played an important role in the development of theoretical models in demography and anthropology, and more recently in reproductive biology. In classic demographic transition theory, for example, it was often assumed that "pre-transition" societies were characterized by uniformly high fertility rates, which provided the starting point for the

recent secular decline in fertility (Coale, 1986). Several ecological anthropologists, in contrast, have suggested that preindustrial societies, especially unacculturated hunter-gatherers, tend to regulate their reproductive output at a relatively low level (Birdsell, 1968; Dumond, 1975;
Harris and Ross, 1987; Cohen, 1989). It has even been claimed that an
earlier, "stone-age" demographic transition toward *higher* birth and
death rates was associated with the emergence of agriculture and settled
village life during the early Neolithic period (Handwerker, 1983; Roth,
1985). Recently, several reproductive physiologists have suggested that
birth-spacing patterns in preindustrial societies may shed light on the
biology of reproduction in the human species as a whole (Short, 1976;
Howie and McNeilly, 1982), a prospect first raised by Louis Henry 40
years ago (Henry, 1953a). Although all these theoretical concerns are relevant to the present book, this last theme will be explored in the greatest detail.

NATURAL FERTILITY

Since the remainder of this book will be concerned primarily with
what demographers call *natural fertility*, it is important to examine this
concept in detail at the outset. Although the phrase "natural fertility"
was used by Raymond Pearl at least as early as 1939, its technical definition was supplied by Louis Henry (1953b, 1961). According to Henry,
natural fertility is

> fertility which exists or has existed in the absence of deliberate birth con-
> trol. The adjective "natural" is admittedly not ideal but we prefer it to
> "physiological" since the factors affecting natural fertility are not solely
> physiological. Social factors may also play a part—sexual taboos, for
> example, during lactation. Some of these factors may result in a reduction
> of fertility but this cannot be considered a form of birth control. *Control
> may be said to exist when the behavior of the couple is bound to the number of
> children already born and is modified when this number reaches the maximum
> which the couple does not want to exceed.* It is not the case [that control exists]
> for a taboo concerning lactation, which is independent of the number of
> children already born (Henry, 1961:81; emphasis added).

Several aspects of this definition are worthy of note. First, natural fertility implies the absence of *deliberate* control. Individuals and couples do
any number of things that limit their reproductive output without
intending to do so, and such behaviors are not inconsistent with natur-

al fertility. Control must also be *parity-dependent*—that is, it must be "bound to the number of children already born." (In demography, parity refers to the number of live-born offspring an individual has produced by a given time, usually the time of interview.) Thus, for control to exist, couples must have a prior notion of how many children represent a desirable family size, and they must modify their reproductive behavior as they approach or attain that number. For demographers, the parity dependence of fertility control is its salient characteristic.

As Henry emphasizes, natural fertility is not determined solely by physiological factors. Behavior, customary practices, and social institutions still have an influence on natural fertility—they simply do not act in parity-dependent fashion, nor are they generally intended to limit fertility. To illustrate this point, Henry uses the example of what is now called *postpartum abstinence*, a cultural proscription against sexual intercourse before the most recently born child has been weaned. Rules of postpartum abstinence are quite widespread among natural-fertility populations, especially in parts of West Africa (Lesthaeghe et al., 1981). (How often these rules are obeyed is a separate issue.) Although such rules undoubtedly act to limit fertility in at least some settings, two considerations suggest that their presence is generally consistent with natural fertility. First, they do not seem to be practiced with the intention of limiting fertility; rather, the preponderance of ethnographic evidence suggests that they are practiced to safeguard the health of the mother and her nursing child (Singarimbun and Manning, 1976; Cantrelle and Ferry, 1979; Gray, 1981; van de Walle and van de Walle, 1993). Second, to my knowledge there is no evidence that postpartum taboos ever operate in parity-dependent fashion, that is, that they change in efficacy as parity increases.

Henry (1953b) originally used the concept of natural fertility as an ideal type to serve as a basis for formulating mathematical models of the reproductive process. Only later did he begin to write as if populations living under a regime of natural fertility actually exist (Henry, 1961). Whether natural-fertility populations exist or have ever existed is a difficult question to answer for two reasons. First, the concept depends upon the intentions of the individual actors (controlled fertility implies deliberate control), which may be difficult to discern. Insight into intentions is especially difficult when researchers are dealing with historical data, but as cultural anthropologists are painfully aware, even the word of living informants is not proof against misstatements of intention. Second, natural fertility involves a logical negative (the *absence* of parity-dependent changes in behavior) and therefore may be unprovable in any given case.

Statistical methods have been developed for inferring parity-dependent changes in reproductive behavior from age patterns of births (e.g., Coale and Trussell, 1974, 1978; Broström, 1983; David et al., 1988), but these methods may be misleading for several reasons. First, the failure of such methods to show parity dependence may reflect a lack of statistical power (i.e., samples are too small or variation too great) rather than a genuine absence of parity dependence. In addition, nondeliberate (even nonbehavioral) factors may exhibit parity dependence. For example, there is a small but real risk of pelvic inflammatory disease and subsequent sterility as a result of *puerperal sepsis,* or infection of the female genital tract during childbirth (see Chapter 10). Although the cumulative incidence of such infection increases with parity (Trussell and Wilson, 1985), no one would call this a method of deliberate fertility control. To take another example, some societies have rules of *terminal abstinence,* which proscribe sexual intercourse once a woman has attained a certain age or life status, such as grandmotherhood. Like postpartum abstinence, such rules need not be inconsistent with natural fertility. However, the probability of becoming a grandmother is clearly parity-dependent: the more children you have, the more children's children you are likely to have. From a purely statistical point of view, it may be impossible to distinguish such inadvertent parity dependence from deliberate fertility control.

Terminal abstinence is an example of what demographers call *stopping behavior* (behavior that permanently terminates reproduction), whereas postpartum abstinence is an example of *spacing behavior* (behavior that increases the time between successive births). As Knodel (1983) has emphasized, Henry's definition of controlled fertility implies that stopping behavior is the dominant mode of fertility control where such control exists.[2] Indeed, according to Henry, control exists precisely "when the behavior of the couple is . . . modified when [the number of children already born] reaches the maximum which the couple does not want to exceed." However, as the example of terminal abstinence shows, stopping behavior by itself is not necessarily synonymous with controlled fertility. Moreover, it is possible in principle that spacing behavior could be used as a means of deliberate control. The relationship of spacing and stopping behavior to Henry's definition of natural and controlled fertility is shown in Table 1.1. As this table indicates, the existence of deliberate spacing behavior intended to limit family size in the *absence* of deliberate, parity-dependent stopping behavior may or may not imply fertility control under Henry's definition. My own inclination is to consider such spacing behavior as implying control, especially if the degree of deliberate spacing increases with parity.

Table 1.1. Natural and Controlled Fertility as Defined by the Occurrence of Deliberate Spacing and Stopping Behavior

Deliberate spacing behavior	Deliberate Stopping Behavior		
	absent	without intent to limit family size	with intent to limit family size
absent	natural fertility	natural fertility	controlled fertility
without intent to limit family size	natural fertility	natural fertility	controlled fertility
with intent to limit family size	ambiguous	ambiguous	controlled fertility

Modified from Knodel (1983)

In view of all these uncertainties and ambiguities, what can we say about the distribution of natural fertility among human populations? How widespread has natural fertility been over human history and pre-history? The consensus among demographers interested in the fertility patterns of the past seems to be that natural fertility was virtually universal before the modern, sustained decline in fertility that first began in western Europe during the late eighteenth and early nineteenth centuries (Leridon, 1977; Coale, 1986; Knodel and van de Walle, 1986; Wilson et al., 1988; Yu, 1991). Although "social group forerunners" of fertility control may have existed in Europe before that time, e.g., among the Genevan bourgeoisie or the Milanese aristocracy in the late seventeenth to early eighteenth centuries (Henry, 1956; Livi-Bacci, 1986), they were few in number and thinly scattered. The ethnographic evidence on contemporary preindustrial societies tends to support this generalization (Caldwell and Caldwell, 1981). Although deliberate birth-spacing practices are very widespread in the ethnographic literature, the evidence suggests, as already noted, that they are intended to protect the mother and already-born child, not to regulate fertility. Deliberate stopping behaviors, in contrast, are evidently much rarer in preindustrial societies and, as far as we know, are never parity-dependent.

The conclusion that most preindustrial societies existed in a state of natural fertility runs counter to a widespread, often deeply held conviction among anthropologists that effective behavioral control of fertility was a common feature of such societies, especially among hunter-gatherers (Birdsell, 1968; Harris and Ross, 1987; Cohen, 1989; cf. Early, 1985).

Such views seem to be based, not on firm demographic evidence, but rather on a belief that humans tend to generate enormous "excess" fertility that, if unchecked by artificial means, would cause ecologically ruinous population growth. From the vantage point of comparative zoology, however, humans appear to be remarkably infecund creatures, as will be argued in more detail below. Moreover, recent demographic studies have identified a widespread phenomenon that may tend to enforce a long-term balance between births and early childhood deaths in preindustrial populations: High fertility (especially that associated with short birth intervals) actually appears to induce higher offspring mortality without any deliberate intervention on the part of the parents (see Chapter 12). This effect seems to be attributable primarily to competition among offspring for breast milk, which is critical to young children in preindustrial settings for both nutritional and immunological reasons (Chapter 8). The modern period of explosive population growth, which is indeed proving to be ecologically disastrous in many parts of the world, has occurred not only because of the persistence of high fertility in many regions, but more particularly because of changes in medical services or public health programs that have radically altered preexisting patterns of infant and childhood mortality. The high levels of fertility observed in undisturbed preindustrial populations have not, for the most part, been associated with rapid population growth because they are countered by high (and highly correlated) mortality.

In the end, the question of whether most preindustrial societies exhibit natural fertility is an empirical one. But certain minimal conditions must be met before fertility control can be said to exist, and these require attention in empirical studies. As Ansley Coale once noted, effective fertility control reaches high prevalences within a population only when a large fraction of couples (1) believes that control is possible, (2) believes that control is morally acceptable, (3) believes that control is desirable, and (4) gains access to effective methods of control (Coale, 1973). (Elements 1–3 make up what Coale terms a "conscious calculus of control.") For control to conform with Henry's definition, we must add another stipulation: (5) that couples exert control in accordance with their achieved parity, which in turn implies that they have conscious norms about what represents a desirable family size. It should be possible for the ethnographer to examine whether any or all of these conditions exist in a particular society, but to date there has been surprisingly little research of this sort in contemporary preindustrial societies.

The ethnographic literature does contain numerous anecdotal references to traditional methods of contraception and induced abortion (Devereux, 1955; Himes, 1963; McLaren, 1990). Although a catalog of these methods may suggest that at least some couples in some preindustrial societies believed that fertility control was possible, it tells us nothing whatsoever about whether any of the other conditions for control were met. In particular, it does not tell us whether such methods were effective, widely used, or used in a way that varied with achieved parity. Although statistical techniques are available for estimating the prevalence and effectiveness of contraception and abortion from community-level surveys (Trussell et al., 1990), to my knowledge such techniques have never been applied to any traditional form of birth control in any population of anthropological interest. Without such estimates, descriptions of traditional methods of birth control do not by themselves demonstrate the existence of controlled fertility in Henry's sense of the term.[3]

In this connection, it is important to recall that most of the decline in European fertility occurred during the nineteenth century, well before the invention of modern methods of contraception and the availability of safe procedures for inducing abortion (Demeny, 1975; Knodel, 1986). Traditional methods, therefore, must have played an important role in fertility control in at least one region during one historical period. Demeny (1975) has argued that the existence of potentially effective traditional contraceptive methods, combined with an absence of evidence for their widespread use in Europe *before* the modern fertility decline, indicates that couples in earlier periods had little motivation to practice control. However, Knodel (1986) has pointed to several lines of evidence suggesting that many births were unwanted, especially by the women involved, and that "at least latent motivation to reduce fertility existed." First, in premodern Europe, couples do not appear to have initiated stopping behavior in response to infant and childhood mortality (Knodel, 1978a, 1988). That is, couples whose children all survived continued to reproduce just as long as did couples whose children had died early. Second, during the sustained fertility transition, marital and nonmarital fertility appear to have declined more or less simultaneously in most European countries (Shorter et al., 1971). Yet it seems unlikely that unwed women would have had the same motivation to bear children as married women. Finally, the large number of foundlings apparently deserted by their parents during the eighteenth and nineteenth centuries testifies to a widespread condition of unwanted fertility (Langer, 1974).

But if methods of control and the motivation to use them coexisted in premodern Europe, it must be concluded that knowledge of *effective* methods (such as condoms) was confined to a small minority and diffused more widely only in recent centuries (Watkins, 1989). Two obvious exceptions to this generalization are withdrawal and abstinence, both of which have undoubtedly been recognized as contraceptive methods in all societies. However, those practices are, in Knodel's (1986) words, "uniquely costly," requiring considerable self-control to forego short-term pleasure for the sake of long-term family-building goals, especially when the vagaries of infant and childhood survival rendered such goals difficult to attain with any precision.

Stone (1977) has argued that deliberate fertility control was simply "unthinkable" in preindustrial Europe; that is, the reproductive process was not viewed as something susceptible to outside intervention. My own ethnographic experience in rural Papua New Guinea reinforces this impression: When we attempted to elicit information about what subjects regarded as an ideal family size, the great majority of respondents clearly considered the question meaningless. Of the small minority who provided an answer, virtually every woman gave the number of offspring she had already produced, whereas men usually said "as many as possible." (For similar results, see Caldwell and Caldwell, 1981; Lesthaeghe et al., 1981.) As Grimm and Diderot put it in the early nineteenth century, just as the sustained decline in French fertility was first getting underway, "The act of propagation is so much in conformity with the wish of Nature, and she invites to perform it by such a powerful, such a repeated, such a constant attraction, that it is impossible for the largest number of people to evade it" (Grimm and Diderot, 1813:318–319). If members of the educated elite, among whom the modern fertility decline began, believed this, could the bulk of the populace have thought otherwise?

On balance, I agree with Pat and John Caldwell that

> There is as yet little satisfactory evidence of conscious limitation of family size in traditional societies. There are widespread findings of knowledge and use of indigenous contraceptive practices and these have frequently been cited in recent years to suggest that the limitation of family size has long been an aim of many traditional societies. The case has not been proven and may well be wrong (Polgar, 1972). The practices usually seem to be aimed at preventing conceptions at certain times (or from certain unions) that are undesirable, rather than at limiting the ultimate size of the family. Such undesired conceptions may be those that could result from premarital or extra-marital relations; where a woman is too young or the relationship is incestuous; where insufficient time has passed since the last birth; or where a woman has reached a stage in life where either her age or a circumstance such as achieving grand-maternal status means that reproduction should cease (Caldwell and Caldwell, 1981:73).

If, as argued here, deliberate fertility control in Henry's sense of the term was not characteristic of most preindustrial societies, it should not be concluded that fertility in such societies was unregulated. As this book is intended to show, limiting factors act in every population to restrain fertility. Such factors are entirely consistent with natural fertility as long as they are not deliberately practiced to limit fertility and do not act in parity-dependent fashion.

THE HUMAN REPRODUCTIVE PATTERN

The rest of this book will concentrate on what studies of natural fertility tell us about the extent and causes of fertility variation in the human species. Since subsequent chapters will stress variability, it is useful to step back first and take a broader, comparative look at general patterns of human reproduction, such as might be done by a zoologist interested in a range of animal species. Is there in fact a general pattern of reproduction that distinguishes us from other species? The answer is obviously yes if we restrict attention to controlled-fertility populations, for what other species uses artificial contraception and abortion in the wild? But what about natural fertility? Is there any sense in which our preindustrial ancestors can be said to have shared a common, distinctive pattern of reproduction?

Despite all the variation to be described in later chapters, it is still meaningful to speak of *the* human reproductive pattern, the main features of which are summarized in Table 1.2. Although as a whole this pattern is uniquely human, certain aspects of it can be viewed as extreme examples of more general primate characteristics. Indeed, recent studies of wild populations of gorillas, chimpanzees, and orangutans (Tutin, 1980; Harcourt et al., 1980; Tutin and McGinnis, 1981; Stewart, 1988; Galdikas and Wood, 1990) have revealed reproductive patterns that are in many respects startlingly similar to those observed

Table 1.2. Elements of the General Pattern of Natural fertility in Humans

---

late sexual maturation
iteroparity[a]
low fecundability[b]
long gestation period
high intrauterine mortality
small "litter" size
prolonged parental care (especially prolonged lactation)
long birth intervals
long reproductive span
low total reproductive output
long postreproductive life for females

---

[a]Serial reproduction, i.e., production of several offspring in succession, as opposed to semelparity or the production of all offspring at the same time
[b]The probability of conception during a month of exposure to regular sexual intercourse

in humans under a regime of natural fertility. And certain aspects of the pattern summarized in Table 1.2, such as iteroparity, are shared by most mammalian species.

Developmentally, humans are characterized by a long period of juvenile dependency, slow physical growth, and onset of sexual maturity that is delayed into the teens, combined with an unusually long life span, even by primate standards. Thus, reproduction begins quite late and continues over several decades. The female reproductive span may terminate several decades before death, a characteristic that appears to be more or less unique to humans (Graham, 1981).[4]

Across the reproductive span, women produce a series of "litters," each normally of the minimum possible size, namely one.[5] Pregnancies are unusually long, and each live birth is followed by a period of prolonged dependence of the juvenile on its parents. One particularly striking form of parental care is breastfeeding, which (assuming the child survives) may last several years in many preindustrial societies. One effect of breastfeeding is a long period following each birth during which the mother is unlikely to conceive again (see Chapter 8). Long gestations, prolonged breastfeeding, and low fecundability all combine in humans to produce unusually long intervals between successive live births, sometimes as long as three or more years. (Macaques, in contrast, normally reproduce every year; small rodents may produce several litters within a single year.) Humans also seem to suffer an unusually high risk of losing pregnancies, perhaps as high as 50 percent, which also tends to lengthen birth intervals (see Chapter 6). The net effect of all these characteristics—late sexual maturation, litter size of one, long birth intervals—is that humans are among the least fecund and least fertile of all organisms. Only in exceptional cases do women produce more than 15 children over their entire life span, a meager achievement when compared with the hundreds of millions of offspring produced by a single female oyster over a few years of adult life (Williams, 1975).

In part, the slow pace of childbearing characteristic of humans is simply a reflection of the fact that we are large creatures (Figure 1.1). The conspicuousness of elephants, whales, and redwood trees should not blind us to this fact: consider the abundance of bats, beetles, and bacteria. Numerous studies have established an association across animal species between overall body size and both the rate of physical development and reproductive capacity (Table 1.3). Thus, mammals with large adult body mass, such as humans, tend to reach physical maturity later, have longer pregnancies, produce smaller litters, have fewer births per year (and thus longer birth intervals), nurse their offspring longer, and live longer than do smaller mammals. As Table 1.3 shows, the same trends are present, and in fact usually accentuated, when

Table 1.3.   Some Interspecific Associations between Life History
Variables and Adult Female Body Mass (in grams)[a]

Taxa	Dependent variable	N	Intercept	Slope	r^2
Mammals	Litter size	250	1.11	−0.30	0.25
Mammals	Gestation length	84	4.18	0.26	0.75
Primates	Gestation length	16	4.55	0.16	0.66
Mammals	Births per year	23	0.23	−0.33	0.98
Mammals	Birth interval	67	1.94	0.19	0.51
Primates	Birth interval	62	3.37	0.37	0.56
Mammals	Age at weaning	98	3.33	0.05	—
Primates	Age at weaning	47	1.00	0.56	0.63
Mammals	Age at maturity	56	5.61	0.29	0.57
Primates	Age at maturity	15	6.77	0.32	0.74
Mammals	Average life span	—	7.62	0.17	0.56
Primates	Average life span	14	8.46	0.24	0.77

[a]Results of least-squares or major-axis regression using the model:
log(dependent variable) = intercept + slope * log(adult female body mass). Only
regressions with a slope significantly different from zero at the 0.05 level or
better are reported. A positive slope indicates that the associated dependent
variable increases with adult body size, a negative slope indicates that it
decreases with adult body size. The value of r^2 indicates the proportion of the
variation in the dependent variable explained by the regression model.
From Peters (1983), Calder (1984), Harvey and Clutton-Brock (1985)

attention is restricted to primate species. However, variation in body
size alone does not explain all the differences in reproduction that exist
among humans and their nearest primate relatives, since wild chimpanzees and orangutans, both of which are smaller than humans,
appear to have even longer birth intervals than we do (Galdikas and
Wood, 1990).

In trying to understand the human reproductive pattern from an ecological perspective, it is important to view it as one aspect of the human
life course as a whole (Figure 1.2). In particular, age patterns of reproduction ultimately cannot be understood apart from age patterns of
somatic growth and the risk of death. In very general terms, an organism can partition incoming food resources in four ways (Sibly and
Calow, 1986; Stearns, 1992): it can put them into *maintenance*, that is, it
can support basic metabolic processes while maintaining tissue integri-
ty against trauma or infection; it can put them into *growth* or the con-
struction of new tissue; it can put them into *reproduction;* or it can simply lose them through *excretion,* including both heat loss and the
elimination of indigestible materials and other by-products of metabolism. Since a given quantum of resource cannot be committed to more
than one of these processes, all four compete with each other for finite

resources. From an ecological perspective, the individual life course represents the organism's attempt to resolve this competition in a way that maximizes its genetic fitness.

Like other primates, humans minimize the competition for resources by separating the periods of growth and reproduction. For much of the first two decades of life, human children grow more or less rapidly but are unable to produce gametes. Significantly, the final period of rapid somatic growth, the *adolescent growth spurt,* coincides with the initial switching-on of the reproductive system; indeed, as discussed in Chapter 9, the same underlying physiological cues appear to induce both sets of changes.

By its nature, the need for maintenance cannot be avoided at any age. The competition between growth and maintenance is reflected in the elevated death rates of early childhood, when the highest growth velocities occur. It is now widely recognized that *growth faltering* (i.e., the failure of a child's growth to keep up with the curve characteristic of the population in which it lives) is an important predictor of an elevated risk of death associated with malnutrition (Chen et al., 1980; Heywood, 1982). The fact that growth faltering often precedes death is an indication that the organism preferentially shunts resources to maintenance at the expense of growth when there is too little food to support both; the fact that death often follows growth faltering indicates that this strategy is not always successful. The competition between maintenance and reproduction during adult (i.e., post-growth) life is sometimes manifested as the *maternal-depletion syndrome,* a decline in maternal nutritional status attributable to the metabolic burden of reproduction, especially prolonged lactation (Adair et al., 1983; Pebley and DaVanzo, 1988; Miller and Huss-Ashmore, 1989).[6] Where food resources are unusually limited, this depletion may not be made up before the next round of reproduction, resulting in a cumulative decline in maternal condition over the course of reproductive life (Tracer, 1991; Wood, 1992). Significantly, maternal depletion seems to be most severe as a result of pregnancies occurring in adolescence (Riley, 1990), when growth is still an important competitor for resources.

The progressive loss of reproductive capacity at later adult ages, which will be considered in detail in Chapters 7, 10, and 11, coincides with an accelerating increase in the risk of death. There are two possible reasons for this association: (1) reproduction itself entails physiological "costs" that compromise both future reproduction and future survival (as in the maternal depletion syndrome), or (2) senescent changes at the cell and tissue level simultaneously affect the capacity for reproduction and the risk of death. As we argue elsewhere (Wood et al., 1994b), the second of these possibilities now appears to be more likely, consistent with the observation that similarly accelerating mortality occurs in both males and females despite the far higher physiological costs imposed upon women by reproduction.

In sum, humans are large, slow-growing mammals with comparatively low fertility and mortality. They generally produce a small number of offspring but tend to invest heavily in each of them—prolonged breastfeeding in particular being a metabolically expensive form of parental care. As numerous ecologists have noted, such life history traits are generally found in species occupying stable, productive environments, such as tropical rain forests or savannas. Primates in general display these same life history traits, but humans are an extreme example of the general primate pattern, although perhaps no more extreme than our closest living relatives, the great apes.

WHY THE EMPHASIS ON FEMALES?

A final issue must be addressed at the outset of this book. A glance at the following chapters will reveal that most of them are devoted to processes occurring in women. Thus, considerable attention is paid to ovarian cycles, pregnancy, breastfeeding, and fetal loss. Only one chapter emphasizes male reproduction. From a genetic point of view, of course, males and females are of approximately equal importance in reproduction. Why, then, the emphasis on women?

There is a good reason and a bad reason for this emphasis. The bad reason is that demographers interested in fertility have traditionally adopted a single-sex, female-dominated perspective, largely as a matter of convenience. It happens that two-sex models of population dynamics are very much more complicated mathematically than single-sex models, a situation referred to as "the demographer's two sex problem" (Pollak, 1990). It is therefore convenient to ignore one sex in most analyses. Since maternity is generally easier to ascertain than paternity, at least when purely demographic data are used, it is most convenient to focus on females.

The *good* reason for emphasizing the female role in reproduction has to do with our interest in the temporal dynamics of reproduction. Given that interest, it makes sense to concentrate on those processes that have the greatest impact on the overall pace of childbearing, on what biochemists would call the *rate-limiting steps* in reproduction (see Chapter 3). In most dioecious species (species with two separate sexes), the major rate-limiting steps involved in reproduction occur in the female. In humans, as in other mammals, these steps include pregnancy itself, as well as the period of infecundity associated with lactation. In addition, since the production of eggs is discontinuous and episodic (unlike sperm production), the female cycle places an irreducible upper limit on the rate of conception. As will be argued in Chapter 3, variation in the rate-limiting steps usually has a greater impact on overall fertility variation than does variation in other aspects of the reproductive process. Thus, in attempting to understand variation in fertility, it makes sense to concentrate on factors affecting women. This is not to say that factors affecting men have no effect on fertility variation; indeed, in Chapter 11 we will try to assess their effect. However, the pace of childbearing—and variation in that pace—is largely set by the female partner.

NOTES

1.   The *reproductive span* is defined as the ages during which individuals are capable of childbearing, usually taken to be ages 15 to 49 for women (Pressat, 1985). These boundaries are somewhat arbitrary since reproduction can occur before age 15, and very occasionally women give birth after age 50. However, reproduction outside this age range is sufficiently uncommon in most human populations that ages 15 to 49 are not unreasonable limits for the female reproductive span. Males, in contrast, can and do reproduce well past the age of 49.

2.   Control can also take the form of *delaying behavior,* i.e., behavior aimed at delaying the first birth but not avoiding it altogether. In general, widespread delaying behavior, as currently exists in the United States, is typical of the more advanced stages of the transition to controlled fertility and is thus less relevant to the boundary between controlled and natural fertility. By its nature, moreover, delaying behavior cannot be parity-dependent.

3.   I agree with Knodel (1986) that induced abortion was unlikely ever to have been important before the early modern period since "it was such a dangerous procedure that women probably resorted to it only in cases of extreme need, and mostly outside of wedlock." No traditional chemical abortifacient has ever been shown to work, and pregnancies are now known to be far more resistant to physical trauma than was previously believed. Any method, such as jumping or piling stones on the pregnant woman's abdomen, would be as likely to kill the woman or render her sterile as to terminate the pregnancy. Infanticide, incidentally, although sometimes discussed as if it were a fertility control mechanism, actually represents *mortality* control. Since it does not prevent a live birth but intervenes only after birth, it is not relevant to the argument presented here. I might mention, however, that the same points made with respect to traditional methods of birth control—that there are no reliable estimates of use-prevalence or use-effectiveness, and no evidence that they are parity-dependent—could be made about infanticide as well.

4.   An exhaustive literature search has uncovered only five verifiable cases of menopause, i.e., complete cessation of reproductive function, in female nonhuman primates. These five cases consist of four chimpanzees and one macaque, all of them living in zoos or laboratories and all undergoing menopause at ages

much greater than normally attained in the wild (Hodgen et al., 1977; Gould et al., 1981).

5. Marmosets and tamarins are the only primates that routinely produce multiple births. Multiple births do, of course, occur in humans, but at very low frequencies. Twin births (both monozygotic and dizygotic) range from about 64 per 10,000 births in Japan to 449 per 10,000 in Nigeria (Morton et al., 1967); most populations appear to cluster near the lower end of this range (Propping and Krüger, 1976). Triplets and higher-order births are much less common; very approximately, if b is the frequency of twins, then b^2 is the frequency of triplets, b^3 the frequency of quadruplets, etc. (Vogel and Motulsky, 1986:209). Because of their rarity in humans, multiple births will be ignored in this book. In effect, then, "birth" will mean a live birth regardless of its multiplicity.

6. Lactation is metabolically costly. Widdowson (1976) estimates that the 36,000 kcal stored by the average, well-nourished western woman during pregnancy is sufficient to provide only about one-third of the energy required to support four to five months of lactation. The National Academy of Sciences recommends that lactating women consume 500 kcal/day over their normal daily requirements (Worthington-Roberts et al., 1985). This recommendation is based on the assumption that the average woman gains about 3.5 kg of fat during pregnancy that can be mobilized to provide the 300 kg/day necessary to produce 850 ml/day of milk for approximately three months. But the typical woman in a natural-fertility setting is unlikely to conform to this paradigm of the average, well-nourished western woman. She often begins pregnancy with lower nutritional reserves; she gains little, if any, fat during pregnancy; she may lactate for more than three years, not three to five months; she often does not supplement her own diet during lactation (and indeed may be subject to special dietary restrictions while she is lactating); and her milk provides virtually all the nutriment required by her child for at least the first six months postpartum. The net effect of these differences is that the nutritional status of women in traditional societies often declines sharply over the course of lactation (Bongaarts and Delgado, 1979; Miller and Huss-Ashmore, 1989).

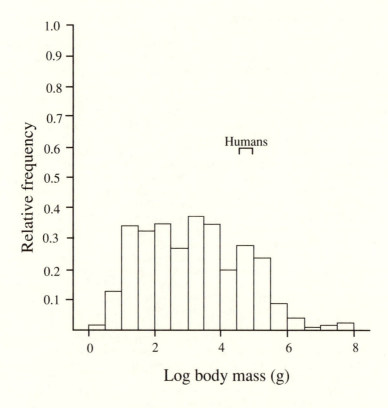

Figure 1.1. Distribution of the average body mass (log₁₀ scale) in adult females of 383 species of mammals. Also shown is the range of variation in mean or median adult female body mass in contemporary human populations. Note that while mammals are comparatively large animals, humans are comparatively large mammals. (Data on humans are from Eveleth and Tanner, 1976. Data on mammals are from Eisenberg, 1981: Appendix 1; for monomorphic species, either the species mean or, when the mean was unavailable, the midpoint of the species range was used; for strongly dimorphic species, the female mean or the midpoint of the female range was used.)

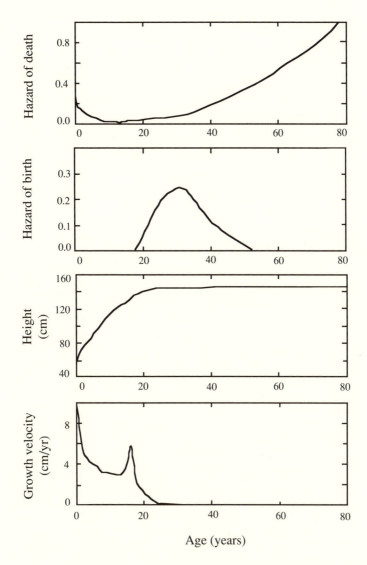

Figure 1.2. Four fundamental aspects of the biology of the female life course in humans (from top): age-specific mortality, fertility, body size, and growth rate. Shown are smoothed curves fit to data on a tribal population in highland Papua New Guinea (Wood, 1980, 1986), thus accounting for the comparatively small adult stature, slow growth, and late adolescent growth spurt.

2

Patterns of Natural Fertility

To suggest, as in the previous chapter, that all human societies share a pattern of low reproductive output, even where effective methods of birth control are unavailable, is not to say that all societies share exactly the *same* reproductive output. We therefore need to address several empirical questions about fertility: What levels of reproduction characterize natural-fertility populations, and how do those levels differ from those of controlled-fertility groups? Do natural-fertility populations differ from each other, or is there a single pattern of reproduction that can be said to characterize them all? In particular, given that such populations lack access to modern methods of birth control, do they all reproduce near the maximum level permitted by biology? And is there variation in reproductive output *within* natural-fertility populations, or are individuals in such populations more or less equivalent with respect to reproduction? Having seen one maternity history from such a group, have we in effect seen them all?

But before we can describe levels of natural fertility, we need some way to measure them. That is, we need standardized, readily interpretable indices of reproductive performance that can be compared across populations. In the next section we develop a basic tool-kit of such indices.

We also need to say a word about the available sources of data on natural fertility, for such data are not easy to come by. The data used by most demographers are often provided by large-scale, institutionalized systems, such as national censuses and birth registries, but populations with well-developed statistical systems of this sort are also likely to have access to modern methods of family limitation and are thus unlikely to be natural-fertility populations. Consequently, many demographers have turned to historical data sources, such as church records. In fact, Louis Henry was a pioneer not only in the study of natural fertility but also in the method of historical "family reconstitution" using parish records of baptisms, burials, and marriages. Unfortunately, as illuminating as this approach has been, it is applicable only to those parts of the world that maintained such records in the past—and, by and large, this has meant Europe. (Japan instituted civil registration of vital events in the seventeenth century, but this material has been studied only recently.) Thus, much work on natural fertility has had an unfortunate "Eurocentric" cast.

Recent advances in survey methodology offer a solution to this problem. Many investigators in demography, anthropology, epidemiology, and related fields have begun to conduct specialized surveys designed to provide information on contemporary populations that approximate conditions of natural fertility. Admittedly, with a few notable exceptions, such as the World Fertility Survey and the Demographic and Health Survey, the data provided by most of these surveys are based upon small samples. Moreover, because surveys usually involve subject recall of past events, they are likely to be less accurate than data from parish or civil registries, which are updated as the relevant events occur. Nonetheless, these surveys are often more representative than parish records, and they are not limited to the narrow kinds of information recorded by parish priests. For example, such surveys can include questions about desired family size or traditional methods of birth control, and survey scientists can even collect biological samples (e.g., of serum, urine, or saliva) for physiological analysis. Such surveys are likely to become increasingly important in the future study of natural fertility.

MEASURES OF FERTILITY

By demographic convention, most fertility indices are defined with reference to women rather than men or even couples. Although this can sometimes be misleading, it is consistent with the point made in Chapter 1, that the "rate-limiting steps" in reproduction most often occur in women. In addition, all the indices discussed here pertain strictly to live births; that is, they exclude stillbirths and miscarriages. For general introductions to fertility indices, much more comprehensive than those presented here, see Campbell (1983), Palmore and Gardner (1983), or Newell (1988).

All fertility indices are computed from finite samples of births and thus represent estimates with some degree of sampling error. As with most demographic measures, very little is known about the sampling properties of fertility indices, except under very restrictive conditions (see below). At present, the best approach to assessing the sampling error of these measures is to use a resampling scheme such as the jack-knife or bootstrap (Efron, 1982).

Fertility indices can be classified as either *period* or *cohort* measures. Period measures are defined with reference to a group of women who exist at a single period of time, although they may be of widely differing ages; strictly, such measures reflect the experience of those women at that time and no other. Cohort measures, in contrast, refer to a group of women followed over an extended period of time, usually many years. The members of a cohort are chosen so they share some characteristic at the outset of the period of follow-up—e.g., all women born during 1920–1924 (an example of a *birth cohort*) or all women married in 1956 (a *marriage cohort*). Since this book is especially concerned with the ways in which individual women time reproductive events over the course of their lives, cohort measures would seem to be more useful for our purposes. Unfortunately, cohort measures have far more severe data requirements than do period measures, simply because they require us to track women over a long period of time. Because of data limitations, we are often forced to make inferences about life-course processes from period measures. As we shall see, such inferences are valid only when fertility levels in the general population are unchanging. In such a situation, but only in such a situation, some simple and straightforward relationships between period and cohort measures exist, as described later in this chapter. When change is known to have occurred, or when we have no information on past fertility levels, extreme caution should be used in making inferences about the reproductive experience of any cohort from period measures.

Period Measures of Fertility

The starting point for all the period measures presented here is the annual *age-specific fertility rate* (ASFR) of women ages a to $a + k$ years. This rate, which I will write as $f(a)$, is estimated as

$$\hat{f}(a) = \frac{\text{observed number of births to women ages } [a, a + k)}{\text{observed number of woman-years at ages } [a, a + k)} \tag{2.1}$$

The hat ($\hat{\ }$) over $f(a)$ signifies that this is an estimate of the "true" value of $f(a)$ based upon a finite sample. The curious brackets in the expression [*a,a* + *k*), where *k* > 0, refer to an interval that contains age *a* and everything up to age *a* + *k*, but not *a* + *k* itself—the interval is said to be "inclusive" of *a* but "exclusive" of *a* + *k*. If we express the interval in terms of a woman's age at her last birthday, then the interval spans ages *a* to *a* + *k* − 1. For example, the age interval [20,25) can be referred to as ages 20 to 24.

The concept of an observed woman-year at ages [*a,a* + *k*) is best explained by an example. Suppose we observe a population for exactly five years; further suppose that one particular woman has her twentieth birthday on the very first day of the observation period and survives (and remains under observation) for the full five-year period. Then that woman provides exactly five woman-years of observation between ages 20 and 25. But suppose that another woman has her twentieth birthday on the same day but dies (or migrates out of the study area) on her twenty-third birthday; she contributes only three woman-years of observation. Or suppose another woman was age 18 at the outset of the study but had her twentieth birthday exactly two years into the study. If she remained in the study until its end, that woman would contribute two woman-years to the 15–19 ASFR and three to the 20–24 ASFR.

In words, *f*(*a*) is the mean number of offspring born in a year to a woman between ages *a* and *a* + *k*. By demographic convention the value of *k*, which determines the width of the age interval, is almost always set equal to five years. Thus, we usually compute ASFRs for women ages 15–19, 20–24, and so on up to 45–49 years, since we assume that females younger than 15 and older than 50 have negligible chances of reproducing. But note that there is nothing magical about *k* = 5; other interval widths are equally valid.

If all women have the same probability of reproducing at ages *a* to *a* + *k*, and if that probability does not change between *a* and *a* + *k*, then equation 2.1 is the maximum likelihood estimator for the rate parameter of an exponential survival model (see Chapter 3), and its large-sample standard error is estimated by the square root of

$$\hat{V}ar[\hat{f}(a)] = \frac{\hat{f}(a)}{\sum_{i=1}^{n} \left\{ 1 - \exp[-\hat{f}(a) \times T_i] \right\}} \tag{2.2}$$

In this expression, *n* is the number of women observed in the age interval, T_i is the number of woman-years of observation contributed to the sample by the *i*th woman, and exp(*x*) denotes e^x where e is the base of the natural logarithms (approximately equal to 2.71828). If the conditions of constancy and homogeneity do not obtain—and they almost certainly never do—this simple expression is not applicable, and some resampling procedure may be necessary to estimate the standard error.

Age-specific fertility rates are useful for comparing different populations because, unlike many other fertility measures, they are unaffected by the age structure of the populations being compared. However, they involve a somewhat crude specification of the women who are actually exposed to the risk of childbearing at each age. Not all women aged, say, 20 to 24 are equally likely to reproduce during a given period of time. In particular, unmarried women in most societies are far less likely than married women to give birth (see Chapter 11). Thus, a more refined fertility measure is the annual *age-specific marital fertility rate* (ASMFR) of women at ages a to $a + k$, written $r(a)$ and estimated as

$$\hat{r}(a) = \frac{\text{observed number of legitimate births to women ages } [a, a + k)}{\text{observed number of married woman-years at ages } [a, a + k)} \quad (2.3)$$

Under constant and homogeneous marital fertility, the standard error of the ASMFR can be estimated by substituting $\hat{r}(a)$ for $\hat{f}(a)$ in equation 2.2 and summing over married women.

ASMFRs preserve the most desirable property of ASFRs in that they are not affected by the population's age structure. In the absence of illegitimate births, ASMFRs and ASFRs are related as $f(a) = g(a)r(a)$, where $g(a)$ is the proportion of women ages $[a, a + k)$ who are married. Since $g(a)$ falls between zero and one, the relationship $r(a) \geq f(a)$ must always hold.

An important single-number measure of total reproductive output that can be computed from a set of ASFRs is the *total fertility rate* (TFR). The TFR is defined as the expected number of offspring ever born to a randomly selected woman who survives to the end of the reproductive span (i.e., to menopause or to some suitably advanced age, such as 50 years), given that the current age-specific fertility rates remain constant. The TFR is estimated as

$$\hat{\text{TFR}} = k \sum_{a=0}^{\infty} \hat{f}(a) \quad (2.4)$$

that is, the sum of all the ASFRs multiplied by the length of the age interval.[1] In addition to its simplicity, the TFR has two major attractions as a single-number measure of fertility: (1) it is a pure fertility measure, uninfluenced by the age and sex composition of the population or by mortality, and (2) it is one of the few aggregate-level period measures of fertility that can be interpreted in terms of an individual woman's expected lifetime reproductive experience. This latter feature is especially important for researchers interested in the linkage between physiology and behavior at the individual level and population dynamics at the aggregate level.

A similar measure with similar virtues is the *total marital fertility rate* (TMFR), which can be estimated from a schedule of ASMFRs:

$$\widehat{\text{TMFR}} = k \sum_{a=0}^{\infty} \hat{r}(a) \tag{2.5}$$

This index is an estimate of the expected number of offspring ever born to a randomly selected woman in the population, given that she survives to the end of the reproductive span, that current marital fertility rates remain constant, *and* that she is continuously married throughout the reproductive span. The TMFR can be thought of as equivalent to the TFR corrected for the fertility-reducing effects of nonmarriage (see Chapter 3). Because of this correction the TMFR must always be greater than or (far less likely) equal to the TFR.

Cohort Measures of Fertility

Since high-quality prospective data are rarely available for natural-fertility populations, we restrict attention here to cohort measures that can be computed from *retrospective* data, such as those provided by maternity histories. The first of these is $F(a)$, the *cumulative fertility* of women at age a, equal to the mean number of offspring ever born to women who are age a at the time of interview. A special case of cumulative fertility is the *mean completed family size* (MCFS), equal to the cumulative fertility of women who have reached the end of the reproductive span.

At each age, we can look not only at the *mean* number of children ever born, but also at the total distribution of children among women in the age interval. This *age-specific parity distribution* is given by the values of $p_a(x)$, the fraction of women age a who have produced x offspring by that age. If a is a postreproductive age, then $p_a(x)$ gives the distribution of completed family sizes. In that case, we drop the subscript for age and write $p(x)$.

Another useful cohort measure is α_x, the *parity progression ratio* (PPR) of order x, defined as the proportion of women with x children who go on to produce at least one more child. This ratio can be estimated from retrospective data on completed family size as

$$\hat{\alpha}_x = \frac{\sum\limits_{i=x+1}^{\infty} \hat{p}(i)}{\sum\limits_{i=x}^{\infty} \hat{p}(i)} \tag{2.6}$$

Parity progression ratios are useful for gauging the effect of achieved parity on the likelihood of further reproduction. Note that

$$\text{MCFS} = \alpha_0 + \alpha_0\alpha_1 + \alpha_0\alpha_1\alpha_2 + \text{ L} \tag{2.7}$$

In addition, the fraction of women who never produce a live birth despite surviving to menopause $(1 - \alpha_0)$ is commonly used as an estimate of *primary sterility* (see Chapter 10).

The parity progression ratios considered in this chapter are all retrospective measures based on completed fertility. Recent work has also been done on age-specific parity progression ratios, which can be calculated for any age during the reproductive span (Feeney, 1983). It appears that such measures may provide sensitive indicators of the presence of parity-dependent changes in reproduction (David et al., 1988).

Relationships between Period and Cohort Measures

If the level and age pattern of fertility in the population we are studying have not changed in recent decades, then the age-specific period rates experienced by women now should be the same as those experienced in the past. This suggests several relationships that can be expected to hold when fertility is constant. For example,

$$F(a) = k\sum_{i=0}^{a-k} f(i) \tag{2.8}$$

That is, cumulative fertility at age a is just the sum of the ASFRs up to age a. Similarly,

$$\text{MCFS} = k\sum_{a=0}^{\infty} f(a) = \text{TFR} \tag{2.9}$$

In principle, then, when fertility is constant the mean completed family size and the total fertility rate should be equal. If they are not, there may be reason to suspect that fertility has changed in the recent past. However, if the mortality of women is selective with respect to fertility, then the MCFS may be a biased estimate of the TFR even when fertility is not changing. For example, if reproduction itself exposes the mother to an elevated risk of death—as is certainly the case in many less-developed countries—then the MCFS may be biased downward, simply because women of low fertility have a better chance of surviving long enough to be in the sample. Alternatively, if healthier women with low risks of death are also better able to reproduce, then the MCFS may yield an upwardly biased estimate of the TFR. If such biases are not taken into account, we may be misled into believing, incorrectly, that fertility has recently changed.

LEVELS OF NATURAL FERTILITY

We can now use the measures developed above to examine empirical patterns of natural fertility. The first set of questions that needs to be addressed has to do with the overall level of reproduction in natural-fertility populations. Do such populations all reproduce at essentially the same level, or are there important differences among them? And how do the fertility levels of natural-fertility populations compare to those of controlled-fertility populations? To answer these questions, it is convenient to use a single-number summary statistic such as the TFR or MCFS. In a recent paper we have summarized the data needed to compute those two measures for a large sample of natural-fertility populations (Campbell and Wood, 1988). We were able to find reliable data for 70 populations from all parts of the world, populations that include hunter-gatherers, tribal horticulturalists, and peasant agriculturalists, both recent and historical (see Campbell and Wood, 1988:App. A, for a complete listing of populations). The data on these populations came primarily from field studies by demographers, anthropologists, and epidemiologists, or from family reconstitution studies by historical demographers. Seven criteria had to be met before a given data set was included in the sample: (1) The population in question had to be relatively unacculturated, and there had to be no evidence of a major change in fertility during the fifty years preceding the study period. (2) If the data pertained to cohort fertility, maternity histories had to be available for at least 50 women of postreproductive age. (3) If the data pertained to period fertility, the age-specific fertility rates from which the TFR was computed had to be based on the direct registration of at least 200 births per year. (4) Demographic analysis had to be a major focus of the study in question, not an afterthought. (5) For contemporary field studies, we required some indication that care was taken to probe for completeness of reporting. (6) Data based on unreliable methods were excluded; an example of such a method would be the estimation of fertility rates from subject-recalled genealogies, which systematically underrepresent low-fertility families (Norton, 1980). And (7), since we were concerned primarily with normal, nonpathological variation in natural fertility, we excluded populations believed to have high prevalences of pathological sterility—specifically, populations with primary sterility rates, as estimated by $(1 - \alpha_0)$, greater than 0.15. Such populations tend to cluster in sub-Saharan Africa, especially East and Central Africa, and in Micronesia.

Depending upon the sort of data available, we computed either the total fertility rate or the mean completed family size. Since we restricted attention to populations in which fertility had apparently not changed appreciably during the period preceding the time of observation, both measures should provide estimates of total fertility. For simplicity, then, I will refer to both kinds of measures as TFRs. This usage, of course, ignores other possible biases introduced by using the MCFS as an estimate of total fertility, for example, the effects of selective maternal mortality. Such biases are often difficult to assess from the sort of data available to us.

Figure 2.1 shows the distribution of TFRs among the 70 natural-fertility populations that met all our criteria. Also shown for comparison is the distribution of TFRs (in the strict sense) among 70 modern, controlled-fertility populations selected at random from recent issues of the United Nations *Demographic Yearbook.*

Are the levels of natural fertility shown in Figure 2.1 high or low? If we use the controlled-fertility populations as a standard for comparison, the natural-fertility TFRs are, not surprisingly, rather consistently high. The mean TFRs for the two distributions are 2.6 and 6.1, respectively, which represents a considerable difference. There is, moreover, virtually no overlap between the two distributions. Indeed, the degree of overlap is exaggerated by the inclusion of several transitional countries in the controlled-fertility distribution—countries such as Taiwan in 1966 and Ireland in 1973 that had experienced substantial fertility declines before the period in question but that clearly had not yet completed their fertility transitions.

But is that the appropriate standard for comparison? All the natural-fertility populations in Figure 2.1 could be considered to have *low* fertility rates in that they are all well below the maximum biological capacity for reproduction in humans. Assume for the moment that every woman experiences a reproductive span that begins at age 15 and ends at age 45, a constant gestation of nine months, a period of postpartum anovulation of about 1.5 months, a lag of three months to next conception, and roughly two months added to each birth interval by fetal loss. As subsequent chapters will show, these values are not biologically implausible, and yet they still leave enough time during the reproductive span to fit in 23 to 24 full-term pregnancies. As a group, even the most fertile of the natural-fertility populations in Figure 2.1 falls well short of that maximum level of reproduction, although the occasional woman in some populations may actually exceed it. In the aggregate, no natural-fertility population reproduces at its maximum biological potential. In other words, factors operate in every population to limit the fertility achieved by its members.

Perhaps the most important conclusion to be drawn from Figure 2.1 is that it is impossible to characterize fertility in preindustrial societies as being either uniformly high or low. There exists a remarkable amount of variation in the level of natural fertility from one population to another. The natural-fertility TFRs show an almost threefold range of variation, from about 3.5 to 9.8, and the variance of the natural-fertility distribution is more than three times that of the controlled-fertility distribution. (And again, the variability of the controlled-fertility groups has been inflated by the presence of transitional populations.) Simply stated, there is no "typical" level of natural fertility.

Admittedly, the natural-fertility distribution includes groups of widely varying subsistence practices, habitats, and social formations, ranging from nomadic hunter-gatherers to densely settled peasant agricultural communities. It might be thought that the apparent heterogeneity in natural-fertility levels might simply reflect the fact that we have thrown together a hodgepodge of groups. However, when the data were grouped into three categories based on subsistence (peasants, tribal horticulturalists, and hunter-gatherers), we found no significant interclass difference in either the mean or the variance in total fertility. When we performed a controlled comparison of hunter-gatherers and all other groups combined, we again found no difference in fertility. Thus, the oft-repeated claim that hunter-gatherers have uniformly low fertility in comparison with other preindustrial societies has no basis in fact.[2] Evidently, such simple categorizations are not helpful in explaining variation in natural fertility.

It is, however, possible to characterize the extreme tails of the distribution in Figure 2.1 in a general way. Note that about 90 percent of the variation in natural-fertility TFRs falls between four and eight. As it happens, all the populations in our sample with TFRs over eight can be considered colonizing populations, that is, populations undergoing rapid expansion after moving into new habitats containing few or no human competitors. These include the 1921–1930 marriage cohort of Hutterites, an Anabaptist sect of the North American Plains that grew from 443 individuals in 1880 to 8,542 individuals 70 years later (Eaton and Mayer, 1953); the Mormons during their phase of westward expansion in the nineteenth century (Skolnick et al., 1978; Mineau et al., 1979); and rural Anglo- and French-Canadians during the early eighteenth century (Henripin, 1954a, 1954b). Also included are the Yanomama Indians of Venezuela and Brazil, who have sustained an average annual population growth rate of about 0.005 to 0.01 over the past several decades while expanding to fill an ecological vacuum left by the postcontact depopulation of neighboring tribes (Neel and Weiss, 1975). By contrast, none of the groups with TFRs less than eight can reasonably be classified as colonizing populations. These findings suggest that there may be widespread, density-dependent damping of fertility in preindustrial societies, damping that is released upon entrance into a new, "empty" habitat. Alternatively, low-fertility colonizing populations may tend to die out rapidly, as predicted by biogeographical theory (MacArthur and Wilson, 1967:68–93), thus leaving behind few cases to study.

At the other extreme, natural-fertility populations with TFRs much below four are likely to have high prevalences of pathological sterility and, indeed, are likely to be located in sub-Saharan Africa's "infertile crescent" (Belsey, 1979; Caldwell and Caldwell, 1983; Frank, 1983). We deliberately tried to exclude such populations from the present sample. Had we included them, they would have extended and heavily dominated the lower tail of the distribution. But, as Bentley et al. (1993) have pointed out, several of the populations that we did include in the sample may have had fairly high levels of sterility—for example, the Greenland Inuit in the early 1950s (Malaurie et al., 1952) and the !Kung San in the late 1960s (Howell, 1979)—even if they did not exceed our arbitrary cutoff point of $(1 - \alpha_0) > 0.15$.

To summarize, total fertility rates in preindustrial societies tend to vary between about four and eight, with a mean of about six. Groups that fall outside that range are likely to have unusual ecological or epidemiological characteristics. Although all preindustrial societies have fertility levels well below the theoretical biological maximum, almost all have levels well above those observed in populations with high use-prevalences of modern contraception, induced abortion, and surgical sterilization. Beyond these generalities, it is impossible to point to a single, typical level of fertility that characterizes all traditional societies. Nor is it possible to sort such societies into simple categories, each with a single level of fertility. These findings do not mean that variation in natural fertility is inherently inexplicable; they simply suggest that we need a more fine-grained approach for explaining such variation. We return to this problem in the next chapter.

INTRAPOPULATION VARIATION IN NATURAL FERTILITY

We should not be misled by the TFR or MCFS into thinking that all or even most couples *within* each population reproduce at or near the average level. In fact, every human population displays internal variation in completed fertility. For example, in each of the four cases shown in Figure 2.2, different women had from zero births up to 10 births or more, despite the fact that they all survived to menopause. When total fertility is modest, as in the Gainj and !Kung cases, it is possible to find women who produce more than twice the mean number of births. When fertility is higher, as among the Amish, at least a few women with 12 or more births can be expected. Note, however, that essentially all women in these populations still reproduce well below any biological maximum.

The Gainj, !Kung, and Amish are all natural-fertility populations. In contrast, the 1928–1930 Norwegian marriage cohort, also shown in Figure 2.2, was well on its way to achieving effective fertility control. Even in Norway, however, there was a marked degree of reproductive variability. Although most women in that cohort had completed family sizes falling in the range of one to five births, the distribution of family sizes displays a long upper tail, with some women experiencing 12 or more births. This upper tail may partly reflect couples who chose not to practice contraception, perhaps for religious reasons; it may also reflect some degree of contraceptive failure.

To make a point that will become crucial in the next chapter, the variation in natural fertility found within populations (as well as among populations) must necessarily be associated with variation in at least some of the principal determinants of fertility. For example, Figure 2.3 summarizes variation in some of the major components of fertility in a rural French community in the seventeenth and eighteenth centuries. In this natural-fertility population, the estimated variance in completed family size exceeds 13, which is quite high. Associated with this high variability is considerable variation in the age at which women marry for the first time and, to a somewhat lesser degree, in the age at which they produce their last live birth. Since in most preindustrial societies regular sexual relations are normally confined to marriage, the ages of women at first marriage and last birth together define the length of the effective reproductive span. In addition, there is a remarkable degree of variation in the length of birth intervals in this population, ranging from less than one year to more than four years (Figure 2.3, bottom panel). Clearly, variation in the length of the reproductive span and variation in the length of birth intervals within the reproductive span must, in combination, determine variation in completed family size.

As later chapters will show, genuine differences among couples in reproductive physiology and behavior probably account for a large fraction of the observed variation in completed fertility and its components. It is important to bear in mind, however, that a certain amount of this variation is purely random or *stochastic* in nature. For example, couples with identical monthly probabilities of conception can wait anywhere from one to twelve months or more before actually conceiving (see Chapter 7). Similarly, women with the same risk of losing a pregnancy can vary appreciably in terms of the actual number of losses that occur before the next successful pregnancy (Chapter 6). This sort of stochastic variation in reproduction is universal and irreducible. Therefore, in order to demonstrate that systematic differences are really affecting reproduction, it is first necessary to show that the observed variance in fertility is significantly greater than that expected from purely random variation.

Two simple probability models allow us to make a preliminary judgment on this score (Wood, 1987b). First, imagine an idealized population of women, all of whom survive to menopause, and all of whom are subject to exactly the same probabilities of reproducing at each age during the reproductive span. If successive births to each woman are independently distributed through time, then completed family size should vary among these women as a Poisson random variable. That is, the probability that a woman selected at random from this population has a completed family size of n live births is

$$p(n) = \frac{\lambda^n}{n!} e^{-\lambda} \tag{2.10}$$

where $n!$ is a quantity called "n factorial," equal to $n(n - 1)(n - 2) \cdots (3)(2)(1)$. This model has a single constant or *parameter* $\lambda > 0$; its mean and variance are both equal to λ.

This admittedly oversimplified model implies, among other things, that there are no systematic differences in fertility among women in this population. In contrast, imagine a more complicated population made up of several such homogeneous subpopulations with varying values of λ. This second case represents a simple form of reproductive heterogeneity in which the total population contains several reproductively homogeneous subgroups that differ from each other. Under certain assumptions about the distribution of λ among subgroups,[3] this model yields a negative binomial distribution of completed family size:

$$p(n) = \frac{(\theta + n - 1)!}{(\theta - 1)! n!} \left(\frac{\gamma}{1 + \gamma} \right)^n \left(1 - \frac{\gamma}{1 + \gamma} \right)^\theta \tag{2.11}$$

This distribution has two parameters, θ and $\gamma > 0$, with a mean of $\theta\gamma$ and a variance of $\theta\gamma(1 + \gamma)$. Since γ is strictly positive, the variance of this distribution must be greater than the mean, in contrast to the Poisson case in which the mean and variance are always equal.

In view of the extreme simplicity of these models, it is perhaps surprising that they fit a wide range of data on human fertility (Brass, 1958; Wood, 1987b). For example, the Gainj and !Kung distributions in Figure 2.2 are well-described by a Poisson distribution, whereas those in the Norwegian and Amish cases approximate a negative binomial. Note that the distinction between Poisson and negative binomial reproduction does not necessarily parallel that between natural and controlled fertility. Gainj and !Kung fertility are both Poisson and natural, Norwegian fertility is negative binomial and controlled—but Amish fertility is negative binomial and natural. Doubtless, the differential use of contraception is one important source of the high variance in completed family sizes observed in Norway, but as the Amish case shows, substantial variation can also exist in the absence of contraception.

As these cases suggest, the relative fit of our two models can be revealing about the presence or absence of reproductive heterogeneity. In the case of the French data summarized in Figure 2.3, the fitted Poisson distribution is too clumped around its mean and does not have enough individuals in its lower and upper tails (Figure 2.4, left). In other words, more heterogeneity in completed fertility exists than is allowed under the Poisson model. The negative binomial distribution provides a much better fit (Figure 2.4, right).[4] We can conclude, therefore, that there is prima facie evidence for systematic heterogeneity in fertility in this population. In other words, something is going on that demands explanation.

Since the Gainj and !Kung are both small-scale societies with little socioeconomic differentiation among households, whereas France and Norway are both state-level, class-organized societies, it is not surprising that substantial heterogeneity exists in the latter but not in the former cases. The existence of significant reproductive heterogeneity among the Amish is, frankly, more puzzling.

PATTERNS IN PARITY PROGRESSION RATIOS

Parity progression ratios summarize precisely the same data as the distribution of completed family size, but do so in a way that brings out different features. For example, the essential difference between natural- and controlled-fertility populations is neatly illustrated by their respective PPRs (Figure 2.5). Under natural fertility, the PPRs decline slowly with parity, describing a curve that is everywhere *convex upward*; controlled-fertility PPRs, in contrast, decline more rapidly and are *concave upward*. This difference reflects the fact that most fertility control is strongly parity-dependent and is dominated by stopping behavior.

There are also clear distinctions between natural-fertility populations with high TFRs (Figure 2.6, top) and those that have low TFRs because of effective birth spacing (Figure 2.6, bottom). In both cases, the PPRs are still convex upward, but the low-TFR curves decline much more rapidly—although not nearly as rapidly as under controlled fertility. The high-TFR curves are, in contrast, much flatter.

Of course, low natural fertility can result not only from long birth intervals, but also from high prevalences of pathological sterility. However, populations with widespread sterility (Figure 2.7, top) show PPR curves quite distinct from those of groups with long birth intervals (Figure 2.6, bottom). The high-sterility curves start at a comparatively low level, reflecting widespread primary sterility, but they then remain much flatter and do not display the rapid decline associated with long birth intervals. One interpretation of the curves in the top panel of Figure 2.7 is that there is a high and more or less constant risk of becoming sterile before each parity progression in these populations; if that risk were removed, these populations would presumably be high-TFR groups. Under this interpretation, if the PPRs at each parity were standardized by dividing by α_0, the curves in Figures 2.6 and 2.7 (top) should be similar. In fact, this seems to be the case (Figure 2.7, bottom panel).

THE AGE PATTERN OF NATURAL FERTILITY

We have seen that there is marked variation in the level of natural fertility within and among human populations. Despite this fact, the *age pattern* of natural fertility is surprisingly consistent. The top left panel of Figure 2.8 shows age-specific marital fertility rates for several populations spanning the observed range of variation in natural fertility; the lower left panel shows the same rates standardized so the rate for age 20–24 is equal to one (i.e., the rate at each age is divided by the rate at 20–24). Clearly, when this sort of correction is made for the *level* of reproduction, the age pattern of marital fertility becomes virtually invariant across natural-fertility populations. The curve of fertility declines monotonically from a peak in the early twenties, nearing zero by the late forties, and is convex upward at all ages.[5] Moreover, this age pattern stands in clear contrast to what is observed in controlled-fertility populations, where fertility rates drop more rapidly with age and are concave upward at all ages (Figure 2.8, right-hand panels). This difference almost certainly reflects the fact that most couples using deliberate fertility control attain their desired family size fairly early in marriage and then terminate their reproduction through contraception or surgical sterilization.

This explanation is supported by an observation first made by Henry (1976). When data on marital fertility are broken down not only by the

wife's age at each birth but also by her age at marriage, another difference between natural and controlled fertility becomes evident. In the
natural-fertility case (Figure 2.9, top panel), the resulting curves overlap
broadly and do not differ in any clear or consistent fashion. Under controlled fertility, in contrast, marital fertility rates are staggered by the
duration of marriage: the rates for longer durations (i.e., women marrying at earlier ages) drop off earlier than those for shorter durations (Figure 2.9, bottom panel).

Henry (1976) interprets this difference in the following way: Consider an idealized natural-fertility population in which marital fertility is
independent of the age at marriage and is constant from age 20 to age
40 but zero thereafter (Figure 2.10, solid line). Let us now introduce fertility control into this population, under the assumption that most couples want only a limited number of children, say, two or three. As long
as these couples have not yet had the two or three children they desire,
they will not practice birth control and their fertility will be about the
same as under the earlier natural-fertility regime. From the second or
third birth on, however, they will effectively terminate their reproduction—or, if contraception is imperfect, at least their fertility will be far
lower than it would have been in the complete absence of control. Thus,
under controlled fertility we should observe patterns analogous to the
broken curves in Figure 2.10. For two ages at marriage separated by five
years we have two staggered curves, themselves separated by five years;
in both instances the desired family size is attained early in marriage,
but the age of the wife at which it is attained differs by five years on
average between the two groups. Thus, concludes Henry (1976:93), "a
highly simplified model brings out the essential elements of the changes
observed when the shift from natural fertility to controlled fertility
occurs, i.e., the changes in the form of the curve and the appearance of
a marked influence of age at marriage."

This simple model underscores Henry's belief that fertility control is
used primarily for stopping rather than spacing births. It also reveals
one of his original motives for studying natural-fertility populations: he
believed that, under controlled fertility, the demographic impact of
purely age-related changes in reproductive physiology would be
masked by the effect of marital duration, but that the marital duration
effect would be entirely absent under a regime of natural fertility. However, as more recent analyses have shown, there is a measurable though
very small effect of marital duration on rates of natural marital fertility
(Page, 1977; Henry, 1979; Mineau and Trussell, 1982; Rodriguez and Cleland, 1988). Presumably this effect does not reflect fertility control but is
the joint outcome of two separate processes: (1) an increase in the cumulative incidence of pathological sterility with increasing duration of

exposure to intercourse (see Chapter 10), and (2) an apparently universal decline in coital frequency with increasing marital duration (see Chapter 7). Whatever its source, the observed influence of the duration of marriage on natural fertility is much less than in controlled-fertility cases.

The adjusted age curve of natural marital fertility shown in Figure 2.8 (lower left panel) has been found in almost all human populations lacking access to modern contraceptives and induced abortion, or at least in almost all such populations for which the necessary data exist. The few exceptions include parts of Nepal and highland Papua New Guinea, where marital fertility rates at age 20–24 are lower than at 25–29 (Knodel, 1983; Wood, 1987a). Both these areas are characterized by late sexual maturation (Wood et al., 1985a; Becker, 1993), which tends to lower fertility during the early twenties. Nonetheless, the populations that deviate from the usual age pattern of natural fertility appear to represent a distinct minority of all natural-fertility populations, and they deviate mostly at younger ages, not throughout the entire reproductive span (see Knodel, 1983, for other exceptions).

A MODEL OF NATURAL FERTILITY AND FERTILITY CONTROL

In one of his earliest papers on the subject, Henry (1961) developed a standard age curve of natural fertility based on the average marital fertility rates from several populations; this standard curve looks very much like the ones in the lower left-hand panel of Figure 2.8. He also suggested that observed departures from the standard curve may provide evidence of deliberate fertility control, although he developed no statistical methodology for implementing this idea. Coale (1971) suggested that a vector describing the effects of birth control (which, like Henry, he took to be synonymous with deliberate, parity-dependent stopping behavior) could be used in conjunction with Henry's standard curve to predict the age pattern of controlled fertility. He also noted that the age patterns of *departures* from natural fertility were very similar in all the controlled-fertility populations he examined. A few years later, Coale and Trussell (1974) incorporated these ideas into a full-scale model of marital fertility. This model provides the basis for a statistical method that has been widely used in an attempt to detect deliberate fertility control. Because many demographers still consider it the method of choice for analyzing marital fertility rates, Coale and Trussell's model is worth examining in some detail.

The Coale-Trussell model (as it is usually called) has four basic elements: (1) an underlying age pattern of natural marital fertility, (2) a scale parameter that determines the *level* of underlying natural fertility, (3) an age pattern of departures from natural fertility attributable to deliberate control, and (4) another scale parameter that determines the extent of deliberate control. Elements (1) and (3) are assumed to be universal and are given by two empirically derived age schedules. Only elements (2) and (4) are actually estimated for any particular population.

The specification chosen by Coale and Trussell for their model is

$$r(a) = n(a)M \exp [mv(a)] \qquad (2.12)$$

Under this model, the predicted marital fertility rate $r(a)$ at ages a to $a +$ k is determined by $n(a)$, the standard natural-fertility rate at those ages, multiplied by M, the scale parameter setting the level of natural fertility, and then discounted by an exponential term containing m, the scale parameter determining the degree of fertility control, and $v(a)$, the standard effect of control at those ages. If $m = 0$ (i.e., no fertility control), then $\exp[mv(a)] = 1$, and the model reduces to one of pure natural fertility. Finding significant nonzero values of m is thus a principal goal of applications of the Coale-Trussell model.

As noted before, the schedules of $n(a)$ and $v(a)$ are taken as given. Henry (1961) originally examined data from 13 populations to form his standard $n(a)$ schedule; Coale and Trussell (1974) rejected three of them because of uncertain data quality. The 10 remaining populations on which they based their own schedule of $n(a)$ values are listed in Table 2.1, along with the data used to compute the schedule. Given this schedule, they then estimated $v(a)$ values from data on 43 developed countries reported in the United Nations *Demographic Yearbook* of 1965. The resulting values of $n(a)$ and $v(a)$ are listed in Table 2.2 and plotted in Figure 2.11. The curve of natural marital fertility is very much like what we have already seen, save that Coale and Trussell include ages less than 20 years. The curve of departures from natural fertility attributable to deliberate control increases steadily in absolute value with maternal age, as would be expected if such control were principally a matter of stopping behavior.

Given these two schedules, we can now fit the model to any observed set of ASMFRs. For example, taking logs and rearranging equation 2.12, we obtain

$$\log\left[\frac{r(a)}{n(a)}\right] = \log M + mv(a) \qquad (2.13)$$

which is a simple linear equation that can be fit by standard least-squares regression.[6] Alternatively, Broström (1983, 1985) has derived maximum likelihood estimators of M and m on the assumption that marital fertility varies at each age as a Poisson random variable. As we have seen, this is tantamount to assuming that fertility is homogeneous at each age. Leaving aside whether this assumption is likely to be universally true, Broström's approach is preferable because of the desirable sampling properties of maximum likelihood estimates, and because it leads to a natural test of the goodness-of-fit of the model based on the likelihood ratio statistic (see Chapter 3). Broström (1983) has also derived the 95 percent confidence ellipses around the joint maximum likelihood estimates of M and m (Figure 2.12). These ellipses can be interpreted as standard 95 percent confidence intervals, except that they surround a point defined by two joint estimates rather than a single estimate.

It would seem, then, that we have a powerful tool for detecting the presence of fertility control, and for estimating what natural fertility would be in the absence of such control. Or do we? Despite the widespread use of the Coale-Trussell model, it is subject to numerous criticisms. Blake, for example, has argued that the model places too great a burden on a single measure of fertility: "To judge such finely tuned degrees of intent from age-specific marital fertility rates has seemed to be a supreme act of faith" (Blake, 1985:394). As noted in the previous chapter, detection of parity-dependent *changes* in reproduction is insufficient by itself to demonstrate deliberate *control* of fertility. And as Trussell himself has emphasized, the model assumes that deliberate control always takes the form of stopping behavior, not spacing behavior; in the presence of deliberate spacing, the estimates of M and m can be negatively confounded (Trussell, 1979). Such confounding could explain cases in which the level of underlying natural fertility appears to drop as fertility control increases, e.g., Sweden from the 1870s to the 1950s (Trussell, 1979).

In addition, for purely statistical reasons there is a *positive* confounding of the estimates of M and m under Broström's maximum likelihood method, because the covariance of the estimates is positive and is also large relative to the sampling variances of the estimates (Broström, 1985). This confounding is reflected in the positive tilt of the joint confidence ellipses in Figure 2.12. In view of this fact, one should use caution in interpreting the sorts of findings shown in Figure 2.13, in which underlying natural fertility seems to increase steadily as fertility control increases. This pattern, which has been found in several cases, is often

Table 2.1. Data Used in Calculating the Coale-Trussell Standard Schedule of Natural Fertility

		Age of Women (Years)						
Population	Female cohort	15–19	20–24	25–29	30–34	35–39	40–44	45–49
Norway	marriages 1874–1876							
woman-years		1,536	16,211	33,465	38,361	36,271	36,676	35,839
births		388	6,437	12,712	13,072	10,488	6,576	1,473
Hutterites	marriages pre–1921							
woman-years		79	758	875	849	827	756	700
births		41	360	395	361	309	155	20
Hutterites	marriages 1921–1930							
woman-years		53	637	807	824	793	563	237
births		33	350	405	368	322	133	15
Canada	marriages 1700–1730							
woman-years		102	462	667	656	580	473	402
births		50	235	331	317	238	109	12
Geneva bourgeoisie	wives of men born pre–1600							
woman-years		91	232	312	303	280	236	213
births		24	90	113	99	77	29	4

Geneva bourgeoisie	wives of men born 1600–1649							
woman-years		98	265	334	322	307	292	246
births		41	139	162	138	88	41	4
Crulai (France)	marriages 1674–1742							
woman-years		66	306	599	633	588	506	400
births		21	128	257	225	172	72	4
Taiwan	women born around 1900							
woman-years		2,700	3,525	3,525	3,650	3,650	3,650	3,650
births		831	1,288	1,180	1,119	960	417	29
Sotteville (France)	marriages 1760–1790							
woman-years		1	28	100	158	142	120	105
births		1	14	44	68	42	15	1
Europeans of Tunis	marriages 1840–1859							
woman-years		185	431	502	463	344	294	224
births		76	202	216	186	112	56	3

From Statistique Générale de la France (1907), Eaton and Mayer (1953), Henripin (1954b), Henry (1956), Gautier and Henry (1958), Tuan (1958), Girard (1959), Ganiage (1960)

43

Table 2.2.   The Coale-Trussell Standard Schedules of $n(a)$ and $v(a)$ Values

	Age of Women (Years)						
	15–19	20–24	25–29	30–34	35–39	40–44	45–49
$n(a)$	0.411	0.460	0.431	0.395	0.322	0.167	0.024
$v(a)$	0.000	0.000	−0.279	−0.667	−1.042	−1.414	−1.671

From Coale and Trussell (1978)

interpreted as showing that the effectiveness of breastfeeding as a traditional birth-spacing mechanism declines during the onset of fertility control (e.g., Knodel, 1988). It is striking, however, that the apparent trajectory of changes in M and m in Figure 2.13 lies right along the main axes of all the joint confidence ellipses, precisely as would be expected if the trajectory were really an artifact of statistical confounding.

In many applications, the temporal trends in M and m simply make no sense whatsoever. An example from a Mexican-American community is shown in Figure 2.14. Despite the absence of any evidence for effective fertility control before about 1950, the estimates from the first half of the twentieth century seem to imply both a high level of underlying natural fertility *and* significant control. Moreover, between the 1920s and 1930s there seems to have been a significant increase in control but no change in natural fertility. In the 1950s, however, underlying natural fertility drops precipitously, while the degree of control slips downward toward its earlier level. Finally, starting in the 1960s we see a sustained decline in both natural fertility and fertility control. In fact, what we are probably seeing are multiple forms of confounding in different periods, rendering the apparent "trends" meaningless. In addition, we are seeing poor model fit: in fully half the decades considered, the model was rejected by the likelihood ratio test.

Unfortunately, very few studies based on the Coale-Trussell model ever present statistical tests of goodness-of-fit.[7] Poor fit can arise from several errors of model specification. For example, the Coale-Trussell model was found to fit poorly to data on the Gainj of highland New Guinea because the underlying "universal" schedule of natural fertility was inappropriate (Wood, 1987a). As mentioned earlier, marital fertility rates from highland New Guinea are often found to depart from the standard fertility schedule, not because of the presence of deliberate control but because of late sexual maturation. There is no reason to expect the schedule of $v(a)$ values to be any more universal; indeed, those values work rather poorly at higher ages even for the populations used to estimate them in the first place (Coale and Trussell, 1974).

When Broström's maximum likelihood analysis is performed, another reason for poor model specification is the assumption that reproduction at each age is Poisson. Any reproductive heterogeneity net of age violates this assumption. Unfortunately, the Poisson is the only available one-parameter model of the birth process; all others involve fitting two or more parameters, which requires more data than just age-specific marital fertility rates.

Finally, Wilson et al. (1988) have questioned the choice of data sets used to estimate the standard schedule of natural fertility in the first place. As already noted, Coale and Trussell selected 10 from the original 13 tabulated by Henry (1961), discarding three because of questionable data quality. Wilson and his colleagues went back to the original data sources for those 10 populations and discovered some interesting anomalies. Two of the 10 schedules selected by Coale and Trussell (Sotteville-les-Rouen and Crulai) had been adjusted by Henry; whatever the original intent of those adjustments, their effect was to make the data conform more closely to the standard curve of natural fertility.[8] Another schedule, that for nineteenth-century Norway, has been shown by subsequent research to be inaccurate (Backer, 1965). This is particularly unfortunate since the Norwegian schedule was based on by far the largest number of births (see Table 2.1). Except for Taiwan, all the other schedules represent small samples.

A concern about small sample size led Wilson and his colleagues to perform an enlightening set of analyses: they applied the Coale-Trussell model in turn to each of the 10 populations used by Coale and Trussell for estimating the standard schedule of natural fertility (Wilson et al., 1988). Their results are shown in Figure 2.15. Clearly these populations are a heterogeneous lot, with estimates of m ranging from −0.3 to 0.4 and M from 0.65 to 1.35. Three of these purportedly natural-fertility populations (Norway, Geneva 1600–1649, Taiwan) even have values of m that differ significantly from zero at the 0.05 level. And two others (Canada and the Hutterites pre-1921) have values that come very close to doing so. Some of the remaining populations, such as Sotteville-les-Rouen, probably fail to reject the value $m = 0$ only because their samples are too small to provide any statistical power.

Wilson and his colleagues do not report the goodness-of-fit of these analyses. To assess fit, it is important to recognize that the data being analyzed have also been used in calculating the standard schedule, thus distorting the degrees of freedom involved in the test. To correct for this bias, I redid the maximum likelihood analyses of each of the 10 standard schedules after first deleting that particular schedule from the standard set. (In fact this correction makes little difference.) These analyses showed that three of the ten standard populations (Hutterites 1921–

1930, Norway, and Taiwan) reject the model at the 0.05 level. Thus, the model does not fit fully 30 percent of the populations used to formulate it in the first place.

In sum, all the populations chosen as the basis of the standard schedule of natural fertility can be faulted on the grounds of data quality, significant departures from natural fertility, small sample sizes, or poor model fit. In light of these and the many other problems that arise in fitting and interpreting the Coale-Trussell model, it should be used, if at all, strictly for exploratory analyses.

SUMMARY

Returning to the question posed at the beginning of this chapter, it is most certainly not true that "when you've seen one maternity history from a natural-fertility population, you've seen them all." There is enormous variation in the level of natural fertility, both within and among human populations. This finding should be a cause for rejoicing, for such variation provides the raw material for analyzing the principal determinants of human reproduction. Although it might be thought that our task of explaining patterns of natural fertility would be easier if natural-fertility populations were in fact homogeneous, variability is a source of information on causality. Had it turned out that all natural-fertility populations were the same, and that all individuals within such populations reproduced at the same level, there would have been absolutely nothing we could learn about the factors influencing fertility. Only differences in effects reveal something about the differential contribution of causes.

This chapter has emphasized constancy as well as variability in the observed patterns of natural fertility. The age pattern of natural fertility, for example, is remarkably constant. Similarly, a few easily interpreted patterns emerge by viewing parity progression ratios from natural-fertility populations. Such regularity offers the hope that variation in natural fertility is not entirely unconstrained, but that a delimitable number of systematic factors that are amenable to analysis dominate the variability. In the next chapter we attempt to identify the most important of those factors.

NOTES

1.   Summation from zero to infinity should cause no alarm. Before the age of first possible reproduction, all values of $f(a)$ are equal to zero, just as they are for all ages after menopause. Thus, summation needs to be done only for values of a that fall within the reproductive span.

2. Because our sample of 70 natural-fertility populations included only 10 hunter-gatherer groups, it could reasonably be argued that we simply had insufficient statistical power to detect any real difference. Since the original analyses, our attention has been drawn to demographic data on four more hunter-gatherer groups: the Ayoreo of Bolivia (TFR = 6.2), the Agta of the Philippines (TFR = 6.5), the Aka of the Central African Republic (TFR = 6.2), and the Ache of Paraguay (TFR = 7.7–8.3) (Bugos and McCarthy, 1984; Goodman et al., 1985; Hewlett, 1987; Hill and Kaplan, 1987). Unfortunately, none of those studies meets our minimum requirements for sample size. However, even if all four groups are added to the analysis, our conclusion about the absence of an effect of subsistence practices is not altered in the slightest. Bentley et al. (1993) have redone our analyses using a slightly different data set and have found a borderline-significant difference between peasant agriculturalists and other groups, but that difference vanishes when our more stringent sample-size criteria are retained. More to the point, subsistence only explains about 8 percent of the observed variation in TFRs, even in the sample of Bentley et al. (1993).

3. Specifically, λ is assumed to be distributed among subgroups as a gamma random variable (Johnson and Kotz, 1969:124–125).

4. The fit of the negative binomial to these data is, of course, not perfect. Aside from sampling error, the primary reason for this lack of fit is that many more couples than expected have had *no* births. That is, the negative binomial model does not allow for a special category of infertile couples. Two methods have been suggested to deal with this problem. First, Golbeck (1981) treats completed fertility as a mixture of two separate negative binomial random variables, one corresponding to subfertile couples, the other to normal but still heterogeneous couples. Second, Chakraborty (1989) has developed maximum likelihood estimators for a *truncated* distribution, one in which the zero class has been deleted. Golbeck and Chakraborty's methods both provide excellent fits to the data used in Figure 2.4.

5. Marital fertility rates are less consistent before age 20. In some populations the rate at ages 15–19 is higher or about equal to that at age 20, in others it is markedly lower. This lack of consistency appears to reflect the variable effects of adolescent subfecundity (see Chapter 9).

6. Wilson et al. (1988) suggest using weighted regression to reduce the effects of small samples at some ages, as well as the tendency for residuals to be large at extreme ages.

7. Coale and Trussell (1978) advocated using an arbitrary upper cutoff point for the mean squared error as a rule of thumb in judging goodness-of-fit, a practice that has been widely adopted. However, as Coale and Trussell point out, this ad hoc method has no real grounding in statistical theory.

8. Henry (1961) mentions that some adjustments had been made but does not make explicit what they were or why he made them. Since most of these adjustments were made for women at young ages when few births occur, Wilson et al. (1988) speculate that Henry may have been trying to smooth random fluctuations.

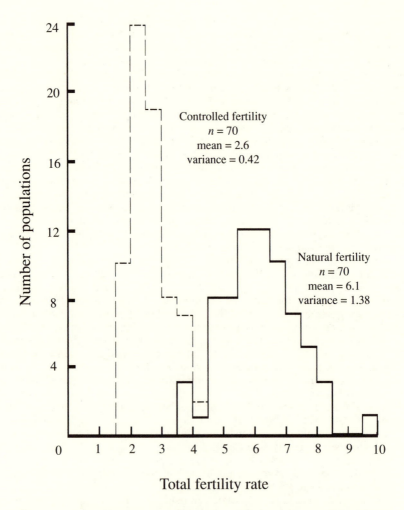

Figure 2.1 Distribution of total fertility rates or mean completed family sizes in 70 natural-fertility populations *(solid histogram)* and 70 controlled-fertility populations *(broken histogram)*. Populations known to have high prevalences of pathological sterility are excluded. (For original data sources, see Campbell and Wood, 1988)

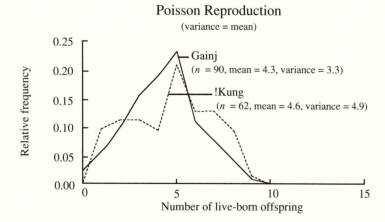

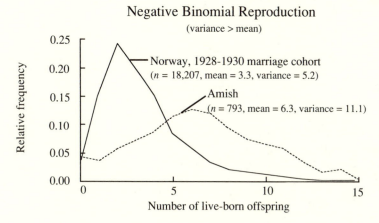

Figure 2.2 Distribution of completed family sizes reported by postre-productive-age women in four populations: the Gainj of Papua New Guinea, the !Kung San of Botswana, Norway (1928-1930 marriage cohort), and the Amish of Holmes County, Ohio. These distributions fall into two general categories, named for the probability models that most closely approximate them: *Poisson reproduction*, in which the mean and variance of completed family size are approximately equal, and *negative binomial reproduction*, in which the variance exceeds the mean, often by a substantial amount. (Data from Cross and McKusick, 1970; Henry, 1976; Howell, 1979; Wood et al., 1985a)

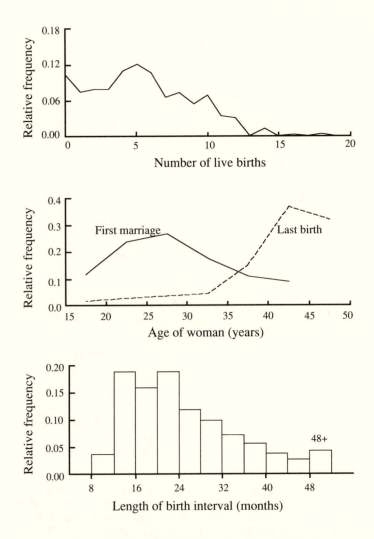

Figure 2.3. Reproductive variability in Tourouvre-au-Perche, France, marriages of 1665-1714: *(top)* distribution of completed family size among women who survived to age 50 (mean = 6.0, variance = 13.1); *(middle)* distribution of the ages of women at first marriage and last birth; *(bottom)* distribution of completed intervals between first and second births. (Redrawn from Bongaarts, 1983; data from Charbonneau, 1970)

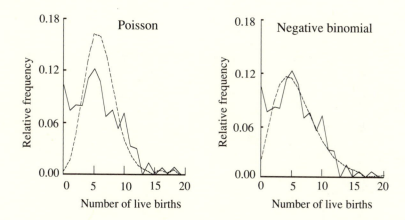

Figure 2.4. Fit of two probability models to data on completed family size, Tourouvre-au-Perche, France, marriages of 1665-1714: *(left)* Poisson model; *(right)* negative binomial model. In each panel, the solid line represents the observed distribution; the broken line is the theoretical distribution fit by the method of moments. (Data from Charbonneau, 1970)

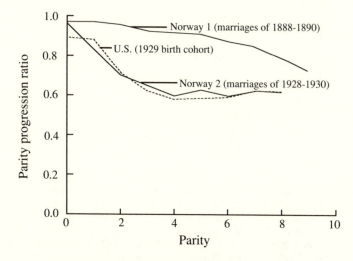

Figure 2.5. Distribution of parity progression ratios under a regime of natural fertility (Norway 1) versus controlled fertility (Norway 2 and the United States). (Data from Henry, 1976)

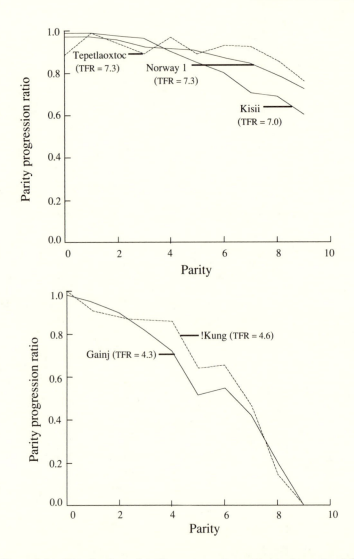

Figure 2.6. Distribution of parity progression ratios under natural fertility: *(top)* populations with high natural fertility, *(bottom)* populations with low natural fertility primarily attributable to long birth intervals. The Kisii are located in Kenya; Tepetlaoxtoc is a rural *municipio* in Mexico. (Data from Brass, 1958; Henry, 1976; Howell, 1979; Golbeck, 1981; Wood et al., 1985a)

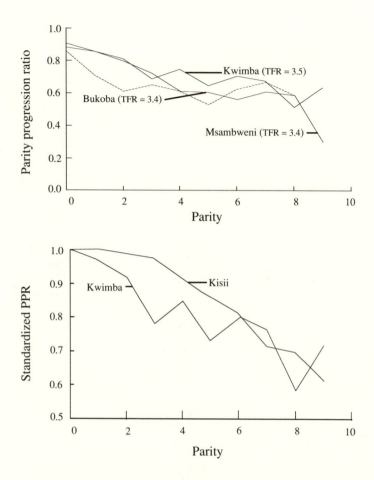

Figure 2.7. *(Top)* distribution of parity progression ratios under natural fertility in the presence of widespread pathological sterility. Kwimba and Bukoba are in Tanzania, Msambweni in Kenya. All three districts are in a region of East Africa known to have high prevalences of sterility, apparently associated with gonorrhea and chlamydia infections. *(Bottom)* distribution of standardized parity progression ratios in two East African populations, one of which has high levels of pathological sterility (Kwimba) and the other of which does not (Kisii). Standardized PPRs are equal to a_x/a_0 ($x = 0,1,2,...$). (Data from Brass, 1958)

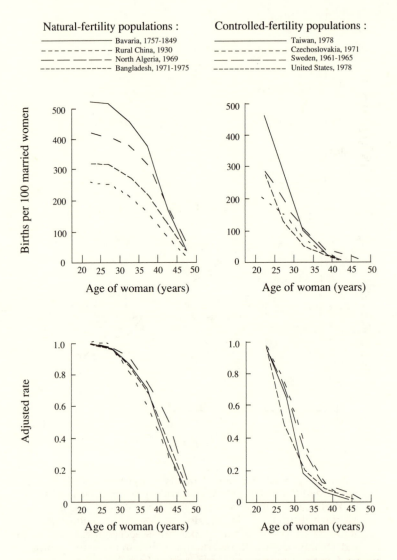

Figure 2.8.   Age-specific marital fertility rates for several natural-fertility populations *(left-hand panels)* and controlled-fertility populations *(right-hand panels)*. Adjusted rates are the raw rates at each
age divided by the rate at age 20-24. (Redrawn from Knodel, 1983;
Wood, 1989)

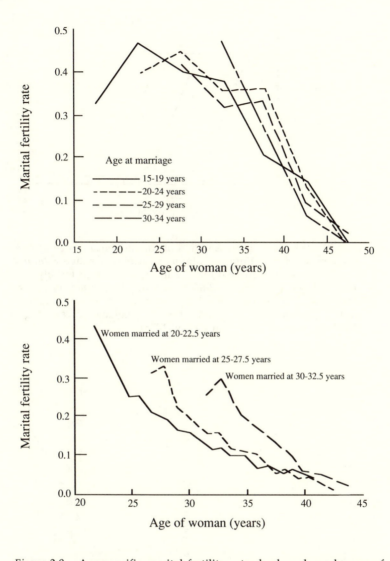

Figure 2.9. Age-specific marital fertility rates broken down by age of wife at marriage: *(top)* a natural fertility case (Crulai, Normandy, late seventeenth to early eighteenth centuries); *(bottom)* a controlled-fertility case (Great Britain, twentieth century). (Redrawn from Henry, 1976)

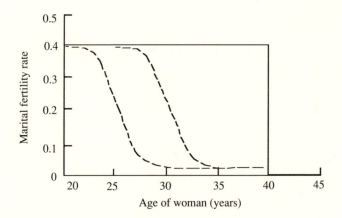

Figure 2.10. Schematic representation of the influence of age of marriage on marital fertility rates under controlled fertility. *Solid line,* the distribution of marital fertility in the absence of fertility control, all ages at marriage. *Broken lines,* the distribution of marital fertility in the presence of effective control for women marrying at two different ages: 20 years *(left-hand curve)* and 25 years *(right-hand curve).* (Redrawn from Henry, 1976)

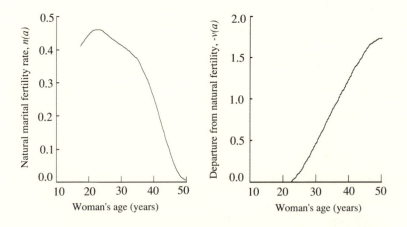

Figure 2.11. The Coale-Trussell standard values of *n(a)* *(left)* and *v(a)* *(right).* Note that –*v(a)* is plotted. Curves have been smoothed by fitting spline functions. (Values from Coale and Trussell, 1978)

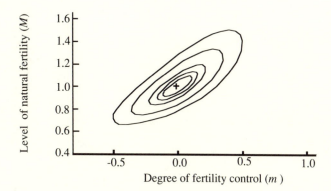

Figure 2.12. The Coale-Trussell model: 95 percent confidence ellipses around the point $m = 0$ and $M = 1$ for sample sizes equal to (reading from the outermost to innermost circles) 100, 200, 500, 1000, and 2000 births. Derived by the maximum likelihood method of Broström (1985). (Redrawn from Wilson et al., 1988)

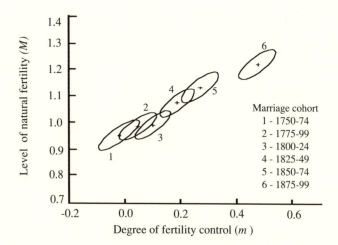

Figure 2.13. The Coale-Trussell model: maximum likelihood estimates of M and m and 95 percent confidence ellipses for 14 German villages, by marriage cohort. (Redrawn from Wilson et al., 1988)

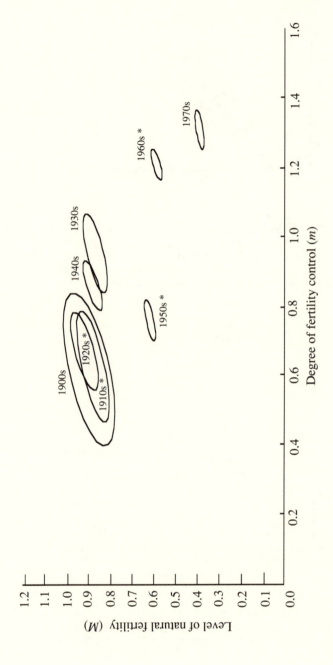

Figure 2.14   The Coale-Trussell model: maximum likelihood estimates of M and m and 95 percent confidence ellipses for Mexican-American women of Laredo, Texas, by period. Decades with a significantly poor model fit ($P < 0.05$) according to the likelihood ratio test of Broström (1988) are marked by an asterisk. (Data supplied by A. V. Buchanan and K. M. Weiss)

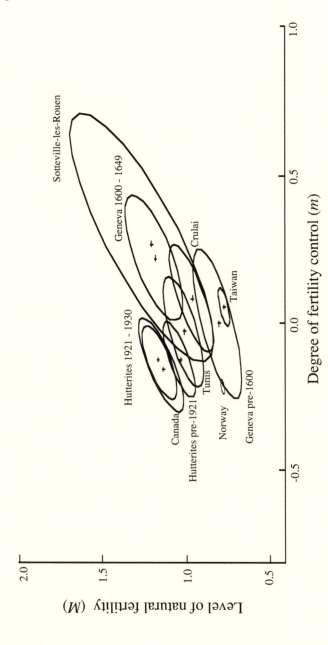

Figure 2.15. The Coale-Trussell model: maximum likelihood estimates of *M* and *m* and 95 percent confidence ellipses for the 10 data sets used to calculate the Coale-Trussell standard *n(a)* schedule. See Table 2.1 for sample sizes. (Redrawn from Wilson et al., 1988)

3

Frameworks for Analyzing the Determinants of Natural Fertility

In the previous chapter, we reviewed evidence for variation in natural fertility, both within and among human populations. One conclusion of that review was that all human populations reproduce at levels far below their biological capacity: in all societies factors that limit reproduction exist. Another conclusion was that simple categorizations of human populations do little to explain interpopulation variation, and by their nature such categorizations tell us nothing about *intra*-population variation. This does not mean, as we have noted, that fertility variation is inexplicable, but only that a more fine-grained approach is needed. In the present chapter, we will review some methods that have been developed in recent years for isolating the sources of variation in natural fertility.

In trying to understand the logic behind these methods, it may be helpful to think in terms of a concrete physical metaphor, namely a series of biochemical reactions that transform some specific input substance into a particular end-product or output—for example, the metabolic reactions that convert starches into simple sugars in the digestive tract. Figure 3.1 schematizes one such series of reactions, involving four conversions (steps 1–4) and three intermediate products (B, C, and D). The input, output, and intermediate products can be thought of as distinct but linked *states* making up the process as a whole.

Suppose we measure each part of the process in one particular person and discover that the conversions take place at differing rates, as indicated in the figure. Clearly the overall speed at which the final output is produced is determined primarily by its slowest step, in this case step 3. For obvious reasons this is known as the *rate-limiting step* (RLS).

The amount of time spent in each state is inversely related to the rate of the conversion immediately following that state.[1] For example, very little time is spent in state D because D is converted almost immediately into the final output E. In general, the greatest time is spent in the state immediately preceding the RLS, in this case state C.

Now suppose we measure the same process in another person and find that the conversion of initial input to final output occurs much more rapidly. Barring the possibility of a difference in the actual sequence of intermediate steps, this observed difference in the overall conversion rate must be caused by a difference in the rate of at least one of the component steps. By the same token, a given step can contribute to the difference between individuals if and only if its rate differs between them. This is as true of the RLS as of any other step. Moreover, the RLS in one individual may or may not be the RLS in the other. If, however, the same step is the RLS in both, and if the speed of the RLS does differ between them, then this difference is likely to have a larger effect than any other step on interindividual variation in the overall rate.

There is a further point to be made. Suppose we discover that the difference between our subjects is caused entirely by a difference in the rate at which C is converted into D. Ultimately there may be many possible explanations for this difference—genetic, nutritional, or whatever. No matter what the ultimate cause, however, it is relevant to the problem at hand only insofar as it affects the conversion of C to D. More generally, any factor, whatever its nature, can influence the rate at which the final output is produced only if it affects one or more of the intermediate steps. Thinking in these terms thus provides a convenient starting point or framework for the analysis of *any* factor acting on the process.

We can think of the intermediate steps as a series of *limiting factors* that place a ceiling on the total output produced per unit of time. Some limiting factors are more powerful than others—that is, they slow down production of output more than other steps do—and the RLS is by definition the most powerful of all. Variation among individuals in the rate of output must ultimately be traceable to variation in one or more limiting factors. And the same can be said of variation from time to time within the same individual.

Since differences among populations are built up from differences among individuals, all that has been said here is applicable to interpopulation variation as well. However, just as the limiting factors that are important for any one individual need not be important for variation among individuals, so too important sources of interindividual (but *intra*-population) variation need not have anything to do with variation among populations. The sources of variation at these different levels are effectively decoupled, at least in principle.

What does all this have to do with fertility? A great deal, as it happens, for it is possible to think of the reproductive process as a series of discrete states and conversions among states that occur at variable rates, much as shown in Figure 3.1. Moreover, reproduction is under sufficiently tight biological control that essentially the same states and conversions occur in all humans. Only the *rates* of conversion, or equivalently the amounts of time spent in each state, vary among individuals or populations. If we can identify the relevant states, as well as the conversions that link them, we will have made an important first step toward understanding fertility variation.

THE PACE OF CHILDBEARING

To develop this logic further, consider differences in the average rate of reproduction in some "archetypal" human populations, such as those shown in Figure 3.2. Although these examples are artificial, the "modern developed" case might be taken to represent recent experience in the United States, Europe, or Japan; "historical European" might be seventeenth-century England or France; and "nonwestern developing" might be India or Pakistan during the 1960s. The other two examples represent extreme simplifications of specific populations: the "Hutterites" case is based on the Anabaptist sect mentioned in Chapter 2, who for a time during the 1940s and 1950s attained the highest level of natural fertility ever recorded (Eaton and Mayer, 1953); the "Gainj" example is based on a small tribal population in highland New Guinea known to have a comparatively low level of natural fertility (Wood et al., 1985a, 1985b).

Figure 3.2 breaks down each of these examples into a series of states and times spent in those states. This scheme is logically very similar to what was shown in Figure 3.1, except that certain states, such as pregnancy, are repeatable. The "input" is now a cohort of females who are both premenarcheal and unmarried, and the ultimate "output" of interest is the total number of live births produced across the reproductive span. In other words, the final output of this process is just what the total fertility rate is intended to measure.[2]

As a standard of comparison, Figure 3.2 also shows one possible version of the maximal rate of childbearing attainable by humans, one in which all women first become pregnant at age 15, lose no pregnancies, ovulate within 1.5 months of each birth, and then wait another three months until their next conception. As we saw in Chapter 2, no known population reproduces at a level anywhere near this theoretical maximum, although individual women may come close and occasionally even surpass it. In all populations, various limiting factors serve to reduce achieved fertility to a level far below the maximum.

The categorization of states in Figure 3.2 suggests what these limiting factors are and how they limit fertility in various settings. In the "modern developed" example, several years are lost to reproduction prior to marriage. In addition, since female fecundity declines precipitously during the last few years before menopause (see Chapter 10), time is also lost at the end of the reproductive span. In this case, there is also a fairly long lag between marriage and the first conception—or rather the first fertile conception, since only those conceptions that produce live births are shown in the figure. Although ovulation resumes fairly rapidly following each birth, there is another long lag between return of ovulation and second fertile conception—a long *fecund waiting time* to next conception, to use demographic jargon.[3] Finally, there is an extraordinarily long lag following the second birth, one that continues to the onset of permanent sterility. Thus, in the modern developed case a low total reproductive output (TFR = 2) is achieved primarily through long periods of nonreproduction within marriage, periods that are at least potentially fecund; these long periods in turn reflect the effects of modern contraception and induced abortion.

The historical European pattern is markedly different: although marriage is considerably later, this delay is more than offset by an absence of the long, potentially fecund waiting times typical of developed societies. Most "nonwestern developing" populations, in contrast, are characterized by earlier marriage, which acts to raise total reproductive output, but also by a much later postpartum return of ovulation—an effect now known to be caused by prolonged breastfeeding (see Chapter 8).

And so on for the other populations: the Hutterites marry somewhat later but have much shorter postpartum anovulation, whereas the Gainj marry at a comparatively early age but experience extremely prolonged postpartum anovulation; they also display a surprisingly long lag between marriage and the first fertile conception, apparently because of a combination of prolonged adolescent subfecundity and irregular cohabitation of spouses during the early years of marriage. The net effect of these differences is that the Hutterites have an extremely high TFR and the Gainj a much lower one.

For the moment, however, the important consideration is not what is going on within any particular population, but rather the fact that we can characterize the fertility experience of a wide variety of human populations—populations that span the range of known variation in total fertility rates—in terms of a small number of events (marriage, conception, birth, etc.) and states (single, pregnant, anovulatory because of breastfeeding, permanently sterile, etc.). We can also learn something about the effectiveness of each reproductive state as a limiting factor, and how each limiting factor varies among populations. Looking at the nonwestern developing case, for example, it is obvious that permanent sterility is an important limiting factor. Indeed, if women in this population did not become permanently sterile before age 50, they could squeeze approximately three more births into their reproductive spans, all other things being equal. One way to express the fertility-reducing effect of permanent sterility quantitatively is as the percentage of the total potential reproductive span that women of a particular population spend in the permanently sterile state (Figure 3.3). In the example shown here, permanent sterility is somewhat more effective than nonmarriage as a limiting factor but not quite as effective as breastfeeding, which is the RLS.

Table 3.1 compares the time spent in various reproductive states among all our archetypal populations. It also provides a measure of interpopulation variability in the time spent in each state, namely, the *coefficient of variation* (the standard deviation of the percentage of time spent in each state divided by the corresponding mean). Restricting our attention to the natural-fertility examples, it appears that marriage and postpartum anovulation are the most variable states and thus contribute most to interpopulation variation in total reproductive output. Note that marriage is important despite the fact that in some of the populations (e.g., the nonwestern developing case) it is a rather weak limiting factor. The important thing here is not its effectiveness within any one population, but its variability across populations.

The next most variable state appears to be pregnancy. The time spent in that state, however, is the outcome of all the other limiting factors since the time spent in pregnancy is directly associated with the total fertility rate. (The higher the total fertility rate, the larger the number of pregnancies experienced by the average woman and the longer the time spent in the pregnant state.) Thus, in looking at pregnancies we are really looking at the end-product of the process and not at an intermediate step.

Compared with marriage and postpartum anovulation, both the potentially fecundable state and permanent sterility appear to contribute little to interpopulation variation in fertility—despite the fact that permanent sterility is a powerful limiting factor within every population. Again the important factor in explaining interpopulation fertility varia-

Table 3.1. Approximate Percentage of the Potential Reproductive Span (Ages 15–50) Spent in Various Reproductive States by Women in Selected Types of Societies

			Reproductive State			
Society	Unmarried	Postpartum anovulatory	Fecundable	Permanently sterile	Pregnant
Modern developed	20	1	51	24	4
Historical European	31	15	16	25	13
Nonwestern developing	10	30	19	26	15
Hutterites	23	13	21	24	19
Gainj	14	34	17	27	8
Coefficient of variation[a]	48	46	12	5	33

[a] Excluding modern developed
Modified from Bongaarts and Potter (1983)

tion is not effectiveness within a single population but variability across populations.

The examples given thus far are not based on genuine data; they are merely intended to illustrate the logic that will be applied in the rest of this chapter. Despite the artificial nature of these examples, however, we can still draw several important conclusions from them. First, variation in reproductive output is wholly attributable to differences in the timing of reproductive events and the time spent in various reproductive states—in other words, the *pace of childbearing*. Although every human population is unique in the precise way it times reproduction, all populations work with the same basic components of fertility, i.e., with the same events and states. It is these shared components, not the similarities or differences in the actual pace of childbearing, that permit us to analyze the sources of interpopulation variation. In addition, although there is no necessary relationship between either the causes or the levels of intra- and interpopulation variation, both forms of variation are built up from precisely the same components. This fact greatly simplifies the analytical task ahead of us.

All that remains to be done is to develop a slightly more detailed, biologically realistic list of fertility components and devise methods for estimating their effects using real data.

THE PROXIMATE DETERMINANTS OF FERTILITY

In the remainder of this chapter I adopt an approach that has proven quite popular among demographers in recent years, although it is still unfamiliar to most reproductive biologists. This is the so-called proximate determinants approach, originally formulated in a pioneering paper by Kingsley Davis and Judith Blake (1956) and more recently refined and championed by John Bongaarts (1976, 1978, 1982b) and others (e.g., Hobcraft and Little, 1984; Palloni, 1984; Wood, 1990).

This approach can be described as follows. In trying to account for variation in fertility, one might be tempted to catalog every single factor potentially affecting reproduction, including socioeconomic variables such as religious affiliation or wife's educational background, behavioral variables such as coital frequency, and a host of physiological variables ranging from maternal nutritional status to the frequency of ovulation or the viability of sperm. Such a catalog, even if possible to compile, would tell us little about which of these variables are likely to be most important in explaining fertility variation. Fortunately, because reproduction is such a regular, systematic process, it is possible to compile a much shorter list of the variables that have a direct effect on fertility and that must always be operating at some level if reproduction is to occur at all. These variables are known as the *proximate determinants of fertility.*

If any other factor is to affect fertility, it must do so via its effects on one or more of the proximate determinants. For example, a couple's religious affiliation (which is not a proximate determinant) can affect fertility only if it affects some more proximate factor, such as the frequency of sexual intercourse. Thus, the proximate determinants mediate the effects of more remote influences on fertility and, for that reason, are often referred to as *intermediate fertility variables.*

Figure 3.4 is a schematic representation of this logic. The figure makes the additional point that there can be secondary feedback effects of fertility itself on both the proximate determinants and more remote influences, as well as feedback effects of proximate determinants on remote influences. For example, if maternal nutritional status is a remote influence, repeated rounds of pregnancy and lactation may act as nutritional drains on the mother and thus worsen her nutritional condition. Although such feedback effects may well be crucial in regulating fertility, we ignore them for present purposes and concentrate instead on the direct effects of the proximate determinants.

One great advantage of thinking in terms of the proximate determinants is that it is fairly easy to provide an exhaustive list of them. The first attempt to compile such a list—indeed, the first application of this logic—was by Davis and Blake in 1956 (Table 3.2). It is worth examining their list in some detail for both historical and conceptual reasons.

As these authors put it,

> The process of reproduction involves three necessary steps sufficiently obvious to be generally recognized in human culture: (1) intercourse, (2) conception, and (3) gestation and parturition. In analyzing cultural influences on fertility, one may well start with the factors directly connected with these three steps. Such factors would be those through which, and only through which, cultural conditions *can* affect fertility. For this reason, by way of convenience, they can be called the "intermediate variables" and can be presented schematically [as in Table 3.2]. . . . It is clear that *any* cultural factor that affects fertility must do so in some way classifiable under one or another of our eleven intermediate variables. Hence the latter provide a framework in terms of which the relevance of cultural factors to fertility can be judged. . . . It is also clear that *each* of the eleven variables may have a negative (minus) or positive (plus) effect on fertility (Davis and Blake, 1956:211–213; emphasis in original).

Table 3.2. Davis and Blake's Intermediate Fertility Variables

---

I. *Factors affecting exposure to intercourse*
  A. Those governing the formation and dissolution of unions in the reproductive period
    1. Age of entry into sexual unions
    2. Permanent celibacy: proportion of women never entering sexual unions
    3. Amount of reproductive period spent after or between unions:
      a. when unions are broken by divorce, separation, or desertion
      b. when unions are broken by death of husband
  B. Those governing the exposure to intercourse within unions
    4. Voluntary abstinence
    5. Involuntary abstinence (from impotence, illness, temporary separations)
    6. Coital frequency (excluding periods of abstinence)

II. *Factors affecting exposure to conception*
    7. Fecundity or infecundity, as affected by involuntary causes
    8. Use or nonuse of contraception:
      a. by mechanical and chemical means
      b. by other means (the rhythm method, withdrawal, etc.)
    9. Fecundity or infecundity, as affected by voluntary causes (sterilization, subincision, medical treatment, etc.)

III. *Factors affecting gestation and successful parturition*
    10. Fetal mortality from involuntary causes
    11. Fetal mortality from voluntary causes

From Davis and Blake (1956)

This paragraph clearly contains the essence of the logic developed here, save that we now wish to gauge the influence of all possible factors, not just cultural ones.

The ideas formulated by Davis and Blake lay fallow for more than two decades until they were picked up again by Bongaarts in his famous 1978 paper. Bongaarts's most important contribution in this area was the development of a quantitative method for assessing the impact of each proximate determinant, a method that we will review below. In addition, he reorganized and simplified Davis and Blake's determinants, as shown in Table 3.3. A principal aim of this reorganization was to incorporate Louis Henry's distinction between natural and controlled fertility. Thus, contraception and induced abortion are included as "deliberate marital fertility control factors," with the stipulation that they operate strictly in parity-dependent fashion.

Table 3.3.   Bongaarts's Proximate Determinants of Fertility

I. *Exposure factors*
   1. Proportion married

II. *Deliberate marital fertility control factors*
   2. Contraception
   3. Induced abortion

III. *Natural marital fertility factors*
   4. Lactational infecundability
   5. Frequency of intercourse
   6. Sterility
   7. Spontaneous intrauterine mortality
   8. Duration of the fertile period

From Bongaarts (1978)

Bongaarts describes his list of determinants as follows:

1. Proportions married: This variable is intended to measure the proportion of women of reproductive age that engages in sexual intercourse regularly. All women living in sexual unions should theoretically be included, but to circumvent difficult measurement problems, the present analysis deals only with the childbearing of women living in stable sexual unions, such as formal marriages and consensual unions. For convenience, the term "marriage" is used to refer to all such unions.

2. Contraception: Any deliberate parity-dependent practice—including abstention and sterilization—undertaken to reduce the risk of conception is considered contraception. Thus defined, the absence of contraception and induced abortion implies the existence of natural fertility.

3. Induced abortion: This variable includes any practice that deliberately interrupts the normal course of gestation.

4. Lactational infecundability: Following a pregnancy a woman remains infecundable (i.e., unable to conceive) until the normal pattern of ovulation and menstruation is restored. The duration of the period of infecundity is a function of the duration and intensity of lactation.

5. Frequency of intercourse: This variable measures normal variations in the rate of intercourse, including those due to temporary separation or illness. Excluded is the effect of voluntary abstinence—total or periodic—to avoid pregnancy.

6. Sterility: Women are sterile before menarche, the beginning of the menstrual function, and after menopause, but a couple may become sterile before the woman reaches menopause for reasons other than contraceptive sterilization.

Table 3.4. The Proximate Determinants of Natural Fertility

I. *Exposure factors*
  1. Age at marriage or entry into sexual union
  2. Age at menarche
  3. Age at menopause
  4. Age at onset of pathological sterility (if earlier than menopause)

II. *Susceptibility factors*
  5. Duration of lactational infecundability
  6. Duration of the fecund waiting time to conception (determined by the following fecundability factors):
     a. frequency of insemination
     b. length of ovarian cycles
     c. proportion of cycles ovulatory
     d. duration of the fertile period, given ovulation[a]
     e. probability of conception from a single insemination in the fertile period
  7. Probability of fetal loss[b]
  8. Length of the nonsusceptible period associated with each fetal loss[c]
  9. Length of gestation resulting in live birth

---

[a] The fertile period is the mid-portion of the cycle when insemination has some nonzero probability of resulting in conception.
[b] Following World Health Organization recommendations (WHO, 1977), fetal loss refers to all spontaneous abortions and thus includes both miscarriages and stillbirths.
[c] The nonsusceptible period associated with a fetal loss includes the truncated gestation ending in loss, plus any residual period of infecundability following the loss.
Modified from Campbell and Wood (1988)

  7. Spontaneous intrauterine mortality: A proportion of all conceptions does not result in a live birth because some pregnancies end in a spontaneous abortion or stillbirth.
  8. Duration of the fertile period: A woman is able to conceive for only a short period . . . in the middle of the menstrual cycle when ovulation takes place (Bongaarts, 1978:107).

In this book, I will adopt a slightly different set of proximate determinants, one that my collaborators and I have developed over the course of our own research on fertility (Table 3.4). This set of determinants builds upon that of Bongaarts but differs from his in two important respects. First, because much of the material developed in subsequent chapters is physiological in nature, a more detailed breakdown of the biological determinants of fertility is necessary. Second, because this book is concerned overwhelmingly with natural fertility, deliberate, parity-specific control factors can be ignored.

In addition, there is a deeper disagreement over the ways in which these lists deal with fertility control. Bongaarts begins with a strong presumption that *all* deliberate fertility control is parity-dependent, and then he builds that presumption into his list of determinants. In this view neither contraception nor induced abortion is ever used as a spacing mechanism (as opposed to a stopping mechanism). Clearly, this assumption may not be correct. To circumvent this problem, we prefer to exclude *all* forms of deliberate control from our list. We do this not because we believe that such control is never important, even in supposedly natural-fertility settings, but because we do not want to prejudice the question of how control operates. Any form of control—whether deliberate or not, parity-dependent or not—can be thought of as operating via one or more of our determinants. Induced abortion, for example, can be viewed as an artificial elevation of the probability of fetal loss; similarly, barrier methods of contraception operate by modifying the probability of conception per insemination, the contraceptive pill affects the probability that a given cycle is ovulatory, and so on for any other form of deliberate fertility control. Thus, our list of determinants establishes a set of baselines against which to gauge the effects of deliberate control, no matter how it is practiced.

Returning to Table 3.4, we classify the proximate determinants of natural fertility into *exposure factors,* which determine if there can be any nonzero probability of conception, and *susceptibility factors,* which govern the conditional probability of successful reproduction given that exposure occurs. One might think of the exposure factors (menarche, marriage, sterility, and menopause) as a series of on-off switches, all of which must be in the proper position for reproduction to occur at all. In reality, none of these factors operates in such a black-and-white manner. Menarche, for example, is followed by a period of subfecundity before the establishment of regular ovarian function, whereas menopause is preceded by a period of declining ovarian function (see Chapter 9). In our scheme, these complications can be handled in terms of age- or time-dependent variation in the susceptibility factors.

If the exposure factors are on-off switches, then the susceptibility factors are rheostats that modify the intensity of reproduction once it has been switched on. Most of the susceptibility factors should be self-explanatory, but some require comment. The *fecund waiting time to conception* refers to the lag between a woman's entrance into the fecund state (in which she is potentially able to conceive) and her next conception. This lag results from the simple fact that conception is never inevitable, not even if a viable sperm and egg are simultaneously present in the female reproductive tract. To express the uncertainty inherent in conception, demographers use the term *fecundability,* defined as the probability that a fecund couple will conceive during a month of exposure to unprotected intercourse. Fecundability determines the expected fecund waiting time to conception: in a homogeneous population, the mean fecund wait is equal to the inverse of fecundability (see Chapter 7). Thus, if fecundability is 0.25, the mean waiting time to conception is four months.

Fecundability is determined by several behavioral and physiological factors, only two of which (the frequency of unprotected intercourse or insemination and the length of the fertile period) are explicitly recognized by Bongaarts. In addition to those two factors, fecundability is affected by the total length of the female cycle, simply because cycle length determines the number of ovulations that can occur during any month of exposure. In addition, a given cycle may or may not make an egg available for fertilization; thus, fecundability can be affected by the proportion of cycles that are *ovulatory*. And even if ovulation occurs and insemination (deposition of sperm in the upper vaginal canal) takes place within the fertile period associated with that ovulation, conception is still not inevitable. Thus, we need to take the probability of conception from a single insemination within the fertile period into account when analyzing fecundability.

With regard to the nonfecundability factors in Table 3.4, natural fertility is clearly influenced by the prevalence and age distribution of intrauterine mortality or *fetal loss,* and by the length of gestation. If the gestation ends in fetal loss, then gestation length affects the total *nonsusceptible* or *infecundable period* associated with intrauterine death. Even when a pregnancy results in a live birth, variation in gestation length could in principle induce fertility variation. To take an extreme example, if gestation could be shortened by half without increasing the risk of reproductive failure, far more live births could be packed into a woman's reproductive career. In reality, of course, such shortening is impossible.

The list of determinants in Table 3.4 will act as our guide throughout the remainder of this book. Our ultimate goal will be to assess the contributions of all these determinants to fertility variation within and among human populations. Before pursuing that goal, however, we need to return to Bongaarts's list of determinants (Table 3.3) as a springboard for a discussion of his method for partitioning the sources of fertility variation.

BONGAARTS'S DECOMPOSITION OF THE PROXIMATE DETERMINANTS

To measure the fertility-reducing effect of each proximate determinant, Bongaarts begins with the TFR associated with a particular population and then sequentially adjusts it upwards, using appropriate data, to estimate what reproductive output would have been in the absence of the determinant's effect. For this purpose, Bongaarts (1978) uses several indices of reproductive performance in addition to the TFR (Figure 3.5):

TMFR = the total marital fertility rate, equal to the expected TFR of women who experience the population's observed age-specific marital fertility rates, but who all marry before age 15 and remain married throughout reproductive life; thus, the TMFR represents the TFR adjusted to remove the effects of nonmarriage.

TNM = the total *natural* marital fertility rate, equal to the expected TMFR of women who use no artificial contraception or induced abortion; thus, the TNM represents the TMFR adjusted to remove the effects of deliberate fertility control.

TF = the total fecundity rate, equal to the expected TNM of women who do not nurse any of their children; thus, the TF represents the TNM adjusted to remove the effects of breastfeeding.

Associated with each sequential adjustment (from TFR to TMFR to TNM to TF) is a coefficient that measures the fertility-reducing effect of the relevant proximate determinant:

C_m = the coefficient of nonmarriage
C_c = the coefficient of deliberate fertility control (i.e., contraception and induced abortion)[4]
C_i = the coefficient of lactational infecundability

The relationships among coefficients and indices are illustrated in Figure 3.5. Note that each coefficient is constructed so that it falls between zero and one, and the larger the coefficient the *smaller* the fertility-reducing effect of the associated proximate determinant. Thus, when a given C-value equals one, the associated determinant has no effect on fertility.

The problem now is how to estimate the C-coefficients from empirical data. The coefficient of nonmarriage is fairly straightforward. From Figure 3.5, $C_m = \text{TFR}/\text{TMFR}$, which can be rewritten

$$C_m = \frac{\sum \text{ASFR}}{\sum \text{ASMFR}} = \frac{\sum_a g(a)r(a)}{\sum_a r(a)} \tag{3.1}$$

where $g(a)$ is the proportion of women already married at age a, and the other quantities are as defined in Chapter 2. Thus, C_m can be estimated directly from standard demographic indices available for a wide variety of populations.

Figure 3.6 shows the relationships observed among C_m and the TFR and TMFR for several human populations. Developed nations as a rule show smaller values of C_m and hence larger fertility-reducing effects of nonmarriage than do most developing nations. Two countries (Bangladesh and Nepal) show almost no effect of nonmarriage (C_m > 0.85). In those countries virtually all women marry early and either remain married or, if their marriages do end, remarry rapidly; very little of their potential reproductive span is spent in the single state.

If all the fertility variation among human groups were caused by differences in marriage patterns, there would be no variation in TMFR values left over after partialling out the C_m coefficient—that is, all the populations in Figure 3.6 would fall along a single horizontal line. This plainly is not the case: one or more additional proximate determinants must be acting to induce fertility variation *within* marriage.

The coefficient of deliberate fertility control (C_c) is more difficult to estimate. Since we are primarily interested in natural-fertility populations, for which $C_c = 1$ by definition, the reader is referred to Bongaarts (1978, 1982b) for a detailed description of the estimation procedure. It is worth noting, however, that the procedure requires some strong assumptions about contraceptive use-effectiveness, the prevalence of "natural" sterility, the age distribution of fecundity, and the number of births averted by induced abortion.

Figure 3.7 shows the relationships among the C_c coefficient and the TMFR and TNM indices. Developed nations have much lower C_c values than do developing nations, which reflects their higher use-prevalences of contraception and induced abortion. Interestingly, they also tend to have higher TNM values once the effects of deliberate control are removed. That is, the scatter of points in Figure 3.7 is not random but shows a decreasing linear trend, with developed counties clustering at the upper left end of the line and developing countries at the lower right. This pattern almost certainly reflects the shift toward less intensive breastfeeding and earlier weaning often found during the course of economic development (Coale et al., 1979; Romaniuk, 1980, 1981). At any rate, a considerable scatter of points remains along the vertical axis in Figure 3.6, and much of this scatter may be attributable to the effects of breastfeeding.

In assessing the fertility-reducing effects of breastfeeding, account must be taken not only of the prevalence of breastfeeding but also of the average amount of time that it adds to the birth interval. Figure 3.8 dissects a typical completed birth interval into several major components: the postpartum infecundable period (or the time to first postpartum ovulation), the fecund waiting time to conception, the time added to the birth interval by fetal loss, and the gestation ending in the next live birth. In estimating C_i Bongaarts assumes that breastfeeding affects only the first of these components, the postpartum infecundable period.[5] In the absence of breastfeeding, this period is typically one to two months long (Habicht et al., 1985; Jones, 1988a), but with lactation its duration takes on some value i determined by the intensity and duration of suck- ling. Assuming that the other components of the birth interval take on the "typical" values shown in Figure 3.8, the average birth interval would be 20 months long without breastfeeding and 18.5 + i months long with breastfeeding. If i is known, then the fertility-reducing effect of lactation is simply

$$C_i = \frac{20}{18.5 + i} \tag{3.2}$$

For most populations, unfortunately, the actual value of i is unknown. However, aggregate data from several studies suggest that i can be esti- mated from the mean duration of breastfeeding in months (B), some- thing that is much more easily measured, using the regression equation

$$\hat{i} = 1.753 \exp(0.1396B - 0.001872B^2) \tag{3.3}$$

(Bongaarts and Potter, 1983:25).[6] The value of i estimated from equation 3.3 can then be substituted in equation 3.2 to estimate C_i.

Once the effects of breastfeeding have been removed by this method, very little residual variation in total fecundity remains (Figure 3.9). With the exception of one outlier (Finland in 1971), all TF values fall between 12 and 18 and cluster around a mean of 15.4. According to Bongaarts this finding indicates that the remaining proximate determinants (fre- quency of intercourse, sterility, fetal loss, and the duration of the fertile period) contribute little if anything to interpopulation variation in fertil- ity; consequently, he does not develop methods for estimating their effects. In this view, once the effects of nonmarriage, deliberate control, and breastfeeding have been removed, women in *all* populations should produce just over 15 children during their reproductive spans, assum- ing that they survive to menopause. This interpretation does not, how- ever, mean that 15.4 births is the theoretical maximum fertility that could be achieved in the absence of all limiting factors. Although coital frequency, fetal loss, sterility, and the length of the fertile period may cause little variation *among* populations, they are all powerful limiting factors *within* populations. Bongaarts's analysis does not allow estima- tion of such intrapopulation effects.

Table 3.5 presents estimates of the C-coefficients for various popula- tions, both recent and historical. Several patterns are readily discernible.

In developing countries, for example, nonmarriage is a strong limiting factor (as indicated by the mean value of C_m) but not a very important cause of interpopulation variation (as indicated by the coefficient of variation); differential contraception and induced abortion are the most important sources of variation among developing countries. Not sur-

Table 3.5. Estimates of the Coefficients of the Proximate Determinants and the Total Fecundity Rate (TF) for Selected Populations

Population	C_m	C_c	C_i	TF
DEVELOPING COUNTRIES				
Bangladesh, 1975	0.853	0.929	0.539	14.8
Colombia, 1976	0.578	0.646	0.841	14.6
Costa Rica, 1976	0.571	0.406	0.905	17.6
Dominican Republic, 1975	0.601	0.692	0.860	16.4
Guatemala, 1972	0.724	0.972	0.612	16.4
Hong Kong, 1978	0.496	0.331	0.930	14.8
Indonesia, 1976	0.706	0.756	0.577	15.2
Jamaica, 1976	0.541	0.637	0.879	14.3
Jordan, 1976	0.745	0.782	0.800	15.9
Kenya, 1976	0.768	0.976	0.673	15.9
South Korea, 1970	0.580	0.631	0.658	16.5
Lebanon, 1976	0.576	0.686	0.780	15.5
Malaysia, 1974	0.607	0.697	0.897	12.5
Mexico, 1976	0.610	0.731	0.841	15.3
Nepal, 1976	0.852	0.980	0.550	13.9
Pakistan, 1975	0.785	0.955	0.642	14.6
Panama, 1976	0.640	0.475	0.879	17.1
Peru, 1977	0.573	0.739	0.759	15.9
Philippines, 1976	0.613	0.705	0.759	15.3
Sri Lanka, 1975	0.513	0.710	0.608	15.9
Syria, 1973	0.730	0.793	0.730	16.6
Thailand, 1975	0.628	0.676	0.660	16.8
Turkey, 1968	0.760	0.698	0.730	14.5
Mean	0.654	0.722	0.744	15.5
Coefficient of variation	0.160	0.237	0.165	0.08
DEVELOPED COUNTRIES				
Denmark, 1970	0.555	0.257	(0.930)†	13.4
Finland, 1971	0.514	0.152	(0.930)	22.2
France, 1972	0.519	0.311	(0.930)	14.7
Hungary, 1966	0.617	0.184	(0.930)	17.0
Poland, 1972	0.437	0.362	(0.930)	14.2
United Kingdom, 1967	0.609	0.258	(0.930)	16.3
United States, 1967	0.631	0.254	(0.930)	15.7
Yugoslavia, 1970	0.572	0.273	(0.930)	14.5
Mean	0.557	0.256	(0.930)	15.1‡
Coefficient of variation	0.117	0.258	—	0.08‡

Table 3.5.   Continued

Population	C_m	C_c	C_i	TF
HISTORICAL EUROPEAN (OR EUROPEAN-DERIVED) POPULATIONS				
Bavaria, 1700–1850	(0.374)	(1.000)	0.856	13.9
Crulai, 1674–1742	0.566	(1.000)	0.673	14.7
Grafenhausen, 1700–1850	(0.442)	(1.000)	0.671	16.0
Hutterites, 1921–1930	0.733	(1.000)	0.816	15.9
Ile de France, 1740–1778	0.505	(1.000)	0.712	17.0
Oschelbron, 1700–1850	(0.477)	(1.000)	0.727	14.6
Quebec, 1700–1730	0.629	(1.000)	0.810	15.7
Tourouvre, 1664–1714	0.591	(1.000)	0.749	13.6
Waldeck, 1700–1850	(0.442)	(1.000)	0.676	14.8
Werdum, 1700–1850	0.403	(1.000)	0.640	14.7
Mean	0.516	(1.000)	0.733	15.9
Coefficient of variation	0.218	—	0.100	0.07

† Figures in parentheses are approximate
‡ Excluding Finland
Modified from Bongaarts and Potter (1983)

prisingly, in developed countries deliberate control is both the most important limiting factor and the most potent source of interpopulation variation. In historical populations, by contrast, nonmarriage is most important from both points of view. Note, however, that in historical populations as well as developing countries the effect of breastfeeding both as a limiting factor and as a source of variation is far from negligible.

AN ALTERNATIVE APPROACH

Because Bongaarts's analysis is easy to apply to aggregate data, it has been widely used in recent years (see, for example, Cleland and Chidambaram, 1981; Casterline et al., 1983; Bongaarts et al., 1984; Kalule-Sabiti, 1984; Bentley, 1985; Singh et al., 1985; Khalifa, 1986; Ross et al., 1986; Wilmsen, 1986). Although the method has proven useful in disaggregating fertility variation in a general way, it has several limitations, chief among which are the following:

1. The method is not statistical. That is, it is not based on an underlying probability model and does not permit either formal estimation or the calculation of error terms for the derived coefficients. As a consequence it is extremely difficult to assess the reliability of particular estimates of the C-coefficients or gauge the model's fit.

2. The estimation procedures involve a large number of simplifying assumptions. In the case of the index of deliberate control, some of those assumptions have already been mentioned. They are also fairly obvious in the case of the index of lactational infecundability. As Figure 3.8 shows, strong assumptions are made about the "typical" length and invariance of birth-interval components, many of which are in fact not well known for most populations and all of which are likely to be influenced by maternal age, parity, and a host of other factors. Since these birth-interval components are determined by several of the proximate determinants *not* measured in Bongaarts's analysis (coital frequency, duration of the fertile period, and fetal loss), the assumption that these components are invariant would seem to take for granted something that the analysis is intended to demonstrate. This becomes a particular problem when i (the average duration of lactational infecundability) is estimated simply by subtracting 18.5 months from the observed mean birth interval (Bentley, 1985; Ross et al., 1986).

3. The method provides no basis for estimating the effects of more remote influences on fertility.

4. Since the method is designed to be used strictly with aggregate data, it is based on means rather than distributions and makes no connection between the micro (individual) and macro (population) levels of analysis. This fact makes it very difficult to assess the impact of compositional differences among populations (e.g., differences in age structure) on differences in the C-coefficients. The one exception is C_m, which is standardized by age.

5. Because the effect of each proximate determinant is removed sequentially, there is no basis for investigating the interactions of the various determinants.

6. Application of the method to World Fertility Survey data has not replicated the earlier finding that total fecundity rates tend to cluster around 15.4; instead, they indicate substantial residual variation in total fecundity (Cleland and Chidambaram, 1981; Casterline et al., 1983). This finding suggests that the other proximate determinants—those not explicitly entered into Bongaarts's analysis—do indeed cause some interpopulation variation in fertility; yet the method provides no basis for assessing those effects.

I do not mean to suggest that Bongaarts is unaware of these problems. The fact that he has gone on to develop alternative approaches clearly shows that he does not consider his original analysis to be the final word on fertility analysis. Other approaches based on the same fundamental logic have also been developed recently by Hobcraft and Little (1984) and Palloni (1984), among others.

None of these analyses, however, answers what I believe to be the most fundamental criticism of the original Bongaarts model: *it is not a model of the family-building process as it operates in real time.* Fate does not deal women 15.4 children at the beginning of their reproductive careers and then take away a few because of nonmarriage, a few more because of contraception and induced abortion, and a few more because of breastfeeding. It would seem that any model based upon such logic could provide little insight into the actual behavioral and physiological mechanisms regulating the reproductive process, and even less insight into the operation of more remote influences on fertility.

How, then, does family-building occur? Fundamentally it is a time-dependent process: women produce a certain number of children by the end of their reproductive lifetimes because of the way in which they time (deliberately or fortuitously) various reproductive events over the course of their lives. A more realistic model of the reproductive process would thus be based on the timing of various reproductive events and on the intervals between those events—in other words, it would be a model of *birth spacing.* As it happens, one such model can be constructed that meets all the objections to Bongaarts's method.[7]

The entire female reproductive life course can be envisioned as a series of time intervals (Figure 3.10). As subsequent chapters will show, each of *our* proximate determinants (Table 3.4) can be related to the distribution of one or more of the time segments in this figure. Most of the linkages are obvious. Lactational infecundability is coextensive with the time until first postpartum ovulation; the fecundability factors jointly determine the distribution of intervals from marriage to first conception, as well as from the first postpartum ovulation until the next conception; the probability of fetal loss, in conjunction with the nonsusceptible period associated with each loss, determines the distribution of times added to the birth interval by fetal death—and so on for all the other proximate determinants. *The proximate determinants exert a direct effect on fertility precisely because they determine the distribution of these time intervals.*

In this scheme, variation in fertility arises through variation in the distribution of time segments. Such variation can occur along any of three dimensions: among populations, among individuals within the same population, and within particular individuals (e.g., with age, parity, and marital status). One virtue of the scheme in Figure 3.10 is that it can accommodate variation along any or all these dimensions.

Implementation of this analytical framework requires two things: (1) methods for observing the reproductive events that delimit birth-interval segments, and (2) statistical techniques for making inferences about the distribution of the time intervals separating those events. Our ability to observe reproductive events in natural-fertility settings has improved dramatically in recent years because of new techniques for measuring the concentration of reproductive hormones in small samples of blood, urine, and saliva (Campbell, 1985, 1988; Ellison, 1988). Using well-established hormonal criteria, along with appropriate demographic data, we can now track women prospectively as they move from one birth-interval component to the next. As later chapters will show, it is now possible to determine fairly precisely from serial samples when a woman resumes ovulating during the postpartum period (Díaz et al., 1982; Howie and McNeilly, 1982; Brown et al., 1985; McNeilly, 1993), when a conception or a fetal loss occurs (Armstrong et al., 1984; Wilcox et al., 1985, 1988), or when a woman becomes menopausal (Richardson al., 1987). It is also possible to track the ovarian cycle itself with an exactitude unimaginable a few years ago (Campbell, 1985; Ellison, 1990). As detailed throughout this book, these new endocrinological techniques promise to revolutionize our understanding of the dynamics of the reproductive process (see the appendix for further discussion).

The other thing that analyzing the reproductive process in terms of birth-spacing patterns requires is the right kind of statistical methodology. Since much of that methodology is likely to be unfamiliar to many readers, the next section discusses it in detail.

ANALYTICAL ISSUES IN THE STUDY OF BIRTH-INTERVAL COMPONENTS

When analyzing birth-spacing patterns, the variable of interest is usually the duration of some particular birth-interval component, such as gestation, the interval from marriage to first conception, the period of lactational infecundability, or indeed the duration of the total birth interval itself. For at least three reasons, more traditional forms of statistical analysis, such as linear regression, are either inappropriate for or unenlightening about these *duration variables*. First, traditional methods are not usually based on *etiologic models,* that is, models intended to capture first principles about the underlying biological mechanisms. Rather, they involve *empirical models* whose principal aim is to achieve goodness-of-fit. Such methods by their nature provide little insight into causation. What we require is an analytical framework that will, at one and the same time, permit realistic modeling of the underlying birth-spacing processes *and* statistical estimation of the effects of interest. Second, traditional methods are poor at handling *time-varying covariates,* i.e., explanatory variables whose values change through time. For example, if we wish to gauge the impact of nutritional condition on a woman's length of lactational infecundability, traditional methods will not allow us to take into account any changes that may occur in her nutritional status over the course of the postpartum period. Such time-varying effects are common in fertility analysis. Third, traditional methods are unable to deal with the biases introduced by *censoring*, a problem that affects all birth-interval data. Any observation on the length of some birth-interval component is "completed" only if that component comes to an end while the woman is under observation. For some women, however, we may never observe completion, either because our study terminates before the component ends or because the subject is lost to follow-up through death, migration out of the study area, or withdrawal from the study. Observations that are begun but not completed during the study period are called *right-censored observations.*[8]

The general principle of right-censoring in prospective studies is illustrated in Figure 3.11, which shows observations on six subjects. Only observations on subjects 1, 2, and 4 are completed. All others are censored—subjects 3 and 5 because of termination of the study, subject 6 because she dropped out before the study was completed. Right-censoring can also occur in retrospective studies. For example, we may wish to draw inferences about birth spacing from maternity histories. Each woman in the study who is still of reproductive age may be able to provide completed information on the length of all intervals between previous births, but not on the current interval (that is, the interval opened by her most recent birth). In this case, the current birth interval is often called a *straddling interval* because it straddles the time of interview. Obviously, information on the straddling interval is right-censored.

The naive approach to censored observations is simply to ignore them, but doing so introduces systematic bias into our estimates because *observations censored by termination of the study are longer on average than uncensored observations.* This is easiest to see in observations that begin at the same time, for example, those for subjects 4 and 5 in Figure 3.11. In this instance, the observation censored by termination of the study obviously must be longer than the one completed during the study, even though we cannot know how long it actually is. (The more general case, in which observations begin at varying times, has been worked out in detail by Sheps and Menken [1972].) Another way to think about the bias introduced by ignoring censored observations is in terms of the *total period of exposure* experienced by the sample as a whole before completion of the birth-interval component of interest. By excluding censored observations from the analysis, we underestimate this period of exposure and thereby overestimate the likelihood of a component ending after a certain time. That is, we infer that the average length of the component is less than it actually is.

An alternative body of statistical techniques has been developed in the past two decades to handle censoring and other problems associated with duration variables, and these techniques are starting to find widespread application in fertility analysis. The relevant area of statistics is known as *hazards analysis, survival analysis,* or *event history analysis.* Several excellent overviews of these methods are now available: Allison (1984) and Lee (1992) are good introductions, and Blossfeld et al. (1989) and Yamaguchi (1991) provide comprehensive and up-to-date reviews at an intermediate level of difficulty. More advanced treatments are given by Elandt-Johnson and Johnson (1980), Kalbfleisch and Prentice (1980), Nelson (1982), Cox and Oakes (1984), and Lancaster (1990). A review of applications in human population biology, with many parallels to the present book, is given by Wood et al. (1992). In the rest of this chapter, I hope to provide enough detail about hazards analysis to illustrate two of its primary attractions for the study of human fertility: (1) it provides a general framework within which the etiologic processes underlying birth-interval dynamics can be modeled, and (2) it is associated with a surprisingly general procedure that permits efficient, unbiased estimation of the models of interest, even in the face of multiple forms of censoring.

A Brief Course in Hazards Analysis

Suppose we wish to characterize the timing of some important reproductive event, such as a conception or the onset of first menses during the postpartum period. We can do so in terms of a random duration variable *T*, which is the length of the time interval beginning with an individual's entry into a state of exposure to the risk of the event and ending in the event itself. For example, *T* might be a woman's age at the event of interest or the time since she experienced some prior event, such as entry into marriage or her previous birth. We will use *t* to denote specific values of *T*; obviously those values are strictly nonnegative.

The distribution of events in time is a direct reflection of the dynamic behavior of *T*, which can be described in various ways.[9] For example, we may wish to know the "risk" (in some general sense) that an event will occur at time *t*, given that it has not occurred previously. This risk may take the form of either a probability or an incidence rate. If the latter, the risk is known as the *hazard* or *hazard rate* of the event at *t*, a quantity so fundamental to this form of analysis that the whole field is often named for it. Although the hazard itself cannot be observed in any direct way, its implicit value determines the occurrence and timing of the events we do observe. Indeed, the principal goal of hazards analysis is to make inferences about the underlying hazard from observations on the timing of events. In addition to the hazard, we may wish to examine the distribution of times to the occurrence of the event. As for any continuous random variable, this distribution is supplied by the *probability density function* (PDF) of *T*. This sort of density function is the subject of our inferences when, for example, we inspect a histogram. In hazards analysis, the PDF is only partially observable: to what times should we assign censored cases? However, the methods described below permit estimation of the entire PDF even in the face of censoring. Once we know the PDF, we can calculate the mean time to the event, its variance, or anything else we may wish to know about *T*. Finally, it is often useful to know the probability that an event has *not yet occurred* by time *t*, a probability given by the *survival function* at *t*. Other quantities and functions occasionally crop up in hazards analysis, but these three—the hazard, the probability density function, and the survival function—are the ones most commonly seen.

When our attention is restricted to continuous-time models, the formal definitions of these three functions are as follows. First, the instantaneous hazard at time *t* is given by

$$h(t) = \lim_{\Delta t \to 0} \left[\frac{\text{Prob}(t \le T < t + \Delta t | t \le T)}{\Delta t} \right] \tag{3.4}$$

At the limit, this expression is not guaranteed to fall within the interval [0,1]. Although it will always be nonnegative, it has no upper bound and is highly dependent upon the time units chosen for the analysis. (The hazard that a particular woman will have a child in the next nanosecond is typically much less than the hazard that she will have one in the next decade.) Consequently, $h(t)$ is a rate rather than a probability—specifically, the rate at which events occur at exact time (or age) *t*, given that they have not occurred before *t*. This rate can always be converted into a proper probability, $q(t)$, of the event occurring during some small subinterval $[t - \frac{1}{2}\delta t, t + \frac{1}{2}\delta t]$ around *t* by solving

$$q(t) = 1 - \exp\left[-\int_{t-\frac{1}{2}\delta t}^{t+\frac{1}{2}\delta t} h(y)dy \right] \tag{3.5}$$

More simply, if $h(t)$ can be assumed to be constant over the subinterval, then $q(t) = 1 - \exp[-h(t) \cdot \delta t]$. For $h(t)$ in the interval $[0,\infty)$, both these expressions constrain $q(t)$ to fall between zero and one.

The probability density function of *T*, which tells us the distribution of time intervals to the event of interest, is defined as

$$f(t) = \lim_{\Delta t \to 0} \left[\frac{\text{Prob}(t \le T < t + \Delta t)}{\Delta t} \right] \tag{3.6}$$

Note that the sole difference between the hazard and the PDF is that the former is conditional on survival to *t*. Loosely speaking, *f(t)* tells us something about the probability of an event occurring at *t* as opposed to some other time, whereas *h(t)* tells us how likely the event is to occur at *t* given that it has not already taken place.

Finally, the survival function is given by

$$S(t) = \text{Prob}(T > t) = \int_t^\infty f(y)dy \tag{3.7}$$

S(t) is the probability that the time to the event of interest is *at least t* time units long. Since *T* cannot take on negative values, it follows that *S(0)* = 1. In addition, *S(t)* is monotonically nonincreasing with *t*, i.e., it can only go down or remain the same as *t* increases. Its lower limit can never be less than zero.

The hazard, density, and survival functions are related to each other in the following ways:

$$f(t) = \frac{-dS(t)}{dt} \tag{3.8}$$

$$h(t) = \frac{-d[\ln S(t)]}{dt} = \frac{f(t)}{S(t)} \tag{3.9}$$

$$S(t) = \exp[-\int_0^t h(y)dy] \tag{3.10}$$

$$f(t) = h(t) \exp[-\int_0^t h(y)dy] \tag{3.11}$$

These relationships will prove useful at several points in what follows.

In formulating a theoretical model for some reproductive event or birth-interval component, we attempt to specify some reasonable mathematical form for either *h(t)*, *f(t)*, or *S(t)*. In other words, we construct a *parametric model* by writing down an equation for one of these functions, an equation we hope will capture at least the main features of what we know about the dynamic behavior of *T*. Many well-characterized, "standard" parametric hazards models (e.g., the exponential, Weibull, Gompertz, lognormal, quadratic, log-logistic, gamma, or Rayleigh) may be appropriate for a wide variety of reproductive processes. Cohen and Whitten (1988) provide a useful catalog of such models, with detailed discussions of their genesis, properties, and estimation. Because of their flexibility, it is often possible to draw upon these standard models in specific applications of hazards analysis, and indeed any model with a right-skew, strictly nonnegative PDF is theoretically permissible. In general, however, it is much more enlightening (and difficult!) to develop

new models—etiologic models—that are specifically tailored to the biological processes of interest. For reproductive processes, it is usually
most intuitive to construct an etiologic model in terms of its hazard.
However, because of the mathematical relationships among $h(t)$, $f(t)$, and
$S(t)$, specification of a model for any one function is equivalent to specification of the other two.

Likelihood Estimation for Hazards Models

Although hazards models can be fit by a variety of methods, perhaps
the most powerful of these is *maximum likelihood estimation*. When computing maximum likelihood estimates for any class of models, we begin
by writing down the probability of obtaining each observation in a randomly drawn sample on the assumption that the model being fitted is
correct—i.e., that it is an accurate portrayal of the process that gave rise
to the observations. This probability is called the *likelihood* of the observation. As initially formulated, the likelihood is not an actual number
but rather an algebraic expression containing constants, known as *parameters*, whose numerical values are as yet unknown. If the observations
are independent—that is, if selection of one subject does not influence
the chance of any other being selected—then the *total likelihood* or *likelihood function* of the complete set of observations is simply the product
of the individual likelihoods. This is because the probability of a series
of independent events is equal to the product of the probabilities of the
events considered singly. Maximum likelihood estimates (MLEs) are
obtained by finding the numerical values of the model's parameters that
maximize the likelihood function. Thus, according to the maximum likelihood principle, the best-fitting version of a particular model is the one
that makes the data in hand most likely to have occurred. Maximization
of the likelihood function can sometimes be done with pencil and paper
but more often requires a computer. MLEs have several desirable statistical properties, including unbiasedness, efficiency, and normality.
Although, strictly speaking, these are all *asymptotic* properties (assured
of holding only in very large samples), they are often realized to a close
approximation in quite small samples.

Given a parametric hazards model, there is a remarkably general and
straightforward maximum likelihood procedure for data sets that
include right-censored observations. (More complex forms of censoring
can also be accommodated, as discussed by Wood et al. [1992].) The
likelihood of obtaining an uncensored observation equal to some value
t_i is simply $f(t_i)$, the PDF at t_i. In contrast, the likelihood of obtaining an

observation right-censored at time t_i is $S(t_i)$, the probability that the event takes place sometime *after* the censoring point. Thus, the total likelihood of obtaining a data set of n observations, allowing for right-censoring, can be written

$$L = \prod_{i=1}^{n} [f(t_i)]^{d_i} [S(t_i)]^{1-d_i} \qquad (3.12)$$

where t_i is the *i*th observed time to the event of interest or to censoring, whichever occurs first, and d_i is the associated *censoring index* (0 if the observation is censored, 1 if uncensored).[10] Since $f(t) = h(t)S(t)$ by equation 3.9, we can rewrite this expression as

$$L = \prod_{i=1}^{n} [h(t_i)]^{d_i} \cdot \exp\left[-\int_0^{t_i} h(y)dy\right] \qquad (3.13)$$

which is the form usually encountered in texts (e.g., Cox and Oakes, 1984:32). To obtain MLEs of the model's parameters, we simply substitute the model-derived expressions for $f(t)$ and $S(t)$ into equation 3.12 or for $h(t)$ into equation 3.13 and maximize L numerically using a standard computer algorithm. The useful series of books on *Numerical Recipes* (e.g., Press et al., 1989) provides computer code for many such maximization algorithms, written in several different programming languages.

In addition to the parameter estimates themselves, it is helpful to have some measure of how close the estimates are likely to come to the true parameter values, assuming, of course, that the model we have chosen is the correct one. That measure is provided by the *sampling variance* of the estimate, or more usually by its square root, the *standard error*. Since MLEs are asymptotically normal, we can say based upon normal-distribution theory that ± two standard errors provides an approximate 95 percent confidence interval for the estimate.

Standard errors for hazards models can be found by the same procedure used in any form of maximum likelihood estimation. First we compute the *information matrix* associated with the model. Let $\theta = (\theta_1, \theta_2, \ldots, \theta_k)$ be a vector of the k parameters of the model we wish to fit. The information matrix is then given by

$$\mathbf{I} = \begin{bmatrix} E\left(\dfrac{-\partial^2 \ln L}{\partial \theta_1^{\,2}}\right) & \cdots & E\left(\dfrac{-\partial^2 \ln L}{\partial \theta_1 \theta_k}\right) \\ \vdots & \cdot & \vdots \\ E\left(\dfrac{-\partial^2 \ln L}{\partial \theta_k \theta_1}\right) & \cdots & E\left(\dfrac{-\partial^2 \ln L}{\partial \theta_k^{\,2}}\right) \end{bmatrix} \qquad (3.14)$$

where the expectations are evaluated at the true value of θ. The estimated *asymptotic variance-covariance matrix* is then

$$\hat{\mathbf{V}} = \left. \mathbf{I}^{-1} \right|_{\theta=\hat{\theta}} \tag{3.15}$$

where $\hat{\theta}$ is the vector of MLEs of the model's parameters. The square
roots of the numbers along the principal diagonal of $\hat{\mathbf{V}}$ provide the standard errors of the estimated parameters. Nelson (1982) provides a useful discussion of numerical methods for approximating $\mathbf{I}$.

An Example

The principles discussed thus far can be illustrated with an example
drawn from fertility analysis. For various reasons, it may be of interest
to ask how likely a couple is to conceive during some fixed period of
regular sexual relations. Since pregnancy is not inevitable, even if both
partners are willing and able to conceive, they can expect to wait some
time before a conception actually occurs. But how long are they likely
to wait? This question is of theoretical interest precisely because the
fecund waiting time to conception is a major component of every birth
interval (see Figure 3.10). To answer the question we need to estimate
the couple's fecundability, since its value determines their expected
waiting time to conception. But fecundability is unobservable. Even
when the most advanced medical techniques are used, there is no way
we can simply examine a couple to determine their chances of conceiving. However, although fecundability has been treated as a probability
since it was first defined by Gini (1924), this probability has a one-to-one
relationship with an associated hazard via equation 3.5. This insight
immediately suggests that fecundability can be estimated from observations on the waiting time to conception through the application of
appropriate hazards models.

Not just any observations will do for this purpose. The use of data
from populations where contraception is widely practiced is problematic for two reasons. First, if we include couples who have used contraceptives, it is essential that we collect information on exactly *when* the
contraceptives were used so we can exclude those intervals from the
period of genuine exposure to the risk of conception. Such precise timing information, however, is very difficult to obtain. Second, if we
exclude such couples and restrict attention to individuals who never use
contraceptives, we are likely to be dealing with a highly unrepresentative subsample of the population. In particular, couples not using contraceptives may be selected for at least a self-assessment of subfecundi-

ty. It is therefore necessary to draw our data from a natural-fertility population. In addition, it is important to select a population without widespread pathological sterility, which can cause confounding problems similar to those associated with contraception. Since pathological sterility is often caused by certain sexually transmitted diseases (see Chapter 10), this restriction is tantamount to saying that we should work only with populations that have fairly conservative sexual mores. Having chosen an appropriate population, it is necessary to impose further restrictions. Most important, since breastfeeding is associated with a variable period of infecundability following childbirth, it is difficult to know when exposure to the risk of conception begins during the postpartum period in higher-order birth intervals. The usual way to avoid this problem is to deal strictly with the first birth interval—that is, the interval from marriage to the first conception. But this last restriction means that we must avoid populations in which premarital conceptions are common.

Miraculously, it is possible to find samples that meet all these criteria. Figure 3.12 shows the distribution of time intervals between marriage and the first fertile conception from an Old Order Amish community in rural Pennsylvania. Because of the possibility of pregnancy loss, the first fertile conception is not necessarily equivalent to the first conception ever; hence, these data allow us only to estimate what demographers call *effective fecundability,* which can be regarded as a lower-bound estimate of fecundability in the strict sense (see Chapter 7). Note that the distribution in Figure 3.12 is incomplete because of the existence of 21 right-censored cases. Some of these represent newlyweds who have not yet had time to conceive, but others represent couples who have been married five years or more without achieving a successful pregnancy. We return later to the question of whether the latter couples are actually capable of conceiving.

As discussed in Chapter 7, demographers delight in agonizing over what types of hazards models are best for estimating fecundability from data such as these. Here we begin with an absurdly simple model. Assume that all the couples in our sample have exactly the same value of fecundability, and that this value does not change through time— what can aptly be called the *homogeneous model* of fecundability. Under the homogeneous model, the monthly hazard of conception is itself a constant, which I will call λ. This model yields the following hazard, PDF, and survival function:

$$h(t) = \lambda, \lambda > 0 \tag{3.16}$$

$$f(t) = \lambda e^{-\lambda t} \tag{3.17}$$

$$S(t) = e^{-\lambda t} \tag{3.18}$$

Because its survival function takes on a simple negative exponential form, this model is often referred to as the *exponential survival model* or simply the *exponential model*. From equation 3.13, the likelihood function for this model is

$$L = \prod_{i=1}^{n} \lambda^{d_i} e^{-\lambda t_i} \tag{3.19}$$

This likelihood, which is extremely easy to maximize, yields an estimate of $\lambda = 0.067 \pm 0.001$ and an expected waiting time to first fertile conception of $1/\hat{\lambda} = 14.97 \pm 0.26$ months.[11] Using equation 3.5, the estimate of effective fecundability provided by this model is 0.065. As shown in Chapter 7, estimates of effective fecundability in most natural-fertility populations range from about 0.15 to 0.30. By comparison, the Amish estimate appears to be extraordinarily low. Thus, our analysis raises an interesting question: is there something unusual about Amish fertility, or is our model a bad one?

Goodness-of-Fit

Unfortunately, hazards analysis has been slow to develop methods for the assessment of goodness-of-fit. In many standard analyses, such as least-squares regression, it is simple to define residuals and investigate their behavior. Hazards analysis is fundamentally different: what is predicted is a hazard rate, which by its nature cannot be observed on an individual. How, then, can we assess the closeness of our predictions to our observations? Several different approaches to this problem have been proposed, none of which has received universal approval (for discussions, see Heckman and Walker, 1987; Wu, 1990). Here I review three methods selected for their broad applicability: (1) a very simple, intuitively appealing graphical method, (2) a more complex graphical method that is not at all intuitive but is ultimately more informative, and (3) a method based on likelihood theory that can be used to compare the fit of two competing models whenever the parameters of one model are a subset of those of the other. Our analysis of Amish fecundability can be used to illustrate all three methods.

The simplest and most appealing way to assess how well a hazards model fits is to compare the survival function predicted by the model with the survival function estimated from the same data by the maximum likelihood method of Kaplan and Meier (1958). Kaplan-Meier estimates (as they are called) are "semiparametric": they make no assumptions about the overall shape of the survival function, but rather approximate it as a piecewise exponential. That is, the hazard is assumed to be constant between each pair of successive uncensored events; the more closely spaced such events are, the closer these estimates come to the true survival function.

Kaplan-Meier estimates are computed as follows. Let $t_1 < t_2 < \cdots < t_k$ be the distinct, uncensored waiting times we observe. Suppose that w_j waiting times end in an uncensored event at t_j and that c_j waiting times are censored during $[t_j, t_{j+1})$ $(j = 1, \ldots, k)$. Then the size of the *risk set* at t_j (i.e., the pool of remaining subjects who have not yet experienced the event) is

$$R_j = \sum_{i=j}^{k} (w_i + c_i) \tag{3.20}$$

The Kaplan-Meier estimator of $S(t)$ is then

$$\hat{S}(t) = \prod_{j|t_j<t} \left[\frac{(R_j - w_j)}{R_j} \right] \tag{3.21}$$

with asymptotic sampling variance

$$\text{Vâr}[\hat{S}(t)] = [\hat{S}(t)]^2 \sum_{j|t_j<t} \left[\frac{w_j}{R_j(R_j - w_j)} \right] \tag{3.22}$$

The estimates of $S(t)$ provided by equation 3.21 are the closest we can ever get to the true survival function of the process under investigation without actually knowing what that function is. Therefore, if the fit of our parametric model is good, the survival function generated by the model should correspond closely to the Kaplan-Meier estimates of the survival function.

The left-hand panel of Figure 3.13 shows the Kaplan-Meier estimates and the model-predicted survival curve for our model of constant, homogeneous fecundability as applied to the Amish data. Clearly, the fit of this model is atrocious: almost every segment of the fitted curve is well over two standard errors away from the Kaplan-Meier estimates. Our earlier question—is there something wrong with the model or something unusual about the Amish?—has thus been answered: something is seriously wrong with the model.

However, since the left-hand panel of Figure 3.13 tells us something about *why* the model is bad, the exercise has not been a complete waste of time. Note that the Kaplan-Meier survival curve drops off more rapidly than predicted by the model during the first few months following marriage but then declines more slowly after about 10 months or so. According to equation 3.9, the rate of decline in the survival function (or the log of the survival function, which amounts to the same thing) is inversely related to the magnitude of the hazard. Thus, the contrast between the Kaplan-Meier estimates and the fitted survival function suggests that the fecundability of couples who have not yet conceived is relatively high immediately after marriage but then declines at later marital durations.

There are two possible explanations for such a change. First, the fecundability of each couple may actually lessen as time goes by, perhaps because sexual intercourse becomes less frequent. Second, couples may simply differ from each other in their fecundabilities, presumably because of differences in either reproductive physiology or sexual behavior. Under this second hypothesis of *heterogeneous fecundability*, the early, rapid decay of the survival function reflects a disproportionate number of high-fecundability couples dropping out of the sample: the higher the fecundability, the more likely a couple is to conceive during any month of exposure. As this process continues, the *remaining* sample of couples who have not yet conceived becomes increasingly selected for low fecundability. As a result of this selection process, the survival curve tends to bottom out at higher marital durations.

Neither of these hypotheses is consistent with our model of constant, homogeneous fecundability, and both probably have some merit. However, analyses summarized in Chapter 7 suggest that the second hypothesis—that fecundability varies among couples—is better supported by the available evidence from other populations. To explore this hypothesis, consider the simplest model of heterogeneous fecundability imaginable: suppose that the population is divided into two homogeneous fractions, one with some constant, nonzero fecundability and the other completely sterile (i.e., with a fecundability of zero). In general, the overall, pooled hazard for such a "two-point" distribution of subpopulation hazards is just the weighted average $\bar{h}(t) = \rho(t)h_1(t) + [1 - \rho(t)]h_2(t)$, where $h_1(t)$ and $h_2(t)$ are the hazards in the two subpopulations and $\rho(t)$ is the fraction of remaining couples at time t who belong to the first subpopulation. In our new model, we want the hazard in the first subpopulation to be some constant, which we will again call λ. Since the couples in the second subpopulation are all sterile, their hazard of a fertile conception is by definition zero. Some fairly straightforward algebra shows the survival function in the *pooled* population to be

$$\bar{S}(t) = \rho(0)S_1(t) + [1 - \rho(0)]S_2(t) = 1 - \rho(0)(1 - e^{-\lambda t}) \qquad (3.23)$$

The corresponding density function of waiting times to the first fertile conception in the pooled population is

$$\bar{f}(t) = \rho(0)\lambda e^{-\lambda t} \tag{3.24}$$

The likelihood function for this model follows immediately from equation 3.12.

The right-hand panel of Figure 3.13 shows the fit of this model to the Amish data. Although the fit is not perfect, it is unquestionably better than that of our original model. (Note that the departure of the Kaplan-Meier estimates from the fitted survival curve of the new model suggests that there may still be some residual heterogeneity among the non-sterile couples.) The estimated value of $\rho(0)$ is 0.934 ± 0.004. By definition, the complement of $\rho(0)$ is equal to the fraction of couples who are sterile at the time of marriage. Thus, it appears that something like 7 percent of Amish couples are unable to conceive from the very outset of married life, a figure consistent with estimates for other natural-fertility populations of European origin (see Chapter 10). Since $\hat{\lambda} = 0.146 \pm 0.012$ under the new model, the effective fecundability of Amish couples who are *not* sterile at the time of marriage is about 0.136 according to equation 3.5, and their waiting time to a fertile conception is only about seven months. In other words, the failure of our original model to consider the possibility that couples may vary in their fecundability led us to underestimate fecundability by more than half. The new estimate is much more in line with the values of effective fecundability estimated for other populations (see Chapter 7).

A second way to compare the fit of our two models is by examining their *martingale residuals* (Barlow and Prentice, 1988; Therneau et al., 1990). This new approach to assessing the fit of parametric hazards models is yet to be applied very widely. The martingale residual, so-called because its behavior is predicted by a mathematical theory developed for the class of stochastic processes known as martingales, is defined for the *i*th observation as

$$\varepsilon_i = d_i - \int_0^{t_i} \hat{h}(y)dy = d_i + \ln \hat{S}(t_i) \tag{3.25}$$

where d_i is the censoring index. This quantity can be interpreted as the difference between the number of events actually *observed* over an exposure of duration t_i, and the number *expected* if the model is correct. Martingale residuals can be used for model diagnostics in much the same way as standard residuals are used in least-squares regression. If our model has been specified correctly, the martingale residuals associated with it should have a mean of zero and a standard deviation of one, and they should follow a highly left-skew distribution with an upper bound of one and no lower limit. More precisely, the expected asymptotic cumulative distribution of the residuals is approximated by

$$\text{Prob}(\varepsilon_i \leq x) = \begin{cases} \exp(x - 1), & -\infty < x \leq 1 \\ 0, & x > 1 \end{cases} \tag{3.26}$$

This expectation provides one of the most powerful ways currently available for assessing the fit of a parametric hazards model. One drawback of this approach is that the theory underlying it is exceedingly complicated, and it is difficult to form an intuitive understanding of how these residuals ought to behave. Thankfully, the approach can be used to good effect even without such an understanding.

Figure 3.14 shows the cumulative distribution of martingale residuals for our two competing models of Amish fecundability, the first of which assumes that all couples have the same fecundability (Figure 3.14, left), the second allowing for some fraction of sterile couples (Figure 3.14, right). In each panel the smooth curve is the expected distribution of the residuals, which is identical for the two models (note the difference of scale along the x-axis). Once again, the simpler model provides an unambiguously bad fit, as indicated by the long lower tail of the observed residuals. The model that allows for sterile couples, in contrast, yields remarkably well-behaved residuals.

A final approach to assessing model fit capitalizes on the fact that our first model is *nested* within the second one; i.e., the single parameter of the homogeneous fecundability model is a *proper subset* of the two parameters of the heterogeneous fecundability model. Stated differently, we can reduce the second model to the first one by assuming that there are no sterile couples, thus forcing $\rho(0)$ to equal one. The first model is therefore less general than the second, which merely states that $\rho(0)$ takes on some value that may or may not equal one. When models are nested, the likelihood at the maximum for the less general model (the one with fewer parameters) can be compared with that for the more general model in terms of the likelihood ratio Λ, defined as

$$\Lambda = \frac{L(\text{less general model})}{L(\text{more general model})} \tag{3.27}$$

where the L values are computed using the MLEs of their respective parameters. Under the null hypothesis that the less general model is correct, $-2 \ln \Lambda$ is asymptotically distributed as a chi-square random variable with degrees of freedom equal to the difference in the number of parameters estimated for each model, i.e., one for the present test (Kendall and Stuart, 1979). From the Amish data we obtain $-2 \ln \Lambda = 92.076$, which is highly significant ($P < 0.0001$). Thus, we gain a substantial improvement in model fit by allowing for some sterile couples, and we can conclude that at least a few Amish couples are likely to be sterile at the time of marriage. It is important to emphasize that the likelihood ratio test, like other goodness-of-fit tests, can be used only with nested models. It is meaningless otherwise.

Since all three approaches to the assessment of fit—Kaplan-Meier estimates, martingale residuals, and likelihood ratio tests—are theoretically sound, and since each is revealing in its own way, I suggest using them all when applying parametric hazards models. In the present case, the three approaches lead to the same qualitative conclusion: the two-point model of heterogeneous fecundability gives a far better fit to the Amish data than does the homogeneous model. Therefore, it appears likely that at least a few Amish couples are subfecund and perhaps even sterile from the very beginning of their marriages. But the Kaplan-Meier estimates also tell us that the two-point model is not perfect—perhaps because it still assumes that fecundability is homogeneous in the nonsterile couples, and because it does not allow any couples to become sterile after marriage has begun. Chapter 10 presents a more elaborate model that captures these additional biological complexities.

Hazards Analysis with Covariates

One advantage of thinking of reproductive events in terms of their hazards is that hazards analysis provides a powerful set of methods for investigating the effects of more remote influences (see Cox and Oakes, 1984). For example, in trying to explain the observed variation in the waiting time to first fertile conception among the Amish, we might wish to investigate the influence of the couple's age, the wife's nutritional status, environmental exposures that predispose to fetal loss, or any of a number of additional variables on the underlying hazard of a fertile conception. A common approach to this sort of problem is to use the *proportional hazards model*, which assumes that the hazard for individual i depends not only on t but also on a vector x_i of covariates or explanatory variables, such that

$$h_i(t; x_i) = h_c(t) \exp(\beta' x_i) \tag{3.28}$$

In this expression, $h_c(t)$ is an underlying "common" hazard function (one shared by all subjects) and β is a vector of coefficients determining the relative impact of the covariates in x_i (β' is the transpose of β). Rearranging,

$$\ln\left[\frac{h_i(t; x_i)}{h_c(t)}\right] = \beta' x_i = \beta_1 x_{i1} + \beta_2 x_{i2} + \cdots + \beta_n x_{in} \tag{3.29}$$

which looks very much like a standard multiple linear regression model. However, this model differs from the usual regression model in two important respects: it lacks an intercept, and it has no random error term. Estimation error in equation 3.29 is, in effect, absorbed by the $h_c(t)$ term. Estimation and testing procedures have been developed not only for this particular model, but for many other specifications of the relationship between the hazard and the covariates (see Cox and Oakes, 1984; Blossfeld et al., 1989). These procedures provide a powerful way to assess the effects of more remote influences on reproductive events.

If a parametric form for $h_c(t)$ is provided, equation 3.28 can easily be estimated by the maximum likelihood methods described above. However, many applications of the proportional hazards model use a semi-parametric approach developed by Cox (1972) based upon so-called partial likelihood techniques. These methods are semiparametric because they do not require specification of a precise mathematical form for $h_c(t)$. For example, even in the absence of an acceptable model for the underlying hazard of a fertile conception, we can estimate the effect of the female partner's age at marriage on the waiting time to a fertile conception from the Amish data using such methods. According to the semiparametric proportional hazards model, the estimated regression coefficient for the main effect of age is $\hat{\beta} = 0.268 \pm 0.185$ (which is not significant) while the coefficient for the square of age is $\beta_2 = -0.006 \pm 0.003$ (which lies at the borderline of significance at the $\alpha = 0.05$ level).[12] Note from equation 3.28 that e^β acts multiplicatively on the hazard. Since a negative β produces a value of e^β that is *less* than one, it *reduces* the hazard and thus *increases* the expected waiting time to a fertile conception. (Conversely, a positive β decreases the expected waiting time.) Some simple algebra shows that $\exp(\hat{\beta}_1 \cdot 40 + \hat{\beta}_2 \cdot 40^2)/\exp(\hat{\beta}_1 \cdot 20 + \hat{\beta}_2 \cdot 20^2) = 0.159$, suggesting that the effective fecundability of a 40-year-old Amish woman is only about 16 percent as great as that of her 20-year-old counterpart, all other things being equal.

Because of their apparent generality, partial likelihood methods have become extremely popular for fitting proportional hazards models to data on reproduction (see, for example, Foster et al., 1986; Anderton et al., 1987; Jones, 1988a, 1988b; Ford et al., 1989). Note, however, that the moment we loosen the assumption of proportional hazards, the partial likelihood method becomes inappropriate. Thus, if we wish to investigate nonproportional effects (e.g., covariates that act additively on the hazard), we cannot adopt the semiparametric approach. In addition, even in proportional hazards models for which the semiparametric method is appropriate, partial likelihood estimators are not guaranteed to have all the desirable statistical properties of MLEs. Specifically, they are not guaranteed to be asymptotically efficient, unbiased, or normal.

Often they come very close, but their sampling properties must be reinvestigated for every separate application rather than taken for granted (for a detailed discussion of this issue, see Tuma and Hannan, 1984:239–247).

Both parametric and semiparametric hazards models provide a straightforward way to deal with *time-varying covariates,* explanatory variables that change in value over the period of observation. Examples of time-varying covariates that might arise in fertility analysis include serum hormone concentrations in a prospective study of pregnancy loss or seasonal variation in maternal nutritional status in a prospective study of ovarian function, since these predictor variables can fluctuate within an individual subject while she is under observation. As noted before, traditional regression analysis cannot cope with explanatory variables whose values change through time, but hazards analysis can. If, for instance, we wish to generalize the proportional hazards model, we can write

$$h_i(t) = h_c(t) \exp [\boldsymbol{\beta}'(t)\boldsymbol{x}_i(t)] \tag{3.30}$$

where $x_i(t)$ is a vector of covariates specific to time t. We then have a choice: we can either write a functional form for $x(t)$, in which case we will generally need to provide observed values of $x_i(0)$ for the ith subject, or we can use serial observations on $x_i(t)$ at differing values of t, in which case more complex forms of bookkeeping will have to be incorporated into the data set (see Lancaster, 1990: Chapter 3). In either instance, the hazard function can readily be substituted in equation 3.13 to yield MLEs of the beta coefficients, as long as we have a parametric expression for $h_c(t)$.

The Untoward Effects of Uncontrolled Heterogeneity

Samples are often heterogeneous with respect to factors affecting the reproductive process. For example, women may vary in their risks of experiencing a fetal loss or couples may have differing fecundabilities. If not controlled analytically, such heterogeneity can introduce surprising artifacts and biases into our hazards analyses (Vaupel and Yashin, 1985a, 1985b; Trussell and Rodriguez, 1990). This problem can be illustrated by two examples drawn from the study of natural fertility.

In natural-fertility populations, often only a weak relationship appears between a woman's achieved parity and the length of her next birth interval (Figure 3.15, top panel). Since parity is highly correlated with maternal age, this finding would seem to imply that birth spacing is only weakly affected by the mother's age, an unexpected conclusion to say the least. But in fact this apparent result is caused by mixing data on women of different fecundities. When the data are disaggregated by the final reproductive output of the women involved, a very different pattern emerges (Figure 3.15, bottom panel). Now birth intervals are seen to increase steadily with parity, rising most sharply as the final completed family size is approached. As I argue in Chapter 10, this apparent effect of parity is almost certainly an indirect effect of maternal age.

Why was this pattern not observed in the pooled data set? The answer is that the pooled sample becomes increasingly dominated at higher parities by women with high fecundity (that is, short birth intervals). Low-fecundity women, who tend to have comparatively long birth intervals, never attain higher parities; they thus contribute observations only at lower parities. If there were no relationship between parity and birth spacing, this heterogeneity/selectivity process would, as an artifact, make it look as if birth intervals were actually *decreasing* as parity increases. As it is, the true increasing relationship is countered by an artifactual decreasing relationship to produce the more or less horizontal line in the top panel of Figure 3.15.

This example poses a problem for prospective studies of birth spacing in contemporary populations, for in such cases we cannot know what the ultimate reproductive output of a woman will be. However, in a study of the reproductive patterns of Agta hunter-gatherers in the Philippines, Goodman et al. (1985) show that age-adjusted cumulative fertility can be used in a multiple regression model to correct for this artifact. And their results conform to expectation: without the correction birth intervals appear to decline slightly at higher parities; with the correction they clearly increase. Another approach is to use the length of the previous birth interval as a proxy measure for the unobservable heterogeneity in fecundity: the higher the fecundity, the shorter the previous interval will tend to be.

The second example has to do with an apparent correlation in the length of birth intervals across parities in the same woman. When pooled data are analyzed, it often appears that successive birth intervals are highly correlated in length. A naive interpretation of this finding is that something in the first interval has a direct causal effect on the length of the next interval. But in fact this need not be the case. Sheps and Menken (1973:73–74) have shown that even when successive birth intervals in a given woman are completely independent of each other, an apparent correlation can arise when data are pooled across women of differing fecundities.

In verbal terms, what their model shows is this: Suppose that birth intervals are independent, but that each woman draws all her intervals from a single probability density function. Suppose too that this PDF differs among women: in high-fecundity women the PDF will have a low mean, in low-fecundity women a high mean. But for each woman, successive intervals are still independent. Despite this independence, a woman with a short first interval will tend to have a short second interval, simply because she has drawn both intervals from a PDF with a low mean—which is simply another way of saying that she is a high-fecundity woman. Conversely, low-fecundity women will tend to have long first birth intervals followed by long second birth intervals, not because the intervals are affecting each other but merely because these women are of low fecundity. If we look solely at women of the same overall fecundity, we see no correlation among intervals. A correlation appears only when we pool across several different fecundity levels.

This kind of heterogeneity artifact has proven to be quite common in hazards analysis, indeed in all forms of demographic analysis (for excellent discussions, see Vaupel and Yashin, 1985a, 1985b; Trussell and Rodriguez, 1990). In some instances, the artifact arises from heterogeneity in a variable that has in fact been measured, in which case the error is usually easy to correct by including the variable as a covariate, for example, in a proportional hazards model. A much more difficult problem arises when the artifact is caused by heterogeneity in some factor z that has not been measured directly and that may even be inherently unobservable. In recent years, two general strategies have emerged for controlling for such "hidden" heterogeneity. Both strategies try to adjust for the heterogeneity by including a theoretical distribution of z-values in the model: the first allows the distribution of unobserved heterogeneity to be unspecified (Heckman and Singer, 1984), whereas the second requires specification of a parametric distribution for z while permitting $h_C(t)$ to be nonparametric (Vaupel et al., 1979). Unfortunately, it has proven impossible for both the hazard and the heterogeneity distribution to be nonparametric at the same time. Thus, under the second strategy, greater flexibility in the form of the common hazard is purchased at the expense of requiring an exact specification for the distribution of z. It is sometimes possible to come up with a specification of the z-distribution based on first principles, for example, by modeling its biological sources (Weiss, 1990). Often, however, too little prior theory exists about the relevant sources of heterogeneity to permit this sort of etiologic modeling. Several attempts have been made to develop more general methods based upon highly flexible parametric specifications of the PDF describing the unobserved heterogeneity (e.g., Heckman and Walker, 1987; Vaupel, 1990a). For example, several authors (Clayton, 1978; Vaupel et al., 1979; Oakes, 1982; Wild, 1983; Clayton and Cuzick, 1985; Gage, 1989; Vaupel, 1990a, 1990b) have used a gamma density function for the distribution of z. Unfortunately, Heckman and Singer (1984) have shown that results using this strategy can be very sensitive to the choice of a distribution for z. At the same time, Trussell and Richards (1985) have shown that the results of the Heckman-Singer strategy, in which the distribution of z is nonparametric, can also be very sensitive to the choice of a parametric form for the *hazard*. In the final analysis, there are only two ways around this problem: (1) figure out how to measure z, in which case it will no longer be an unobservable; or (2) develop enough *theory* to allow specification of a realistic parametric form for both $h_c(t)$ and the distribution of z.

In classic fertility analysis, fecundity has been the sine qua non of unobservables; it has therefore been considered by many demographers to be little more than a nuisance, an annoying creature whose sole aim in life is to confound analyses of the social factors affecting fertility. If this book accomplishes nothing else, I hope it will convince its readers that that view is naive and unsatisfying. Using modern endocrinological techniques, we can now observe the various components of fecundity quite directly. And by developing etiologic hazards models, we may actually be able to say quite a lot from a theoretical perspective about how the components of fecundity ought to behave. From this perspective, it is time for fecundity to take its proper place within fertility analysis as something interesting in its own right.

SUMMARY

As I have tried to show, it is useful to concentrate our analyses, at least initially, on the proximate determinants of fertility because all other factors that influence reproduction must act through those determinants, and because the proximate determinants have theoretical linkages to reproductive performance that are comparatively easy to model and to estimate. The most natural way to model those linkages is in terms of a small set of reproductive events and the time intervals separating those events. Hazards analysis provides the ideal statistical framework within which models of the birth-spacing processes and the birth-interval components that are responsible for fertility variation can be developed and tested. Once appropriate models have been constructed, general likelihood methods exist for estimating the model parameters and their associated standard errors. Many readily available statistical packages, such as SAS, SYSTAT, BMDP, GLIM, and SPSSX, now include modules for performing a wide variety of hazards analyses, and special-purpose packages like RATE and CTM permit more advanced applications (for descriptions of RATE and CTM, see Allison, 1984:79; Heckman and Walker, 1987). In sum, hazards analysis provides a general, practical, and easily implemented framework for developing, fitting, and testing biologically meaningful models of the proximate determinants. And hazards models with covariates provide a straightforward way to incorporate the effects of more remote influences on fertility.

The remainder of this book is in effect an elaboration of the logic developed in this chapter, particularly the claims that the proximate determinants of fertility operate via birth-interval components, and that specification of birth-interval models is a useful way to assess the contribution of the proximate determinants to fertility variation. The next several chapters review each proximate determinant in turn; what might be called their "natural history" is described in some detail, and an attempt is made to relate them to the relevant birth-interval components. The final chapter provides a preliminary assessment of the joint contributions of all the proximate determinants to variation in natural human fertility.

Before we can formulate realistic models of the proximate determinants, however, we need to review certain basic facts about the biology of reproduction. The next two chapters provide the necessary biological foundation for etiologic hazards models of the reproductive process.

NOTES

1. This is simplest to show in the *homogeneous case*, in which all items in a given state are converted into the next state at exactly the same rate. If $t(i)$ is the average amount of time spent in the ith state, and $\lambda(i)$ is the rate of conversion from the ith to the $(i + 1)$th state, then $t(i) = 1/\lambda(i)$. Things are more complicated in the *heterogeneous case*, that is, when the various items in a given state are converted at different rates. Then we have, very approximately, $t(i) = 1/E_i(\lambda) + k_i \mathrm{Var}_i(\lambda)$, where $E_i(\lambda)$ is the mean or expected rate of conversion to the $(i + 1)$th state, $\mathrm{Var}_i(\lambda)$ is the variance in rates of conversion, and k_i is a positive constant whose value is determined by the specific pattern of variation in conversion rates. Since k is always greater than zero, heterogeneity of rates lengthens the average time spent in the state.

2. For simplicity, Figure 3.2 assumes that the potential reproductive span begins at age 15 and ends at age 50, all women survive to age 50, all women marry only once and remain married throughout reproductive life, and all reproduction occurs within marriage.

3. Fecund waiting time is something of a misnomer in the present case, since the waiting period includes any time added to the birth interval by pregnancy loss; thus, not all of this interval may be genuinely fecund—for example, the gestation period ending in fetal death.

4. In his original analysis, Bongaarts (1978) recognized two separate coefficients, C_c for contraception and C_a for induced abortion. In most empirical applications, however, it is not possible to estimate those coefficients separately, but only to estimate their product, $C_c C_a$. As a consequence—and in view of the fact that we are overwhelmingly interested in natural fertility anyway—I have combined both determinants into a single deliberate control factor.

5. Although this assumption may appear reasonable, recent research has shown that prolonged breastfeeding can also affect the length of the subsequent fecund waiting time to conception (see Chapter 8).

6. This regression involved 46 populations and yielded an r^2 of 0.96. It should be noted that the curve described by equation 3.3 is non-monotonic and will produce declining values of i once B has surpassed 37.3 months. Thus, the equation tends to produce biologically implausible results at longer durations of breastfeeding.

7. It should be emphasized that Bongaarts (1976) has made fundamental contributions to the development of this alternative model.

8. Observations can also be *left-censored*. That is, a particular birth-interval component may already be underway when the study begins, so that although we may know when the component ends, we do not know when it started. In general, it is more difficult to correct for left-censoring than right-censoring (Wood et al., 1992).

9. For simplicity I will treat T as if it is measured on a continuous scale, which it usually is.

10. Strictly speaking, this likelihood function is correct only in the case of *random censoring* in which the expected time remaining to the ith subject, conditional on survival to t_i, is not affected by the fact that censoring occurs at t_i. Otherwise, the fact of censoring itself contains information about the underlying hazard, and more complicated likelihood functions are required. Throughout this discussion, I assume random censoring.

11. Throughout the text, point estimates will be presented ± one standard error and marked by a "hat" (ˆ).

12. In view of the well-established impact of age on female fecundability discussed in Chapter 7, it may seem surprising that a less ambiguous age effect was not found in this Amish sample. The problem, however, is not so much that age is unimportant but that there is an extremely narrow age-range of first marriages among the Old Order Amish: approximately 90 percent of all women in the present sample married for the first time between the ages of 19 and 26 years. Thus, even if age were an important and powerful covariate, there would be too little variation to permit estimation of its effects.

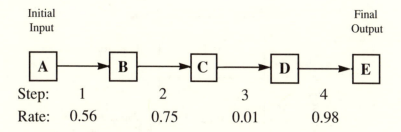

Figure 3.1. An idealized series of metabolic reactions converting an initial input substance (A) into a final output product (E). This process involves four distinct conversions (steps 1–4) and three intermediate products (B, C, and D). The four conversions occur at widely varying rates, expressed as the fraction of a given product converted into the next product in the series per unit time.

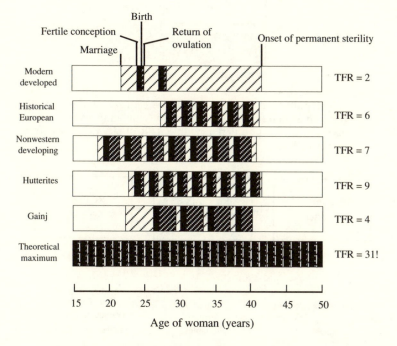

Figure 3.2. The average timing of reproductive events in selected types of societies. This figure is highly schematic and is intended for illustrative purposes only. (Modified from Bongaarts and Potter, 1983)

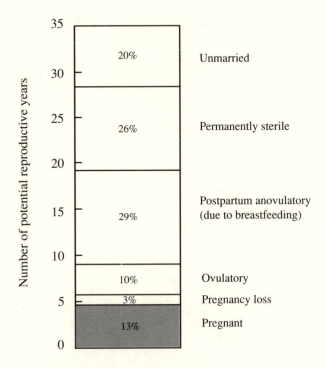

Figure 3.3. Hypothetical example of the percentage of the potential reproductive span (ages 15–50) spent by women in various reproductive states in a population with a TFR of six. (Modified from Bongaarts, 1976)

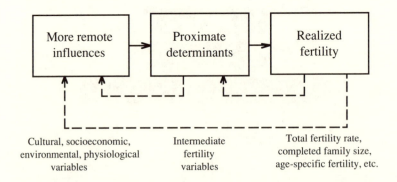

Figure 3.4. The relationship of the proximate determinants and more remote influences to realized fertility. *Solid arrows,* primary causal effects; *broken arrows,* secondary feedback relationships. (Modified from Bongaarts, 1978)

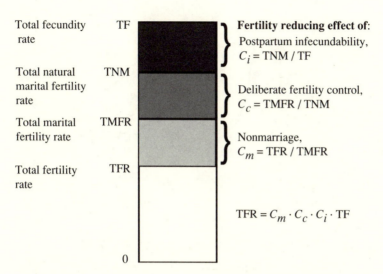

Figure 3.5. Relationships among the fertility-reducing effects of the proximate determinants and various measures of fertility. (Redrawn from Bongaarts, 1982)

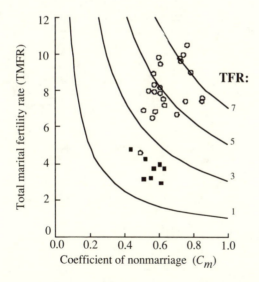

Figure 3.6. Total fertility rate (TFR) as a function of the total marital fertility rate and the coefficient of nonmarriage. For illustration, several developing countries *(open symbols)* and developed countries *(closed symbols)* are plotted (data from 1966–1978). (Redrawn from Bongaarts, 1978)

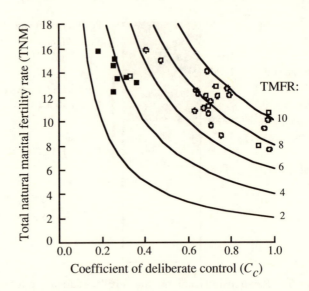

Figure 3.7. Total marital fertility rate (TMFR) as a function of the total natural marital fertility rate and the coefficient of deliberate fertility control. For illustration, several developing countries *(open symbols)* and developed countries *(closed symbols)* are plotted (data from 1966–1978). (Redrawn from Bongaarts, 1978)

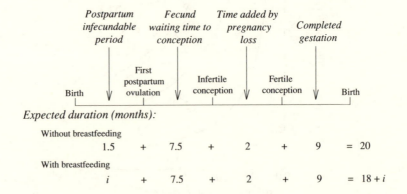

Figure 3.8. The major components of the completed birth interval, with Bongaarts's estimates of their typical length without and with prolonged breastfeeding. The variable i is the average duration of lactational anovulation.

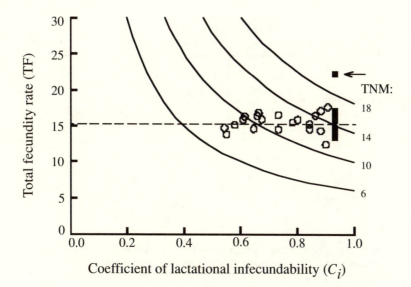

Coefficient of lactational infecundability (C_i)

Figure 3.9. Total natural marital fertility rate (TNM) as a function of the total fecundity rate and the coefficient of lactational infecundability. For illustration, several developing countries *(open symbols)* and developed countries *(closed symbols)* are plotted (data from 1966–1978). The fact that all the developed countries have the same C_i reflects Bongaarts's assumption that the average duration of lactational infecundability is three months in all such countries. The *broken line* is the estimated mean total fecundity rate for all 31 countries minus one outlier (Finland, indicated by *arrow*); this mean is equal to 15.4 ± 0.2. (Data from Bongaarts and Potter, 1983)

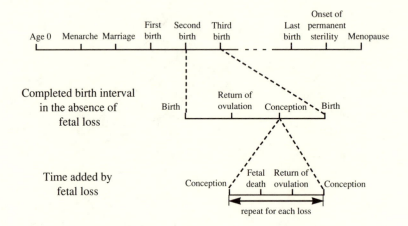

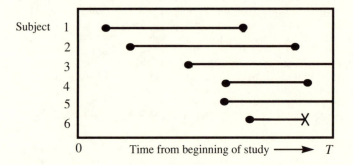

Figure 3.10. The female reproductive life course viewed as a series of time intervals. (Modified from Bongaarts and Potter, 1983)

Figure 3.11. An illustration of right-censoring in a prospective study of birth interval dynamics. Each circle represents a birth. Two circles joined by a horizontal line are successive births to the same woman; in such cases, the length of the line is proportional to the total duration of the birth interval. Lines that terminate at the end of the study (time *T*) represent censored observations; their length is proportional to the time from the last birth to censoring. For subject 6, the X indicates the time at which this particular woman dropped out of the study. Her observation was thus censored before the end of the study.

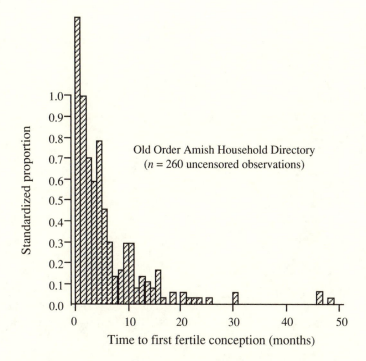

Figure 3.12. Distribution of 260 completed intervals from marriage to first fertile conception in an Old Order Amish community in rural Pennsylvania (first marriages only). Not shown are 21 right-censored observations. One apparent premarital conception and four very long (>60 months) uncensored observations have been excluded from this data set; in all five cases misrecording is suspected, but the data have not yet been field-checked. All data were drawn from a household directory compiled by the Old Order Amish. To avoid the bias toward higher fertility caused by ascertainment of marriages through offspring, a marriage was included only if at least one surviving spouse was recorded. (From Wood et al., 1992)

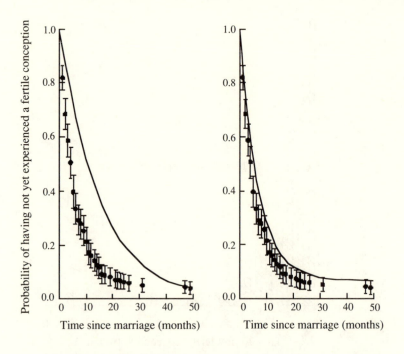

Figure 3.13. Comparison of Kaplan-Meier estimates *(closed circles)* and model-predicted survival functions *(smooth curves)* for two models fitted to the waiting time from marriage to first fertile conception among the Old Order Amish: the homogeneous model of fecund-ability *(left)* and the two-point model with homogeneous fecund and sterile subpopulations *(right)*. The Kaplan-Meier estimates are presented ± two standard errors; the number of estimates is equal to the number of distinct uncensored waiting times. The degree to which the Kaplan-Meier estimates and the model-predicted curve correspond is an indication of how well each model fits the data. (Redrawn from Wood et al., 1992)

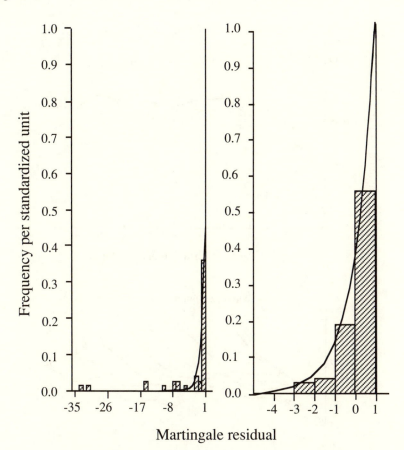

Martingale residual

Figure 3.14. Distribution of martingale residuals for two models fitted to the waiting time from marriage to first fertile conception among the Old Order Amish: the homogeneous model *(left)* and the two-point fecundability/sterility model *(right)*. The smooth curve in each case is a unit exponential with a mean of zero and a variance of one, the expected distribution of residuals on the hypothesis that the model is correct. Note the large number of residuals generated by the homogeneous model whose absolute values are substantially greater than expected. (Redrawn from Wood et al., 1992)

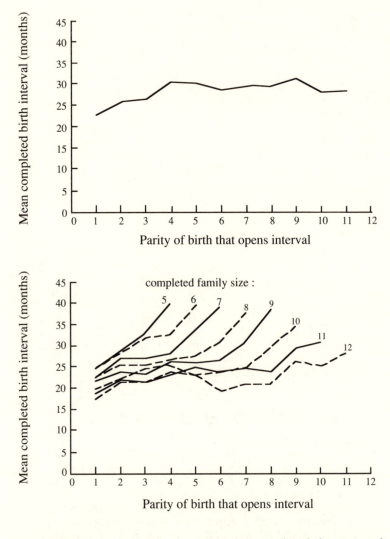

Figure 3.15. The relationship of mean birth interval and the parity of the birth that opens the interval, rural France, marriages of 1675–1780: *(top)* data pooled across all completed family sizes, *(bottom)* data broken down by completed family size. Only families of completed size 5+ are represented. (Data from Leridon, 1967)

PART II

The Proximate Determinants
of Natural Fertility

4

Ovarian Cycles and the Fertile Period

In common with the females of most other mammalian species, women experience regularly recurring cycles in their ability to conceive and establish a pregnancy—loosely called "menstrual cycles" in humans because periodic menstrual bleeding is their most conspicuous feature. Although many tissues in the body participate in these cycles, including parts of the brain and uterus, the principal actor in many respects is the ovary (Figure 4.1). Hence, physiologists commonly refer to these cycles as *ovarian cycles*. A normal ovarian cycle must occur in order for conception and pregnancy to be possible. Moreover, even when a normal cycle does occur, there is only a comparatively brief period during the midcycle, known as the *fertile period*, when there is any chance at all that insemination will result in conception. Thus, the length of the ovarian cycle itself and the length of the fertile period within each cycle place an irreducible upper limit on the rate of conception and ultimately on the pace of childbearing.

From a biometric point of view, components of the ovarian cycle can induce variation in fertility only insofar as they vary themselves. The question of variation in cycles and their components, and the possible contribution of such variation to fertility differentials, will be taken up in the second part of this chapter. First, however, it will be necessary to outline the basic physiology of the ovarian cycle. Although the first part of this chapter is motivated primarily by a need to provide some basic background for discussing the biometry of cycles, study of the ovarian cycle and its regulation is in its own right one of the most important and fascinating areas of reproductive biology.

115

MAJOR FEATURES OF THE HUMAN OVARIAN CYCLE

Figure 4.2 shows, in schematic form, some of the major landmarks of the normal ovarian cycle as it might be experienced by a woman of mid-reproductive age, say, in her late twenties or early thirties. It is customary to think of a given cycle as beginning on the first day of *menses* or menstrual bleeding, which normally lasts some three to five days. On average there is a lag of approximately 28–29 days from one menstrual onset to the next, although as detailed below, there is so much variation in this number both among women and within individual women that it is meaningless to speak of the "typical" length of a cycle as if it were a characteristic of the species as a whole.

If the cycle is normal, ovulation will occur at some point during its mid-portion—that is, the ovary will release a single mature egg or ovum that will then be available for fertilization. Ovulation divides the cycle into two distinct phases, the *follicular* (or preovulatory) *phase* and the *luteal* (or postovulatory) *phase.* (The reasons for these particular labels will become clear later.) The fertile period, which obviously can occur only in ovulatory cycles, overlaps the follicular and luteal phases. Ovulation takes place, very approximately, at the midpoint of the fertile period. As we shall see, there are major difficulties in estimating the total length of the fertile period; however, for the moment we can say that it lasts a few to several days, whatever that may mean.

Associated with the ovulatory cycle are regular, measurable changes in the concentrations of several hormones in the woman's peripheral circulation (Figure 4.2, panels b and c). As discussed below, these hormones play a critical role in the regulation of the ovarian cycle. For the present, however, we are interested in them primarily as markers of the various phases of the cycle and of ovulation itself. For a more detailed discussion of the basic biology of hormones, and of how we go about measuring their concentrations, the reader should consult the appendix to this book.[1]

The hormones of interest here fall into two classes: the *pituitary gonadotropins* and the *ovarian steroids.* The pituitary gonadotropins, *luteinizing hormone* (LH) and *follicle-stimulating hormone* (FSH), are peptide (protein) hormones of moderate molecular weight secreted by the anterior portion of the pituitary gland, which lies just below the brain. Both LH and FSH begin the cycle at fairly low concentrations, although there is a slight elevation in FSH in the late luteal and early follicular phase (Figure 4.2, panel b). The levels of LH and FSH remain low throughout most of the follicular phase. A day or so before ovulation, however, both gonadotropins display a sudden and dramatic increase in concentration, known as the preovulatory LH/FSH surge. As we shall see, this surge in LH and FSH is essential for ovulation to occur. The surge itself is quite transitory: by the time of ovulation, both hormones are dropping back to low levels, where they remain throughout the luteal phase.

The principal ovarian steroids of interest here are *estradiol* and *progesterone.* Steroids are hormones derived from cholesterol (see below). All the steroids are linked by complex pathways of interconversion, which can take place in various tissues, including parts of the ovaries. Estradiol is one of a subgroup of steroids known as the estrogens. Although estradiol is the most important of the estrogens, at least in terms of ovarian function, it is sometimes easier to measure the concentration of total estrogen rather than that of estradiol alone. (The two concentrations are highly correlated.) Progesterone, an earlier molecule in the pathway of steroid synthesis, is the most important of the progestins.

The concentrations of both estradiol and progesterone in the bloodstream undergo distinctive patterns of variation over the course of the cycle (Figure 4.2, panel c). Beginning at a very low level in the early follicular phase, estradiol increases in approximately exponential fashion up to the time of the LH/FSH surge. The level of estrogen then crashes at about the time of ovulation, but it displays a broad secondary peak during the mid-luteal phase. In contrast, progesterone remains at a low, usually undetectable level throughout the follicular phase. At about the time of ovulation, its concentration begins to increase, reaching a high point in the mid-luteal phase and then dropping back to undetectable levels by the end of the luteal phase if conception does not occur. Indeed, in hormonal terms it is precisely this postovulatory elevation in progesterone secretion by the ovary that defines the luteal phase.

These cyclical changes in steroid concentrations parallel physical changes within the ovary itself, particularly changes in the size and internal complexity of the *follicle,* the tiny capsule of cells that contains and nurtures the developing egg (Figure 4.3). Estradiol is secreted by the follicle, and its rising concentration parallels growth in follicle size across the follicular phase. Indeed, it is precisely because the preovulatory phase of the cycle is dominated by maturation of the follicle and increasing follicular secretion of estradiol that it is known as the follicular phase.

After ovulation, when the egg is released into the fallopian tube or oviduct, the follicle collapses, regresses, and then redifferentiates into a specialized gland called the *corpus luteum* (lit. "yellow body"), which is devoted primarily to the secretion of progesterone (some of which is secondarily converted to estradiol within the ovary). If fertilization and implantation of the egg do not occur during the cycle in question, the corpus luteum undergoes a final regression (*luteolysis*) after about two weeks and remains in the ovary as a whitish patch of scar tissue (the *corpus albicans* or "white body"). If a pregnancy intervenes, however, the life of the corpus luteum is extended, as discussed in Chapter 5. It is the corpus luteum that is responsible for the luteal phase rise in progesterone concentration and that gives the luteal phase its name.

Using radioimmunoassay and related techniques, it is now possible to measure the concentration of LH, FSH, estradiol, and progesterone even in very small quantities of serum collected by venepuncture from the peripheral bloodstream (for discussions of basic endocrinological techniques, see Campbell, 1985, 1988; Chard, 1987). This fact makes these hormones the markers of choice for detecting ovulation and for monitoring the ovarian cycle, both in the laboratory and in the field. These hormones or their metabolites can also be measured in urine (Table 4.1), and the urinary concentrations prove to be highly correlated with the corresponding serum concentrations.[2] In addition, it has recently been shown that estradiol and progesterone are present in saliva at concentrations that correlate well with their serum concentrations (Riad-Fahmy et al., 1987; Ellison, 1988; Lenton et al., 1988). (Since the higher molecular weight and hydrophilic nature of the gonadotropins prevent them from diffusing across the cell membranes of the salivary glands, those peptide hormones will never be measurable in saliva.) Because it is much less invasive to collect repeated salivary samples from a woman than to collect a large number of serial blood samples, a great deal of effort has been made in recent years to develop reliable, inexpensive radioimmunoassays for salivary steroids. Such assays are already finding important applications in anthropological field studies of ovarian function (Ellison, 1988).

Another marker of the ovarian cycle needs to be mentioned for the sake of completeness, namely, change in *basal body temperature* (BBT) or resting oral temperature.[3] At about the time of ovulation, associated with the beginning of the luteal phase rise in progesterone, some cycles in some women are characterized by a more or less sharp increase in BBT (Figure 4.2, panel d). Unfortunately this rise in temperature is of very small magnitude, amounting to only about 0.5°C, and thus is near the usual limits of detection for most commercially available oral thermometers. To make matters worse, BBT itself is quite labile, fluctuating by several degrees over the course of a normal day; for that reason, BBT is usually taken as early as possible in the morning, preferably after awakening but before rising. And even if measured reliably, a slight rise in BBT may just as easily indicate a mild fever as ovulation. Studies have revealed serious problems with the validity of using BBT as an indicator of ovulation (Johansson et al., 1972; Hilgers and Bailey, 1980).

Table 4.1. Correspondence between Major Ovarian Steroids and Gonadotropins in Serum and Their Urinary Metabolites

Serum	Urine
Estradiol-17β	Total urinary estrogen
Estradiol-17β	Estradiol glucuronide
Progesterone	Pregnanediol glucuronide
17-OH progesterone	Pregnanetriol glucuronide
LH	LH
FSH	FSH

From Rebar (1991)

Nonetheless, before the development of reliable assays for LH, FSH, progesterone, and estradiol, BBT was practically the only noninvasive method available for detecting ovulation. Its importance, in other words, is largely historical, and we will come across several early studies that made use of BBT.

It is important to mention one other physical effect of the hormonal changes described above, having to do with the cycle's most conspicuous marker, menstruation (Brenner and Maslar, 1988). Both before and after ovulation, under the influence first of rising levels of estradiol during the late follicular phase and then of progesterone throughout the luteal phase, marked histological changes occur in the tissues lining the uterus (Figure 4.4). During the follicular phase, rising estrogen stimulates the growth and vascularization of the *endometrium* (the innermost lining of the uterine cavity) and thickening of the *myometrium* (the layer of smooth muscle underlying the endometrium). It also induces endometrial cells to produce receptors for progesterone. During the luteal phase, progesterone acts upon the estrogen-primed endometrium to convert it into secretory tissue: special glands develop within the endometrium that secrete fluids rich in glycogen and various enzymes into the uterine cavity. At the same time, the arteries feeding the endometrium link to form blood-engorged pools or *anastomoses* (Figure 4.4). All these changes prepare the uterus to accept and nourish the implanting embryo, should fertilization occur. If fertilization does not occur, however, luteolysis will ensue, progesterone and estradiol will be cleared from the bloodstream, and the endometrium will be sloughed off the uterine wall. It is the loss of this blood-rich endometrial material that constitutes menstruation.

Proliferation of the endometrium is only one of many changes throughout the female body induced cyclically by estradiol and progesterone in preparation for a pregnancy. Others are listed in Table 4.2.

Table 4.2.   Cyclic Effects of Estradiol and Progesterone on the Female Body
in Preparation for a Possible Pregnancy

ESTRADIOL
1. Stimulates growth of follicle and ovary
2. Stimulates growth of smooth muscle and epithelial linings of reproductive
   tract. In addition,
   a. Fallopian tubes: increases contractions and ciliary activity
   b. Uterus: increases myometrial contractions; stimulates secretion of
      abundant, clear cervical mucus; prepares endometrium for progesterone's
      actions by inducing progesterone receptors
3. Stimulates enlargement of breast tissues, especially ducts and fat deposition
4. Stimulates anterior pituitary to secrete prolactin, but inhibits prolactin's
   milk-inducing action on the breasts (see Chapter 8)
5. Stimulates retention of fluid
6. Has feedback effects on hypothalamus and anterior pituitary (see text)

PROGESTERONE
1. Stimulates endometrial glands to secrete glycogen and enzymes
2. Induces thick, sticky cervical mucus
3. Decreases contractions of uterine tubes and myometrium
4. Stimulates enlargement of breast tissues, especially glandular tissue
5. Inhibits milk-inducing effects of prolactin (see Chapter 8)
6. Has feedback effects on hypothalamus and anterior pituitary (see text)

Modified from Vander et al. (1990)

The patterns shown in Figure 4.2 are highly idealized, and it is important to see how they actually look in the real world. Figure 4.5 shows
these same patterns in five successive ovulatory cycles experienced by a
healthy, mid-reproductive age Australian woman, based upon daily
measurement of BBT and daily collection of serum and urine.[4] It is
immediately apparent that, despite some variation, the basic hormonal
patterns are fairly consistent across cycles. BBT, on the other hand, is so
variable as to be almost useless; in cycles 2 and 5, for example, we
would be hard-pressed to infer from BBT alone that ovulation had
occurred at all, much less when in the cycle it fell.

The cycle-to-cycle variability in hormone concentrations evident in
Figure 4.5 raises an important issue concerning *sampling frequency*. The
ability to detect rapid, transient fluctuations in hormone concentrations
clearly depends upon the frequency with which biological specimens
are collected—or, equivalently, the time interval between collections.
The more often we draw samples, the more likely we are to observe
"micro-level" peaks, valleys, and transitions. Close inspection of Figure
4.5 reveals that much of the apparent cycle-to-cycle variability in hormone levels involves differences in the height of the peak secretion of
those hormones that exhibit periodic surgelike behavior, that is, the

height of the periovulatory LH surge and of the preovulatory peak in follicular estradiol secretion. In cycle 1, for example, the LH surge seems to have reached a much higher level than in cycle 2. However, this difference may be more apparent than real. It may be that, purely by chance, a serum sample was drawn in cycle 1 at a time very near the true peak in LH secretion, whereas in cycle 2 the peak was missed by only a few hours. In general, sharp peaks and narrow valleys are difficult to detect unless very frequent samples are collected.

This point is brought home with startling clarity by Figure 4.6, which contrasts the apparent shape of the LH surge when serum samples are drawn every eight hours versus every five minutes. With more frequent sampling, the smooth, unimodal LH surge vanishes, replaced by a complex series of microscopic pulses in LH secretion. It turns out that the surge does not represent a single, massive dumping of LH into the bloodstream by the pituitary, but merely an increase in the amplitude of these tiny pulses of LH secretion. Recent work has shown that many hormones, especially peptide hormones like LH and FSH, are released in pulsatile fashion, and this discovery has led to a rethinking of the regulation of hormone secretion, a topic to which we return below.

Thus far, we have dealt with the main features of the ovarian cycle primarily in a descriptive way. We now turn to a more detailed discussion of female reproductive cycles from a functional point of view.

THE LIFE CYCLE OF THE FEMALE GERM CELL

The developmental sequence of the egg or female germ cell from its primordial state up to fertilization (if such occurs) is known as *oogenesis* (lit. "egg emergence"). The major stages of oogenesis are summarized in Figure 4.7. The primordial germ cells are first detectable as a distinct cell line well before birth, indeed by about the twenty-fourth day after conception. Remarkably, the germ cells make their first appearance in extra-embryonic tissue, near the junction of the yolk sac and the allantois (Figure 4.8). They then migrate by ameboid action, entering the embryo proper, and by about day 35 of gestation lodge in the genital ridges, embryonic tissues that will eventually become the gonads. For most of the remainder of gestation, the female germ cells (now called *oogonia*) undergo little morphological change but increase greatly in number through *mitosis*, the normal cell division that retains the full complement of 46 chromosomes in each of the daughter cells. Near the end of the second trimester of pregnancy, mitosis stops, and the oogonia enter a phase of active DNA synthesis. The resulting duplication of DNA prepares the oogonia for the first stage of *meiosis*, the reduction cell division that will eventually reduce the number of chromosomes per egg cell by half as a prelude to fertilization. By uniting the chromosomes of egg and sperm, fertilization restores the full chromosomal complement in the newly created embryo (see Chapter 5).

In marked contrast to sperm production in the male, in which meiosis does not begin until puberty, the first stages of meiosis in the female actually begin while she is still in her mother's womb. But once the developing egg cell (now a *primary oocyte*) has completed the first few stages of meiosis, meiosis stops and the egg enters a kind of suspended animation, at least as far as cell division is concerned. Meiosis will not resume until sometime after puberty. In fact, since only a small number of eggs reenter meiosis in each ovarian cycle and normally only one per cycle will complete meiosis, a given egg cell may remain in suspended animation for anywhere from 12 to 50 years or more. It has been suggested that accumulating insults to the chromosomes while in this prolonged state of arrested development are partly responsible for the observed increase with maternal age in the risk of birth defects associated with chromosomal abnormalities, for example, the nondisjunction of chromosome 21 associated with Down's syndrome (Antonarakis et al., 1991).

The physical changes in the germ cells described thus far are paralleled by marked changes in the number of germ cells (Figure 4.9). From only a few thousand primordial germ cells that migrate into the developing gonads, the oogonia proliferate mitotically at a rapid rate, reaching about seven million by the end of the second trimester of pregnancy. Since mitosis ceases at that point, and since each oogonium initiating meiosis will ultimately produce *at most* one mature ovum, the number of germ cells present at this stage represents the maximum number that the female will ever possess.

Once the very first germ cells have entered meiosis at about the sixth month of gestation, the total number of viable germ cells present in the ovary begins a long decline that continues throughout premenopausal life (Figure 4.9). By the beginning of the menopausal transition, after which oogenesis ceases, perhaps only 100–200 germ cells remain, most of them still arrested at the primary oocyte stage. In fact, it now appears that menopause occurs *because* the ovary runs out of eggs (see Chapter 9). The female thus loses several million egg cells over the course of her life, at an average rate of about one every four minutes; current evidence suggests that the rate of loss per germ cell is more or less constant throughout premenopausal life, including prepubescent life, except perhaps for a slight acceleration just before menopause (Richardson et al., 1987).

A moment's reflection will show that only a tiny fraction of oocytes can be lost through ovulation. Even if a woman begins cycling at a fairly young age, say 12 years, and experiences menopause fairly late in life, say at age 50, and even if she experiences a complete ovulatory cycle every single lunar month in between, she still will have shed fewer than 500 mature ova over the course of her lifetime. The vast majority of oocytes never leave the ovary, but rather undergo a process of degeneration and death in situ. The degeneration of the egg and its follicle within the ovary is a process known as *atresia*. Clearly atresia, not ovulation, is the fate of most eggs: many are culled, but few are chosen.

FOLLICULAR DEVELOPMENT AND OVULATION

While still in the ovary, each developing germ cell is contained within its own follicle, and the development of the germ cell is accompanied by follicular growth and internal differentiation, especially just before ovulation (see Figure 4.10 for the various stages of follicular development). In a strict sense, there are two distinct developmental trajectories that the egg and follicle can follow, one leading to ovulation and the other to atresia—and we have seen that atresia is by far the more common path. Although atretic follicles do play a role in the regulation of ovarian cycles (see below), we will concentrate for the moment on those few eggs and follicles that participate in ovulation.

Once the primordial germ cells have lodged in the embryonic gonads, each of them is surrounded by a simple, thin follicular capsule made up of a basement membrane (or *basal lamina*) and a single layer of epitheliumlike cells known as the *granulosa precursors* (Figure 4.11, A–D). After the first stage of meiosis begins and up to the time meiosis is reinitiated (if it ever is), this primordial follicle grows into a primary follicle, increasing in size and complexity. The granulosa precursors differentiate into *granulosa cells*, which increase in number until two or more layers are present. In addition, the basal lamina becomes surrounded by a new kind of specialized squamous cells known as *theca cells* (Figure 4.11, E–F). As we shall see, both the theca and the granulosa cell layers play important roles in the endocrinology of the ovarian cycle.

If a follicle is selected for maturation in some particular cycle, it undergoes rapid and dramatic changes. First, it grows enormously, from <1 mm (about the size of a pinhead) up to more than 20 mm in diameter. Although this growth is partly attributable to an increase in the number of theca and granulosa cells, during the final stages of development the follicle is mostly a hollow, fluid-filled ball (Figure 4.11, G–H).

Even before final maturation of the egg, the granulosa cells secrete a nutritive fluid rich in steroids and proteins known as *antral fluid*. As this fluid accumulates, it forces gaps between the granulosa cells; these gaps eventually join to form a single, continuous space called the *antrum* (Figure 4.11). Once the follicle has reached this final stage of development just before ovulation, it is called a *Graafian follicle*, named for the early Dutch microscopist who first described it.

Within the Graafian follicle, the maturing egg floats attached to the side of the follicle by a thin stalk of granulosa cells (Figures 4.11 and 4.12). Surrounding the egg is a thin, milky layer of residual granulosa cells given the felicitous name *cumulus oophorus* (lit. "cloud of the egg").[5] Separating the egg and cumulus oophorus is a layer of jellylike material known as the *zona pellucida*, made up of glycoproteins (proteins with short side-chains consisting of sugars) secreted by the egg itself (Figure 4.13). By the Graafian follicle stage the egg has increased by about 500 percent over its primary oocyte size; it has also undergone its first, nonreductive meiotic division (Figure 4.7) and has ejected the first polar body into the space between its own cell membrane and the zona pellucida. It will not complete meiosis and release the second polar body until well after ovulation, at about the time of fertilization.

The mature preovulatory follicle is both large and complex in structure, creating special difficulties of communication and interaction among its various parts. This is especially true since the bloodstream is a common medium of communication among tissues, and the interior of the follicle is not vascularized (Figure 4.14). How then do nutrients and chemical messengers like the gonadotropins get into the follicle and eventually to the egg itself, and how do waste products and follicular hormones like estradiol get out? Clearly, the cell membranes and cytoplasm of the granulosa cells must actively transport materials into and out of the interior follicle. This function permits the granulosa cells to act as a selective filter regulating what reaches the developing egg. The granulosa cells thus provide a partial *blood-egg barrier*, shielding the egg from potentially harmful substances such as chemical mutagens. In addition, the granulosa cells play a special role in transporting materials across the zona pellucida by sending thin processes through the zona to interact with microvilli on the cell membrane of the egg (Figure 4.15).

As the time of ovulation nears, the expanded follicle presses tightly against the outer membranes of the ovary, causing a distinct and readily visible bulge in its side known as the *stigma* (Figures 4.11 and 4.16). About half an hour before ovulation, there is a marked reduction in blood flow at the stigma, and the granulosa and theca cells in this region begin to thin out (Figure 4.16, B–C). Minutes before ovulation, when

almost all follicular cells and interstitial ovarian tissues have disappeared near the stigma, the ovum, now free of its stalk of granulosa cells
but still carrying the cumulus oophorus, moves toward the outer wall of
the ovary. Finally the follicle and outer ovarian membranes rupture and
the egg is sent streaming into the woman's abdominal cavity on a trail
of antral fluid (Figure 4.16, D–E). Ovulation is now completed.

At this point we interrupt our narrative of the egg's journey, which
we will take up again in the next chapter. We turn now to a discussion
of the complex system that acts to regulate the remarkable processes we
have been describing.

NEUROENDOCRINE REGULATION OF THE OVARIAN CYCLE

The regularly recurring cycles of follicular development and oocyte
maturation obviously must be under strict physiological control. As it
happens, this control system involves not only the ovary itself, but parts
of the brain and pituitary gland as well. Because of the involvement of
both hormone-secreting and neural tissues in this process, it is common
to speak of it as being *neuroendocrine* in nature.

The part of the brain of most immediate relevance is the *hypothalamus*,
which lies near the base of the midbrain, just in front of the brain stem
(Figure 4.17). The hypothalamus acts as a kind of central coordinating
node for neural impulses linking several parts of the brain; it also plays
a prominent role in regulating various biological rhythms such as the
wake-sleep cycle, as well as the ovarian cycle. The hypothalamus is, in
addition, the most important site of communication between the central
nervous system and the endocrine system. Much of that communication
involves a dialogue between the hypothalamus and the pituitary gland,
which lies just below the hypothalamus, suspended from it by a slender
stalk of neural, connective, and vascular tissue (Figure 4.18). Although
the pituitary has two distinct lobes, we are concerned here primarily
with the more anterior of these, where the gonadotropins LH and FSH
are synthesized by a distinct line of cells known as the *gonadotrophs*.

The hypothalamus contains specialized nerve axons that, in response
to the appropriate neural signals, produce not the neurotransmitters
made by most nerve cells, but rather hormonelike substances known as
neurohormones.[6] These neurohormones are released into a short, dense
bed of capillaries, the *portal system*, linking the hypothalamus and the
anterior pituitary. One particular neurohormone produced by the hypothalamus and carried to the anterior pituitary is *gonadotropin-releasing*

hormone (GnRH).[7] As its name implies, GnRH acts on the gonadotrophs within the pituitary, causing them to increase the rate at which they synthesize and secrete LH and FSH. LH and FSH are in turn carried by the general circulation to the ovary, where they induce *steroidogenesis* (the production of ovarian steroids) and the growth and differentiation of follicle cells.

The hypothalamus, anterior pituitary, and ovary are linked by the hormones they produce to form a single functional unit, the *hypothalamo-pituitary-ovarian axis*. The regulation of ovarian cycles cannot be said to reside in any one part of this system; rather it involves a complex set of feedback relationships that integrates the system as a whole (Figure 4.19).

A principal role played by the ovary in this regulatory system is as a steroid-secreting gland. The ovary synthesizes and secretes estradiol and progesterone under the influence of LH and FSH. It is interesting, from this viewpoint, that LH and FSH appear to act solely on the gonads, in contrast to estradiol and progesterone, which have ramifications in many tissues throughout the body (Table 4.2).

There are three major classes of sex steroids, i.e., steroids secreted by the gonads: (1) the androgens, or so-called male sex steroids, (2) the progestins, and (3) the estrogens, which together with the progestins make up the female sex hormones.[8] As mentioned earlier, steroids are ultimately derived from cholesterol. All three classes of sex steroids are linked by complex pathways of interconversion, and production of any particular steroid usually requires production of several others as intermediate stages (Figure 4.20). For example, synthesis of an estrogen such as estradiol requires prior synthesis of one or more androgens. Therefore, in order to produce estrogens, the ovary must actually produce "male" sex steroids first.

Steroidogenesis thus proceeds by a complex series of stages, which are distributed across several compartments of the body as shown in Figure 4.21. The low levels of LH and FSH produced by the pituitary during the early follicular phase of the cycle are carried by the general circulation to the ovary, in particular to the primary follicle. LH binds to receptors on the surface of the theca cells and stimulates production of two androgenic steroids, androstenedione and testosterone (Figure 4.22). These androgens in turn diffuse across the basal lamina (recall that no blood vessels penetrate the follicle beyond the theca cells) and are absorbed by the granulosa cells. At the same time, FSH binds to the granulosa cells and activates the enzymes of the aromatase complex, which convert androgens to estradiol. Thus, LH and FSH turn on the production of estradiol by the follicle and stimulate further follicular development.

Throughout most of the follicular phase, the estradiol produced by the ovary acts at the level of the hypothalamus and anterior pituitary to restrict secretion of GnRH, LH, and FSH. Thus, there is a *negative feedback control loop* involving the entire hypothalamo-pituitary-ovarian axis that acts to keep LH and FSH at fairly low concentrations (Figure 4.23). Once started, however, estradiol production is *autocatalytic:* the more estradiol the follicle produces, the more it is stimulated to produce. This autocatalytic stimulation of estradiol synthesis is possible because, in the presence of small amounts of LH and FSH, as well as various growth factors produced by the ovary itself, estradiol promotes vascularization of the theca and multiplication of the follicle cells; in other words, estradiol stimulates growth in the very cellular apparatus that produces estradiol. As a result, the concentration of estradiol in the woman's circulation steadily increases over the course of the follicular phase, as has already been described (Figure 4.2).

Somewhere near midcycle—and for reasons that are still obscure—the level of estradiol passes a threshold beyond which the feedback effect of estradiol on GnRH and the gonadotropins becomes positive rather than negative (Figure 4.24). This shift may be mediated by two additional substances produced by the granulosa cells, the peptide hormone *inhibin*, which is known to inhibit FSH synthesis and secretion (Steinberger and Ward, 1988), and the still poorly characterized *gonadotropin-surge attenuating factor* (GnSAF), which appears to modulate the effects of estradiol on LH/FSH secretion in complex ways that are not fully understood (Messinis and Templeton, 1991). It may be that negative feedback requires a particular combination of estradiol, inhibin, and GnSAF, and that the switch from negative to positive feedback represents the withdrawal of at least one of the last three substances, most likely inhibin. But that withdrawal by itself, although it may obliterate positive feedback, would not be enough to establish *negative* feedback. A variety of changes are occurring at midcycle at both the hypothalamic and pituitary levels, however, and any or all of those changes could modulate estradiol feedback on LH secretion. For example, high levels of estradiol tend to increase the number of GnRH receptors on the gonadotrophs, which could serve to increase the amplitude of the LH response to pulsatile GnRH secretion. Whatever its cause, once the switch has occurred the high circulating level of estradiol triggers a sudden, massive increase of LH and FSH in the bloodstream—the LH/FSH surge. This surge is necessary for the final induction of ovulation, which follows the peak of the surge by about 16 hours (WHO, 1980); it is also necessary for conversion of the now-empty follicle into a corpus luteum—hence the name *luteinizing* hormone. Table 4.3 summarizes the principal changes induced in the follicle by the LH/FSH surge.

Table 4.3. Summary of Changes in the Preovulatory Follicle Induced
by the LH/FSH Surge, Leading to Ovulation and Formation
of the Corpus Luteum

---

1. Increase in antral fluid volume and blood flow to the follicle
2. Temporary increase and then inhibition of estradiol secretion
3. Reduction in LH receptors on theca and granulosa cells
4. Stimulation of progesterone secretion by granulosa cells
5. Resumption of meiosis in oocyte
6. Severing of cumulus oophorus cells from other granulosa cells
7. Breakdown of basal lamina between theca and granulosa cells
8. Increased synthesis of prostaglandins and proteolytic enzymes, which break
   down other follicular and ovarian structures

Modified from Baird (1983a), Vander et al. (1990)

There are still many unsolved mysteries concerning the fine details of
this control system. However, intriguing clues have been offered by the
recent discovery that both GnRH and the gonadotropins are released
from the hypothalamus and pituitary in tiny bursts or pulses (see Figure 4.6).[9] The frequency of these pulses is ultimately controlled by a
complex of neurons within the hypothalamus known as the *GnRH pulse
generator,* which times the release of GnRH into the portal system. The
pulse generator is located in the *suprachiasmatic nucleus,* a region of the
hypothalamus lying just above the area where the optic nerves cross
(the optic chiasma in Figures 4.17 and 4.18). When cultured in vitro, the
cells of the pulse generator display an inherent rhythmicity, firing their
action potentials at regular intervals of about two hours. In vivo, this
rhythmicity appears to be subject to modulation by signals from other
neurons and, more importantly, by steroids such as progesterone.

As its name implies, the rhythmicity of the GnRH pulse generator is
translated into pulses of GnRH secretion and thereby into pulses of LH
secretion, which can readily be monitored in the peripheral circulation
(Figure 4.25). During the early follicular phase, LH pulses appear to be
frequent but of low amplitude, maintaining the overall concentration of
LH in the bloodstream at a relatively low level. The midcycle LH surge
is attributable to a marked increase in the amplitude of LH pulses, but
apparently not in their frequency (Figure 4.25, middle panel). The feedback suppression of LH during the luteal phase appears to be fundamentally different from that in the early follicular phase, since it is associated with low frequency (but moderate amplitude) pulses (Figure 4.25,
bottom panels).

These observations can be combined with what is known about
steroid feedback to produce a model of the neuroendocrine control of
the cycle (Figure 4.26). In the early follicular phase, the negative effect

of estradiol and inhibin (and perhaps GnSAF) on LH/FSH secretion appears to operate at the level of the pituitary by somehow making the gonadotrophs less sensitive to GnRH, thus lowering the amplitude but not the frequency of LH/FSH pulses. Whether this negative amplitude modulation by estradiol is reversed in the periovulatory period, or whether the gonadotropin-inhibiting effects of inhibin or GnSAF are simply removed, is not yet known. It is known, however, that high circulating concentrations of progesterone, such as those typical of the luteal phase, directly affect the GnRH pulse generator in such a way as to lower the frequency of the GnRH pulses. It is not clear how this repression of GnRH pulsatility is accomplished, but it may well involve β-*endorphin*, one of the so-called endogenous opiates (biologically produced peptides with effects similar to morphine). β-endorphin is found, among other sites in the central nervous system, in a major group of neurons within the hypothalamus that richly innervate the region where most GnRH-containing neurons are found. Hypothalamic β-endorphin concentrations fluctuate predictably over the course of the cycle, with highest levels in the luteal phase and lowest in the follicular phase. When administered intravenously, β-endorphin suppresses pulsatile gonadotropin secretion and abolishes the preovulatory LH surge. Consistent with this finding, administration of the drug naloxone, which "jams" opiate receptors and thereby blocks β-endorphin action, causes a marked elevation of plasma LH levels. It is now believed that elevated β-endorphin levels reduce the frequency of GnRH pulses within the hypothalamus and are thus a major mediator of the suppressive effects of progesterone on pulsatile LH/FSH secretion during the luteal phase (Crowley et al., 1985; Lincoln et al., 1985).

Whatever the mechanism, LH/FSH secretion is held in check throughout the luteal phase, and further follicular development is postponed until the time of luteolysis, when progesterone is withdrawn. From this perspective, luteolysis may turn out to be the most critical event in the timing of the whole cycle. Unfortunately, luteolysis is still something of a mystery, at least as it occurs in humans and other primates. It is known that the presence of at least some LH, even at the low levels characteristic of the follicular phase, is necessary to maintain the corpus luteum. Thus, experimental injection of LH antagonists (chemicals that block the normal action of LH) is sufficient to cause luteolysis in a variety of species (Niswender and Nett, 1988). In nonprimate mammals such as sheep, luteolysis appears to be induced by a substance secreted by the uterus, namely, the prostaglandin $PGF_{2\alpha}$, which acts to lower the number of LH receptors in the corpus luteum (Behrman et al., 1978). In primates, however, including humans, there is no evidence of uterine involvement. Indeed, women who undergo hysterectomy (removal of the uterus) without oophorectomy (removal of the ovaries) continue to experience all the hormonal changes associated with normal cycles (Beling et al., 1970; Doyle et al., 1971). But *intraovarian* $PGF_{2\alpha}$ may be involved, since infusion of $PGF_{2\alpha}$ into the primate corpus luteum induces luteolysis (Wentz and Jones, 1973; Auletta et al., 1984). Possible effects of ovarian $PGF_{2\alpha}$ on the corpus luteum are summarized in Table 4.4.

A final mystery has to do with *follicular selection,* the process whereby one (and usually only one) follicle is "chosen" each cycle from the pool of developing follicles to go through the final stages of maturation and ovulation (Figure 4.27). The selected follicle, known as the dominant follicle for that particular cycle, thus avoids the common fate of atresia that awaits most follicles. In two recent reviews, Baird (1987) and Greenwald and Terranova (1988) have discussed several aspects of this mystery. First, why do some follicles initiate growth early in the woman's reproductive life, say at age 12 or 13, and some as late as age 50? Current evidence suggests that the difference is purely random, with every primary follicle having the same small and apparently constant probability of initiating development in each cycle. If correct, this view implies a negative exponential decline in the number of primary follicles remaining in the ovary at any age, consistent with the data in Figure 4.9. Although gonadotropins are essential for the final maturation of the follicle and for ovulation, most of the preantral stages appear not to require gonadotropin support (Niswender and Nett, 1988). This explains why follicular loss occurs well before adolescence, when gonadotropin secretion is undetectable (see Chapter 9).

Remarkably, if a growing follicle does go on to final maturation, its total growth period covers about three cycles (approximately 85 days in humans). Therefore, events in earlier cycles may influence the selection of dominant follicles in later cycles. Evidence from rats and hamsters suggests that the FSH surge in the first or second cycle may be critical in selecting the dominant follicle for the third cycle. However, many

Table 4.4.   Possible Mechanisms by Which Ovarian $PGF_{2\alpha}$ May Induce
       Luteolysis

1. Reduced blood flow to the ovary
2. Reduced number of LH receptors in the corpus luteum (CL)
3. Decoupling of the signal transduction system within the CL cell that links LH and steroidogenesis
4. Direct cytotoxic effects within the CL

Adapted from Niswender and Nett (1988)

chemicals stimulate superovulation in humans and other primates, including gonadotropins, antisera to inhibin, and estrogen antagonists (Greenwald and Terranova, 1988). This fact suggests that there may be no single stimulus underlying follicular selection. What seems clear in a wide variety of species, however, is that the final stages of follicular maturation cannot occur in the face of high levels of progesterone. Thus, luteolysis and the length of the luteal phase can be thought of as the ultimate regulator of the pace of follicular maturation.

Just as the cause of follicular selection is unclear, it is not clear what induces atresia. There is no clear-cut evidence that the dominant follicle in any way "shuts down" the other developing follicles, although this possibility cannot be ruled out (Baird, 1987). In humans, one critical factor in atresia appears to be a loss of activity in granulosal aromatase, one of the enzymes involved in converting androgens to estrogens within the follicle. Thus, a primary sign of atresia is failure of estradiol secretion and a resultant failure of follicular autocatalysis. Based on present evidence, however, it is impossible to say whether this failure is an inherent feature of follicles doomed to atresia or is somehow imposed upon the follicle from outside.

Leaving such mysteries aside, Table 4.5 summarizes the major events and processes that appear to be involved in the neuroendocrine regula-

Table 4.5. Summary of the Major Events Involved in the Ovarian Cycle

Day[a]	Major events
1–5	Estradiol and progesterone are low because the previous corpus luteum has completely regressed. *Therefore:* (a) the endometrial lining sloughs, and (b) secretion of FSH is released from inhibition and its serum concentration rises, stimulating several follicles to enlarge
5–15	Serum estradiol rises because of secretion by the dominant follicle
13–14	LH/FSH surge is induced by high serum estradiol. *Therefore:* (a) the oocyte is induced to complete its first meiotic division and undergo cytoplasmic maturation, and (b) the follicle is stimulated to secrete lytic enzymes and prostaglandin
15	Ovulation is caused by follicular enzymes and prostaglandin, which rupture follicular and ovarian tissues
16–24	Corpus luteum forms and secretes progesterone and estradiol, causing serum concentrations of those hormones to rise
25–29	Corpus luteum degenerates. *Therefore:* plasma concentrations of progesterone and estradiol decline. *Therefore:* endometrium begins to slough, and a new cycle begins

[a] These figures are intended to indicate the "typical" timing of events as observed in mid-reproductive age western women
Modified from Vander et al. (1990)

tion of the female cycle. Although the table indicates the "typical" cycle days on which those events and processes occur, considerable variation in timing has been observed within the same woman, as well as among women. We turn next to a detailed discussion of this variability.

INTRAPOPULATION VARIATION IN OVARIAN CYCLES

The processes underlying ovarian cycles would be irrelevant to a study of fertility variation if they did not themselves vary, either among women or within the same woman at different times. We could then treat ovarian function as if it were a perfectly "hard-wired" trait of the species as a whole, profoundly important from a physiological point of view but in no way contributing to fertility differentials. In fact, this is precisely how most demographers tend to view ovarian function as well as other aspects of reproductive biology, insofar as they view them at all. But we now know that cycles are extremely variable within individual women, among women of the same population, and as recent evidence suggests, perhaps even among populations.

One of the largest sets of data on cycle length in normal, healthy western women was compiled by Chiazze et al. (1968) in a multi-center study conducted during the 1960s (Figure 4.28). A major virtue of this data set is that it was collected in a series of populationwide surveys and thus avoids much of the selectivity bias associated with clinical data. The sample includes a total of 30,655 cycles in 2,316 white U.S. and Canadian women ages 15 to 45 years, or an average of about 13 cycles per woman. In other words, the sample combines intersubject variation (including variation by age) with intrasubject variation, if such exists.

As Figure 4.28 (top panel) indicates, substantial variability in cycle lengths is apparent in the total sample. The distribution of cycle lengths is slightly right-skewed, with a single clear-cut mode at about 27 days and a mean and variance of 29.1 and 55.7, respectively. Although this variance is unquestionably large, it tells us nothing about the relative contributions of inter- and intrasubject variation to its value. Moreover, the estimated mean and mode are almost certainly biased downward to at least a slight degree by a disproportionate weighting of shorter cycles. This bias arises because, during an observation period of fixed duration, women with short cycles experience somewhat more cycles and thus contribute more observations to the total data set than do women with longer cycles. This bias can be removed, and the contribution of intersubject variation made explicit, by inspecting the distribution of *mean* cycle length for each subject (Figure 4.28, bottom panel).

As expected, the mean and mode of this second distribution are slightly greater than those of the first one (29.4 versus 29.1 for the mean, and 28 versus 27 for the mode), but the differences are so small as to be trivial. More important, the variance of this new distribution is approximately 80 percent less than the variance of the total distribution, suggesting that a large fraction of the total variability in cycle lengths is attributable to variation from cycle to cycle within individual women. In addition, the much smaller upper tail of the second distribution suggests that long cycles tend to occur sporadically among women. Considerable subject-to-subject variation in cycle length remains, however, with approximately 95 percent of the means falling between 25 and 37 days.

As discussed below, much of the observed intersubject variation in this study is probably attributable to the wide age range of subjects. However, even when attention is restricted to mid-reproductive age women, significant inter-subject variation in cycle length remains. This was shown clearly almost 70 years ago (King, 1926) in a classic study of more than 300 cycles in four healthy women who were ages 21–30 at the beginning of observation (Figure 4.29). Unquestionably, then, variation in cycle length—not to mention deviations from the supposedly "normal" cycle of 28 days—is the rule, not the exception.

This conclusion must be tempered somewhat by noting one shortcoming in the two studies just described. As we shall see, cycles vary in quality (i.e., fecundity) as well as in length, and in fact some of this variation in cycle quality *causes* variation in cycle length. Stated differently, some variation in cycle length is now known to be associated with cycles that fall outside the range of normal ovarian function. Since neither of those earlier studies attempted to detect ovulation or to measure the adequacy of progesterone secretion during the luteal phase, two principal determinants of cycle quality, there is no way to remove subfecund cycles from the samples and thus look strictly at variation in normal cycles. However, more recent work, though involving much smaller sample sizes, has confirmed the general impression of tremendous variability in normal cycles both within and among women (Lenton et al., 1984a, 1984b).

Whenever it has been investigated in humans, age has proved to be a major factor associated with differences in cycle length. The influence of age is perhaps most clearly seen in the massive data compilation of Treloar et al. (1967), who collected prospective menstrual diaries over more than 30 years from a total of 2,702 U.S. women, each followed for an average of 9.6 years (Figure 4.30). As this and other studies have shown, age patterns of ovarian function stand out most clearly when the data are classified by "gynecological age" (i.e., years since menarche for subjects less than 20 years old and years before menopause for those over

40) rather than simple chronological age. This finding is not surprising. It is well known that women vary considerably in the age at which they experience *menarche* and *menopause,* the times of first and last menstrual bleeding, respectively (see Chapter 9). By looking at subjects of chronological age 14, say, we would be mixing together individuals with quite different developmental rates, thus obscuring the effects of developmental age, presumably the age-variable of most direct biological importance. By its nature, gynecological age is almost certainly more closely related to developmental age than is chronological age.

As shown by the Treloar et al. study (Figure 4.30, top panel), cycles are slightly longer on average and more variable in length during the first few years following menarche. After age 20, mean cycle length declines slowly to a low point at or just after age 40, a decline amounting to only about one or two days. During the mid-reproductive years, cycles are comparatively invariant in length, as indicated by their small standard deviations. During the three or four years preceding menopause, however, mean cycle length again increases, reaching 50+ days at about the time of the last menses, and cycles become much more variable (Figure 4.30, top panel). As the bottom panel of Figure 4.30 clearly indicates, these age-related changes in cycle length are attributable almost entirely to changes in the fraction of very long cycles at each age: at perimenarcheal and especially perimenopausal ages, a disproportionately large fraction of cycles is more than 35 or 40 days long. To put it differently, cycles are much more variable in length at extreme ages, with most of the variation taking the form of unusually long cycles.

Figure 4.31 reproduces one of the most remarkable figures ever published on the biometry of ovarian function, also taken from the study of Treloar et al. (1967). This figure presents the only known complete menstrual history for one particular woman, recorded prospectively from menarche to menopause, and it shows the usual age patterns in cycle length with exceptional clarity. Fortunately for us all, this woman seems to have experienced normal patterns of cyclicity throughout her life.[10] During the first few years following menarche at age 14, her cycles are variable in length, and it takes perhaps four or five years for them to settle completely into the normal adult pattern. Between ages 25 and 34 she cycles with remarkable regularity. There is a slight hint of increased variability just before her first fertile conception at age 35, perhaps attributable to the presence of one or more "silent pregnancies" ending in fetal loss and occurring at a time when she was trying to become pregnant. Each successful birth is followed by a brief period of increased variability in cycle length, representing the normal postpartum period during which regular cycles are reestablished (see Chapter 8). Finally, beginning at about age 40, she shows the usual increase in cycle length and in variability of cycles that typifies the perimenopausal years; she then experiences her final menses at age 46.

For women in general, variation in total cycle length appears to be caused primarily by variation in the length of the follicular phase; in contrast, the luteal phase is more nearly invariant (Figure 4.32). As a major study in Japan by Matsumoto et al. (1962, 1963) has revealed, this is true at all ages, including perimenarcheal and perimenopausal ages when very long cycles are most likely to occur (Figure 4.33). The functional significance of this finding is unclear, although it has been suggested that the luteal phase is under more severe functional constraints than the follicular phase because of its role in the maintenance of the early stages of pregnancy (see Chapter 5).

Since the end of the follicular phase and the beginning of the luteal phase are demarcated by ovulation, the data on follicular phase lengths in Figure 4.32 tell us something about variation in the timing of ovulation within the cycle. Although ovulation in mid-reproductive age women appears to occur *on average* at about cycle day 16–17, it is obvious that variability in the length of the follicular phase makes it almost impossible to predict the day of ovulation for any one cycle with any exactitude simply by counting forward from the most recent menses. More complex hormonal methods are needed to predict ovulation (for useful reviews, see Jeffcoate, 1983; Campbell, 1985; Royston, 1991).

There is an interesting statistical artifact having to do with an apparent correlation in the lengths of the follicular and luteal phases. Earlier studies based on the measurement of BBT indicated a substantial and significant *negative* correlation between the two phases: the longer the follicular phase, for example, the shorter the luteal (Matsumoto et al., 1962; Presser, 1974). When more refined but still error-prone physiological indicators of ovulation, such as bioassay and endometrial biopsy, were used, the negative correlation was reduced in magnitude but still persisted (Potter et al., 1967). But as modern hormonal methods for the detection of ovulation became available, permitting the day of ovulation to be pinpointed with considerable accuracy, the correlation in the length of the two phases vanished (McIntosh et al., 1980). From all recent evidence, the two phases vary independently.

Why did the apparent negative correlation arise in the first place? The answer is suggested by Figure 4.34. In this diagram, total cycle length (A–B) is measured by the interval between two readily identifiable markers, the onsets of successive menses. Thus, we can treat total cycle length as fixed and known with practical certainty. The midcycle rise in

BBT, in contrast, can be located with much less certainty, and its position can only be said to be located within a fairly broad range (Figure 4.34, shaded area). However, in describing any particular cycle we are forced to specify the day of the BBT rise, even though recognizing that an element of subjectivity may be involved. Suppose we assign the BBT rise to position 1 (Figure 4.34, left-hand arrow). Because the outer limits of the cycle are fixed, this assignment immediately implies a comparatively short follicular phase and a long luteal phase. On the other hand, had we assigned the BBT rise to position 2 (right-hand arrow), the follicular phase would have appeared long and the luteal phase short. Thus, an apparent but spurious negative association between the two phases arises strictly because of measurement error. This artifact should serve as a reminder that methods and inferences are never entirely separable, and that the former sometimes dictate the latter in discomfiting ways.

There are fairly consistent age patterns in cycle quality as well as cycle length. With respect to quality, cycles can vary in one of two ways, each of major functional significance. First, a cycle can be *anovulatory*—that is, follicular development can be aborted before the final maturation of the egg, and the endometrium sloughed off without a corpus luteum ever being formed. The apparent menses that follows such an anovulatory cycle is actually a form of *estrogen-withdrawal bleeding* rather than the full-scale, normal menses that follows luteolysis and the withdrawal of progesterone. Compared with a normal cycle, the anovulatory cycle is likely to be short; if, however, follicular development is aborted too early for noticeable estrogen-withdrawal bleeding to occur, then a series of such anovulatory cycles may have the appearance of one very long cycle.

The second way in which cycles can vary in quality has to do with the adequacy of the luteal phase to support a pregnancy. Even if ovulation occurs, the ensuing luteal phase may be defective either because the corpus luteum is too short-lived or because it produces too little progesterone (Figure 4.35). In such cases, the secretory endometrial lining of the uterus, which normally undergoes final maturation under the influence of rising progesterone, may be inadequate for the establishment and maintenance of a pregnancy (see Chapter 5). Again, such defective cycles are likely to be somewhat shorter than normal cycles.

Several studies have shown that the probability of a cycle being anovulatory or having an inadequate luteal phase is highest during the first few years following menarche, with a secondary rise during the five or so years preceding menopause (Lenton et al., 1988; Lipson and Ellison, 1992). These aspects of cycle quality underlie the well-known phenomenon of *adolescent subfecundity*, as well as the decline in fecundity during the final years of a woman's reproductive life (see Chapter 9 for details). The data summarized in Figure 4.36, which come from the single largest study of cycle quality ever conducted, show the general trends fairly clearly. It is important to realize, however, that this study used the BBT method to detect ovulation and to determine the length of the luteal phase; since BBT shifts are notoriously difficult to detect, it is likely that this study overestimated the prevalence of anovulation at each age. However, more recent studies based on modern endocrinological methods have verified the same general patterns (Figure 4.37).

An interesting statistical analysis by Lenton et al. (1984a) estimates the contribution of luteal inadequacy to overall variation in the length of the luteal phase. Figure 4.38 (panel a) shows variation in luteal-phase length as indicated by the location of the LH surge in 327 cycles experienced by the British and Swedish women included in their study; panel b shows a normal probability plot of the same data. (A straight line on such a plot would indicate a normal distribution of luteal phases.) The normal probability plot strongly suggests the existence of two distinct subpopulations of cycles, each following its own normal distribution: a

Table 4.6. Observed Distribution of Luteal Phase Lengths in 327 Ovulatory Cycles, and Predicted Distributions from Two-normal Mixture Model Broken Down into Population I ("short luteal phases") and Population II ("normal luteal phases")

Duration of luteal phase (days)	Observed frequency (%)	Predicted Frequency (%)			Estimated % in Pop. I
		Pop. I	Pop. II	Total	
7	0.6	0.6	0.0	0.6	100
8	1.5	1.0	0.0	1.0	100
9	0.6	1.4	0.1	1.5	97
10	1.8	1.2	0.4	1.6	74
11	3.7	0.7	2.4	3.1	22
12	8.6	0.2	8.8	9.0	2
13	18.0	0.1	19.3	19.4	0
14	26.6	0.0	26.2	26.2	0
15	24.2	0.0	21.9	21.9	0
16	10.4	0.0	11.3	11.3	0
17	3.4	0.0	3.6	3.6	0
18	0.6	0.0	0.1	0.1	0

χ^2 for goodness-of-fit = 4.53, df = 7, 0.9 > P > 0.5
log likelihood ratio test statistic for hypothesis that only one subpopulation is represented in the sample = 48.0, df = 2, P < 0.001
Modified from Lenton et al. (1984a)

larger subpopulation with relatively long luteal phases, represented by the higher, more rapidly ascending portion of the plot, and a much smaller one with short luteal phases, represented by the more slowly rising lower tail of the plot. In order to estimate the characteristics of the two subpopulations, Lenton et al. (1984a) fit a mixture model combining two normal distributions to their data (Table 4.6). The fit of the model was excellent, and the analysis firmly rejected the alternative hypothesis that only one normally distributed population was represented in the sample. According to their analysis, Subpopulation I (with short luteal phases) had a mean luteal phase of 9.2 days and made up five percent of the sample, while Subpopulation II (normal luteal phases) had a mean luteal phase of 14.1 days and made up 95 percent of the sample.[11] They were also able to show that the fraction of cycles in Subpopulation I was elevated during adolescence and at perimenopausal ages, consistent with other studies. Unfortunately, because their analysis cannot determine which subpopulation a *particular* luteal phase belongs to, we cannot be certain whether Subpopulation I represents a distinct subpopulation of women or of cycles distributed sporadically among all women.

In addition to age, the effect of which is fairly obvious, genetic factors, nutritional status, and, more recently, activity patterns have all been implicated as possible causes of intrapopulation variation in ovarian cycles (Figure 4.39). A curious but intriguing indication that environmental factors may indeed affect cycle length comes from the study of Sundararaj et al. (1978). Analyzing menstrual diaries maintained by 3,766 U.S. women over a period of more than 10 years, these investigators were able to detect a small but significant seasonal pattern in cycle length variation, with longer cycles occurring from October to January and shorter cycles from June to September (Figure 4.40). Although the magnitude of the estimated seasonal shift in cycle length (about two days over the course of the year) is far too small to have any measurable demographic impact, these results serve to remind us that despite the exquisite physiological control to which cycles are subject, they are far from being absolutely "hard-wired."

INTERPOPULATION VARIATION IN OVARIAN CYCLES

In comparison with the effort expended in recent years to characterize cycles in western women, cycles in women of other populations have been almost completely unstudied, leaving unanswered the question of whether significant interpopulation differences exist in either cycle length or cycle quality. One study of Japanese women, conducted during the late 1950s and early 1960s, revealed cycles that were almost indistinguishable from those of U.S. or European women (see Figure 4.33). Coming as early as they did in the biometric study of ovarian cycles, these findings may have served to deflect attention away from research on interpopulation variation. Despite the unambiguous ethnic difference between Japanese women and most western women, however, it could be argued that these women are all so similar in terms of general socioeconomic background that differences in their cycles should not have been expected in the first place. For a clear-cut answer to the question of interpopulation variation, it would be better to study women who are not only nonwestern but nonindustrial and nonurban as well.

Two recent studies are relevant in this connection. The first involved a sample of 36 rural women in highland Papua New Guinea, known from their serum hormone profiles to be cycling normally (P. Johnson et al., 1987). Analysis of menstrual diaries collected from these women over a 10-week period suggests that they experience cycles that are approximately 40 percent longer on average than those of a sample of U.S. women matched for gynecological age (Figure 4.41). The second study involved 35 mid-reproductive age women from the Ituri Forest of northeastern Zaire in central Africa (Ellison et al., 1989). According to daily salivary progesterone measures collected over a four-month period, these women experienced a significantly higher frequency of anovulatory cycles and, when their cycles were ovulatory, a significantly lower luteal phase rise in progesterone than U.S. women (Figure 4.42). In addition, women in this study exhibited significantly shorter durations of menstrual bleeding than the U.S. controls (Bentley et al., 1990), which may point to suppression of ovarian function. Although there are major (and fundamentally undecidable) questions about possible selectivity bias affecting the results of both these studies, they at least suggest that more research needs to be done on ovarian cycles in nonwestern settings.

The question remains whether any of the variation in ovarian cycles discussed thus far, either within or among populations, is sufficient to have anything other than a trivial effect on fertility at the aggregate level. Such an effect, if it exists, must operate through one of two mechanisms: (1) by modifying the probability of conception during any fixed time period, thus affecting the expected waiting time to conception; or (2) in the case of luteal inadequacy, by modifying the ability to carry a pregnancy full term, thus affecting the time added to each birth interval by fetal loss. This question will be deferred to later chapters, after the analytical machinery for answering it has been developed. For now, it will suffice to say that the age-related changes in ovarian function described above are unquestionably large enough to have important demographic consequences (Weinstein et al., 1993a; Wood, 1989). It is much less clear that the same can be said for the interpopulation differences that have been observed to date (Johnson et al., 1990).

DO WOMEN IN NATURAL-FERTILITY POPULATIONS EXPERIENCE REGULAR OVARIAN CYCLES?

In 1976 the noted reproductive biologist Roger Short made the provocative suggestion that "women may be physiologically ill-adapted to spend the greater part of their reproductive lives having an endless succession of menstrual cycles" (Short, 1976). Elsewhere he remarks that "an excessive number of menstrual cycles is an iatrogenic disorder [i.e., one induced by medical intervention] of communities practicing any form of contraception" (Short, 1984c).[12] Although it may be a dramatic overstatement, Short's assertion is worth considering here because it concerns, first, the distinction between controlled and natural fertility and, second, the key role played by breastfeeding as a determinant of total reproductive output (see Chapter 8).

According to Short's argument, once a woman in a natural-fertility population has commenced reproduction, she is likely to be either pregnant or lactating (or both) and thus acyclic for much of the remainder of her reproductive life. There is, moreover, a consistent balance between the two: women of low fertility experience comparatively few pregnancies but longer periods of lactational infecundability, whereas those of higher fertility experience shorter periods of lactational infecundability per pregnancy but more pregnancies overall. Thus, under natural fertility both sorts of women devote roughly equal amounts of time to reproduction—where "reproduction" is taken to include both pregnancy and lactation. Unless a woman's married life is interrupted by the absence of her husband, by widowhood, or by divorce, there are only a few short intervals during which she is likely to be cycling at all: menarche to marriage, marriage to first fertile conception, and postpartum resumption of menses to the next fertile conception or to menopause, whichever occurs first. Under reasonable assumptions about the distribution of those intervals, Short estimates that the typical woman in most natural-fertility settings is likely to experience fewer than 100, and perhaps fewer than 50, ovarian cycles during her entire lifetime (Short, 1976).

When fertility control is adopted on a large scale, the intervals during which regular cycles can occur are greatly extended. With birth-control methods that act without preventing ovulation (e.g., barrier methods, IUDs, or induced abortion), an increased number of cycles is inevitable. Paradoxically, one widely used method of artificial contraception that completely obliterates ovulation, namely, the combined estrogen-progestin pill, is carefully designed to mimic both the hormonal changes associated with the natural ovarian cycle and menses itself. Thus, we have the situation exemplified by the woman whose menstrual history appeared in Figure 4.31, who experienced a total of 355 complete menstrual cycles during her lifetime.

Unfortunately, no one has ever compiled lifetime data on the number of cycles experienced by women of any natural-fertility population, and by their nature such data would be very difficult to collect. However, it is possible to test Short's idea indirectly using cross-sectional data. In 1983, while working with a natural-fertility population in rural Papua New Guinea, we studied ovarian function in 65 married women, all of whom were known from their serum gonadotropin and steroid profiles to be both postmenarcheal and premenopausal—that is, to be of those ages at which menstrual cycles could potentially occur. In Table 4.7, those 65 women are classified by current reproductive status according to their hormone levels and detailed data on breastfeeding. For women less than 40 years old, it is clear that Short is more or less right: fewer than 15 percent are neither pregnant nor lactating and are thus presum-

Table 4.7. Reproductive Status of 65 Postmenarcheal and Premenopausal Gainj Women Who Were Married and Living with Their Husbands in 1982–1983. Tabulated figures are the age-specific proportions of women in each reproductive category, based on serum hormone profiles and self-reports of lactation (standard errors in parentheses)

Age of woman (years)	N	Pregnant	Lactating, not Pregnant		Not pregnant, not lactating
			Acyclic	*Cyclic*	
20–39	45	.33 (.07)	.33 (.07)	.20 (.06)	.13 (.05)
40–49	20	.10 (.07)	.20 (.09)	.30 (.10)	.40 (.07)

Test of difference between age groups:
log likelihood ratio test statistic = 8.64, df = 3, $P < 0.05$
From Wood (1992)

ably cycling with regularity. Although some 40 percent of all lactating women under 40 have already resumed cycling, their cycles are likely to be both irregular and anovulatory (see Chapter 8). Thus, at any one time only a minority of married women are cycling normally, at least up to age 40. After 40, the situation changes. Now fully 40 percent of all married, premenopausal women are neither pregnant nor lactating, and 60 percent of all lactating women have resumed cycling. This change probably reflects, at least in part, the declining fecundity of women at those ages. It may also indicate a cultural norm of terminal abstinence, although we have no positive evidence of this.

When we examine all reproductive-age women in this population, both married and unmarried, a slightly different picture emerges (Figure 4.43). It now appears that a large fraction, perhaps even a majority, of postmenarcheal, premenopausal women are cycling at any one time. In this particular sample, however, about 86 percent of all currently cycling women were unmarried, either because they had not yet married or because they had been widowed or divorced. This finding suggests that Short's hypothesis needs to be modified to allow for differing patterns of marital formation and dissolution, as well as spousal absence and the occasional taboo against intercourse within marriage. With this important proviso, it would appear that our data support Short's claim that women experience far fewer cycles under natural fertility than under controlled fertility. We estimate that the New Guinea women in our study may experience something on the order of 90 cycles during their lives, roughly the number a western woman might experience in about seven years.

This conclusion raises an interesting question about the neuroendocrine control of the ovarian cycle: why did such a complex regulatory system evolve if it so seldom functions under conditions of natural fertility? The answer, of course, is that it *does* function under such conditions, and at critical times, but it does so in a manner designed to close itself down as rapidly as possible, i.e., through conception, establishment of a pregnancy, and suckling-induced anovulation. Conception is something of a gamble and is by no means inevitable in any cycle of exposure to intercourse. The hypothalamo-pituitary-ovarian axis is designed, therefore, not only to maximize the probability of conception in any one cycle, but also to clear the way as quickly as possible for another opportunity should conception not take place. The distribution of fecund waiting times to conception, those segments of the female life course during which cycles are expected to occur, will be examined in detail in Chapter 7. For now, it is worth noting that the same pattern of sporadic and short periods of reproductive cycling that Short postulated for humans under natural fertility has also been observed in wild populations of our closest living primate relatives, the African apes (Tutin, 1980).

THE FERTILE PERIOD

Even when ovulation occurs, there is only a comparatively brief period during the middle of the cycle when an act of unprotected intercourse or *insemination* has any chance of resulting in a conception. The length of this so-called fertile period is of obvious practical importance for couples who wish to maximize their chances of conception or who wish to avoid it altogether. It is also of considerable theoretical significance because of its effect on fecundability. All things being equal, the shorter the fertile period, the lower the probability of conception during any fixed period of exposure to unprotected intercourse. In the analytical models reviewed in Chapter 7, the length of the fertile period emerges as a surprisingly potent determinant of fecundability. It is unfortunate, therefore, that it has proven so difficult to measure.

In principle, the major factors determining the length of the fertile period have been identified, if not precisely measured (Figure 4.44). The absolute outer limits of the fertile period are set by the maximal fertile life spans of sperm and egg, that is, by the maximum amount of time either type of gamete can survive in the female reproductive tract and still be capable of taking part in fertilization. (For sperm this life span is counted from the time of ejaculation, for eggs from ovulation.) Clearly, if sperm can survive no more than T_s days after ejaculation and still be fertile, then an insemination occurring more than T_s days before ovulation cannot result in fertilization. Similarly, if an egg can survive and be fertilized no more than T_e days after ovulation, no insemination falling more than T_e days after ovulation can result in fertilization. Thus, we would expect the fertile period to be at most $T_s + T_e$ days long.

A complication is introduced by the processes of *sperm capacitation* and *activation*. It is now known that sperm cells must spend some time within the female reproductive tract before they are fully capable of fertilizing an egg (see Chapter 5 for details). In theory this period of capacitation and activation should be subtracted from the total length of the fertile period (Figure 4.44). That is, if capacitation and activation take T_c days, and if an insemination occurs fewer than T_c days before the end of the maximal fertile life span of the egg, the sperm will not yet be capable of fertilizing the egg when the egg loses its own capacity for fertilization. Thus, the fertile period is at most $T_s + T_e - T_c$ days long. Studies of in vitro fertilization (which may or may not be completely relevant to the situation in vivo) suggest that capacitation and activation may take only a few hours (Edwards, 1980), a period of time that appears to be negligible compared to the fertile period as a whole. Thus, capacitation and activation can probably be ignored in statistical analyses of the fertile period.

Bongaarts (1983) has made the important point that estimates of fertile period length, as well as the probability of conception from an insemination in the fertile period, are sensitive to the method by which the day of ovulation is ascertained. Figure 4.45 presents two early sets of empirical estimates (panels a and b), along with Bongaarts's assumptions about what the "true" fertile period looks like (panel c). Panel a is based on the work of Vollman (1953), who analyzed coital records in a group of noncontracepting women, taking cycle day 14 as the presumptive day of ovulation. According to his analysis, the fertile period is as long as 12–14 days, almost half the entire cycle, but the daily probability of conception never rises much above 15 percent. A later analysis by Barrett and Marshall (1969), using the midcycle rise in BBT to indicate ovulation, is shown in panel b of Figure 4.45. According to that study, the fertile period is only about seven days long, and the daily probabilities of conception within that period are consistently higher than in Vollman's study.

How can we account for the large difference in the results of these two studies? As Bongaarts (1983) notes, neither the BBT shift nor cycle day 14 is an error-free indicator of ovulation; but BBT, as imperfect as it is, is certainly more valid than arbitrarily choosing cycle day 14. Even random errors in timing ovulation tend to spread out and lower the estimated conception probabilities, and this effect should be much more severe when using cycle day 14 than when using the BBT method. If the exact day of ovulation were known without error, then Bongaarts would argue that the fertile period should appear as in Figure 4.45, panel c. That is, it would be only about two days long, with a more or less uniform distribution of conception probabilities at a level of about 0.5–0.6.

Bongaarts's point about the "spreading out" of the fertile period associated with measurement error is extremely telling, but I would argue that he is wrong to assume that the true distribution of conception probabilities across the fertile period is uniform. This is because individual gametes differ in their fertile life spans (France, 1981). Thus, in speaking of the "maximal fertile life span of sperm and egg" it is important to bear in mind that only a small fraction of gametes may actually survive to that maximal age, even if there are only random differences in gamete survival. Consequently, the probability of conception from a single insemination almost by necessity will vary over the course of the fertile period. If, for example, an insemination occurs very early in the fertile period, few sperm will survive to ovulation and there will be little chance for conception. Similarly, if insemination occurs very late in the fertile period, the egg is likely not to have survived long enough to be fertilized. Clearly, the chance of fertilization should be highest if insemination occurs around the time of ovulation (or even, in view of capacitation and activation, a little before ovulation) and should drop off more or less rapidly the further from ovulation that insemination occurs. Such a pattern is clearly inconsistent with Bongaarts's hypothesis of a uniform distribution of conception probabilities across the fertile period.

Several attempts have been made to measure the length of the human fertile period, as well as the probability of conception per insemination for each day within the fertile period, most notably by Barrett and Marshall (1969), Barrett (1971), Schwartz et al. (1980), and Royston (1982). Since Royston's analysis was an extension of the previous studies, and since it worked with the largest set of data, including all the data used in the earlier analyses, it is arguably the most comprehensive and successful of those attempts. In addition, it is the only one that allows for variation in the survival of gametes. However, since it timed ovulation by the BBT method, Royston's analysis does not escape Bongaarts's argument about the spurious "spreading out" of the fertile period.

Royston based his analysis on the following hazards model. First, he assumed that the fertile life spans of sperm and egg follow two separate negative exponential distributions of the general form

$$S_i(t) = \exp(-\lambda_i t) \tag{4.1}$$

where $S_i(t)$ is the probability that the gamete in question, either egg or sperm, survives t days or more in the female tract. The hazard rate parameters, λ_e for eggs and λ_s for sperm, determine the daily risk of death (or, more accurately, loss of fertilizing capacity) per gamete. The mean fertile life spans of egg and sperm are then $1/\lambda_e$ and $1/\lambda_s$, respectively, with variances $1/\lambda_e^2$ and $1/\lambda_s^2$.

It is important to understand what this model implies. Although the two types of gametes are allowed to differ in their survival probabilities, all gametes of the same type suffer the same likelihood of death each day. Thus, there is no allowance for systematic heterogeneity among sperm or among eggs either in viability or in fertilizing capacity, and there is no aging of the gametes during their transit through the female reproductive tract. All variation in fertile life spans except that between the sexes is attributable strictly to chance. The distribution of fertile life spans implied by this model is extremely asymmetrical and right-skewed, with a single mode at zero and a long upper tail. As Royston remarks, "There is no evidence available for humans concerning the distribution of sperm and egg lifetimes *in vivo*" (1982:399), an observation that remains true today. Royston's simple model, which makes the fewest possible assumptions about the underlying processes, seems appropriate in the absence of such evidence.

According to Royston's model, the risk of conception from a single insemination on cycle day t relative to the risk from an insemination on t_{ov}, the exact day of ovulation, is

$$a_f(t) = \begin{cases} \exp\left[-\lambda_s(t_{ov} - t)\right], & t < t_{ov} \\ 1, & t = t_{ov} \\ \exp\left[-\lambda_e(t - t_{ov})\right], & t > t_{ov} \end{cases} \tag{4.2}$$

Since the probability of conception from an insemination on the day of ovulation is presumably not equal to one but rather some value a_0 that is less than one, the actual probability of conception per insemination on cycle day t is $a_0 a_f(t)$. In addition, there is some probability a_d that a fertilized egg will not survive long enough for the pregnancy to be detected. Unfortunately, these two probabilities are completely confounded when real data are being used. If we make $K_0 = a_0(1 - a_d)$ and assume that inseminations within the same fertile period are independent of each other, the total probability of a *detectable* conception per cycle is

$$a_f = 1 - \prod_t \left[1 - K_0 a_f(t)\right]^{x(t)} \tag{4.3}$$

where $x(t)$ equals one if unprotected intercourse occurs on cycle day t and zero otherwise. The values of t in this expression are restricted to a single cycle.

Royston applied this model to a sample of 2,604 ovulatory cycles experienced by 653 married British and French women, none of whom was using any form of contraception. On every cycle day, each woman in the sample recorded her own BBT and noted whether intercourse had occurred that day. Ovulation was assumed to precede the rise in BBT by

Table 4.8.   Maximum Likelihood Estimates of the Parameters Affecting the Length of the Fertile Period

Parameter	Estimate	Standard error	95 percent confidence limits
Mean sperm life (days)	1.47	0.20	1.1–1.9
Mean egg life (days)	0.70	0.11	0.5–0.9
K_0	0.48	0.08	0.3–0.6

From Royston (1982)

one day.[13] The outcome of each cycle—i.e., whether or not a pregnancy was detected—was also recorded. In total, 515 detectable pregnancies resulted from these 2,604 cycles.

The results of Royston's analysis are shown in Table 4.8 and Figures 4.46 and 4.47. As expected, the fertile period straddles the day of ovulation, but not symmetrically so. Under the exponential survival model, there is a tiny but finite probability that a gamete will last a very long time, even longer than the cycle itself. Consequently it is impossible to specify an absolute maximal life span for either eggs or sperm, and thus to say absolutely when the fertile period begins or ends. However, we may arbitrarily judge a cycle day to be outside the fertile period whenever the probability of conception per insemination on that day falls below 0.05. By that criterion, it appears that the fertile period begins some four days before ovulation and ends some two days after ovulation, for a total duration of six days (Figure 4.46). The probability of a detectable conception per insemination varies widely over this interval, from a low (by definition) of 1 in 20 to a high just before ovulation of 1 in 2.

Since this analysis is based on the BBT method, it probably overestimates the true length of the fertile period to some degree. In addition, the probabilities in Figure 4.46 are necessarily underestimates of the actual probabilities of conception per insemination, quite possibly by a substantial amount. This is because the analysis confounds the probability of conception itself with the probability that the conception will result in a detectable or diagnosable pregnancy—that is, we can estimate K_0 but not a_0 and a_d separately. The interpretation of Royston's results is thus confounded by an unknown amount of early fetal loss. The profound difficulties involved in estimating the level of such loss will be

Table 4.9.   Estimates of Mean Survival Time (in hours) of Viable Fertile Gametes in the Female Reproductive Tract

Species	Sperm	Eggs
Human	30–45	6–24
Cow	30–48	12–24
Sheep	30–48	15–25
Horse	140–150	15–25
Pig	25–50	10–20
Mouse	6–12	6–15
Rabbit	30–36	6–8

From Johnson and Everitt (1988)

reviewed in Chapter 6. For now it is enough to say that the probabilities in Figure 4.46 should probably be adjusted upward by something on the order of 20–30 percent (Wilcox et al., 1988; Wood, 1989).

Royston's analysis suggests that the survival curves of the male and female gametes are quite different, with far more sperm than eggs surviving as long as three or four days (Figure 4.47). This difference is, of course, responsible for the asymmetry of the fertile period about ovulation. The estimated mean life span of sperm is about 1.5 days, of eggs less than one day (Table 4.8). These figures appear to be consistent with physiological studies of gamete survival in vivo (Table 4.9).

Despite its shortcomings, Royston's study represents the best statistical analysis of the fertile period to date. However, it completely omits one major physiological factor: cyclical changes in cervical mucus. For much of the cycle, the uterine cervix represents an effective barrier to sperm transport. Not only is the cervix itself something of a labyrinth, filled with crypts, folds, and blind alleys, it is usually occluded by a thick, viscous mucus secreted by glands in its lining. This mucus is a complex mixture of glycoproteins and mucopolysaccharides and has been shown to have bacteriocidal properties. Both its bacteriocidal and occlusive functions are probably essential, since intercourse is far from sterile, and since the female reproductive tract ultimately opens the peritoneal cavity to the outside world.[14] Because the mucus barrier persists throughout pregnancy, it also protects the developing fetus from infection.

For most of the ovarian cycle, cervical mucus effectively blocks penetration of sperm into the uterine cavity by forming a dense mesh of fibrils (Figure 4.48, panel a). Under the influence of rising estradiol during the late follicular phase, however, the mucus undergoes a series of marked changes, which become noticeable about two days before ovulation. First, the volume of mucus secreted by the cervix increases more than tenfold. Second, the water content of the mucus rises, with a consequent decline in protein and mucoid concentration. Associated with these changes in chemical composition, the dense mesh of fibrils making up the mucus relaxes, giving way to a more open and penetrable structure (Figure 4.48, panel b). At about the same time, the glycoprotein molecules arrange themselves into long fibers oriented parallel to the main axis of the cervical canal. On insemination, sperm cells become oriented along these fibers and are apparently guided by them through the cervix (Figure 4.48, panel c). Thus, far from being a barrier to sperm, mucus in the periovulatory period becomes a conduit facilitating their movement into the main body of the uterus. A few days after ovulation, the cervical mucus returns to its earlier, occlusive state (Figure 4.48, panel d).

Certain readily discernible physical properties of mucus are so closely associated with these cyclical changes in composition and ultrastructure that they can be used as indicators of ovulation (Figure 4.49). These include increases in mucus volume and viscosity. A few days before ovulation, the mucus becomes not only more copious but also wetter, more slippery, and more elastic (e.g., when suspended between two fingers). This latter property is measured in terms of *spinnbarkeit*, or the distance a string of mucus of known volume can be stretched before breaking. Changes in chemical composition can be monitored by observing the propensity of thin smears of mucus to form glycoprotein and salt crystals when dried, a phenomenon known as *ferning* (Figure 4.50). The degree of ferning is at its highest at about the time of ovulation.

From a functional point of view, changes in cervical mucus are important because of their influence on the fertile period. The amount of sperm penetration into the uterus appears to be well correlated with changing mucus characteristics (Figure 4.49). If the period of effective penetration is shorter than the sum of the fertile life spans of sperm and eggs, then the former is necessarily more important than the latter in determining the length of the fertile period. Despite this possibility, differential sperm penetration has never been included as a factor in any quantitative analysis of the fertile period.

Previous studies, based unfortunately on BBT, suggest that the fertile period parallels changes in cervical mucus quite closely (Figure 4.51). This finding does not imply, however, that analyses of the fertile period based upon gamete survival rather than changes in mucus are necessarily meaningless. It is likely that gamete survival and changes in mucus are themselves highly coevolved, with the former acting as a major determinant (in the evolutionary sense) of the latter. Functionally, it is important for the period of sperm penetration to be as narrow as possible in order to minimize a woman's risk of infection. Natural selection should act to favor occlusion of the cervix whenever the probability of conception on a given day of insemination falls below some threshold value; that probability, in turn, is determined primarily by gamete viability, as assumed in Royston's model. Thus, the two systems should act in concert, making it difficult if not impossible to disentangle their roles in setting the fertile period.

As the preceding paragraphs should indicate, it is difficult enough to measure the characteristics of the fertile period in western women. Large samples and sophisticated analyses are needed, and even then major analytical problems remain, such as controlling for early fetal loss. Not surprisingly, we know nothing whatsoever about interpopulation variation in the fertile period. However, recent mathematical models suggest that, if such variation exists, it may exert a remarkably large influence on birth spacing via its effect on fecundability (see, for example, Potter and Millman, 1986; Wood and Weinstein, 1988). It might be suggested, then, that this difficult but intriguing problem should be a focus of future research on the dynamics of human reproduction.

NOTES

1.   For present purposes, hormones can be defined as molecules whose major function is to transfer information among cells and tissues. Each class of hormones has a very specific molecular structure. Hormones are synthesized by various tissues and secreted into the blood or extracellular fluid, which carries them to specific *target cells* where they produce either permanent or reversible physiological responses. The target cells for any one hormone are characterized by *hormone-specific receptor molecules*, located either on the cell surface or within the cell, that bind that particular hormone. Binding of the hormone produces a *hormone-receptor complex* that initiates a cascading series of molecular changes within the target cell via a molecular *signal transduction* system. Once activated, the signal transduction system may modulate the synthesis or secretion of specific products by the target cell, or it may affect the growth and reproduction of that cell. The dynamics of the formation and dissociation of hormone-receptor complexes are a major determinant of hormone action. Hormone-secreting tissues are often organized into larger organs known as *endocrine glands;* hormone-secreting tissues and their immediate neural connections make up the *endocrine system.* (Target cells, by contrast, are often widely scattered throughout the body.) The study of hormones and hormone receptors, including their genetics, biochemical properties, synthesis, and action, is called *endocrinology.* For further details, see the appendix to this book.

2.   Complications having to do with the timing of collection and the volume of the total void make it somewhat more difficult to use urinary assays in the field; however, these difficulties are not insurmountable.

3.   A final marker of ovulation, involving changes in cervical mucus, is discussed at the end of this chapter.

4.   The single best indicator that all five cycles in Figure 4.5 were ovulatory is the luteal-phase rise in progesterone, as reflected here in urinary pregnanediol excretion. This rise can occur only in the presence of a functioning corpus luteum, which in turn is formed only following ovulation. Rare exceptions to this rule occur in cases of *follicular luteinization syndrome,* in which the developing ovum becomes entrapped within the follicle while the residual follicular tissues undergo the usual changes associated with the formation of a corpus luteum.

5.   Some authors recognize two regions of granulosa cells surrounding the egg: the cumulus oophorus and the corona radiata. In this usage, the corona radiata refers to the cells that lie closest to the egg, whereas the cumulus oophorus refers to the more diffuse cells further away from the egg. In some species of mammals, these two regions are quite distinct and may be functionally differentiated as well. In humans, however, the distinction appears to have little

biological basis; I have therefore grouped both regions under the label "cumulus oophorus."

6. The distinction between neurotransmitter and neurohormone is arbitrary since the same substances can play both parts. It is based on the fact that neurohormones are secreted into the bloodstream to be carried to their target cells, whereas neurotransmitters are secreted into the synapses between adjacent neurons.

7. GnRH is sometimes called *gonadoliberin* or *luteinizing-hormone releasing hormone* (LHRH).

8. A fourth class of steroids, the corticosteroids, is secreted by the cortex of the adrenal gland. The adrenal cortex also secretes androgens.

9. This pattern appears to be functionally important since it is known that the pituitary will not respond to GnRH stimulation if GnRH is administered at a constant rate. Apparently the gonadotrophs "habituate" to such steady GnRH stimulus by losing their GnRH receptors.

10. The identity of this woman is unknown. Even if we were not told that she was born in the United States during the twentieth century, however, it would be fairly obvious that she is not a member of a natural-fertility population. She experiences too long a lag between marriage and first fertile conception, as well as between the end of childbearing and menopause, for us not to suspect deliberate contraception. In addition, the very short interval between each birth and the first postpartum menses tell us that she is almost certainly not practicing prolonged breastfeeding. Given the combination of late marriage, even later commencement of childbearing, early cessation of childbearing, and low total reproductive output shown by this woman, one might guess that she is a career professional. She is, in fact, a physician.

11. The analysis constrained both subpopulations to have the same variance in luteal phase lengths, estimated to be 1.99.

12. Short (1984c) goes on to review studies implicating exposure to recurring cycles as a risk factor for such diseases as endometriosis and cancer of the breast, cervix, and ovarian epithelium. This more epidemiological part of the argument, which lies outside the scope of the present book, has recently been assessed by Pike (1988) and Eaton et al. (1992).

13. In 50 cycles BBT was too variable for reliable timing of the temperature rise; those cycles were excluded from the analysis.

14. Significantly, Sweet et al. (1985) and Cates et al. (1993) report that infection of the endometrium, fallopian tubes, and peritoneal cavity by microorganisms via the cervico-vaginal tract appears to be most likely during menses, when the cervical mucus is washed away.

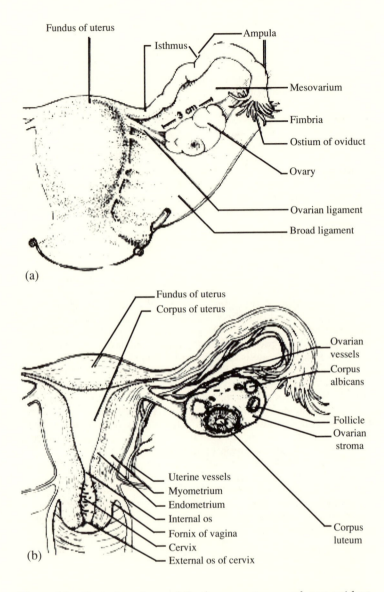

Figure 4.1. Posterior views of the human uterus and one oviduct and ovary: *(a)* intact, *(b)* sectioned. The ovaries, which have been pulled upward and laterally, would normally have their long axes almost vertical. (Redrawn from Johnson and Everitt, 1988)

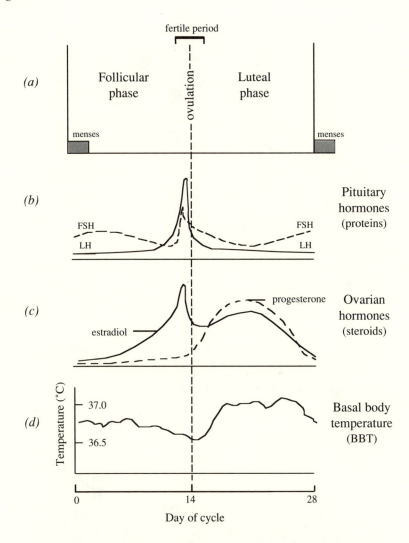

Figure 4.2. Major features of the human ovarian cycle. (a) Principal landmarks and phases of the cycle (shaded bars indicate days of menstruation). (b) Changes in the serum concentration of the pituitary gonadotropins luteinizing hormone (LH) and follicle-stimulating hormone (FSH). (c) Changes in the serum concentration of the ovarian steroids estradiol and progesterone. (d) Changes in basal body temperature.

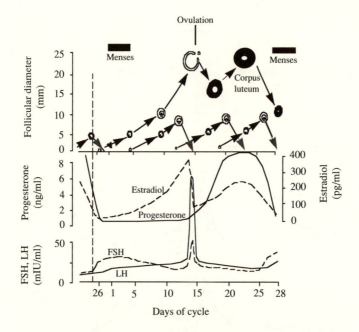

Figure 4.3. Changes in follicle growth and in the concentration of pituitary and ovarian hormones across the ovarian cycle. In the follicular phase, only a single follicle acquires dominance over the other follicles and proceeds to ovulation. The vertical *broken line* indicates the onset of luteolysis in the previous cycle. (Redrawn from Baird, 1984)

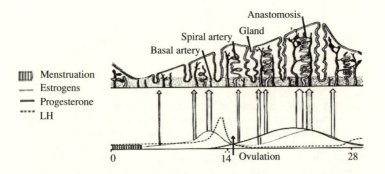

Figure 4.4. Changes in the endometrium during the ovarian cycle. Underlying changes in ovarian steroids are indicated below; thickness of *arrows* indicates strength of action. (Redrawn from Johnson and Everitt, 1988)

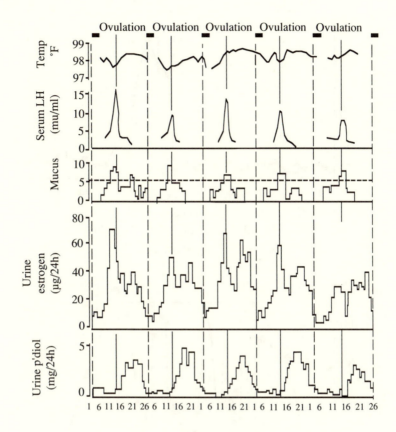

Day of cycle

Figure 4.5. Changes in basal body temperature, serum LH concentration, viscosity of cervical mucus, and total urinary estrogren and urinary pregnanediol excretion in a woman during the course of five successive ovulatory cycles. Pregnanediol is a urinary metabolite of progesterone, and its concentration is highly correlated with the serum level of that steroid. A mucus score of five or more, indicating elastic, wet, and slippery mucus, correlates well with maximal follicular estrogen secretion and the preovulatory LH surge. *Black bars* indicate menstruation. (Redrawn from Short, 1984b; data from J. B. Brown)

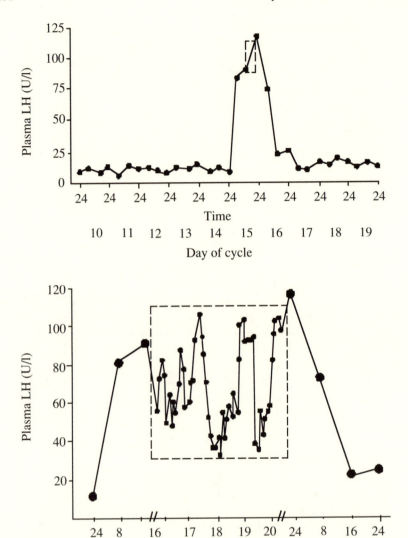

Figure 4.6. Fluctuations in the plasma concentration of LH in the peri-ovulatory segment of the cycle: *(top)* samples collected every eight hours, *(bottom)* samples collected every five minutes during the period indicated by the *broken box*. (Redrawn from Baird, 1983b)

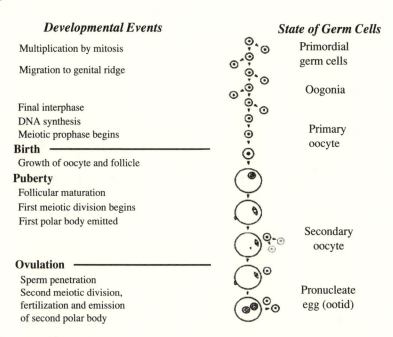

Developmental Events	**State of Germ Cells**
Multiplication by mitosis	Primordial germ cells
Migration to genital ridge	
	Oogonia
Final interphase	
DNA synthesis	
Meiotic prophase begins	Primary oocyte
Birth	
Growth of oocyte and follicle	
Puberty	
Follicular maturation	
First meiotic division begins	
First polar body emitted	
	Secondary oocyte
Ovulation	
Sperm penetration	
Second meiotic division,	Pronucleate
fertilization and emission	egg (ootid)
of second polar body	

Figure 4.7. Schematic representation of the major stages of oogenesis. (Redrawn from Baker, 1982)

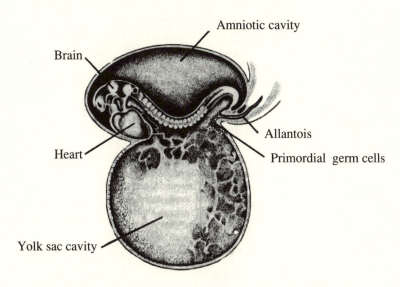

Figure 4.8. The apparent site of origin of primordial germ cells in a 24-day human embryo. (Redrawn from Witschi, 1948; Byskov, 1982)

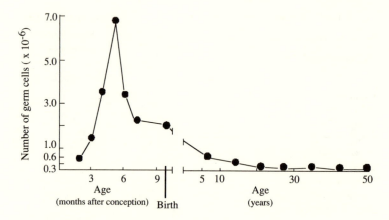

Figure 4.9. Changes in the total number of primary oocytes in the human ovary with increasing age. (After Block, 1952; Baker, 1971)

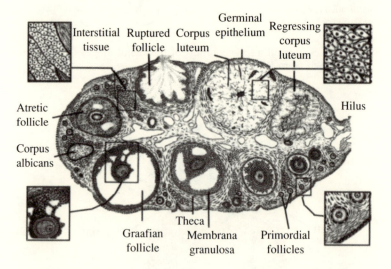

Figure 4.10. Idealized drawing of the structure of the human ovary, showing follicles at various stages of development and the formation and regression of the corpus luteum. Although the Graafian follicle, ruptured follicle, and corpus luteum are not expected to be present at the same time, most stages of follicular development preceding the Graafian follicle, as well as the regressing corpus luteum and corpus albicans, coexist in the adult ovary. (Redrawn from Turner, 1966)

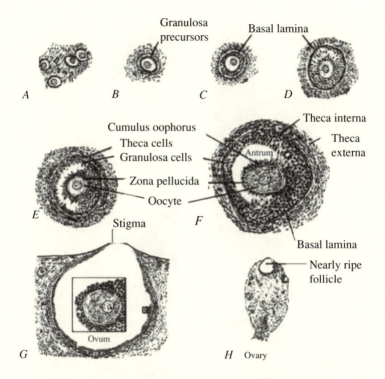

Figure 4.11. Stages in the development of the follicle and oocyte: *(A–D)* primary follicles; *(E–F)* secondary follicles, with the formation of an antrum; *(G–H)* mature or Graafian follicles. (Drawings in *A–F* magnified by 10, but the inset detail of the oocyte is magnified 100 times.) (Redrawn from Carlson, 1988)

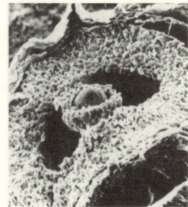

Figure 4.12. Scanning electron micrograph (× 800) of a mature mammalian follicle showing the spherical oocyte *(center)*, surrounded by smaller cells of the cumulus oophorus, projecting into the antrum. (Courtesy of P. Bagavandoss)

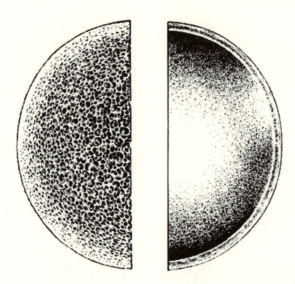

Figure 4.13. Drawing of scanning electron micro-
graphs of the outside *(left)* and inside *(right)* of
the zona pellucida. (Redrawn from Phillips and
Shalgi, 1980)

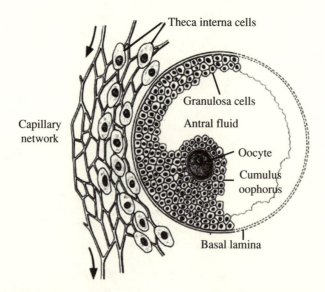

Figure 4.14. Structure of a Graafian follicle showing
how the granulosa cells are deprived of a direct
blood supply by the basal lamina. (Redrawn from
Baird, 1984)

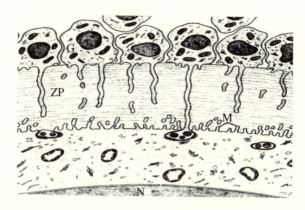

Figure 4.15. Structure of the fully formed zona pellucida (ZP) around an oocyte in a Graafian follicle. The zona is penetrated from one side by processes from the granulosa cells (G) and from the other by microvilli (M) arising from the oocyte. The granulosa cell processes indent the cytoplasm of the oocyte and may provide nutrients and maternal protein. The N marks the oocyte nucleus. (Redrawn from Baker, 1982)

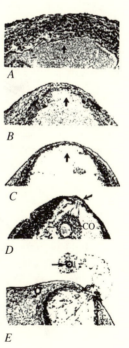

Figure 4.16. Stages in ovulation: *(A)* several hours before ovulation, the granulosa and theca layers are still quite thick; *(B)* a half-hour before ovulation, there is a significant thinning of the follicular cells and connective tissue in the region of the stigma (*arrow*); *(C)* a few minutes before rupture, follicular cells and connective tissue have almost disappeared; *(D)* the ovarian wall ruptures, and the free cumulus oophorus (CO) streams toward the opening; *(E)* ovulation is completed, but the viscous antral fluid still adheres to the site of the rupture. (Redrawn from Vander et al., 1980)

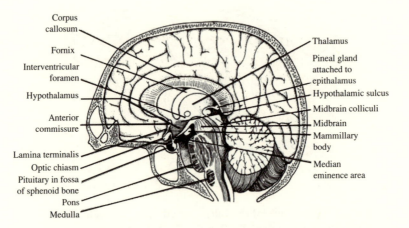

Corpus callosum
Fornix
Interventricular foramen
Hypothalamus
Anterior commissure
Lamina terminalis
Optic chiasm
Pituitary in fossa of sphenoid bone
Pons
Medulla

Thalamus
Pineal gland attached to epithalamus
Hypothalamic sulcus
Midbrain colliculi
Midbrain
Mammillary body
Median eminence area

Figure 4.17.   Sagittal section of the human brain with the pituitary gland attached. Note the comparatively small size of the hypothalamus *(shaded)*. The thalamus and hypothalamus form one wall of the third ventricle of the brain. (Redrawn from Johnson and Everitt, 1988)

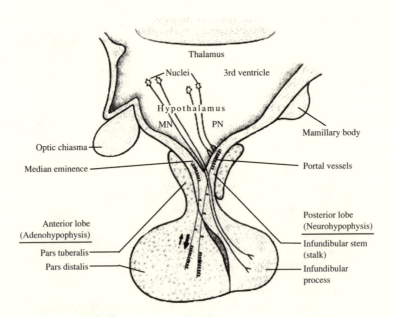

Thalamus
Nuclei   3rd ventricle
Hypothalamus
MN   PN
Optic chiasma
Median eminence

Mamillary body
Portal vessels

Anterior lobe (Adenohypophysis)
Pars tuberalis
Pars distalis

Posterior lobe (Neurohypophysis)
Infundibular stem (stalk)
Infundibular process

Figure 4.18.   Anatomy of the hypothalamo-pituitary axis. *Arrows* designate relative amount of blood flow in each direction in the pituitary stalk. MN, magnocellular neurons; PN, paraventricular neurons. (Redrawn from Karsch, 1984)

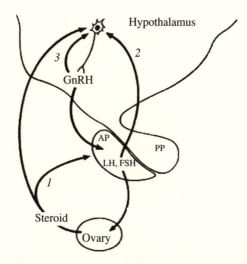

Figure 4.19. Illustration of the three levels of hormonal feedback that integrate the hypothalamo-pituitary-ovarian axis: *(1)* long-loop feedback, *(2)* short-loop feedback, *(3)* ultra-short-loop feedback. GnRH, gonadotropin-releasing hormone; AP, anterior pituitary; PP, posterior pituitary. (Redrawn from Karsch, 1984)

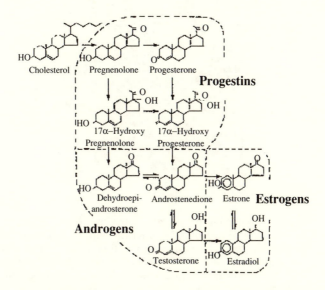

Figure 4.20. Biosynthesis of steroids in the ovary. (Redrawn from Griffin and Ojeda, 1992)

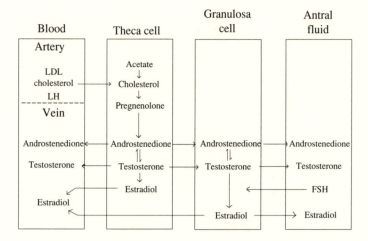

Figure 4.21. Interactions among the bloodstream and various compartments of the ovarian follicle in the biosynthesis of steroids. LDL, low-density lipoproteins. (Redrawn from Baird, 1984)

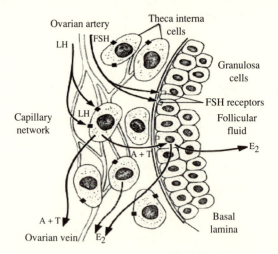

Figure 4.22. The action of the gonadotropins on the follicle in the synthesis of estradiol. LH interacts with receptors on the theca cells (■) to stimulate production of androgens and small amounts of estradiol. FSH activates the aromatase enzyme system in the granulosa cells by interacting with receptors bound to the granulosa cell membranes (□). A, androstenedione; T, testosterone, E_2, estradiol. (Redrawn from Baird, 1984)

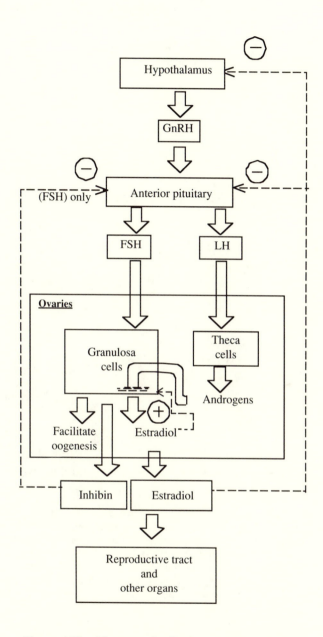

Figure 4.23. Negative feedback control of ovarian function during the early to middle follicular phase. Inhibin is a peptide hormone that appears to inhibit FSH secretion. (Redrawn from Vander et al., 1990)

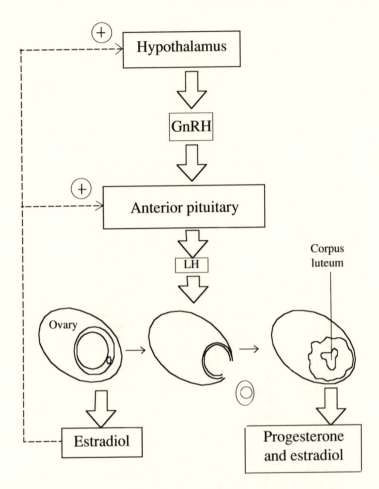

Figure 4.24.   Positive feedback control of ovarian function during the late follicular phase. The dominant follicle secretes
large amounts of estradiol, which acts upon the hypothalamus and anterior pituitary to stimulate the LH surge. The
increase in serum LH then triggers both ovulation and formation of the corpus luteum. These effects of LH are mediated by the granulosa cells. (Redrawn from Vander et al.,
1990)

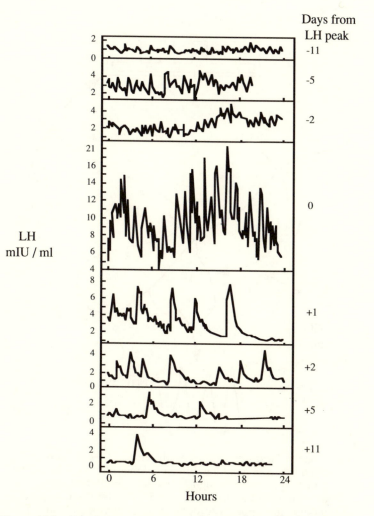

Figure 4.25. Frequency and amplitude of LH pulses on selected cycle days. The days of the cycle are indicated relative to the day of the LH surge (Day 0). During the follicular phase (Days −11 to −2), low-amplitude LH pulses occur almost hourly. On the day of the LH surge, the frequency of the LH pulses is still approximately hourly, but the amplitude is much higher. During the luteal phase, the LH pulse frequency is reduced. (Redrawn from Knobil and Hotchkiss, 1988)

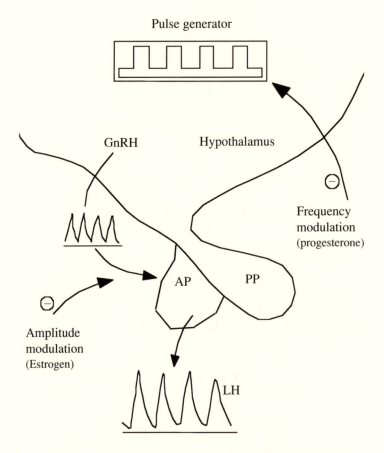

Figure 4.26. Feedback control of pulsatile LH secretion by ovarian hormones. Note the differential modulation of amplitude
and frequency. Although frequency modulation appears to
act at the level of the GnRH pulse generator itself, amplitude
modulation appears to occur at the anterior pituitary, perhaps by making the gonadotrophs more or less sensitive to
GnRH stimulation (e.g., by changing the number of GnRH
receptors on their surface). (Redrawn from Karsch, 1984)

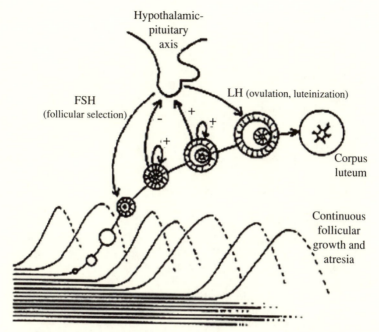

Hypothalamic-
pituitary
axis

FSH
(follicular selection)

LH (ovulation, luteinization)

Corpus
luteum

Continuous
follicular
growth and
atresia

Pool of nongrowing follicles

Figure 4.27. Diagram illustrating the concept of follicular selection. Follicles that go to maturity are "chosen" from a continuously replenished cohort of growing follicles. Most members of the cohort become atretic, but if one should reach a critical stage (with a critical number of gonadotropin receptors) at a point in the cycle when the proper balance of FSH and LH is present in the circulation, it is "rescued" and can attain full development followed by ovulation and formation of a corpus luteum. Approximately 85 days elapse between entry of a follicle into the cohort of growing follicles and eventual ovulation. The interplay of gonadotropins and estradiol necessary to induce these events is shown; – and + indicate negative and positive feedback effects, respectively, of estradiol. (Redrawn from Peters and McNatty, 1980)

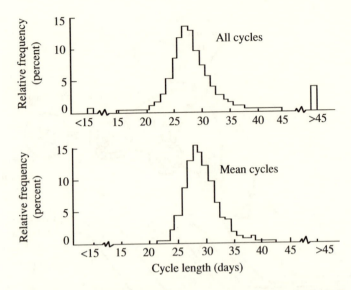

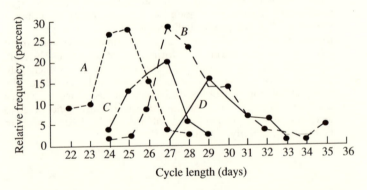

Figure 4.28. Variation in the time between menstrual onsets in a composite sample of 30,655 cycles experienced by 2,316 white North American women ages 15–45 years. *(Top)* distribution of all 30,655 cycle lengths (mean = 29.1 days, variance = 55.7). *(Bottom)* distribution of the mean cycle length for each of the 2,316 subjects (mean = 29.4 days, variance = 10.7). (Data from Chiazze et al., 1968)

Figure 4.29. Individual-level variation in the length of the menstrual cycle in four normal, healthy U.S. women (*A–D*). The ages of the women over the period of observation were 30–39 years for *A*, 29–35 for *B*, 21–27 for *C*, and 23–27 for *D*. None of the women reported any form of dysmenorrhea during the observation period. (Data from King, 1926)

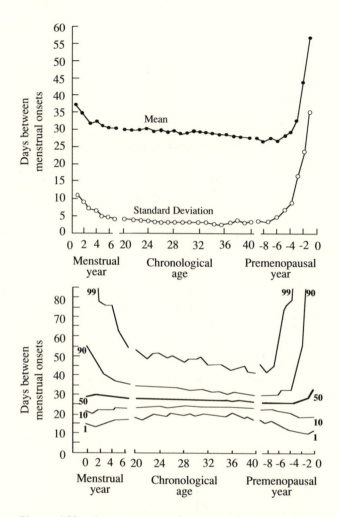

Figure 4.30. Age patterns of ovarian function in U.S. women based on 25,844 woman-years of menstrual experience in 2,702 subjects originally recruited when they were students at the University of Minnesota: *(Top)* mean and standard deviation in cycle length; *(bottom)* first, tenth, fiftieth, ninetieth, and ninety-ninth percentiles of the cycle-length distribution (the fiftieth percentile or median is indicated by *heavy line*). Age is given as years since menarche for the first six years of menstrual life, chronological age for ages 20–40, and years before menopause for the last nine years of menstrual life. (Data from Treloar et al., 1967)

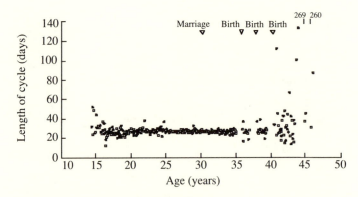

Figure 4.31. One woman's complete history of menstrual experience from menarche to menopause. (Redrawn from Treloar et al., 1967)

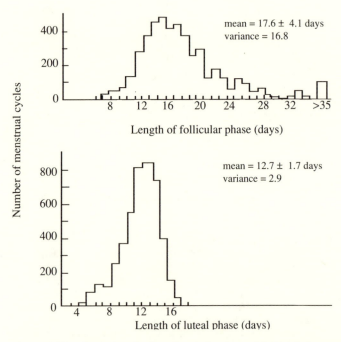

Figure 4.32. Variation in length of the two phases of the ovarian cycle: follicular phase *(top)* and luteal phase *(bottom)*. The follicular phase is longer and more variable than the luteal phase, but the difference in length is slightly exaggerated by the fact that these figures are based on the BBT method. (Data from Döring, 1963)

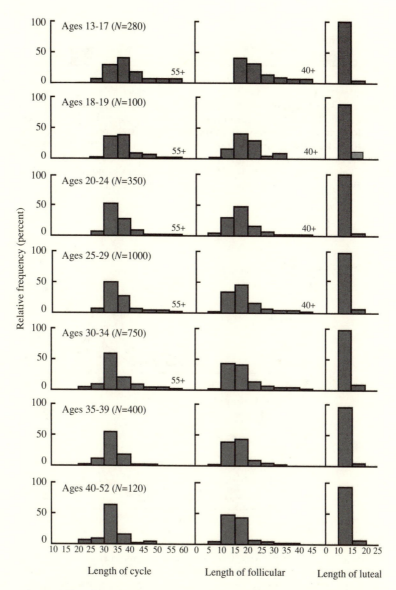

Figure 4.33. Distribution in the length of the total cycle, follicular phase, and luteal phase at different ages in Japanese women. Note that the follicular phase is more variable than the luteal phase at each age. Ovulation was timed by the BBT method. (Data from Matsumoto et al., 1962)

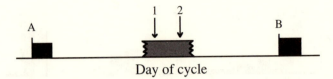

Day of cycle

Figure 4.34. Diagram explaining the spurious correlation between phases of the cycle when an error-prone method like BBT is used to determine the day of ovulation. *Black bars* represent menses; thus, the line segment *A–B* is the length of the cycle. The *shaded area* is the portion of the midcycle during which the BBT rise is likely to occur; points 1 and 2 are separate days on which the BBT rise might be judged to occur. (See text for details)

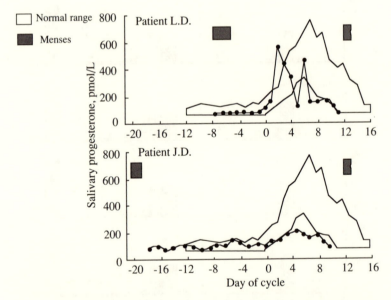

Figure 4.35. Salivary progesterone concentrations in two patients presenting at infertility clinics with luteal phase abnormalities, compared with the range (mean ± one standard deviation) in normal women. Patient L.D. exhibits a short or inadequate luteal phase; patient J.D. has an insufficient or hypofunctional corpus luteam. Cycles are dated from the first day of progesterone elevation. (Redrawn from Riad-Fahmy, 1984)

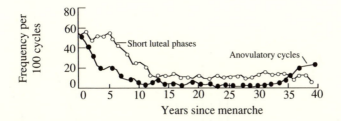

Figure 4.36. Age-related variation in cycle quality based on 31,645 cycles in 656 U.S. women: estimated frequency of anovulatory cycles (*closed circles*) and cycles with luteal phases ≤ 11 days long (*open circles*). Ovulation was detected by the BBT method, and the frequency of short luteal phases was estimated for apparently ovulatory cycles only. (Redrawn from Vollman, 1977)

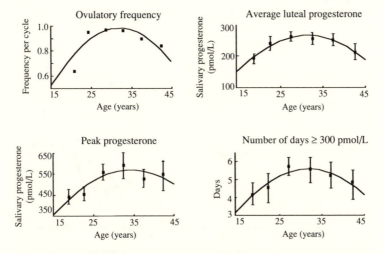

Figure 4.37. Age patterns of ovulation and luteal function, as indicated by salivary progesterone measures collected from 111 healthy U.S. women ages 18–44. All subjects experienced regular menstrual cycles, were within 20 percent of ideal body weight, reported no weight loss/gain or vigorous exercise during the study period, reported no use of oral contraceptives or other steroid medications, and were neither pregnant nor lactating. Daily saliva samples were collected over one complete menstrual cycle. *Solid curves* are quadratic regressions fit to five-year age-group means. (Figure generously provided by Peter Ellison)

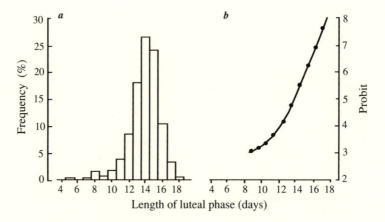

Figure 4.38. Frequency distribution *(a)* and normal probability plot *(b)* of luteal phase length in 327 apparently ovulatory cycles experienced by British and Swedish women. Ovulation and length of the luteal phase were assessed according to the presence or absence of an LH surge. (Redrawn from Lenton et al., 1984b)

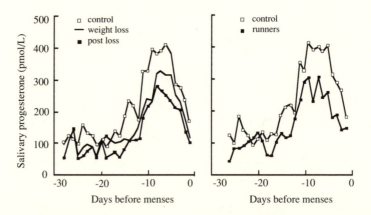

Figure 4.39. Effects of diet and exercise on luteal function. *(Left)* mean daily salivary progesterone profiles from 19 cycles in normally menstruating U.S. women, 14 cycles of weight loss in dieting women, and 9 cycles following a month of weight loss in dieting women. *(Right)* mean daily salivary progesterone profiles in a group of recreational runners and nonexercising controls. Both panels are aligned with day of onset of next menses. (Redrawn from Ellison and Lager, 1986; Lager and Ellison, 1987)

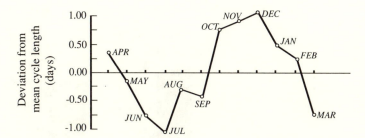

Figure 4.40. Estimated seasonal variation in mean menstrual interval, based on 38,194 woman-years of menstrual experience in 3,766 U.S. women. The overall mean menstrual interval has been age-adjusted. (Redrawn from Sundararaj et al., 1978)

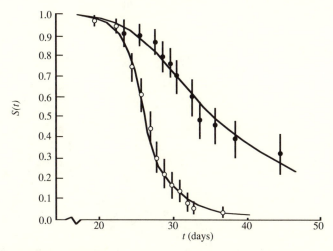

Figure 4.41. Kaplan-Meier estimates of the survival function $S(t)$, the probability that an ovarian cycle is longer than t days. *Closed circles* represent 36 women from Papua New Guinea, *open circles* a synthetic sample of 36 U.S. women matched with the New Guinea subjects for gynecological age. Standard errors of the estimates are also shown. The number of points for each group corresponds to the number of distinct cycle lengths observed. Smooth curves represent the regression of ln [logit $S(t)$ + 5] on ln t for each group ($r^2 = 0.959$ for the New Guinea women, 0.954 for the U.S. women). Interpolation from the regression lines yeilds an estimated median cycle length (and 95 percent confidence interval) of 36.0 (32.8–39.8) for the New Guinea women and 26.1 (23.1–29.3) for the U.S. women. (Redrawn from P. Johnson et al., 1987)

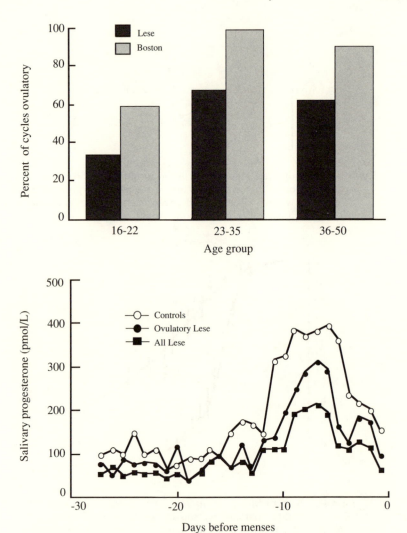

Figure 4.42. Luteal suppression among the Lese, an agricultural population in rural Zaire: *(top)* ovulatory frequency in 35 Lese women
monitored for four months, compared with that in Boston women
of similar age; *(bottom)* mean daily salivary progesterone profiles
for all 35 Lese subjects, for a subset of 20 ovulatory Lese women,
and for a sample of similarly aged Boston controls. (Redrawn from
Ellison et al., 1989)

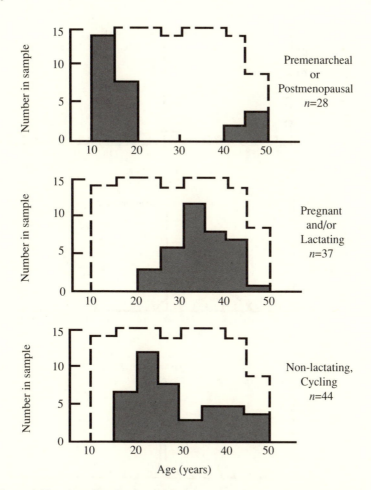

Figure 4.43. Age distribution of Gainj women by reproductive status. Not shown are two pathologically sterile subjects, both age 35–39. Broken line is the distribution of the total sample (n = 111). (Redrawn from Johnson et al., 1990)

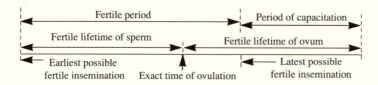

Figure 4.44. Factors determining the length of the fertile period. (Redrawn from Bongaarts, 1983)

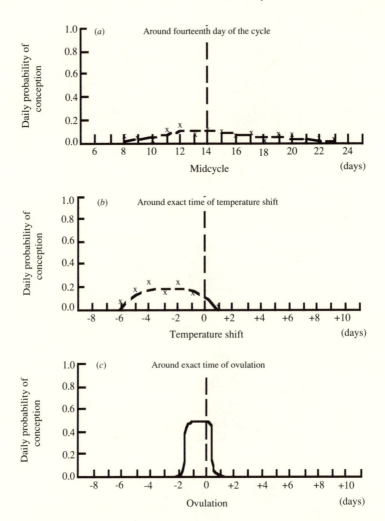

Figure 4.45. Estimated daily probabilities of conception from a single insemination relative to (*a*) day 14 of the cycle counted forward from the onset of last menses, (*b*) day of midcycle BBT shift, and (*c*) exact day of ovulation. x, point estimates; *dashes*, smoothed curve fit to point estimates; *solid line*, presumed smoothed curve if true day of ovulation were known. (Redrawn from Bongaarts and Potter, 1983)

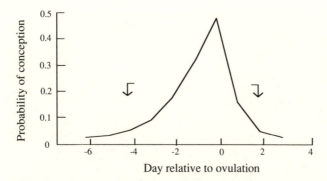

Figure 4.46. Maximum likelihood estimates of the probability that a single insemination on a given cycle day will result in a detectable pregnancy. Timing of ovulation was estimated by the BBT method. Since K_o was allowed to vary with the female partner's age, the results shown here are adjusted to age 32 years, the mean age of women in the sample. In addition, the probabilities of conception have been rigidly translated leftward by one-sixth of a day as an approximate correction for sperm capacitation and activation. *Arrows* indicate days on which the probability of conception reaches 0.05. (Redrawn from Royston, 1982)

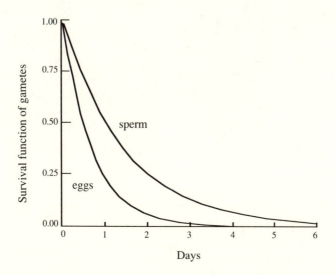

Figure 4.47. Survival functions for the fertile life spans of male and female gametes estimated by the analysis of Royston (1982)

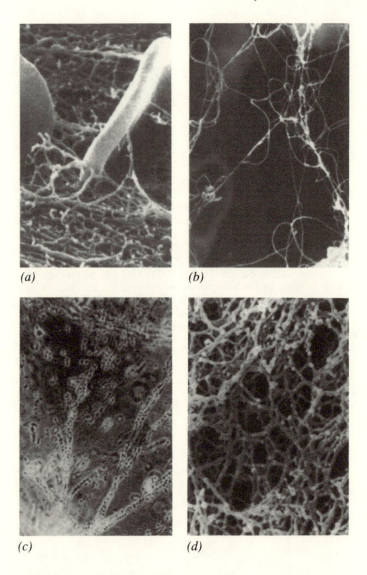

Figure 4.48.   Cervical mucus *(a)* during menstruation (note red
    blood cells caught in mucus strands) (× 10,000) and *(b)* on
    the day of ovulation (× 5,700). *(c)* Sperm migrating through
    the cervical mucus (note alignment along mucus fibrils). *(d)*
    Cervical mucus toward the end of the cycle (× 10,400).
    (From Hogarth, 1978)

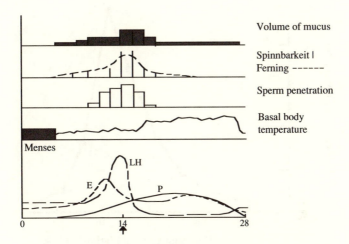

Figure 4.49. Changes in the cervical mucus at various days of the cycle (*arrow* denotes time of ovulation). Properties changing under the influence of high estradiol (E) and low progesterone (P) include (*from top*) volume of mucus, spinnbarkeit (stretchability or elasticity of mucus), ferning (crystallization pattern of dried mucus), and penetration of mucus by sperm. Note that luteal estradiol does not induce changes because of elevated progesterone. (Redrawn from Johnson and Everitt, 1988)

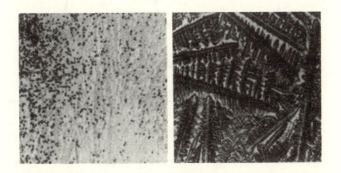

Figure 4.50. (*Left*) absence of ferning in a cervical smear obtained on cycle day 5 from a normally cycling woman (× 88). (*Right*) ferning in a cervical smear obtained from the same woman just before ovulation (× 88). (From Rebar, 1991)

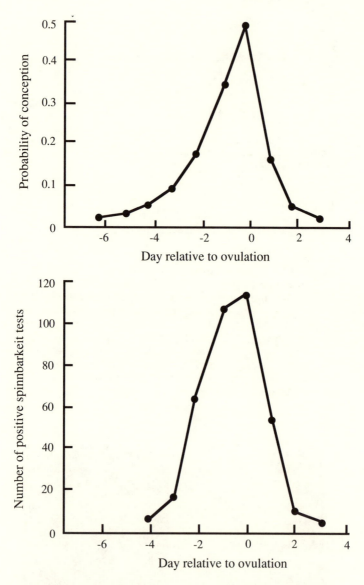

Figure 4.51. Temporal relationship between the probability of conception per insemination *(top)* and the frequency of positive spinnbarkeit tests *(bottom)* for each day of the fertile period. Day of ovulation was determined by the BBT method. (Data from Döring, 1963; Royston, 1982)

5

Conception, Implantation, and Pregnancy

In this chapter, we resume the saga of the egg cell, last encountered at ovulation. We follow the egg and its descendants through fertilization, establishment of a pregnancy, early embryonic development, and eventual parturition. In the final section of the chapter, we turn to biometrical aspects of pregnancy, with special attention to variation in *gestation*, or the length of pregnancy, since such variation can (at least in principle) cause variation in birth intervals. Attention will be restricted in this chapter to pregnancies resulting in live birth; pregnancies terminated by fetal loss are discussed in the following chapter.

Figure 5.1 shows the location within the upper regions of the female reproductive tract of some of the major events covered in this chapter. The timing of those events is described in Table 5.1. At ovulation, the mature ovum is shed into the woman's peritoneal cavity, not into the reproductive tract per se. The ovum must therefore be actively located and brought into the reproductive tract, specifically into one of the *fallopian tubes* or *oviducts* of the uterus. Around the time of ovulation, the rising concentration of estradiol secreted by the preovulatory follicle induces the fingerlike projections (*fimbria*) at the distal end of the oviduct to move, causing them to sweep the surface of the associated ovary. At the same time, the cilia lining the ampullary segment of the oviduct are activated, setting up a strong current of fluid moving from the distal toward the proximal segments of the oviduct (Figure 5.2). Together, these actions rescue the egg from the peritoneal cavity and move it down the oviduct toward the uterus. Since the egg has no motive power of its own, and since the oviduct is a complex maze of tis-

Table 5.1. Timing of Events (in days) Associated with the Establishment of Pregnancy, Early Development, and Gestation

Species	Stage of Early Development				Entry into uterus	Implantation	Total gestation[a]
	2-cell	4-cell	16-cell	Blastocyst			
Human†	1.5	2	3	4	2–3	8–13	252–274
Rhesus monkey†	1.1–2.0	1.0–2.2	4–6	7–8	3	9–11	159–174
Rat‡	1–2	2–3	4	4.5	3	5	20–22
Mouse†	0.9–1.0	1.6–2.1	2.5–2.9	2.8–3.4	3	4	19–20
Rabbit†	0.9–1.0	1.0–1.3	1.7–2.0	3.1–4.0	2.5–4.0	7–8	30–32
Horse†	1	1.2–1.5	4.1–4.2	6	4–5	28	335–345
Cow†	1	2–3	4	7–8	3–4	30–35	275–290
Sheep†	1	1.8	3	6–7	2–4	15–16	145–155
Pig†	0.6–0.8	1.0–3.1	3.3–5.0	5–6	2.0–2.5	11	112–115
Guinea pig‡	1–2	1.2–3.1	4.5	4.8	3.5	6	63–70
Ferret‡	2.1–3.0	2.7–3.1	4–5	4.5–6.0	5–6	7–8	42
Mink‡	3	3–4	5–6	8–10	8	25	42–52
Cat‡	1.7–2.1	3	4	5–6	4–8	13–14	52–65

[a] Fertilization to birth
† Days from ovulation
‡ Days from coitus
From McLaren (1982)

sue (Figure 5.3), the movements of the oviductal cilia are essential to ensure that the egg reaches the uterus.

If fertilization occurs, it generally takes place in the distal segment or *ampulla* of the oviduct.[1] Before passing into the main uterine cavity, the fertilized egg (now called a *zygote*) undergoes several rounds of cell division, the so-called cleavage divisions (see below). When the zygote reaches the proximal end of the oviduct, it is a solid ball of 8–16 cells known as a *morula* (lit. "mulberry," an apt description of its shape). By the time it passes into the main body of the uterus, the zygote has grown into a hollow ball of 50–60 cells called a *blastocyst*. Some ten days after fertilization, the blastocyst imbeds itself in the endometrium of the uterus by a process called *implantation*. At implantation, we can say that a pregnancy has been established.

Contrary to previous views, which saw the egg as playing a rather passive role in fertilization, the process actually involves complex interactions between egg and sperm. Before these interactions can be discussed, it will be necessary to describe the principal features of the sperm cell and to outline the major stages of *spermatogenesis*, the process by which these remarkable cells are generated.

THE SPERM CELL AND SPERMATOGENESIS

The mature spermatozoan is one of the most specialized cells of the human body, a highly refined system for delivering the male genetic material to the egg. It is the only human cell designed to function outside the body of the person who produces it. The mature sperm has two major components, the head and the tail (Figure 5.4). As shown in Figure 5.5, the head is filled almost entirely by the nucleus of the cell, which at maturity contains the *haploid complement* (half the normal number) of chromosomes, and by a structure called the *acrosome* or *acrosomal vesicle*, a bag of enzymes whose role in fertilization will be described below. As will be seen, the outer membrane of the sperm head, in combination with the inner and outer acrosomal membranes, are also important in fertilization, as is the *equatorial segment*, the region where the acrosome attaches to the base of the sperm head.

The tail of the sperm has three segments, the mid-, principal, and end pieces (Figure 5.4). Running down the center of the tail for its full length are two parallel microtubules, whose lateral movements relative to each other provide the sideways whipping motion characteristic of the sperm tail during its final stages of existence, motion that is responsible for the sperm's *motility*. Energy is conveyed to these structures by a radially arrayed set of microtubules and denser fibers, which in turn receive chemical energy from the sheath of mitochondria that makes up the midpiece of the tail (Figure 5.6).

In the course of spermatogenesis, these highly specialized sperm cells must be created from comparatively undifferentiated precursors, the *spermatogonia* or male primary germ cells. All except the final stages of spermatogenesis take place in the *testes* or male gonads. Each of the paired testes is divided into two functional compartments, the *seminiferous tubules*, where actual sperm production takes place, and the *interstitial tissue* separating the seminiferous tubules (Figure 5.7). The interstitial tissue is made up of connective tissue and *Leydig cells*, specialized endocrine cells that secrete the principal male sex steroid, testosterone. The seminiferous tubules are surrounded by a basal lamina and a thin layer of smooth muscle cells, which impart peristaltic movement to the tubules. Lining the inside of the tubules are two populations of cells, the germ cells themselves, and the *Sertoli cells*. Each Sertoli cell extends from the basal lamina to the tubule's lumen, the fluid-filled central channel where maturing sperm cells are deposited.

Each testis contains several seminiferous tubules with a combined length of about 250 m; obviously, the tubules must be packed into the testis in a highly convoluted way (Figure 5.8). Each seminiferous tubule empties its contents into a dense mesh of interconnected channels called the *rete testis*. These in turn are connected by several small ducts, the *efferent ductules*, which pass through the fibrous outer covering of the testis and empty into a single, large duct contained within the *epididymis*. The epididymis is draped over the testis and connects it to the *vas deferens*, a large duct that conveys the sperm to the outside world. Arrayed along the vas deferens are several glands, including the seminal vesicles, the prostate gland, and the bulbourethral glands; these glands secrete a semiliquid medium that, when mixed with sperm, makes up the *semen*. The secretions of these glands contain a variety of substances, including nutrients for the sperm cells, buffers to protect the sperm against acidic secretions in the female tract, and prostaglandins, the functions of which are uncertain but may include stimulation of vaginal and uterine contractions which help propel the sperm to the fallopian tubes.[2] The sperm themselves make up only a small fraction of the semen.

Returning to the seminiferous tubules, the primary site of spermatogenesis, the male germ cells can be observed to pass through several distinct stages of development (Figure 5.9). As in the case of female germ cells, the primordial male germ cells are first detectable during intrauterine life. Originally arising in extraembryonic tissue, these primordial germ cells migrate to the structures that will eventually give rise to the testes. Once the testes have differentiated, the germ cells (now called *spermatogonia*) are lodged in the seminiferous tubules between the basal lamina and the Sertoli cells (Figure 5.10). The Sertoli cells envelope the spermatogonia, overlap with each other, and form tight junctions that seal off the interior of the seminiferous tubule. Since the spermatogonia lie outside this *blood-testis barrier,* the sperm cells that descend from them must actively move across it with the help of the Sertoli cells in order to reach the lumen of the seminiferous tubules.

Unlike oogonia, which undergo mitosis and even begin meiosis in utero, the spermatogonia lie quiescent until puberty. Only then, under the influence of rising pituitary gonadotropins, do they enter mitosis (Figure 5.9). For the first several rounds of mitosis, the spermatogonia differentiate little and remain outside the blood-testis barrier. Eventually, however, some of the daughter cells produced by this clonal proliferation differentiate into *primary spermatocytes,* which move across the blood-testis barrier, where they enter the first stage of meiosis. Each original spermatogonium thus gives rise to a large clone of germ cells that begin differentiating into functional sperm. Unlike the primary oocyte, which yields at most one mature egg, each spermatogonium may have dozens of descendants.

If all the cells descended from each spermatogonium were to differentiate into mature sperm, the supply of spermatogonia within the testes would eventually be exhausted, like the supply of primary oocytes at menopause. However, this does not happen. Every few round of mitosis, one daughter cell reverts to a less differentiated spermatogonium stage, from which it can produce further clones of germ cells. In this way, the testis is able to produce many billions of sperm cells throughout adult life from the comparatively small number of spermatogonia originally present at puberty, thus providing a practically inexhaustible supply of gametes.

As it enters meiosis, the primary spermatocyte is a comparatively large cell. Successive meiotic divisions produce progressively smaller daughter cells, two *secondary spermatocytes* followed by four *spermatids* (Figure 5.9). These daughter cells remain connected by cytoplasmic bridges until the final stages of spermatogenesis, when virtually all the cytoplasm is jettisoned and left behind as a residual body (Figures 5.9 and 5.10). Each of the four spermatids will normally develop into a mature spermatozoan—again contrasting with oogenesis, in which all but one of the four products of meiosis are extruded as polar bodies, which are normally unfertilizable. During the progression from primary spermatocyte to spermatid, the developing germ cell is moved by the Sertoli cells away from the outer wall of the seminiferous tubules toward the lumen (Figure 5.10). At the same time, the cell becomes increasingly differentiated (Figure 5.11). The head and tail become more prominent, the Golgi apparatus gives rise to the acrosome, and the mitochondria, initially scattered throughout the cytoplasm, migrate and organize themselves around the base of the developing tail. Finally, the almost-mature spermatozoa are released into the lumen of the tubules. The Sertoli cells play an important part in this entire process, helping spermatocytes cross the blood-testis barrier and move toward the lumen, providing nutrients and growth factors, guiding the remodeling of the spermatids, and secreting the fluid that fills the lumen and carries the sperm toward the epididymis.

It takes about 64 days for male germ cells to pass from the primary spermatocyte stage to ultimate release into the lumen of the seminiferous tubule. All segments of the tubule do not, however, undergo these various stages synchronously; otherwise, sperm would become available in discrete bursts every 64 days. Instead, adjacent portions of the tubule are staggered: one portion may be at the primary spermatocyte stage, the next at the secondary spermatocyte stage, and so on. In most mammals, this staggering gives rise to regular waves of spermatogenesis that can be traced all the way down the length of the tubule. In humans, however, the situation is more complex because different "wedges" of a single cross-section of the tubule can be at different stages of spermatogenesis; thus, continuous waves along the entire tubule are not observed. However, "wavelets" can be observed in humans along restricted subportions of each tubule, with a wavelength of about 4 mm and a duration of 16 days per wave.

Movement of the sperm from the seminiferous tubules to the epididymis, where they are temporarily stored, does not occur as a result of the sperm's own motility; in fact, the sperm will not become fully motile until after they enter the female tract (see below). Rather, the movement of sperm results from peristalsis of the seminiferous tubules themselves, as well as pressure from the continuous secretion of fluid into the lumen of the tubules by the Sertoli cells.

Although sperm appear to be morphologically mature once they reach the epididymis, they are not yet capable of taking part in fertilization. During roughly 12 days of passage through the epididymis, the sperm undergo further maturation, including small changes in morphology and metabolism. The epididymis also absorbs fluid from its own lumen, thereby concentrating the sperm by approximately a hundredfold. If ejaculation occurs, this dense bolus of sperm is moved along the vas deferens by peristalsis, becoming mixed in the process with other components of semen. If ejaculation does not take place, the sperm are gradually resorbed and destroyed.

After ejaculation, sperm must undergo further maturational changes in the female reproductive tract. That such changes are necessary is indicated by an observation from the field of in vitro fertilization: if freshly ejaculated sperm are placed in the immediate vicinity of a cultured egg cell, they will not attach to the egg's outer covering or otherwise initiate fertilization. Rather, the sperm must be incubated for 1–2 hours in a medium that mimics the uterine environment before they can take part in fertilization. Three final stages of maturation must occur within the female tract: *capacitation, activation,* and the *acrosome reaction.* Capacitation, which prepares the sperm to bind to the zona pellucida, occurs during the transit from the vaginal canal to the site of fertilization in the oviduct. Activation and the acrosomal reaction, during which the sperm's full motility is switched on and the sperm head is prepared for tunneling through the zona pellucida, occur more or less simultaneously upon binding to the zona. These final preparatory steps are induced by chemicals produced by the female (see below for details).

As in the case of oogenesis, the neuroendocrine control of spermatogenesis involves an integrated hypothalamo-pituitary-gonadal axis, with principal players already familiar from Chapter 4: the GnRH pulse generator, hypothalamic GnRH, pituitary LH and FSH, and gonadal steroids. Now, however, the principal steroid of interest is testosterone rather than estradiol or progesterone. Testosterone appears to be the single most important hormonal prerequisite for spermatogenesis, and severe testosterone deficiency is invariably associated with male sterility. Testosterone is produced by the Leydig cells of the interstitial tissue and exerts its effects on spermatogenesis via nearby Sertoli cells. The concentration of testosterone within the adult testis is approximately one hundred times greater than in the general circulation.

The synthesis and secretion of testosterone are under hypothalamo-pituitary control (Figure 5.12). In response to the pulsatile secretion of GnRH, the pituitary releases LH and FSH. LH and FSH have distinct effects on the testes, with LH acting predominantly on the Leydig cells to stimulate testosterone secretion and FSH acting on the Sertoli cells to stimulate the differentiation and division of germ cells. However, LH can be thought of as having an essential but *indirect* effect on the Sertoli cells, since testosterone is required for spermatogenesis. Finally, both the Leydig cells and the Sertoli cells have important negative feedback effects on the hypothalamus and pituitary. The Sertoli cells secrete inhibin, which inhibits FSH secretion, while the testosterone produced by the Leydig cells inhibits LH secretion in two ways: (1) by making the pituitary less responsive to GnRH and thereby decreasing the amplitude of LH pulses, and (2) by acting at the level of the hypothalamus to decrease the frequency of GnRH pulses (Figure 5.12).

The neuroendocrine regulation of male reproductive function is thus dominated by negative feedback—in contrast to the system in females, which alternates between negative and positive feedback over the course of the cycle. As a result of negative feedback, male reproductive function is not notably cyclic. Instead, the total amounts of GnRH, LH, FSH, and testosterone secreted, and the total numbers of sperm produced, are maintained within a fairly narrow range of variation. This is not to say, however, that any or all of these aspects of male reproductive function are absolutely constant. Sexual arousal, for example, leads to increased production of all the hormones involved in the regulatory system, and several aspects of testicular function are known to vary by season (Tjoa et al., 1982; Levine et al., 1990; Levine, 1991, 1994). At present, however, much less is known about heterogeneity of reproductive function in males than in females, primarily because males lack an obvious external marker such as menses by which such heterogeneity can be assessed.

Although spermatogenesis has several points of similarity to oogenesis, there are some critical differences in timing that have important implications for the unequal contributions of male and female physiology to the pace of childbearing. Most critical is the fact that sperm production is continuous rather than cyclic. This, combined with the fact that several hundred million sperm cells are normally produced by the adult male each day (in contrast to the single mature egg produced in a normal ovarian cycle of 28–30 days), makes oogenesis rather than spermatogenesis the rate-limiting step in conception. In addition, since a clone of spermatogonia is constantly being set aside to support further mitosis, males do not run out of germ cells in the way females do. Thus, although male reproductive capacity may decline somewhat at higher ages, there is no irreversible cessation of reproductive function in males that is at all comparable to the menopause (despite fashionable talk about the "male menopause"). This, too, makes the timing of reproductive events over the female life course of far greater significance than that over the male life course.

A final distinction between sperm and egg is one of size: the sperm cell becomes progressively smaller during the course of gametogenesis, whereas the egg grows dramatically larger. As a result, the egg is approximately one thousand times greater in volume than the sperm. Not surprisingly, this massive difference in size reflects important differences in function. Sperm must move large distances, and small size is a virtue in sperm transport. Small size means, however, that virtually all that the sperm contributes to the zygote is nuclear DNA (it also contributes one centriole that will generate the spindle for the first mitotic cell division). In addition to DNA, the egg must provide the materials necessary for the first several days of embryonic development, until the embryo implants and is able to obtain nutrients from the maternal bloodstream. The egg therefore has a huge cytoplasmic storehouse accumulated during oogenesis, containing, in addition to nutrients, large quantities of enzymes, messenger RNA, transfer RNA, ribosomes, and mitochondria.[3]

FERTILIZATION

Fertilization, the fusing of sperm and egg and the merging of their haploid chromosome complements, accomplishes three things: (1) restoration of the full (*diploid*) complement of 46 chromosomes, (2) determination of the chromosomal sex of the zygote, and (3) initiation of *embryogenesis*, i.e., formation of the basic body plan of the embryo. In addition, two dangers must be avoided during fertilization, either of which could lead to failure of embryonic development and thus a wasted opportunity for reproduction: (1) inappropriate fertilization by sperm of another species, and (2) *polyspermy*, or fertilization of an egg by more than one sperm. Both these errors usually result in an unbalanced chromosome complement, aberrations of mitosis, and early embryonic death. Recent studies show that the zona pellucida, the capsule of gelatinlike material surrounding the mature egg, plays an unexpectedly important part in circumventing both dangers.

Upon insemination, sperm are deposited in the upper reaches of the vaginal canal. The site of fertilization, usually the ampulla of the oviduct, is some 15–20 cm away from the site of deposition, a tremendous distance from the perspective of a cell as small as the spermatozoan. Given the limited life span of sperm after insemination (see Chapter 4) and the slow rate at which sperm swim (about 2–4 mm/min), it is unlikely that many would reach the egg alive if they had to arrive under their own power. Moreover, although most sperm do not reach the oviduct for an hour or more, a few sperm are detectable there as soon as five minutes after insemination (Table 5.2).[4] It is clearly impossible for sperm to cover such a distance by itself in that short a time.

Recent evidence suggests that sperm are transported to the oviduct by the female through several mechanisms. First, some unidentified component of seminal fluid (perhaps the prostaglandins) appears to stimulate contractions of the upper vagina, which may help propel sperm into the cervical canal. There, as described in Chapter 4, sperm are guided by the cervical mucus into the main cavity of the uterus. It has been suggested that the spasmodic contractions of the uterine smooth muscles

Table 5.2. Interspecific Variation in the Number of Sperm Ejaculated, Number of Sperm Reaching the Site of Fertilization, and Time between Coitus or Artificial Insemination and Arrival of Sperm at the Site of Fertilization[a]

Species	Site of sperm deposition	Mean no. of sperm per ejaculate (millions)	Mean no. of sperm at site of fertilization	Time to arrival of sperm at site of fertilization
Human	vagina	280	200	5–68 min
Mouse	uterus	50	<100	15 min
Rat	uterus	58	500	15–30 min
Rabbit	vagina	280	250–500	2–60 min
Guinea pig	vagina & uterus	80	25–50	15 min
Cow	vagina	3000	few	2–13 min
Sheep	vagina	1000	600–700	6 min–5 hr
Pig	uterus	8000	1000	15 min

[a] In all cases, the site of fertilization is the ampulla of the oviduct.
From Harper (1982)

experienced during orgasm may facilitate sperm transport across the uterine cavity, although there is no evidence that those inseminations accompanied by female orgasm have a higher probability of conception than those that are not. In the oviducts, movement of the ductal cilia becomes important; rings of contraction divide the ducts into compartments among which currents of different directions can be set up, thus distributing the sperm along the oviduct even as they move the egg toward the uterus. The activity of the sperm themselves are probably of minor importance in sperm transport until the sperm reach the ampulla, where they orient themselves against the ciliary current. Because of this positive "rheotactic" response on the part of the sperm, downward currents in the ampulla serve as an effective orienting stimulus to point the sperm toward the egg. Recent experimental evidence (Ralt et al., 1991) suggests that sperm may also respond to chemical attractors associated with the egg (or rather, with the antral fluid still surrounding the egg), although the functional significance of the chemotactic response, as well as the chemical responsible for it, remains uncertain.

Of the millions of sperm ejaculated, only a few hundred make it to the ampulla of the oviduct. Interestingly, while there is considerable variation among species of mammals in the number of sperm per ejaculate, there is far less variation in the number of sperm at the site of fertilization (Table 5.2). In general, the volume of the ejaculate is correlated with the size and complexity of the female tract. For example, pigs, which have one of the most massive ejaculates known, also possess uterine horns that are among the longest of any mammalian species; moreover, fertilization occurs at several sites along the full length of the pig uterus, and sperm must be evenly (and densely) distributed along this length. Evidently, the size of the ejaculate has been adjusted during evolution to compensate for the difficulties of maneuvering the female tract.

Once the sperm have reached the ampulla, fertilization proceeds in several stages (Figure 5.13, A–B). First, a single sperm binds to the outside of the zona pellucida, a process that appears to be highly species-specific. If a cultured egg cell is stripped of its zona, sperm from many different species can fuse with the egg cell membrane and be incorporated into the egg. If the zona is left intact, however, only sperm of the same species will bind to the zona. There must be some highly specific molecule that recognizes the appropriate type of sperm, thus providing the principal defense against fertilization by sperm of the wrong species.

Upon binding to the zona, the sperm cell becomes activated and undergoes the acrosome reaction (Figure 5.14). Little is known about the process of activation in mammals, but it may be similar to what has been observed in sea urchins (Tombes and Shapiro, 1985): a sudden elevation in intracellular pH activates the enzyme ATPase in the midpiece of the sperm, causing a rapid turnover of ATP and a 50 percent increase in mitochondrial respiration; most of the resulting energy is used for flagellar motility. Consistent with this, activation in mammals is accompanied by a marked increase in the movement of the sperm tail.

Considerably more is known about the acrosome reaction (Figure 5.15). As a first response, the plasma membrane of the sperm head fuses at many points with the outer acrosomal membrane, causing the covering of the acrosome to disintegrate and exposing the inner acrosomal membrane. Embedded within this membrane are several enzymes, including hyaluronidase, neuraminidase, and a trypsinlike protein called acrosin. These enzymes digest the zona material immediately in front of the sperm head and, in combination with the whiplike movement of the sperm tail, facilitate passage of the sperm through the zona (Figure 5.16). The fact that the slit in the zona through which the sperm passes is no wider than the sperm head itself suggests that the enzymes involved in digesting the zona are firmly bound to the inner acrosomal membrane and do not diffuse more widely through the zona material.

After passing through the zona pellucida, the sperm enters the perivitelline space separating the zona and egg (Figure 5.13, C–D). The sperm then binds to receptors on the surface of the egg plasma membrane (Figure 5.13, E–H). Apparently the sperm proteins recognized by these receptors are clustered most densely on the equatorial segment of the sperm head. Consequently, even though initial binding may occur at the tip of the head, the sperm becomes most firmly attached when bound at the equatorial segment. Upon binding, the egg sends out microvilli that engulf the sperm head, and the plasma membranes of the egg and sperm fuse (Figure 5.17).

Binding of the sperm to the egg's outer membrane triggers an important series of responses known as the *cortical reaction* (Figure 5.13, E–F). Lying just below the surface of the egg are thousands of tiny vesicles, the *cortical granules,* which were derived during oogenesis from the Golgi apparatus of the primary oocyte and are thus homologous to the acrosome of the sperm (Nicosia et al., 1977). Binding of the sperm to the egg plasma membrane releases a burst of calcium ions from internal stores in the egg cytoplasm, causing the cortical granules to move toward the plasma membrane and fuse with it. Membrane fusion is propagated as a wave, emanating from the point of sperm-egg fusion and moving out over the entire surface of the egg. As a result of fusion, the granules open and spill their contents into the perivitelline space. These cortical contents, a mixture of proteinases, peroxidases, and other hydrolytic enzymes, digest the sperm receptors on the egg's plasma membrane and induce the *zona reaction* (see below). Both these actions set up blocks to polyspermy.

The release of calcium ions caused by binding of the sperm not only sets off the cortical reaction but also reinitiates meiosis in the egg. It apparently does so by abolishing the activity of a protein, *cytostatic factor,* in the egg's cytoplasm (Meyerhoff and Masui, 1977). Cytostatic factor, produced during oogenesis, effectively freezes the egg at the second meiotic interphase. By deactivating this protein, binding of the sperm allows the egg to complete meiosis.

The final stages of fertilization proceed rapidly (Figure 5.13, I–N). As the sperm is incorporated into the egg cytoplasm, the dense mass of DNA packed into its nucleus swells to form the *male pronucleus.* At about the same time, the egg completes its final meiotic division, ejects the second polar body, and forms the *female pronucleus.* The centriole that was originally located at the base of the sperm tail (the proximal centriole in Figure 5.5) generates an aster around the male pronucleus, sending out microtubules that attach to the female pronucleus and draw the two pronuclei toward each other. The centriole divides, a mitotic spindle is formed, and the now-fertilized egg undergoes its first cell division.

It has only been in the last decade that some of the molecular details of this remarkable process have been revealed. It turns out that a single molecular constituent of the zona pellucida plays a pivotal rote at several stages of fertilization. The zona is now known to consist primarily of three glycoproteins, denoted ZP1, ZP2, and ZP3 (Nakano, 1989). All three are linked together in the zona in long, multistranded filaments, as shown in Figure 5.18. Competitive binding experiments show clearly that ZP3 is the molecule that binds the sperm head to the zona (Figure 5.19). More specifically, one or more of the sugar side chains of ZP3 is responsible for the binding, for binding does not occur when ZP3 is stripped of these carbohydrates (Figure 5.19). More recent research has suggested that it is the so-called O-linked oligosaccharides (sugars attached to the amino acids serine and threonine) that bind sperm (Wassarman et al., 1989). In addition, it now appears that the same sugar chains, in combination with the polypeptide component of ZP3, induce the acrosome reaction (Wassarman, 1987). Thus, a remarkable economy is achieved: the same molecule screens out inappropriate sperm, binds appropriate sperm, and prepares the sperm head to tunnel through the zona.

Capacitation is now thought to be a process in which the specific molecules on the sperm head that are recognized by ZP3 are made available for binding (Figure 5.20). The sperm head appears to have a protein, glycosyltransferase, embedded in its outer membrane with its active site oriented outward (Shur and Hall, 1982a, 1982b). When exposed, this active site is capable of recognizing and binding with terminal N-acetylglucosamine residues, such as those on the O-linked oligosaccharides of ZP3. However, at the time of ejaculation, this active site is blocked by a "coating factor" containing an N-acetylglucosamine moiety. (This coating factor appears to be supplied by the male during the course of spermatogenesis.) It has recently been suggested that, during capacitation, enzymes within the uterine secretions release the coating factor, thus exposing the site at which glycosyltransferase can bind with ZP3 (Lopez et al., 1985). This hypothesis would explain the inability of uncapacitated sperm to bind to the zona and undergo the acrosome reaction.

The block to polyspermy induced by the cortical reaction also involves ZP3, as well as the other zona glycoproteins. Various enzymes (proteinases, glycosidases, and peroxidases) released into the perivitelline space by the egg's cortical granules are absorbed by the highly porous zona. There, they appear to have two important effects. First, they "harden" the zona, perhaps by increasing the number of cross-linkages among its glycoprotein filaments, and thus render the zona resistant to penetration by other sperm. In addition, cortical enzymes, especially the glycosidases, appear to shear off the sugar side chains of ZP3 that serve as sperm head receptors. As a result, once the cortical and zona reactions have occurred, additional sperm are unable to bind to the zona, and polyspermy is averted.

In sum, contrary to earlier views that treated the zona pellucida as an inert, rather boring substance that does little more than provide a "bumper" for the egg and early embryo, the zona now appears to play an extraordinarily active role in fertilization.

IMPLANTATION AND EMBRYOGENESIS

Fertilization initiates a series of mitotic cell divisions, known as *cleavage*, which transform the fertilized egg into a multicellular organism, capable of undergoing differentiation but as yet undifferentiated. Cleavage divisions occur at intervals of about 12–24 hours as the zygote passes down the oviduct. Each cleavage occurs more or less simultaneously in all cells of the zygote, producing, in sequence, a 2-cell, 4-cell, 8-cell, and finally a 16-cell embryo. An important aspect of cleavage is that little if any new cytoplasm is produced between rounds of mitosis. Thus, the enormous cytoplasm of the egg is gradually parcelled out into several smaller cells, known as *blastomeres.* Experiments on invertebrates suggest that the ratio of cytoplasm to nucleus is critical in determining when cleavage begins and when it ends (Longo, 1987).

The zygote's metabolic rate, as measured by oxygen uptake or carbon dioxide output, increases little during early cleavage; in addition, the early zygote does not appear to produce new RNA, despite the fact that protein synthesis has already begun (Figure 5.21). These observations suggest that the early embryo, up to the morula or even blastocyst stage, draws upon existing stores provided by the egg for its energy needs and for the molecular machinery required for DNA and protein synthesis. The zygote's own genome is probably not transcribed until about the time of implantation (McLaren, 1981).

The cells produced by the first three cleavage divisions form a loose aggregate within the zona pellucida. During the 8-cell stage, however, this aggregate suddenly draws together, maximizing the amount of contact among the blastomeres and forming a compact ball of cells (Figure 5.22). This process of *compaction,* caused by the formation of numerous tight junctions among the blastomeres, sets the stage for the first differentiation of embryonic tissues. The tight junctions effectively seal off the interior of the zygote, and the interior cells produced by the next several rounds of mitosis begin to differentiate from the external cells. In addition, a hollow cavity called the *blastocoel* begins to form in the interior of the embryo. By the time the embryo passes into the main body of the uterus, it consists of two distinct tissues, a hollow ball of cells called the *trophoblast,* which surrounds and defines the blastocoel, and a rather unprepossessing lump of cells stuck to one side of the inner trophoblast, the *inner cell mass* (ICM). Unpromising though it may appear, the ICM contains the cells that will give rise to the embryo. (The entire ICM consists initially of eight cells, about three of which will actually produce the embryo.) The trophoblast, in contrast, produces the placenta and other extraembryonic tissues. Thus, the very first differentiation that takes place in human development divides the embryo proper from the tissues that will communicate between the embryo and its mother.

About day five after conception, the embryo "hatches" from the zona pellucida. The region of the trophoblast farthest from the ICM secretes an enzyme (strypsin) that digests a hole in the zona, through which the embryo escapes (Figure 5.23). Since the zona prevents attachment of the embryo to the endometrium, hatching from the zona is a prerequisite for implantation; indeed, one probable function of the zona is to minimize the risk of implantation in inappropriate maternal tissue such as the epithelium of the oviducts.

Approximately 8–10 days after fertilization, the "hatched" embryo undergoes implantation. First, the embryo attaches to the uterine endometrium, usually at the upper portion of the back wall of the uterine cavity (Figure 5.24, top). Enzymes produced by the trophoblast then digest away the endometrial epithelium, and the embryo burrows into the blood-rich tissues of the underlying endometrium (Figure 5.24, bottom). During this process, the trophoblast differentiates into two distinct layers, an inner layer of epithelial cells called the *cytotrophoblast,* and a more diffuse layer of multinucleated tissue without cell membranes called the *syncytiotrophoblast* (Figure 5.25). The syncytiotrophoblast is the part of the embryo that invades the endometrium most vigorously, digesting away the endometrial tissues and opening up the maternal capillaries to form pools of maternal blood. The syncytiotrophoblast thus prepares the interface where maternal and fetal tissues will meet.

Implantation takes place at about the time that the corpus luteum would have started regressing had conception not occurred. Since luteolysis causes the loss of the endometrium, the very tissue in which the embryo is implanted, luteal regression during a cycle of conception would be disastrous. How is it averted?

The answer seems to be that the embryo itself "rescues" the corpus luteum from luteolysis. Beginning at about the time of implantation, when the embryo first gains access to the maternal bloodstream, the trophoblast of the blastocyst secretes the peptide hormone *human chorionic gonadotropin* (hCG), and it is this hormone that causes persistence of the corpus luteum. This fact is demonstrated by experiments in which the activity of hCG is blocked by highly specific anti-hCG antibodies, invariably inducing luteal regression (Figure 5.26). Biochemically, hCG is very similar to LH, and it strongly stimulates steroid secretion by the corpus luteum. Thus, the signal that maintains luteal function (and thus pregnancy) comes from embryonic tissue.

In terms of measurement, hCG is important because it is the earliest reliable biochemical indicator of pregnancy. Recently, extremely sensitive and specific hCG assays have been developed for use with serum or urine (Armstrong et al., 1984).[5] With the advent of these assays, hCG has become the hormone of choice not only in the clinical detection of pregnancy, but also in studies of fecundability and early fetal loss (see Chapter 6).

From the mother's point of view the implanting embryo is, in effect, a tissue transplant. Since it expresses its own genome, 50 percent of which is nonmaternal, the embryo will inevitably expose the maternal bloodstream to foreign antigens. It is known, for example, that histocompatibility antigens are present on the syncytiotrophoblast soon after implantation (Hogarth, 1982; Claman, 1993). What is to prevent the mother's immune system from recognizing those antigens as foreign and attacking the embryo? Clearly, the maternal immune system is not globally suppressed during pregnancy, and maternal IgG antibodies are capable of crossing the placenta and entering the fetal bloodstream. And in some conditions, such as Rh incompatibility, maternal antibodies may recognize fetal antigens and attack fetal tissues. Why, then, do maternal antibodies not recognize the histocompatibility antigens expressed by the fetus as if they were foreign matter?

Two factors seem to prevent this. First, the amniotic fluid appears to contain large quantities of soluble fetal antigens; if maternal antibodies cross the outer embryonic tissues, they may be "mopped up" by these free antigens before they can harm the embryo. Second, some unknown substance or combination of substances secreted by the embryo appears to act at the syncytiotrophoblast as a local immunosuppressant, perhaps by masking the fetal antigens expressed there. In recent years, attention has focused on a fetal protein called *early pregnancy factor* (EPF) as one possible immunosuppressant (Smart et al., 1981). According to some studies, EPF may be secreted by the embryo as early as 48 hours after conception, which would make it extremely valuable for the detection of early pregnancy. (By contrast, hCG is not detectable in maternal blood or urine until 7–10 days after conception.) Unfortunately, no standardized method for detecting EPF has been developed, and some reproductive endocrinologists have come to believe that EPF does not really exist. The EPF story has been one of the most tantalizing and frustrating developments (or rather lack of developments) in reproductive biology during the past decade. Perhaps future editions of this book will have more to say about it.

Beginning in the third week of pregnancy, two critical embryogenic events occur, *gastrulation* followed by *neurulation*. Gastrulation establishes the basic tissue layers of the embryo, as well as its dorsal-ventral (back to belly) axis, while neurulation establishes the cephalic-caudal (head to tail) axis. Together these events lay down the basic body plan of the embryo.

Just before gastrulation, the implanting blastocyst has the structure shown in Figure 5.27. The inner cell mass has differentiated into two cell layers, the embryonic epiblast and the hypoblast. In addition, an *amniotic cavity* has started to open up between the epiblast and trophoblast. During gastrulation, cells along the longitudinal axis of the embryonic epiblast invaginate and migrate relative to each other to form the *primitive streak*, which will develop into the notochord (Figure 5.28, A). Involuting epiblast cells form a disorganized layer of mesenchyme cells and a more orderly layer of epithelial cells between the remaining epiblast and the hypoblast (Figure 5.28, B). This movement yields the three fundamental *germ cell layers* of the embryo: the *endoderm*, which gives rise to the lining of the digestive tract and other internal organs; the *mesoderm*, which produces skeletal and connective tissues, as well as smooth muscle, pleural, pericardial, and peritoneal tissues; and the *ectoderm*, which produces skin and neural tissues.

The major extraembryonic tissues also differentiate during gastrulation (Figure 5.29). The hypoblast extends and closes in on itself to form the *yolk sac*. In addition, the amniotic cavity grows in size as it fills with a secretion called *amniotic fluid*, which serves as a shock absorber for the embryo while protecting it from desiccation. Meanwhile, the cytotrophoblast, now called the *chorion*, sends out a series of thin projections, the *chorionic villi*, into the space between the syncytiotrophoblast and the maternal tissues (Figure 5.30). As the syncytiotrophoblast digests away the endometrium, including the walls of the endometrial blood vessels, this space becomes filled with maternal blood. Lying within the resulting pools of maternal blood, the chorionic villi become the major site of exchange between the maternal and embryonic circulations. In combination with the *allantois*, which grows out of the lower gut of the embryo, the chorion will give rise to the placenta. Figure 5.31 shows the relationships among these extraembryonic tissues at various stages of pregnancy.

Soon after implantation, the embryo and extraembryonic tissues are overgrown by maternal endometrial tissue, also called *decidual tissue* since it will be lost at birth (Figure 5.32, A). At this stage, chorionic villi are more or less evenly distributed over the entire chorion. Later, as the allantois makes direct contact with the chorion, the chorionic villi become denser in the region of the *decidua basalis*, the endometrial tissue oriented toward the myometrium of the uterus, which is also where the allantois and chorion meet (Figure 5.32, B). In contrast, the chorionic villi of the *decidua capsularis*, the endometrial tissue overlying the implanted zygote, start to thin out as that endometrium becomes more attenuated (Figure 5.32, C). The chorionic villi oriented toward the decidua basalis continue to proliferate as the chorion and allantois fuse, producing the placenta. In humans, as in other primates, the placenta is *hemochorial*, that is, the chorionic villi lie in direct contact with reservoirs of maternal blood (Figure 5.33) This arrangement allows efficient transfer of nutrients and waste products, but it also poses special immunological problems.

After gastrulation, the next major step in embryogenesis is neurulation, which divides the dorsal ectoderm into (1) the epidermis of the skin; (2) the internal neural tube, which will give rise to the central nervous system (brain and spinal chord); and (3) neural crest cells, which migrate into the mesoderm and produce the peripheral nervous system, pigment cells, and other structures. The process of neurulation involves an invagination of the ectoderm, which closes in on itself and pinches off the neural tube (Figure 5.34). This process begins near the midpoint of the longitudinal axis of the embryo and proceeds toward either end (Figure 5.35). Once neurulation is accomplished, the basic tissues and body plan of the embryo are set, and the rest of pregnancy is dominated by *organogenesis*, the differentiation of the various organs of the body from the basic tissues already established.

Discussion of the cellular and genetic mechanisms of embryogenesis and organogenesis is far beyond the scope of this book (for outstanding recent reviews, see Gilbert, 1991; Wilkins, 1993). Essentially, however, the process involves three coordinated series of genetic events: (1) activation in hierarchical order of certain regulatory genes known as *homeotic genes*, which determine the major axes of the overall body plan as well as the axes of secondary structures, such as limbs, the brain, and even teeth; (2) activation of special-purpose genes so the cells in which they are located (and the descendants of those cells) begin to produce the specific proteins that determine their function; and (3) the permanent repression of other genes whose expression would be inappropriate in the particular cell line involved. Processes 2 and 3 often involve chemical induction by adjoining cells or tissues, so contiguous structures have the proper orientation to each other. At the outset of development, the cells of the early embryo are "totipotent," i.e., capable of differentiating into any or all of the tissues present in late pregnancy. During the course of embryonic development, the descendants of those early cells lose their totipotency and become increasingly specialized. A major goal of contemporary developmental biology is to understand the complex process of gene regulation whereby this specialization is achieved.

THE ENDOCRINOLOGY OF PREGNANCY

Pregnancy is characterized by marked changes in the concentration of several hormones in the maternal circulation (Figure 5.36). Secretion of hCG by the trophoblast has already been mentioned. The concentration of hCG goes up rapidly in early pregnancy, reaching a peak at about 60–80 days after the last menses; it then falls again just as rapidly, but it remains detectable throughout pregnancy. Coinciding with the fall in hCG, the placenta starts to secrete increasing quantities of progesterone and estrogen (especially estriol, the major estrogen of pregnancy). The marked increases in steroid concentrations observed during late pregnancy are almost entirely of placental (and therefore fetal) origin, and the concentrations of those steroids are highly correlated with the total size of the placenta (Figure 5.36, top).[6] The corpus luteum remains throughout pregnancy, but it contributes less and less to total steroid secretion. Although removal of the ovaries during the first two months of pregnancy will invariably terminate the pregnancy by disintegrating the endometrium, removal during the last seven months has essentially no effect. By that time, the role of the corpus luteum, so crucial to early pregnancy, has been usurped by the placenta.

Among other things, estriol stimulates the growth of the smooth muscles of the myometrium, the very muscle that will eventually expel the fetus; at the same time, progesterone inhibits the contractility of these muscles so the fetus is not expelled prematurely.

It will be recalled from Chapter 4 that GnRH, LH, and FSH secretion is powerfully inhibited by progesterone in the presence of estrogen, e.g., during the normal luteal phase of the ovarian cycle. Since both steroids are high throughout pregnancy, the hypothalamo-pituitary axis is closed down, and follicular development and ovulation, as well as menstruation, cease for the duration of the pregnancy.

In sum, the endocrine signals that tell the mother she is pregnant, and that she should discontinue cyclic ovarian function and menstruation for the duration, are produced in large part by the embryo itself—trophoblastic hCG during early pregnancy and placental estrogen and progesterone during later pregnancy. The placenta also produces the peptide hormone *human placental lactogen* (hPL). Among other functions, hPL is important in regulating maternal metabolism during pregnancy: it maintains a positive protein balance, mobilizes fats for energy, and stabilizes serum glucose at a comparatively high level to support the fetus; it also probably facilitates development of breast tissue. Thus, hPL "primes" the mother to meet the nutritional needs of the rapidly growing infant, both before and after delivery.

In addition to hormones secreted by the placenta, several hormones are secreted by the mother in amounts that increase over the course of pregnancy. One of the most important of these is *prolactin*, a peptide secreted by the anterior pituitary gland. During pregnancy, the anterior pituitary increases in size by about 40–50 percent; this increase is attributable almost entirely to growth and multiplication of the lactotrophs,

the specific cells that produce prolactin. As detailed in Chapter 8, prolactin appears to collaborate with hPL and the placental steroids to
cause proliferation and differentiation of the milk-secreting glands of the
breast.

THE PHYSIOLOGICAL COSTS OF PREGNANCY

As the previous section has hinted, pregnancy induces profound
changes in maternal physiology, changes that can be thought of as reset-
ting the maternal metabolism to support fetal growth and development.
The pregnant mother is not just eating for two, she is catabolizing and
anabolizing for two as well. Although the fetus is extremely small relative to the mother, its growth rate is extraordinarily high, higher than it
will ever be again (Figure 5.37). This rapid turnover of fetal tissues
places heavy metabolic demands on the mother, reflected in substantial
increases in oxygen consumption that can be measured in various
maternal tissues as pregnancy progresses (Figure 5.38).

These demands require that pregnant women increase their dietary
intake of a variety of nutrients, often by substantial amounts (Table 5.3).
If those requirements are met, normal, healthy western women can
expect to gain an average of about 10.8 kg during pregnancy, about half
of which represents fetal tissue and the rest increased maternal blood
and tissue fluid volume and maternal energy stores (Figure 5.39, top).
These maternal stores consist largely of body fat laid down during pregnancy under the influence of rising progesterone and hPL, although
some increase in lean body mass may also occur (King et al., 1973; King,
1975). These stores are necessary not only for a successful pregnancy,
but also for lactation once the child is born; the energetic costs of lactation are even greater than those of pregnancy itself (Worthington-
Roberts et al., 1985).

This point is important to the study of natural fertility, because
women in rural preindustrial settings cannot always achieve the dietary
intakes needed to meet the costs of pregnancy and lactation. Under
these circumstances, it appears from several studies that the growth of
fetal tissue will be supported at the expense of maternal tissues (Figure
5.39, bottom). Therefore, women may gain little if any weight during the
course of pregnancy, and they may actually lose substantial amounts of
body fat during pregnancy and subsequent lactation (Bongaarts and
Delgado, 1979; Miller and Huss-Ashmore, 1989). If maternal stores are
not replenished before the next pregnancy—and they may well not be
since lactation often continues until the next pregnancy in traditional

Table 5.3. Recommended Dietary Allowances
of Various Nutrients for Pregnant Women,
Expressed as Percentage Increase
over Recommendations for Nonpregnant
Women Ages 23–50

Nutrient	Increase over nonpregnant allowance (%)
Energy	12.5–20
Protein	68
Vitamin A	25
Vitamin D	100
Vitamin E	25
Ascorbic acid	33
Folic acid	100
Niacin	15
Riboflavin	25
Thiamin	40
Vitamin B_6	30
Vitamin B_{12}	33
Calcium	50
Phosphorus	50
Iodine	167
Iron	167–333
Magnesium	50
Zinc	33

Adapted from National Academy of Sciences (1980)

societies—repeated rounds of pregnancy and lactation may lead to a cumulative deterioration in maternal nutritional status, the maternal depletion syndrome discussed in Chapter 1. Thus, under conditions of natural fertility, reproduction may be a major determinant of the health and well-being of women.

PARTURITION

During the last few weeks of pregnancy, the contents of the uterus are shifted downward, and the baby's head (normally) is brought into contact with the cervix. The cervix becomes soft and pliable, changes referred to as the "ripening" of the cervix. Cervical ripening is caused by a breakdown of collagen fibers under the influence of several hormones secreted by the corpus luteum and the uterus, including estrogen, prostaglandins, and the peptide hormone *relaxin*.

These changes prepare the mother for *parturition,* i.e., delivery of the infant and placenta. Parturition is induced by strong, rhythmical contractions of the smooth-muscle cells of the myometrium, which rupture the amniotic sac and expel the baby and placenta. In the nonpregnant myometrium, the smooth-muscle cells have an inherent rhythmicity, but the resulting contractions are weak and uncoordinated. During the course of pregnancy, the smooth-muscle cells grow in size and number under the influence of estriol and their contractility increases, partly because of mechanical stretching by the growing fetus. As labor begins, contractions become coordinated and quite strong, sweeping downward from the top of the uterus at intervals of about 10–15 minutes. Coordination of these contractions is made possible by the formation during pregnancy of numerous gap junctions linking the uterine smooth-muscle cells. These gap junctions, which allow several neighboring cells to be activated by a single neural impulse, increase markedly in number near the end of pregnancy in response to high levels of estrogens in the maternal circulation.

Several questions about the factors that trigger parturition in humans remain unanswered. It is clear that the maternal autonomic nervous system is not directly involved, since severing the autonomic neurons to the uterus does not interfere with delivery. Instead, the trigger seems to be entirely hormonal. Near the time of delivery, the uterus secretes a prostaglandin that is a powerful stimulator of contractions. Increased secretion of this prostaglandin has been observed during delivery itself. Oxytocin, a neuropeptide hormone secreted by the posterior lobe of the pituitary gland, is also a powerful stimulator of contractions; it acts directly on the uterine smooth muscles and also stimulates the uterus to produce more prostaglandins. Oxytocin is released by the posterior pituitary in response to neural signals to the hypothalamus from receptors in the uterus, especially the cervix. During pregnancy, the concentration of oxytocin receptors in the uterus increases as a result of estrogen stimulation.

Acting in opposition to all these contraction stimulants, progesterone strongly inhibits contractions, apparently by desensitizing the myometrium to estrogen, oxytocin, and prostaglandins. Just before parturition, the ratio of progesterone to estrogen in the maternal bloodstream drops precipitously, largely because of a marked increase in estrogen secretion. This change in the progesterone:estrogen ratio appears to be the immediate trigger of birth.

But why is the trigger pulled at the right time, rather than too early or too late? In nonprimate mammals the trigger seems to be fired, oddly enough, by the fetal adrenal cortex. Once the fetal adrenal glands have attained the right level of development, they secrete large quantities of dehydroepiandrosterone and androstenedione, androgens that are pre-

cursors for estrogen (see Chapter 4, Figure 4.20). This may tip the pro-
gesterone:estrogen balance enough to induce contractions. Thus, the
trigger for birth in such species is clearly pulled by the fetus, not the
mother. The evidence for such a mechanism in humans, however, is
recent and somewhat equivocal. As in other species, elevated concentrations of adrenal steroids have been observed in fetal serum and amniotic fluid near the end of human pregnancy. Moreover, fetal adrenal
hypoplasia is sometimes associated with prolonged gestation—but not
necessarily so. Indeed, normal gestation and parturition have been
observed in human fetuses with a congenital *absence* of adrenal glands.
The onset of labor in humans, therefore, does not seem to be entirely
dependent upon maturation of the fetal adrenals. Thus, the question of
whether the mother or fetus plays the larger role in inducing parturition
has not yet been settled for humans.

THE BIOMETRY OF PREGNANCY

We next turn to statistical questions similar to those asked of ovarian
cycles in the final sections of the preceding chapter: how much variation
exists in the timing of pregnancy and parturition, and is this variation
sufficient to induce significant differentials in birth intervals? In other
words, is there important variation, either within or among human populations, in the length of gestation ending in live birth?

The pattern typically observed within a human population is illustrated in Figure 5.40. The distribution of gestations is approximately
normal, deviating slightly from normality by being leptokurtic
("peaked") and somewhat left-skewed. On an absolute time scale, the
amount of variation is small: a variance of 5 implies that about 95 percent of all pregnancies fall within 4½ weeks of the mean. Indeed, the
great majority of gestations in Figure 5.40 fall between 35 and 40
weeks. At the level of birth intervals, which are typically measured
in months or years, this amount of variation must be regarded as
trivial. Thus, variation in the length of pregnancies ending in live
birth is unlikely to be an important source of intrapopulation variation in fertility.

What about interpopulation variation? Typical results are shown
in Figure 5.41 and Table 5.4. In each intergroup contrast illustrated,
there is a significant difference in the distribution of gestations
across groups but the magnitude of the difference is small. For
example, black and white pregnancies in the United States differ by
about 8.5 days on average (Figure 5.41). And gestations among high-
er-class women of Aberdeen are only about 11 days longer than

those of much poorer women in Lahore (Table 5.4). Again, the degree of variation in gestations is minor compared with the amount of variation in the total birth interval.

The conclusion that variation in gestation length is of negligible importance is purely a demographic one. Such variation may, in fact, be of profound importance at the individual level. Note that, in each comparison, women of lower socioeconomic status tend to have a larger fraction of pregnancies ending in preterm births.[7] Numerous epidemiological studies (reviewed by Kline et al., 1989) have established that such births are often associated with low birth weights and elevated risks of infant death. Therefore, when I conclude that variation in gestation length is unimportant, that conclusion pertains strictly to its effects on birth spacing and nothing else. In addition, although the *direct* effect of gestation length on birth-interval variation is minor, gestation may have an important *indirect* effect via its impact on the duration of lactational infecundability. That is, if preterm babies are at elevated risk of early death, mothers of such infants may resume ovulating earlier than they might have because breastfeeding has been terminated by the death of their offspring. In that capacity, however, gestation is operating as a remote influence on fertility, not as a proximate determinant.

It is of interest to ask why variation in the length of gestation is held within such narrow bounds. I suggest that gestation may per-

Table 5.4. Distribution of Gestation Lengths in Three Social Classes of Primiparous Women, Aberdeen, Scotland, 1951–1959 (legitimate births only), and in Women of Lahore, Pakistan, 1963–1965

| | Aberdeen Social Class[a] | | | | | | Lahore | |
| | I-II | | III | | IV-V | | | |
Gestation (weeks)	No.	%	No.	%	No.	%	No.	%
≤35	22	1.7	169	3.0	81	4.4	71	7.1
36–37	40	3.1	238	4.2	86	4.7	164	16.4
38–39	215	16.8	1,007	17.9	374	20.4	367	36.6
40–41	716	55.0	2,913	51.8	882	48.0	285	28.5
42+	289	22.5	1,299	23.1	414	22.5	114	11.4
Median	39.0		39.0		38.8		37.4	

[a] Standard British census designation
Likelihood ratio tests of heterogeneity:
Among classes in Aberdeen: ratio = 39.1, df = 8, $P < 0.001$
Aberdeen (pooled) vs. Lahore: ratio = 3890.1, df = 4, $P < 0.001$
Data from Hytten and Leitch (1964)

haps be subject to fairly severe stabilizing selection. Several studies have indicated a U-shaped relationship between gestation length and perinatal mortality, i.e., the risk of death during the first week of postnatal life (Figure 5.42). Gestations that depart significantly from the mean, either in the direction of longer or shorter durations, are at much higher risk of early infant death than are gestations of more "typical" length. If there is (or ever has been) any additive genetic variance in gestation length, such a relationship would imply natural selection for gestations near the mean. There is, incidentally, an important confounding factor in this scenario: birth weight, which is highly correlated with length of gestation, also has a U-shaped relationship with perinatal mortality (Figure 5.42), and this fact makes it difficult to determine whether it is birth weight or gestation per se that is subject to selection. However, because the two traits are highly correlated, selection for one amounts to selection for the other.

It may be significant from this perspective that gestation length is remarkably invariant across the higher primates (Figure 5.43). The somewhat shorter gestation of chimpanzees, which may reflect their comparatively small body size, appears to be a derived characteristic; the primitive condition for pongids and hominids seems to be a gestation of about 260–265 days, similar to what is observed in humans.[8] Thus, the length of gestation may have been virtually constant over the past several million years of human evolution, consistent with the hypothesis of strong stabilizing selection for this trait.

In sum, despite the obvious centrality of conception and gestation in successful reproduction, and despite the fact that every completed birth interval must include a time segment corresponding to the full length of gestation, the processes described in this chapter appear to be of little consequence for explaining fertility *variation*. Models of birth-interval dynamics must, of course, allow for the gestation that results in live birth, but they can probably treat it as a constant without much loss of realism. The same conclusion does not, however, hold for pregnancies ending in fetal death, for as the next chapter shows, those can be much more variable in length.

NOTES

1. If fertilization does not occur, the egg passes into the uterine cavity and is eventually lost with the endometrial material during the next menses.

2. Prostaglandins were first isolated in semen and were so named because it was thought they were secreted by the prostate gland. In fact, the prostaglan-

dins in the semen are now known to be secreted primarily by the seminal vesicles. Moreover, prostaglandins are produced by many different tissues, including ovarian and uterine tissues in the female (see Chapter 4).

3. Transcription of the embryonic genome does not occur during the first few days of embryogenesis; thus, the machinery for the production of new proteins must be supplied by the egg itself.

4. There is considerable interest in the fact that some sperm arrive much earlier in the oviduct than others, since such variation may indicate important differences in the viability, motility, or fertilizing capacity of sperm. There is even a suggestion that sperm bearing Y chromosomes may travel faster than those bearing X chromosomes, which has been cited to explain the observed excess of male over female zygotes in detectable pregnancies. However, there is little evidence that early arriving sperm are any more likely to take part in fertilization than are late arrivers. Indeed, early arrivers have probably not spent enough time in the female tract for capacitation to have occurred. On the basis of present evidence, early arrivers and late arrivers appear to differ strictly by chance.

5. Sensitivity and specificity are of special concern when working with hCG because (1) hCG levels are low during early pregnancy, and sensitive assay systems are required to detect them; and (2) the structural similarities between hCG and LH cause considerable cross-reactivity between the two unless highly specific antibodies are used.

6. Strictly speaking, the early stages of estriol synthesis do not occur in the placenta, which lacks the proper enzymes, but in the fetal adrenal cortex and liver. In addition, steroidogenesis by the fetus requires maternally derived cholesterol or pregnenolone as a precursor.

7. Preterm births have been defined by the World Health Organization and the International Federation of Gynecology and Obstetrics as pregnancies terminated by live birth before 259 days or 37 completed weeks of gestation (WHO, 1977).

8. The terms "primitive" and "derived" come from the cladistic approach to the reconstruction of evolutionary phylogenies. For a specific group of related species, the primitive condition is the one that appears to have characterized the common ancestor of all the species, whereas a derived character is one that has evolved in only one or a few of the species since the time of the common ancestor.

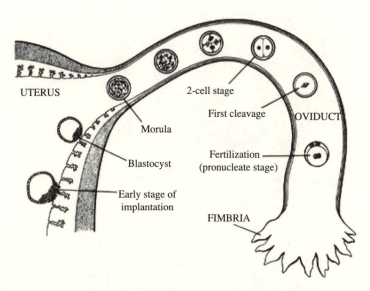

Figure 5.1. Development of the zygote and early embryo in the female reproductive tract, from fertilization to implantation. (Redrawn from Tuchmann-Duplessis et al., 1971; McLaren, 1982)

Figure 5.2. Drawing of a scanning electron micrograph of the cilia lining the oviduct, illustrating the high density of these structures in the ampulla. The comparatively smooth areas are probably mucus-secreting cells. (Redrawn from Hafez, 1975; Harper, 1982.)

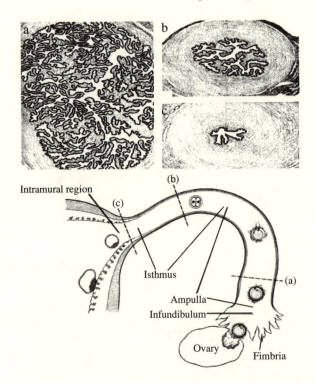

Figure 5.3. (*Top panels*) cross-sections of the oviduct in the region of the ampulla (*a*), the isthmus (*b*), and the intramural part of the tube (*c*) showing differences in the lumen size and complexity. (*Lower panel*) approximate positions of the sections portrayed in (*a*), (*b*), and (*c*). (Redrawn from Bloom and Fawcett, 1969; Harper, 1982)

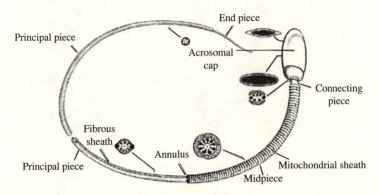

Figure 5.4. Internal structure of a typical mammalian spermatozoan with the cell membrane removed. (Redrawn from Fawcett, 1975; Setchell, 1982)

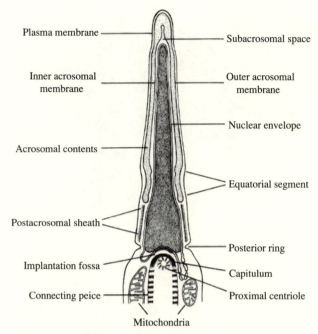

Plasma membrane

Subacrosomal space

Inner acrosomal membrane

Outer acrosomal membrane

Nuclear envelope

Acrosomal contents

Equatorial segment

Postacrosomal sheath

Posterior ring

Implantation fossa

Capitulum

Connecting peice

Proximal centriole

Mitochondria

Figure 5.5. Sagittal section of the head of a primate spermatozoan. The large stippled area is the nucleus, containing the male haploid genetic material. (Redrawn from Fawcett, 1977; Setchell, 1982)

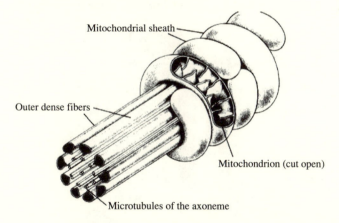

Mitochondrial sheath

Outer dense fibers

Mitochondrion (cut open)

Microtubules of the axoneme

Figure 5.6. The midpiece of a mammalian spermatozoan with the cell membrane removed. Note the spiral, sheathlike arrangement of mitochondria around the proximal end of the midpiece. (Redrawn from Fawcett, 1977; Setchell, 1982)

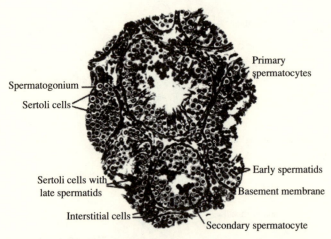

Primary
spermatocytes

Spermatogonium

Sertoli cells

Early spermatids

Sertoli cells with
late spermatids

Basement membrane

Interstitial cells

Secondary spermatocyte

Figure 5.7. Cross-section of a human testis (× 170) showing the general morphology and the arrangement of the germ cells at various stages of spermatogenesis. The large, rounded structures are the seminiferous tubules, the site of sperm production. Lying between the tubules is the interstitial tissue, whose cells (the Leydig cells) secrete the principal male sex steroid, testosterone. (Redrawn from Bloom and Fawcett, 1968; Monesi, 1972)

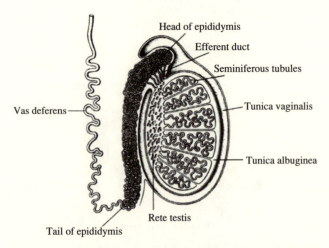

Head of epididymis

Efferent duct

Seminiferous tubules

Vas deferens

Tunica vaginalis

Tunica albuginea

Rete testis

Tail of epididymis

Figure 5.8. Sagittal section of a human testis and epididymis, showing efferent ducts leading from the rete testis to the head of the epididymis, and the tail of the epididymis merging with the vas deferens. (Redrawn from Dym, 1977; Harper, 1982)

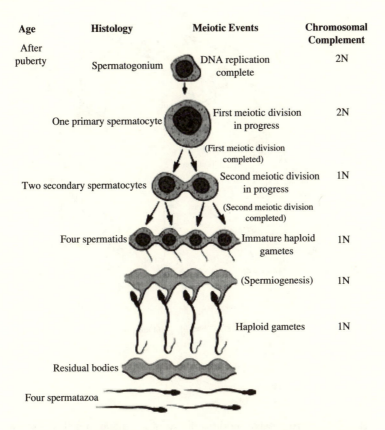

Age	Histology	Meiotic Events	Chromosomal Complement

After puberty — Spermatogonium — DNA replication complete — 2N

One primary spermatocyte — First meiotic division in progress — 2N

(First meiotic division completed)

Two secondary spermatocytes — Second meiotic division in progress — 1N

(Second meiotic division completed)

Four spermatids — Immature haploid gametes — 1N

(Spermiogenesis) — 1N

Haploid gametes — 1N

Residual bodies

Four spermatazoa

Figure 5.9. The major events in spermatogenesis following commitment of the spermatogonial stem cell to meiosis. (Redrawn from Carlson, 1988)

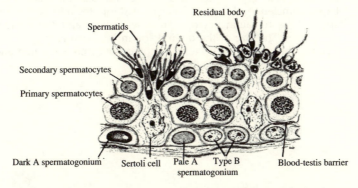

Residual body

Spermatids

Secondary spermatocytes

Primary spermatocytes

Dark A spermatogonium   Sertoli cell   Pale A   Type B   Blood-testis barrier
spermatogonium

Figure 5.10. Partial cross-section of a seminiferous tubule showing the relationship between Sertoli cells and developing sperm cells. (Redrawn from Dym, 1977)

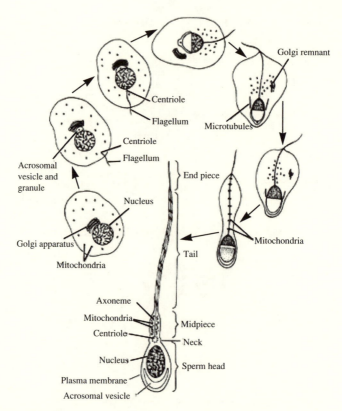

Figure 5.11. Progressive modification of a male sperm cell to form a spermatozoan. The centriole produces a long flagellum at what will be the posterior end of the sperm, and the Golgi apparatus forms the acrosomal vesicle at the future anterior end. The mitochondria *(dots)* collect around the flagellum near the base of the haploid nucleus and become incorporated into the midpiece of the sperm. The remaining cytoplasm is jettisoned, and the nucleus condenses. The size of the mature sperm has been enlarged relative to other portions of the figure. (Redrawn from Clermont and Leblond, 1955; Gilbert, 1988)

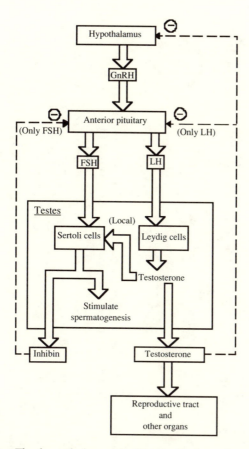

Figure 5.12. The hypothalamo-pituitary-testicular axis and the neuroendocrine control of testicular function. Gonadotropin-releasing
hormone (GnRH) reaches the anterior pituitary via the portal system. Follicle-stimulating hormone (FSH) stimulates the Sertoli
cells; in turn, secretion of FSH is inhibited by inhibin, a peptide
hormone secreted by the Sertoli cells. Similarly, luteinizing hormone (LH) acts upon the Leydig cells, and its secretion is inhibited by testosterone produced by those cells. Testosterone acts locally to stimulate spermatogenesis; it also has wider effects on the
male body, such as inducing the secondary sexual characteristics.
(Redrawn from Vander et al., 1990)

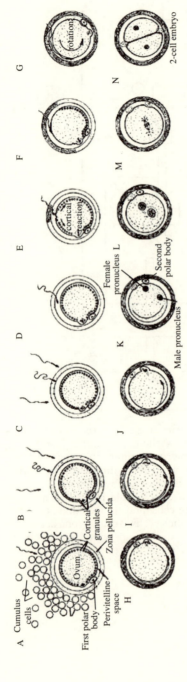

Figure 5.13. Stages of fertilization. (*A*) A sperm passes through the cumulus oophorus. (Note that cumulus cells are omitted in illustrations of later stages.) The egg has shed its first polar body and has formed its second meiotic spindle. (*B*) species-specific binding of sperm to ZP3 molecules on the surface of the zona pellucida. Upon binding, sperm undergoes activation and the acrosomal reaction. (*C*) passage of the sperm through the zona pellucida. (*D*) sperm enters the perivitelline space separating the zona and the egg. (*E*) Binding of the sperm to the cell membrane of the egg, initiation of cortical reaction. As cortical granules release their contents into perivitelline space, adjacent areas of the zona begin to harden (*dark stippling*). (*F*) sperm is fully bound at its equatorial region to the egg membrane; cortical and zona reactions continue. (*G*) whipping motion of the sperm tail causes the egg to rotate relative to the zona as cortical and zona reactions reach their end. (*H–I*) cell membranes of sperm and egg fuse, sperm head is incorporated into the egg cytoplasm. (*J*) sperm head decondenses, egg begins to extrude second polar body. (*K–L*) female and male pronuclei swell; centriole originally at base of sperm tail forms an aster that will bring the two pronuclei together and provide a mitotic spindle. (*M–N*) zygote undergoes first cleavage division to form a two-cell embryo. (Redrawn from Austin and Bishop, 1957; Yanagimachi, 1981; Carlson, 1988)

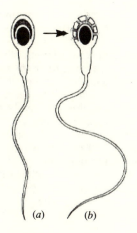

Figure 5.14. Schematized spermatozoan (*a*) before and (*b*) after activation and the acrosome reaction. At activation, the amount of tail movement increases. During the acrosome reaction, multiple fusions occur between the plasma membrane of the sperm head and the outer acrosomal membrane, exposing the acrosomal contents. (Redrawn from Johnson and Everitt, 1988)

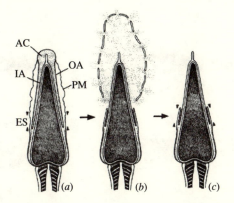

Figure 5.15. Sagittal sections of the sperm head showing stages of the acrosome reaction. (*a*) Intact acrosome (AC) in a nonactivated sperm, showing the typical irregular contour of the plasma membrane over the acrosome. (*b*) As a first reaction response, multiple point fusions develop between the plasma membrane and the outer acrosomal membrane. The unbound fraction of the acrosome contents then swells and escapes through the pores created by the fusion points. (c) The acrosome covering is lost, exposing enzymes bound to the inner acrosomal membrane and leaving behind the equatorial segment. IA, inner acrosomal membrane; OA, outer acrosomal membrane; PM, plasma membrane; ES, equatorial segment (boundaries defined by *arrows*). (Redrawn from Bedford and Cooper, 1978; Bedford, 1982)

Figure 5.16. (*a*) Transmission electron micrograph of the zona
pellucide (× 22,500) showing
the characteristic slit made by a
spermatozoan penetrating it.
Several sections of the sperm
tail (T) can be seen. PVS, perivitelline space between the
zona and the egg. (*b*) Oblique
sections of the sperm head (×
30,000) cleaving its way
through the zona substance
and thus creating a penetration
slit behind it. (Sperm tail is out
of the plane of the section.)
(From Bedford, 1970, 1982)

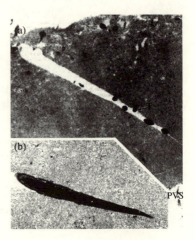

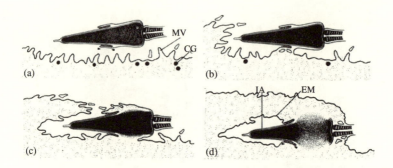

Figure 5.17. Sequence of events involved in the fusion and incorpora-
tion of the fertilizing sperm by the egg. After penetrating the zona
pellucida, the sperm in the perivitelline space (*a*) uses the surface
overlying the equatorial segment (delineated by *arrows*) to establish fusion (*b*) with the egg's plasma membrane. Fusion stimulates
release of the cortical granules (CG), decondensation of the sperm
nucleus begins, and soon thereafter (*c*) the egg cytoplasm engulfs
the front of the sperm head bounded by the inner acrosomal membrane. When this process is completed (*d*), that region of the sperm
nucleus is encased by a vesicle composed of internalized egg
membrane (EM), persisting equatorial segment, and inner acrosomal membrane (IA). MV, microvillus. (Redrawn from Bedford and
Cooper, 1978)

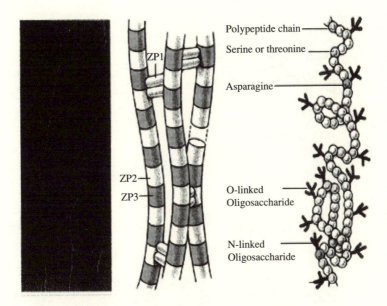

Figure 5.18.   (*Left*) scanning electron micrograph of the glycoprotein
filaments that form the zona pellucida; ZP1, ZP2, and ZP3 are all
glycoproteins (polypeptide chains to which sugar side-chains are
attached). (*Center*) ZP2 and ZP3 combine to form the basic build-
ing block of the filaments, which are linked by ZP1. (*Right*) structure of ZP3, the receptor molecule that binds sperm and induces
the acrosome reaction. The actual binding sites are a subset of the
sugar chains radiating from ZP3's polypeptide backbone, specifically, the O-linked oligosaccharides attached to the amino acids
serine and threonine. (From Wassarman, 1988)

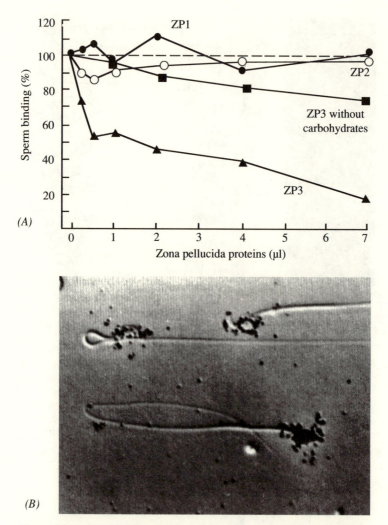

Figure 5.19. Competitive binding of sperm to intact zona pellucidae and free zona glycoproteins. (*A*) inhibition assay showing
the specific decrease of sperm binding to intact zonae when
sperm are incubated with increasingly large amounts of ZP3.
The importance of the oligosaccharide side-chains of ZP3 is
indicated by the lack of inhibition when those carbohydrates
are removed. (*B*) binding of radioactively labeled ZP3 to
capacitated mouse sperm. (After Bleil and Wassarman, 1980,
1986; Florman and Wassarman, 1985)

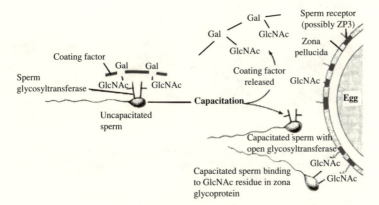

Figure 5.20. A model of capacitation and the recognition of sperm and egg. The sperm cell surface contains a protein, glycosyltransferase, capable of recognizing terminal N-acetylglucosamine residues. In uncapacitated sperm, the binding sites on those sperm proteins are blocked by a coating factor containing such N-acetylglucosamine moieties. Upon capacitation, the coating factors are stripped away by enzymes in the uterine secretions, freeing the active sites on the sperm surface glycosyltransferases. These sperm proteins can now recognize the exposed N-acetylglucosamine residues of the ZP3 glycoprotein on the zona pellucida. (Redrawn from Lopez et al., 1985; Gilbert, 1988)

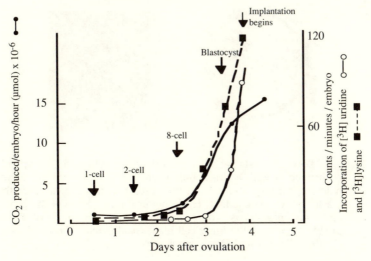

Figure 5.21. CO_2 output and RNA and protein synthesis in the preimplantation mouse embryo. (Redrawn from McLaren, 1982)

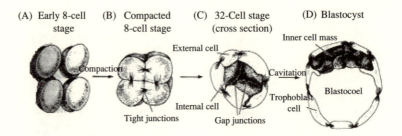

(A) Early 8-cell   (B) Compacted   (C) 32-Cell stage   (D) Blastocyst
    stage   8-cell stage   (cross section)

Figure 5.22. Schematic diagram of compaction and the formation of the blastocyst. (*A,B*) 8-cell embryo; (*C*) morula; (*D*) blastocyst. (Redrawn from Gilbert, 1988)

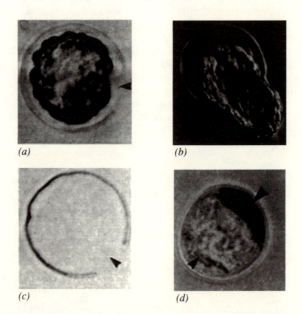

(a)   (b)

(c)   (d)

Figure 5.23. Blastocysts hatching from the zona pellucida. (*a*) Expanded blastocyst with a single hole (*arrow*) in its zona. (*b*) Blastocyst hatching through the hole in the zona. (*c*) Empty zona after the blastocyst has hatched. (*d*) Histochemical localization of the proteolytic enzyme strypsin (*large arrow*) on the membrane of a trophoblast opposite the inner cell mass (*small arrow*). (From Mark et al., 1985; Perona and Wassarman, 1986)

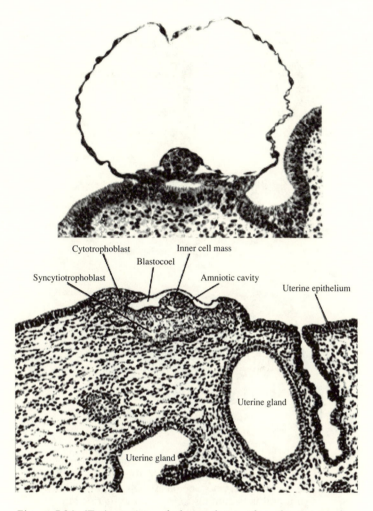

Figure 5.24. (*Top*) section of the embryo of a rhesus monkey adhering to the uterine wall nine days after conception (× 400). The layer of vertically aligned cells on which the (hollow) embryo rests is the epithelium of the endometrium; below the endometrium is the decidual tissue. In the two areas where the embryo and the uterus connect, the trophoblast has started to invade the endometium. (*Bottom*) sections of a 7½-day human embryo implanting in the uterine endometium. The thin membranous portion of the blastocyst wall has collapsed. (From Hamilton et al., 1945; Beer and Billingham, 1974; Renfree, 1982)

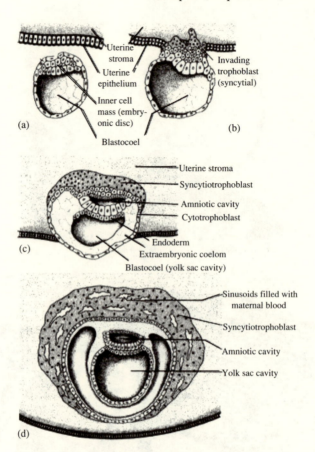

Figure 5.25. The process of implantation. (*a*) The blastocyst has not yet attached to the uterine epithelium. (*b*) The trophoblast has penetrated the epithelium and is beginning to invade the uterine stroma. Note that the trophoblast has started to differentiate into a cytotrophoblast with clear cell membranes and an invasive syncytiotrophoblast, which is multinucleated. (*c*) The blastocyst has sunk further into the stroma, and anamniotic cavity has appeared. The syncytiotrophoblast continues to grow and invade the stroma. (*d*) The endometrial epithelium has grown over the implantation site, so the blastocyst is entirely enclosed in maternal tissue. Irregular spaces (sinusoids) filled with maternal blood have appeared in the syncytiotrophoblast. (Redrawn from Tuchmann-Duplessis et al., 1971; McLaren, 1972)

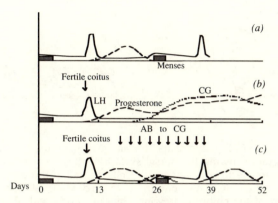

Figure 5.26. Circulating hormone levels during ovarian cycles in a higher primate. (*a*) Nonpregnant cycles; (*b*) a cycle in which conception occurs; (*c*) similar to (*b*) but with injections of highly specific anti-CG antibodies (*arrows*). CG is suppressed, pregnancy loss occurs, and normal cyclicity is restored. CG, chorionic gonadotropin; LH, luteinizing hormone; AB, antibodies. (Redrawn from Johnson and Everitt, 1988)

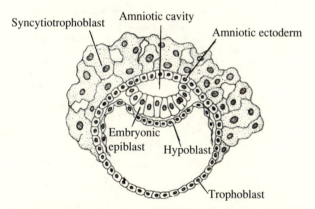

Figure 5.27. Embryonic tissues during implantation. The hypoblast and epiblast have differentiated, and the hypoblast has started to grow around the inside of the trophoblast. (When it completes this process, the hypoblast will form the yolk sac.) The epiblast has split into the amniotic ectoderm, which surrounds the amniotic cavity and will become the amnion, and the embryonic epiblast, which will produce the embryo proper. (Redrawn from Carlson, 1988; Gilbert, 1988)

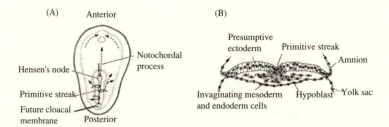

Figure 5.28. Cell movements during gastrulation. (*A*) schematic diagram showing the dorsal surface of the embryonic epiblast (amniotic ectoderm removed). Cells migrating through a structure called Hensen's node travel toward what will become the head of the embryo, drawing Hensen's node with them. This lengthening structure, known as the primitive streak, will become the notochord. The remaining epiblast cells traveling through the streak migrate laterally to become the mesoderm and endoderm precursors. *Dotted lines* indicate internal migrations (i.e., below the epiblast). (*B*) transverse section of the gastrulating embryo. (Redrawn from Langman, 1981)

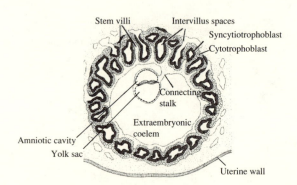

Figure 5.29. Extraembryonic tissue 13 days after conception. The amniotic cavity (between the amniotic ectoderm and the embryonic epiblast) and yolk sac (derived from the hypoblast) have formed. The cytotrophoblast has sent out a series of villus processes into the syncytiotrophoblast; these villi, which will form the placenta, are coming into contact with the blood vessels of the uterine endometrium. The embryo is connected to the trophoblast by a stalk of extraembryonic mesoderm that will shortly carry the fetal blood vessels to the placenta. (Redrawn from Langman, 1981)

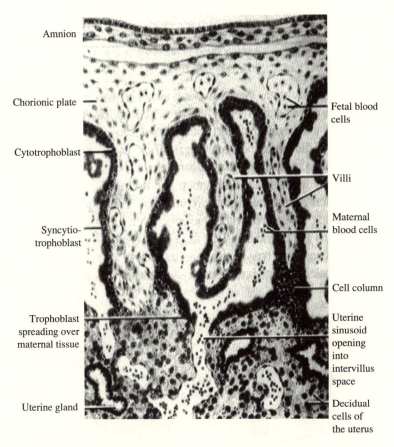

Amnion

Chorionic plate

Cytotrophoblast

Syncytio-
trophoblast

Trophoblast
spreading over
maternal tissue

Uterine gland

Fetal blood
cells

Villi

Maternal
blood cells

Cell column

Uterine
sinusoid
opening
into
intervillus
space

Decidual
cells of
the uterus

Figure 5.30. Relationship of chorionic villi to pools of maternal blood.
(After Hill, 1932)

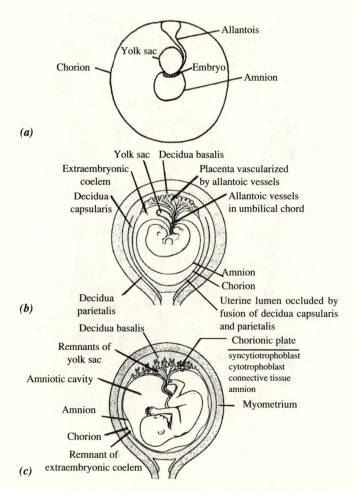

Figure 5.31. Extraembryonic tissues and their relationships
to the embryo and the uterus at various stages of pregnancy: (*a*) about the time of neurulation, (*b*) about 40
days, and (*c*) by the fourth month. Note the gradual loss
of the yolk sac, which has only a minor nutritive function in mammals. Also note that the amnion and chorion fuse in later pregnancy, and that both tend to fuse
with the decidua basalis. The allantois fuses with the
portion of the chorion that lies next to the decidua
basalis, and the amnion and chorion together form the
placenta. The figure in (*a*) is highly schematized and
does not show any maternal tissue. (*b* and *c* redrawn
from Renfree, 1982)

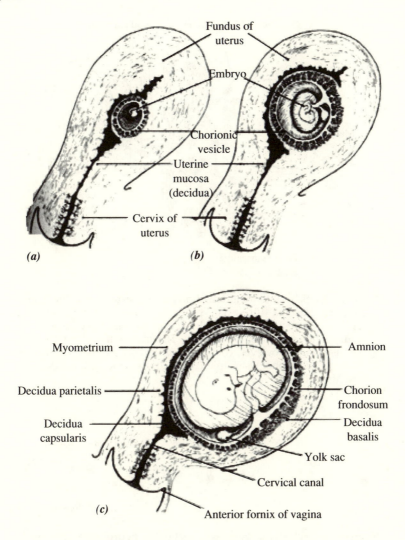

Figure 5.32. The uterus at various stages of pregnancy: *(a)* three weeks post-fertilization, *(b)* five weeks, and *(c)* eight weeks. (Redrawn from Carlson, 1988)

Maternal blood sinus   Umbilical chord

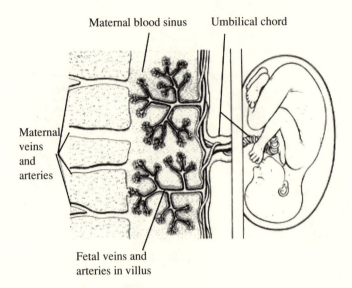

Maternal
veins
and
arteries

Fetal veins and
arteries in villus

Figure 5.33.   Schematic diagram showing the structure of the haemo-
chorial placenta during late pregnancy. The chorionic villi lie in
pools (sinuses) of maternal blood, and the villi have become vascularized by fetal blood vessels. The maternal and fetal circulations do not, however, come into direct contact; they remain separated by trophoblastic tissue. (After Beer and Billingham, 1974;
Renfree, 1982)

Neural plate   Neural fold   Neural crest   Epidermis

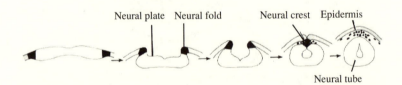

Neural tube

Figure 5.34.   Schematic transverse section of neurulation. Ectodermal
cells are represented as precursors of either the neural crest (*black*),
the neural tube (*white*), or the epidermis (*shaded*). The ectoderm
invaginates at its most dorsal point, forming an outer epidermis
and an inner neural tube connected by neural crest cells. (Redrawn
from Balinsky, 1975; Gilbert, 1988)

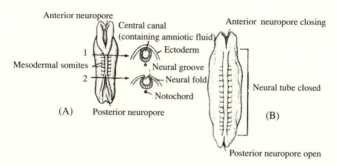

Figure 5.35. Neurulation in a human embryo. (*A*) dorsal view and transverse sections of the early stages of neurulation. (*Arrows 1* and 2 indicate the locations of the transverse sections.) (*B*) dorsal view of a late neurula embryo. The anterior neuropore (future location of the brain) is closing while the posterior neuropore remains open. (Redrawn from Gilbert, 1991)

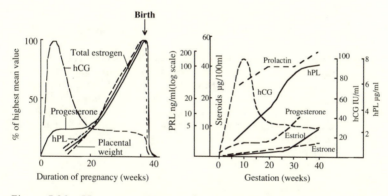

Figure 5.36. Hormone patterns of pregnancy. (*Left*) changes in the concentrations of several hormones in the maternal bloodstream and their relationship to the size of the placenta; all values expressed as a percentage of maximum value (which, with the exception of hCG, occurs in late pregnancy). (*Right*) absolute concentrations of several hormones during pregnancy. In the right panel, total estrogen is broken down into its individual components. The typical maximum weight of the human placenta in late pregnancy is 0.7 kg. hPL, human placental lactogen; hCG, human chorionic gonadotropin; PRL, prolactin. (After Heap and Flint, 1984; Norman and Litwack, 1987; Johnson and Everitt, 1988)

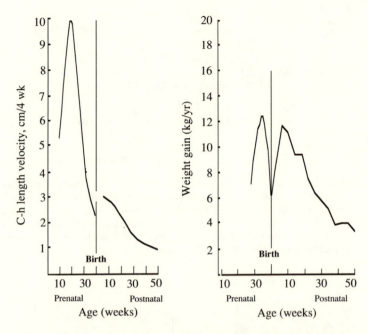

Figure 5.37. Fetal and postnatal growth velocity curves for crown-heel length (*left*) and weight (*right*) in males, mixed longitudinal data. Growth velocity is defined as the absolute increment in length or weight per unit of time. Prenatal curves are based on live-born male children delivered spontaneously before 40 weeks of gestation, a sample that may not be representative of all pregnancies. (Redrawn from Tanner, 1989)

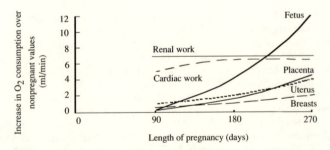

Figure 5.38. Increases over pregnancy in oxygen consumption by various maternal and fetal tissues. (Redrawn from Hytten and Chamberlain, 1980)

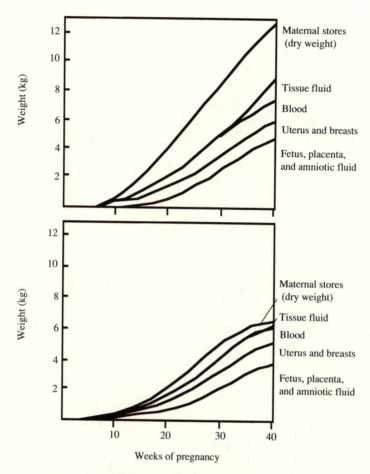

Figure 5.39. Estimated components of weight gain during pregnancy in a normal, healthy western woman (*top*) and an undernourished woman from rural India (*bottom*). Note that dietary restriction in the latter case has a small effect on the fetal component, but a much larger effect on maternal stores. (Redrawn from Hurley, 1980; Worthington-Roberts et al., 1985)

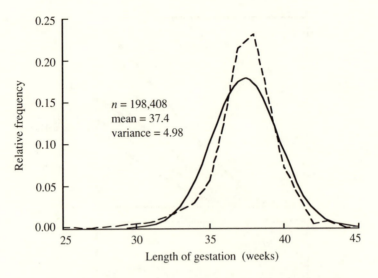

$n = 198,408$
mean $= 37.4$
variance $= 4.98$

Figure 5.40. Distribution of gestation lengths for all live-born white infants, California, 1966. *Broken line*, observed distribution; *smooth curve*, normal distribution fit to the observations by maximum likelihood. (Data from Hammes and Treloar, 1970)

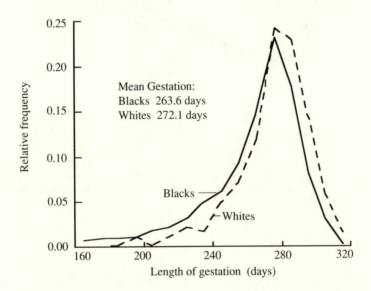

Mean Gestation:
Blacks   263.6 days
Whites   272.1 days

Blacks

Whites

Figure 5.41. Distribution of gestation lengths for all live births among blacks (*solid line*) and whites (*broken line*), United States, 1965. (Data from Henderson and Kay, 1967)

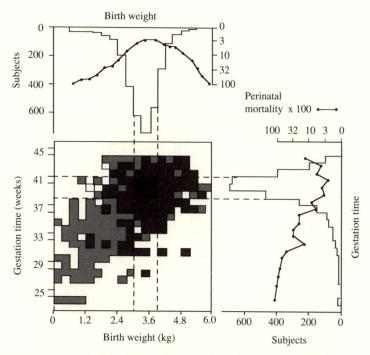

Figure 5.42. Perinatal mortality as a function of birth weight and length of gestation, 3,358 live-born white male infants, Italy, 1959–1960. The total perinatal mortality rate was 91 per 1,000 live births. (Redrawn from Federici and Terrenato, 1982)

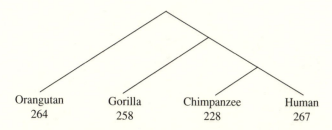

Figure 5.43. Cladogram of evolutionary relationships among the hominoids, with estimated mean live-birth gestation lengths. The gorilla-chimpanzee-human trichotomy, which may be too close to resolve into dichotomies as shown here, is based on several molecular studies. (Data from Harvey and Clutton-Brock, 1985; Napier and Napier, 1985)

6

Fetal Loss

Although it may seem out of place to discuss mortality in a book on fertility, there is one form of death that has a definite and possibly substantial effect on the pace of childbearing: death of the embryo or fetus in utero. All things being equal, the more intrauterine deaths experienced by a woman, the lower her achieved fertility. Depending upon its frequency and when during pregnancy it occurs, intrauterine mortality may act as a powerful limiting factor in reproduction.

Unfortunately, as we shall see, the study of intrauterine mortality in humans is fraught with unavoidable difficulties of measurement and analysis. They are compounded by needless terminological complexities, which we would do well to clear up at the outset. The terms *intrauterine mortality* and *fetal loss* will be used synonymously in this chapter, and solely with reference to naturally occurring death in utero, that is, to *spontaneous* rather than *induced* abortion. (Use of the term *abortion* by itself is ambiguous and should be avoided.) Intrauterine mortality includes any natural death prior to birth, no matter what the stage of gestation, and thus encompasses both *embryonic* and *fetal* death.[1] In clinical practice, it is conventional to use the term *embryo* to refer to the product of conception from about a week after conception up to the end of the eighth week of pregnancy, and the term *fetus* thereafter. In fact, this distinction is arbitrary since it does not correspond to any definite developmental transition, and the term *conceptus* will be used in reference to both phases of development. There is a growing tendency among demographers and epidemiologists to use the terms *fetal death* or *fetal loss* to refer to all forms of intrauterine mortality, a practice that has

239

recently received international approval (Table 6.1). Except where it is likely to cause confusion, this usage will be followed here.

Two other terms sometimes used in connection with fetal loss are *miscarriage* and *stillbirth*. These expressions are meant to apply respectively to deaths that occur before and after the conceptus would have been viable in the outside world, usually taken to occur near the end of the second trimester of pregnancy. As the ability to save premature babies through medical intervention improves, however, this distinction becomes increasingly blurred. For registration purposes, some states in the United States have recently changed the cutoff point for distinguishing the two types of death from 28 to 20 weeks of gestation, while others have not. To make things even more muddled, WHO (1977) has recommended distinguishing miscarriages and stillbirths on the basis of birth weight rather than duration of pregnancy. To avoid this terminological confusion, these terms will not be used here.

Because the risk of fetal death varies considerably by stage of pregnancy, there is need for an expression for the time that has elapsed since conception. Since it is impossible to pinpoint the time of conception with accuracy, however, most studies instead count *gestational age* from the day of onset of last menses, which usually occurs about two weeks before conception (Table 6.1). Although widely adopted, this convention is unfortunate because it can lead to systematic bias in the estimation of

Table 6.1. World Health Organization Definitions of Live Birth, Fetal Death, and Gestational Age

---

Live birth—the complete expulsion or extraction from its mother of a product of conception, irrespective of the duration of pregnancy, which after such separation, breathes or shows any evidence of life, such as beating of heart, pulsation of the umbilical cord, or definite movement of voluntary muscles, whether or not the umbilical cord has been cut, or the placenta is attached; each product of such a birth is considered a live birth.

Fetal death (or loss)—death prior to the complete expulsion or extraction from its mother of a product of conception, irrespective of the duration of pregnancy; the death is indicated by the fact that after such separation the fetus does not breathe or show any other evidence of life, such as beating of the heart or pulsation of the umbilical cord, or definite movement of voluntary muscles.

Gestational age—duration of gestation as measured from the first day of the last normal menstrual period. Gestational age is expressed in completed days or weeks. For example, events occurring 280 to 286 days after the onset of the last menstrual period are considered to have occurred at 40 weeks of gestation.

From WHO (1978)

early fetal loss. In this chapter, I will use the term *corrected gestational age* to refer to conventional gestational age minus two weeks—an imperfect correction, to be sure, but certainly more accurate than dating pregnancy from the previous menses.

Because pregnancies are virtually impossible to detect before implantation, some demographers prefer to ignore preimplantation deaths altogether. They accordingly distinguish *conception* or *fertilization* (production of a fertilized egg) from *impregnation* (establishment of a pregnancy by implantation) and regard pregnancy as beginning only at impregnation. From the standpoint of measurement, there is considerable justification for this view—if a phenomenon is unmeasurable, why attempt to investigate it? However, there is no reason to believe that preimplantation pregnancies are *inherently* undetectable. For example, if reliable assays could be developed for early pregnancy factor (EPF) or any other substance uniquely produced by the early zygote, they would greatly enhance our ability to monitor preimplantation losses (see Chapter 5). Moreover, indirect evidence described below suggests that the number of preimplantation deaths is by no means negligible. Despite the difficulties in detecting them, those deaths may still add time to the birth interval. In view of the real demographic and biological importance of such deaths, and given the likelihood that we may be able to detect at least some of them in the future, there is every reason to begin formulating theoretical models that incorporate preimplantation mortality, even if we are not yet in a position to estimate its frequency with precision. Accordingly, pregnancy will be assumed to begin at conception, not impregnation.

Almost all studies equate the gestational age of the conceptus at death with the duration of gestation at the time the conceptus is expelled from the uterus, ignoring the fact that it may have died well before expulsion. This confusion is unavoidable, except in those rare instances when the precise time of death is known from monitoring fetal heartbeat or when the developmental stage of the conceptus at death can be inferred on morphological grounds. (And even using morphological criteria, the correlation between developmental and gestational ages at death is imperfect.) Unless otherwise stated, "gestational age at death," both corrected and conventional, will actually refer to the duration of gestation at expulsion of the conceptus rather than at death per se.

Finally, it is sometimes necessary to distinguish conceptions by their ultimate outcomes. Following standard demographic usage, *fertile conception* will refer to any conception initiating a pregnancy that ultimately produces a live birth. Although there is no well-established term for conceptions that end in fetal loss, it seems logical (if strange) to call them *infertile conceptions*. An alternative convention was suggested by

Sheps and Menken (1973): fertile conceptions are "*l*-conceptions" because they end in *l*ive births, and infertile conceptions are "*a*-conceptions" because they end in spontaneous *a*bortions. This convention will occasionally be useful in what follows.

SOURCES OF DATA ON FETAL LOSS

Two major methodological problems plague the study of fetal loss: underreporting and selectivity bias. The first arises primarily because of the near impossibility of detecting early pregnancies, the second because of difficulties in choosing a sample of women representative of some larger population with respect to a number of risk factors predisposing to fetal loss. Broadly speaking, studies of fetal loss are of two types, clinical and community-level. Clinical studies are almost always prospective (in which case they are called *clinical follow-up studies*), whereas communitywide studies can be either prospective or retrospective.[2] The design of a particular study largely determines the degree to which it suffers from underreporting and selectivity.

Recruitment of subjects into clinical studies involves women who suspect they may be pregnant and who go to a physician for confirmation. Since a subject may not enter the study until several weeks after conception, this method of recruitment necessarily entails a substantial amount of underreporting of early fetal deaths—often called *subclinical* fetal deaths because they occur before a clinical diagnosis of pregnancy. In addition, the women who go to clinics may be a highly self-selected sample of all women, especially in the rural developing world where the majority of pregnant women may never see a physician. In particular, women who go to clinics early in pregnancy may be precisely those who are most concerned about the outcomes of their pregnancies because of some prior experience with fetal loss or because they are already experiencing some difficulty with their current pregnancies. As with all clinical studies, there is cause for concern about whether the clinical population is at all representative of the larger population that the clinic supposedly serves.

In principle, by using appropriate sampling frames, communitywide studies can largely avoid the problem of selectivity bias. To date, however, such studies have still suffered enormous underreporting of fetal deaths. This problem is especially serious in the case of retrospective studies, which enquire about the number of previous fetal losses in the course of taking a maternity history from each subject. Obviously, no woman can report truthfully on any pregnancy that ended before she became aware of it, and in some parts of the world a woman may not know she is pregnant until many weeks or even months after conception.[3] In addition, recall error may be substantial, affecting even those pregnancies about which she was once aware, since fetal losses, especially early ones, are almost certain to be less memorable than live births.

It should be possible to design a prospective, communitywide study that minimizes both selectivity bias and underreporting. Ideally, such a study would not use pregnancy itself as a criterion for subject recruitment but would recruit from all groups of women. At the beginning of the study, a serum or urine sample would be taken from each subject and assayed for an early indicator of pregnancy such as hCG; each detected pregnancy would then be monitored until it was terminated either by fetal death or live birth. If a larger battery of hormones was assayed, including progesterone, estriol, and hPL, an assessment of gestational age at the time of recruitment could also be made (see Figure 5.36), providing the basis for a hazards analysis of fetal death.

Such a study has never been attempted. In practical terms, the requisite hormone assays are still too expensive and time-consuming for use in large-scale surveys, and reliable standards for estimating gestational age from hormone concentrations in maternal blood or urine have yet to be developed. Those prospective studies of fetal loss that have been conducted on a populationwide basis have involved subjects' self-report of pregnancy—sometimes with medical confirmation, sometimes not—and thus are subject to underreporting. And even the ideal study based on hormone assays would still involve some underreporting of subclinical loss, since we cannot yet detect a pregnancy during its first 7–10 days (see Chapter 5).

In sum, the one unavoidable problem facing all studies of fetal loss is the inability to detect early pregnancies. All data on fetal loss are thus left-truncated and length-biased to some degree—that is, very short gestations ending in loss are not observable—and special techniques must be adopted to correct for this fact. Even when such techniques are used, however, there is no direct way to estimate the probability of fetal loss before the truncation point. In this light, all the empirical studies reviewed below are incomplete.

To end this section on a less pessimistic note, I will argue in what follows that despite these measurement problems, we can still make considerable progress in the study of fetal loss by constructing theoretical models of the processes involved in intrauterine death. Indeed, this is precisely the sort of situation in which such models are most likely to further our understanding. To paraphrase R. A. Fisher (1956), the fact that something is difficult to measure is no reason not to try to think clearly about it.

THE FERTILITY-REDUCING EFFECT OF FETAL LOSS

Fetal loss reduces fertility by lengthening birth intervals (Figure 6.1). In the absence of fetal death, a given completed birth interval would be made up of the following segments: a postpartum infecundable period ending with the first postpartum ovulation, a fecund waiting time to the next conception, and because that conception is fertile, a complete gestation culminating in the next live birth. Each fetal death within the interval adds three more segments of time: an incomplete gestation terminated by death of the conceptus, some residual period of "postpartum" infecundability before the next ovulation, and a fecund waiting time to the next conception. And, of course, the next conception may not be fertile, in which case precisely the same segments will be added to the birth interval again. Clearly, the total fertility-reducing effect of fetal loss depends upon the length of these segments, as well as the number of deaths occurring in each interval.

As has already been noted, the expected fecund waiting time to next conception is determined by fecundability, or the monthly probability of conception. There is no conclusive evidence that fecundability is affected by the outcome of the previous conception (i.e., whether it was fertile or not), or that the length of the fecund wait has any effect on the outcome of the *next* conception (Wilcox et al., 1988). It seems permissible, therefore, to treat fecundability and fetal loss as separate issues, and to postpone discussion of the former to the next chapter.

This leaves three things we need to know in order to assess the impact of fetal loss on fertility: (1) the number of fetal deaths expected to occur in each birth interval, (2) the average length of gestation associated with each fetal death, and (3) the expected duration of residual infecundability following each fetal death. Despite the fact that we have no very exact empirical knowledge about any of these things, it is still possible to make a preliminary assessment of their impact.

THE EXPECTED NUMBER OF FETAL DEATHS
IN EACH BIRTH INTERVAL

To take the simplest case, think of successive conceptions experienced by a particular woman during a single birth interval as a series of "trials," each with some fixed probability q of ending in fetal death and $(1 - q)$ of ending in live birth. If q is constant, and if the trials are independent—that is, if the outcome of any one trial does not influence the outcome of any other—then the number of trials to the first fertile conception is a geometric random variable:

Prob (N losses before next successful pregnancy) = $q^N(1 - q)$ (6.1)

In this case, the expected number of fetal deaths in each birth interval is simply $q/(1 - q)$, with variance $q/(1 - q)^2$. In other words, everything we need to know about the number of fetal deaths in each birth interval is determined by the value of q, the total probability of fetal loss per conception.

In fact, this simple model is probably not a bad first approximation to reality. It contains two principal assumptions: (1) that the probability of fetal loss is constant across conceptions, and (2) that successive conceptions are independent. As long as we are talking about a single birth interval, so that changing maternal age has little effect, the first of these assumptions is probably approximately correct. There are, however, data that seem to contradict the second assumption, and we will discuss that assumption in more detail below. For now, we will take it as given that this model is an adequate description of reality.

What, then, do we know about the value of q? The answer is, first, that we know very little, and second, that what we *think* we know has changed dramatically over the past few decades. Several large-scale, communitywide surveys conducted during the 1930s and 1940s suggested that the probability of loss per conception in mid-reproductive age women is between 0.10 and 0.15 (United Nations, 1954). Those surveys, most of which were retrospective, inevitably involved a substantial amount of underreporting, a fact acknowledged by the original authors. Nonetheless, the estimates from the surveys gained currency, especially in the clinical literature but also in demographic and epidemiological circles.

The earliest indication that these figures were probably gross underestimates emerged from a remarkable study by A. T. Hertig and his colleagues of "34 fertilized human ova, good, bad and indifferent" (Hertig et al., 1959). This study involved 210 married women of proven fertility (i.e., who had produced at least one live-born child) who were about to undergo therapeutic hysterectomy. Hertig asked these women to time sexual intercourse during the cycle immediately preceding the operation in an attempt to maximize their chances of fertilization. Of these 210 women, a subset of 107 who were under the age of 45, whose last cycle had been ovulatory according to the BBT method, who reported at least one act of unprotected intercourse during the periovulatory period, and whose reproductive tracts showed no patent signs of inflammatory disease were selected for the final study. After the hysterectomy, each subject's oviducts and uterus were flushed and an attempt was made to recover her last egg. Thirty-four fertilized eggs were obtained, ranging from a two-cell embryo still in the oviduct to a 17-day embryo that had already implanted. Of these 34 early conceptuses, ten were clearly abnormal and were either already dead or were judged unlikely to survive to term. This study thus implied a probability of *early* fetal loss, 0.294 ± 0.078, that was substantially higher than the then most widely quoted estimates of *total* fetal loss.

The results of this study, which in all likelihood will never be replicated, have been used by several demographers to support the claim that approximately 30 percent of all conceptions end in *preimplantation* death. This conclusion (which Hertig himself did not draw) is unwarranted for several reasons. First, Hertig's sample of fertilized eggs was small and the standard error of the estimated probability of death is large. Since the 95 percent confidence interval around this estimate ranges from about 0.14 to 0.45, it is not even possible to conclude that Hertig's estimate is *necessarily* inconsistent with the earlier estimates.

Second, the estimated probability of loss of 0.294 does not pertain solely to preimplantation mortality. For one thing, some of the recovered conceptuses had already implanted. For another, only one conceptus was definitely known to be dead at the time of recovery, since it showed clear signs of necrotic degradation. Hertig reckoned that at least some of the other abnormal conceptuses might have survived as late as the second or even third trimester of pregnancy.

Finally, it is impossible to decide how representative Hertig's subjects were of any larger population of women. They were comparatively old, ranging from 26 to 43 with an average age of 35.0 years, and therefore were probably somewhat more prone to fetal loss than younger women (see below).[4] Moreover, despite Hertig's claims to the contrary, it is difficult to believe that the pathologies warranting hysterectomy had no influence whatsoever on these women's prospects of a successful pregnancy. Hertig's study, as extraordinary as it was, suffered the same possibility of selectivity bias that haunts all clinical studies.

Nonetheless, Hertig's study is of enormous historical importance, and more recent research has confirmed his general conclusion that spontaneous fetal deaths are more common than originally believed. Estimates of the total probability of fetal loss from several studies conducted since Hertig's work are summarized in Table 6.2. In comparing these estimates it is important to bear in mind that they depend not only on study design, which affects both the reliability of the data and the size of the

Table 6.2.   Estimates of the Total Probability of Fetal Loss per Conception

Type of study	Probability of loss (s.e.)	Number of pregnancies	Source
Survey (retrospective)	0.12 (0.01)	5,673	Leridon, 1976
Clinical (follow-up)	0.24 (0.01)	3,083	French and Bierman, 1962
hCG (follow-up)	0.31 (0.03)	197	Wilcox et al., 1988
hCG (follow-up)	0.43 (0.04)	152	Miller et al., 1980
hCG (follow-up)	0.62 (0.05)	118	Edmonds et al., 1982
EPF (follow-up)	0.89 (0.07)	18	Rolfe, 1982

sample, but also on the method by which pregnancies are ascertained. Some recent studies using hormonal indicators of early pregnancy, such as hCG or EPF, have yielded extremely high estimates of q, from more than 0.6 to almost 0.9 (Edmonds et al., 1982; Rolfe, 1982). While these estimates cannot be dismissed out of hand, there is reason to be skeptical about them. Given the expense of hormone assays, these estimates are almost inevitably based on small, highly select samples. In addition, it appears that existing assays of EPF may entail large numbers of false positive diagnoses of pregnancy, which by their nature inflate the estimate of q (Grudzinskas and Nysenbaum, 1985). Although hCG has become the hormone of choice in studies of early pregnancy loss, it too may involve numerous false positives unless an assay system is used that minimizes cross-reactivity with luteinizing hormone (Figure 6.2). Both hCG and LH consist of two polypeptide subunits, an α-chain and a β-chain. The α-chains of hCG and LH are identical, as are the β-chains except for an extension of 30 amino acids at the carboxyl end of the hCG β-chain. Cross-reactivity of hCG and LH can occur unless the hCG assay specifically recognizes that 30 amino acid tag.

At present, the best estimate of q comes from a study by Wilcox et al. (1988) using a highly sensitive and specific immunoradiometric assay of β-hCG in urine. One example of a hormone profile from a woman who experienced subclinical fetal loss during this study is shown in Figure 6.3. Of a total of 198 pregnancies detected among the 221 women in the study, 62 terminated in fetal loss, giving an estimate of q equal to 0.31 ± 0.03. (Hertig, who died within two years of this study, must have been gratified by this result, since it is almost identical to his own estimate.) Even this study, however, was not immune to selectivity bias, as its authors acknowledge. Moreover, their estimate of q may be a slight underestimate for two reasons: (1) a somewhat conservative criterion was used for diagnosing a pregnancy ending in subclinical loss (at least three consecutive days of elevated hCG), and (2) no pregnancy could be detected before the beginning of hCG secretion by the trophoblast at about the time of implantation. Nonetheless, this study remains the single best investigation of subclinical fetal loss to date.

In sum, various studies have provided estimates of q ranging from as low as 0.1 to as high as 0.9. Our model cannot, of course, decide among these estimates, but it can indicate how much difference they make to the expected number of fetal deaths per birth interval. It is clear from Table 6.3 that the expected number of deaths is very sensitive to the probability of death per conception, with the number of deaths rising sharply as q surpasses 0.5. On current evidence it seems extremely unlikely that as many as nine fetal deaths occur in the typical completed birth interval; thus, the model allows us to rule out estimates of q as high as 0.9. It is difficult, however, to guess how much lower q may be. The estimate of q from the study of Wilcox et al. (1988) suggests that one fetal death occurs on average about every second birth interval.

The model used here assumes that successive conceptions within the birth interval are independent events. Evidence from a number of studies would seem to indicate, however, that the outcome of a woman's previous pregnancy may exert a substantial influence on her current risk of fetal loss (for a recent review, see Kline et al., 1989). One example, taken from the Australian Family Project, is summarized in Table 6.4. As a pioneering analysis by Leridon (1976) makes clear, however, this apparent effect of the previous pregnancy is almost certainly spurious, an artifact of unobserved heterogeneity among women in the underlying risk of loss.

To understand the origin of this artifact, imagine a population made up of four distinct classes of women whose risks of fetal loss are zero, low, medium, and high, respectively. Suppose that the class to which a

Table 6.3. Relationship between q, the Total Probability of Fetal Loss per Conception, and $E(N)$, the Expected Number of Losses Occurring in Each Birth Interval

q	$E(N)$
0.1	0.1
0.3	0.4
0.5	1.0
0.7	2.3
0.9	9.0

Table 6.4. Estimated Probability of Fetal Loss by Pregnancy Order and Number of Previous Losses, Australia, 1986–1987[a]

Pregnancy order	Previous losses	Number at risk	Probability of loss	Standard error
1	0	1,990	0.132	0.008
2	0	1,448	0.121	0.009
	1	228	0.285	0.030
3	0	737	0.138	0.012
	1	301	0.209	0.028
	2	52	0.423	0.077
4	0	310	0.139	0.018
	1	186	0.208	0.033
	2	72	0.230	0.069
	3	22	0.500	0.163

[a] Based on a total of 6,213 pregnancies ascertained retrospectively in a one-in-one-thousand sample of households. Because of rapidly diminishing sample sizes, women with 5 or more pregnancies are excluded from the table.
From Santow and Bracher (1989)

woman belongs does not change, so there is no opportunity for her previous pregnancy outcome to affect future outcomes. At the outset of our study, we have no idea to which class a particular subject belongs. If she experiences a fetal death on her first conception, we know that she cannot belong to the zero-risk class. If she goes on to lose her second conception, we might begin to suspect that she is in the medium- or high-risk class. In general, the more losses a woman experiences, the higher our posterior assessment of her risk. Stated differently, at higher numbers of previous losses, our sample becomes increasingly selected for high-risk subjects. Had we ignored this selection process and pooled data across subjects, it would have appeared that previous losses were actually increasing a woman's current risk of loss.

In analyses designed to correct for this selection process, Leridon (1976), Casterline (1989), and Santow and Bracher (1989) have shown that substantial heterogeneity in the risk of fetal loss does appear to exist among women (Figure 6.4). Once this heterogeneity is taken into account, the apparent effect of previous pregnancy outcome vanishes. In this light, our assumption of independence appears to be justified.

THE RELATIONSHIP BETWEEN MATERNAL AGE
AND THE RISK OF FETAL LOSS

Even if earlier studies underestimated the total risk of fetal loss, they were rather consistent in suggesting that it varies with maternal age. Table 6.5 summarizes data from nine such studies conducted in a variety of locales from North America, Europe, Asia, and the Caribbean. These studies were selected for two reasons: they included reasonably large samples of pregnancies, and they were conducted at periods or in locations that minimized contamination of the results by induced abortion, which is a competing risk. Despite variation among the studies in the overall level of fetal loss, they all indicate a marked rise in the risk of loss after maternal age 30 or 35. Less consistently, they also suggest that there may be an elevated risk of fetal loss associated with teenage pregnancies.

Adjusting these results so they all have the same overall risk of loss (arbitrarily set at 150 per 1,000 pregnancies) and pooling results across studies, we obtain the relationship between maternal age and the risk of fetal loss shown in Figure 6.5. Interestingly, a recent analysis of an enormous data set (1,507 fetal deaths during the first 28 weeks of pregnancy in more than 3.25 million woman-days of observation) shows essentially the same relationship, with the additional finding that the elevated risk for mothers less than 20 years old may involve second- but not first-trimester losses (Figure 6.6). Whatever the actual level of fetal loss, it appears from these studies that the relationship between the risk of loss and maternal age is U-shaped, with a low point during the early to midtwenties and a rapid increase after age 30. If this finding is correct, both the risk of fetal death and the average number of fetal deaths per birth interval would be expected to vary markedly with maternal age.

Several recent studies have suggested, however, that the apparent relationship of fetal loss to maternal age may be partly an artifact of unmeasured heterogeneity, at least in populations with effective fertility control (Resseguie, 1974; Wilcox and Gladen, 1982; Gladen, 1986; Casterline, 1989; Santow and Bracher, 1989). In controlled-fertility settings, most women attain their desired family size well before the later reproductive years. Therefore, the women who are still reproducing at age 40, for example, are likely to be a highly select subpopulation of women; in particular, they may not yet have produced their desired number of children precisely because they have experienced a disproportionate number of pregnancy losses. In other words, women who reproduce at later ages may be selected for a high risk of fetal loss, not because all women at those ages are at high risk, but because those who

Table 6.5. Rate of Fetal Loss per 1,000 Detected Pregnancies, by Age of Mother, Nine Populations in which Induced Abortion Is Not a Competing Risk

Location	Age of Mother (Years)						Source
	<20	20–24	25–29	30–34	35–39	40+	
Hawaii	87	68	81	97	122	207	Yerushalmy et al., 1956
Belfast	137	109	116	163	183	275	Stevenson et al., 1958
United States	122	143	137	155	187	255	Warburton and Fraser, 1964
Punjab	191	121	105	125	171	240	Potter et al., 1965b
United States	103	108	134	163	....	230	...	Shapiro et al., 1970
UK	100	117	141	148	166	255	Naylor, 1974
United States (Amish)	88	77	81	68	112	155	Resseguie, 1974
Paris	129	141	165	193	....	247	...	Leridon, 1976
Martinique	101	97	104	139	139	235	Leridon, 1976

are *not* at high risk have ceased reproducing. Consequently, any attempt to isolate a pure age effect must control for the number of previous losses or, less precisely, the number of previous pregnancies. When such controls are implemented, the rise in the risk of loss with maternal age becomes less marked and may not begin until age 35 or 40 (Table 6.6). It is unlikely, however, that selectivity entirely accounts for the elevation in risk of fetal loss at later reproductive ages, since the same elevation is observed among the Amish and in the rural Punjab, where effective fertility control is minimal (Table 6.5).

A different selectivity process may be operating during the early reproductive years. It has been observed in a variety of cultural settings that girls who undergo early menarche also experience earlier initiation of sexual relations and earlier first pregnancy than later-maturing girls (Udry and Cliquet, 1982; Sandler et al., 1984). Thus, teenage pregnancies may occur in a subpopulation of girls selected for early menarche. Other studies suggest that early menarche may predispose toward a high risk of fetal loss, at least during the first two or three pregnancies (Leistøl, 1980; Casagrande et al., 1982; Wyshak, 1983). Therefore, teenage pregnancies may appear to be at elevated risk of loss, not because all teenage girls who are capable of conceiving are at elevated risk, but merely because the ones who actually do conceive are the early maturers.

On the basis of current evidence, I suspect that selectivity may account for essentially all the apparent effect of very young maternal ages on the risk of loss, but only a fraction of the effect at later maternal ages. At present, however, it is very difficult to say what that fraction may be. At any rate, the findings of all these analyses should be borne in mind when studies of fetal loss are being compared, since they may be based upon samples of women with very different age compositions and pregnancy histories.

LENGTH OF GESTATION ENDING IN FETAL LOSS

Each fetal death adds a gestation of some finite duration to the birth interval. If the death occurs early in pregnancy, the addition may be small, although even subclinical fetal deaths are now known to extend the interval by at least a few days (Figures 6.7, 6.8). If the conceptus lives almost to term before dying, the addition to the birth interval will be substantial. Thus, we need to know something about the distribution of gestations ending in fetal death in order to assess the overall impact of such death on fertility.

Table 6.6. Percentage of Pregnancies Ending in Fetal Death (± standard error) by Pregnancy Order and Maternal Age (number of pregnancies in parentheses)

Pregnancy order	Maternal Age at Loss of Pregnancy (Years)				
	15–19	20–24	25–29	30–34	35–49
1	14.6 ± 0.8 (383)	12.5 ± 1.1 (959)	13.4 ± 1.5 (508)	10.3 ± 2.8 (116)	29.2 ± 9.3 (24)
2–3	33.6 ± 4.2 (125)	16.2 ± 1.2 (971)	14.1 ± 1.0 (1149)	12.4 ± 1.6 (419)	13.7 ± 3.4 (102)
4–5	42.9 ± 18.7 (7)	28.1 ± 3.7 (146)	17.6 ± 2.0 (363)	13.5 ± 2.1 (274)	26.8 ± 4.9 (82)
6+	— (0)	31.3 ± 11.6 (80)	31.3 ± 5.2 (80)	21.2 ± 4.2 (96)	26.0 ± 5.0 (77)
Total	19.6 ± 1.7 (515)	15.4 ± 0.8 (2092)	15.2 ± 0.8 (2100)	13.2 ± 1.1 (905)	22.1 ± 2.5 (285)

From Santow and Bracher (1989)

This problem can be restated in terms of the probability of death at each gestational age, since there is a one-to-one relationship between the distribution of those probabilities and the distribution of gestations ending in death. Figure 6.9 shows three idealized cases. In Case A the probability of fetal loss is high immediately following conception and declines steadily thereafter, implying that most infertile gestations end early in pregnancy. In Case B, the probability of loss is constant throughout pregnancy, giving rise to a distribution of infertile gestations that is somewhat more evenly spread but that still has a single mode immediately following conception (the decline in this distribution with gestational age reflects a diminishing number of pregnancies at risk of loss). Case C, in contrast, has the probability of death rising as gestational age increases, implying that most infertile gestations go almost full-term before ending in death.

All the available evidence suggests that reality is far closer to Case A than to either B or C. Figure 6.10, based on an analysis of 3,083 diagnosed pregnancies registered in Hawaii in 1953–1956, indicates the usual empirical pattern and, at the same time, suggests some of the analytical problems involved in detecting this pattern. The broken line shows the distribution by gestational age of an obsolete measure of fetal loss known as the *fetal death ratio*, equal to the fraction of all observed fetal deaths that occur at a given gestational age. This measure is misleading insofar as it fails to take account of changes in the number of known pregnancies at risk of death at each gestational age. The number of fetal deaths at gestational age 4–7 weeks, for example, appears to be low because only a small fraction of pregnancies have been diagnosed by that time.[5] Similarly, the low uncorrected fetal death rate at very late gestational ages (> 36 weeks) partly reflects right-censoring; that is, it ignores the fact that some pregnancies that might otherwise have been at risk of death have already ended in live birth. In order to reveal the true pattern, shown by the corrected fetal death rates in Figure 6.10, we need to perform a hazards analysis that adjusts for both biases.

The published data on fetal loss by gestational age come in discrete-time form; that is, gestational age at loss is classified into discontinuous, nonoverlapping categories, usually of four-week durations. We therefore need to construct a *life table* of fetal loss (life-table analysis is simply a discrete-time or *interval-censored* form of hazards analysis). More specifically, what is needed is a *double-decrement* life table that allows for two distinct ways to "exit" pregnancy, namely, fetal death and live birth. The life table is such an important tool for hazards analysis that it is worth describing in detail. For simplicity, we will assume that the data to be analyzed come from a community-level survey based on hormonal assays, thus minimizing selectivity and underreporting.

At the beginning of the survey, having selected an appropriate sample of reproductive-age women, we draw one 5–10 ml blood specimen from each subject in the sample. Then, exactly four weeks later, we draw a second, smaller specimen (< 5 ml) from the same subject. At the time of the second draw, we ask each woman if she has produced a live birth in the interval between draws. Back in the lab, we assay the first draw for hCG, progesterone, estriol, and hPL. Together these hormones will tell us whether each woman was pregnant at the beginning of the survey and, if she was, will allow us to estimate the gestational age of the conceptus at the time of the first draw. If the woman is found to have been pregnant on the first draw, we check to see if she reported a live birth between draws. If she did not, we assay the second draw for hCG to learn if she was still pregnant at the time of the second draw. A negative assay response is taken as a sign of fetal loss.

With the information provided by this survey, we can form the following relation:

$$P_2(t + 1) = P_1(t) - D(t) - B(t) - C(t) \tag{6.2}$$

where

$P_k(t)$ = number of pregnancies of gestational age t observed on the kth blood draw (k = 1,2)

$D(t)$ = number of fetal deaths between gestational ages t and $t + 1$

$B(t)$ = number of pregnancies of gestational age t on the first draw terminated by live birth before the second draw

$C(t)$ = number of women with pregnancies of gestational age t on the first draw lost to observation before the second draw.

(Note that t is measured in units of four weeks, the time interval between blood draws.) Since we know $P_k(t)$, $B(t)$, and $C(t)$, we can arrange equation 6.2 to find

$$D(t) = P_1(t) - P_2(t) - B(t) - C(t) \tag{6.3}$$

We can now construct a double-decrement life table as follows. If we assume that births, fetal deaths, and losses of women to follow-up are evenly spread across each interval, the probability of fetal loss between gestational ages t and $t + 1$ is estimated as

$$\hat{q}(t) = \frac{D(t)}{P_1(t) - \frac{1}{2}[B(t) + C(t)]} \tag{6.4}$$

(It is this probability that is plotted as a solid line in Figure 6.10.) Similarly, the estimated probability of a live birth between gestational ages t and $t + 1$ is

$$\hat{v}(t) = \frac{B(t)}{P_1(t) - \dfrac{1}{2}[D(t) + C(t)]} \tag{6.5}$$

The estimated probability of "survival" (in the special sense that the conceptus escapes both death and birth) to gestational age t is then

$$\hat{S} * (t) = \prod_{y=0}^{t-1} [1 - \hat{q}(y) - \hat{v}(y)] \tag{6.6}$$

and the estimated probability of survival in the absence of live birth is

$$\hat{S}(t) = \prod_{y=0}^{t-1} [1 - \hat{q}(y)] \tag{6.7}$$

The distribution of deaths by length of gestation is thus

$$\hat{d}(t) = \hat{S}(t) - \hat{S}(t + 1) \tag{6.8}$$

and the mean length of gestations ending in death is

$$\hat{E}(t) = \frac{\displaystyle\sum_{t=0}^{\infty} t\hat{d}(t)}{\displaystyle\sum_{t=0}^{\infty} \hat{d}(t)} \tag{6.9}$$

This last quantity tells us the expected time added to the birth interval by a single a-gestation.

The fetal life tables that are currently available are not, unfortunately, based on this sort of ideal survey but rather on clinical follow-up studies, and they suffer the usual selectivity and underreporting typical of such studies. No pregnancies in any of these studies were detected earlier than two weeks post-conception, and very few were detected earlier than six weeks. Nonetheless, these life tables have essentially the same logical structure as the life table outlined above. Table 6.7 summarizes the results of five such studies.

All five studies show essentially the same pattern: the risk of death is highest at early gestational ages, drops steadily until about the twenty-second week, is more or less constant from week 22 to week 34, and then rises slightly again at later gestational ages. If we take the results

Table 6.7.   Fetal Life Tables from Five Studies: Estimated Probability of Fetal
Death by Gestational Age

Corrected Gestational Age (weeks)[a]	Estimated Probability of Fetal Death				
	(1)	(2)	(3)	(4)	(5)
2–5	0.108	0.082	(0.016)[b]	0.161	0.061
6–9	0.070	0.067	0.064	0.135	0.049
10–13	0.045	0.028	0.044	0.053	0.025
14–17	0.013	0.011	0.006	0.017	0.011
18–21	0.008	0.009	0.001	0.007	0.008
22–25	0.003	0.002	—	0.004	0.003
26–29	0.003	0.004	—	0.003	0.004
30–33	0.003	0.002	—	0.003	0.003
34–37	0.004	0.007	—	0.004	0.004
38–41	0.007	0.011	—	0.002	0.004
42–45	—	—	—	—	0.010
46+	—	—	—	—	0.018
Sample size[c]	232	175	125	339	159

[a] Time since onset of last menses minus two weeks
[b] Week five only
[c] Total number of fetal deaths recorded
Sources and locations: (1) French and Bierman (1962), Hawaii; (2) Erhardt (1963), New York City; (3) Pettersson (1968), Uppsala County, Sweden; (4) Shapiro et al. (1970), New York City; (5) Taylor (1970), Oakland, California.

of the French-Bierman life table, widely regarded as the most reliable of these studies, and express the probability of death on a log scale, this pattern emerges with particular clarity (Figure 6.11, left-hand panel). Now the decline in the probability of fetal loss before the twenty-fourth week of pregnancy is almost perfectly linear. There appears to be a sharp break-point at about the twenty-fourth week separating two quite different patterns of intrauterine death, which might be called "early" and "late" fetal loss, respectively.

One possible interpretation of this pattern is indicated in the left panel of Figure 6.11, which splits fetal loss into three components: an early excess (A), a constant portion (B), and a late excess (C). Ignoring component C, which in fact represents a very small fraction of all fetal deaths, it looks as if the total pool of conceptuses beginning pregnancy may actually consist of two distinct subpopulations, a "normal" fraction with a low (but nonzero) probability of dying and an "abnormal" fraction with a much higher probability of dying. Both probabilities appear to be constant across gestation. The apparent decline in the probability of death with gestational age is, in this view, attributable to the selective weeding-out of abnormal conceptuses; as gestational age increases, the pool of surviving pregnancies is dominated more and more by normal conceptuses, and the probability of death for the pool as a whole declines. By week 24 virtually all abnormal conceptuses have died, and the probability of death reaches a plateau.[6] The early and late patterns of fetal loss are therefore distinguished primarily by the extent to which the deaths of abnormal conceptuses predominate.

This interpretation suggests the following hazards model of fetal loss, based on a two-point distribution in the risk of loss.[7] The hazard rate for deaths in the total pool of zygotes at gestational age t is made up of two parts:

$$h(t) = p(t)\mu_1 + [1 - p(t)]\mu_2 \tag{6.10}$$

where μ_1 and μ_2 are the constant mortality rates of abnormal and normal zygotes, respectively. The mixing parameter $p(t)$ is the fraction of zygotes left at age t that are abnormal, equal to

$$p(t) = \frac{p(0)S_1(t)}{p(0)S_1(t) + [1 - p(0)]S_2(t)} \tag{6.11}$$

where $S_i(t) = \exp(-\mu_i t)$ is the survival function for the ith subgroup of zygotes ($i = 1, 2$), and $p(0)$ is the fraction of abnormal zygotes at conception. The mean length of gestations ending in loss, conditional on loss occurring before the maximal length of pregnancy T, can be found as

$$E(t\,|\,t < T) = \frac{p(0)[1 - S_1(T)]E_1(t\,|\,t < T) + [1 - p(0)][1 - S_2(T)]E_2(t\,|\,t < T)}{p(0)[1 - S_1(T)] + [1 - p(0)][1 - S_2(T)]} \tag{6.12}$$

where

$$E_1(t\,|\,t < T) = \frac{\int_0^T t f_i(t)dt}{\int_0^T f_i(t)dt} \tag{6.13}$$

and $f_i(t) = \mu_i \cdot \exp(-\mu_i t)$ is the density function for gestations ending in loss in the ith subgroup.

When the French-Bierman data up to the thirty-eighth week of pregnancy are used (thus avoiding the confounding effect of losses involving complications of delivery) and T is set at 45 weeks, the estimates of the model's parameters obtained by nonlinear least-squares are $\hat{p}(0) = 0.285 \pm 0.011$, $\hat{\mu}_1 = 0.169 \pm 0.013$, and $\hat{\mu}_2 = 0.001 \pm 0.001$, suggesting that the risk of death for abnormal conceptuses is about 170 times greater than that for normal conceptuses, and that abnormal zygotes make up between a quarter and a third of all conceptions.[8] Extrapolating from these parameter estimates, the probability of fetal loss during the first two weeks of pregnancy, when deaths are effectively invisible, is 0.083.[9] That is, something like 8 percent of all conceptuses die before a clinical diagnosis of pregnancy occurs. (Moreover, between a quarter and a third of all the conceptuses that *will* die in utero do so during the first two weeks.) If we add this level of subclinical loss to the rest of the French-Bierman life table, the total probability of fetal death over all gestational ages turns out to be approximately 0.3 and the mean duration of gestations ending in death is 7.5 weeks, or about 1¾ thirty-day months.

It is important to understand that these estimates presuppose the correctness of the model. Although it is premature to judge the model on its merits, it does fit the French-Bierman data well (Figure 6.11, right-hand panel), and it yields an estimate of the total probability of fetal loss remarkably close to that provided by the most sensitive and specific of the available hCG assays (Wilcox et al., 1988). In addition, a considerable body of cytological evidence supports the model from an etiologic perspective (see below). Thus, the model is at least biologically plausible, and it provides a basis for estimating the probability of death per conception and the expected duration of gestations ending in death, two important aspects of fetal loss that would otherwise be inestimable.

THE DURATION OF RESIDUAL INFECUNDABILITY FOLLOWING EACH FETAL DEATH

With respect to the final piece of the puzzle, the expected duration of "postpartum" infecundability following fetal death, very little is known. It seems likely that some period of anovulation is associated with all but the earliest losses: as discussed in Chapter 5, hCG maintains the corpus luteum of pregnancy (thus postponing the next cycle), and it takes about 12–15 days to clear hCG from the maternal bloodstream once its concentration has become elevated (Lähteenmäki and Luukkainen, 1978; Cameron and Baird, 1988). Beyond that, we are left with the plausible supposition that the duration of anovulation is somehow related to the stage of pregnancy at which loss occurs: the later the loss, the longer the subsequent anovulation. If a pregnancy is terminated within two weeks of conception, little if any delay in the next ovulation is expected (Wilcox et al., 1988). At the other extreme, when a live birth occurs but the infant is not breastfed, there is an average lag of about 1.5–2.0 months before the mother resumes ovulating (Habicht et al., 1985; Gray et al., 1987; Jones, 1989); if a pregnancy lasts almost full term but then ends in fetal death, a similar period of residual infecundability presumably takes place. Thus, the duration of residual infecundability following a fetal death probably ranges from near zero for early losses up to almost two months for late ones (Figure 6.12). Unfortunately, little more can be said about the details of this relationship. Is it linear or curvilinear? If curvilinear, how does it depart from the linear? The few empirical studies pertaining to these questions are summarized in Table 6.8, and their results are plotted in Figure 6.12. None of these studies yields a very strong statement about the relationship between the length of gestation and subsequent anovulation, except that they probably rule out the lower curve in Figure 6.12. In view of our almost total ignorance, it seems prudent to assume the simplest possible relationship, that is, a linear one. It should be understood, however, how weak a justification this is.

We have already estimated that the mean length of gestations ending in fetal death is approximately 7½ weeks. Under the assumption of a linear relationship, it would appear that the expected duration of infecundability following death is something like 1½ weeks. Summing these two figures, the total nonsusceptible period (gestation plus residual infecundability) associated with a single intrauterine death appears to be nine weeks or 2.1 thirty-day months on average.

Table 6.8. Relationship between Gestational Age at Abortion (spontaneous or induced) and Subsequent Duration of Anovulation, Various Studies

Type of abortion	N	Mean gestation (weeks)	Mean anovulation (weeks)	Source
Vacuum aspiration	13	6.9	4.1	Cameron and Baird, 1988
Vaginal prostaglandin suppository	16	7.0	3.4	Cameron and Baird, 1988
Spontaneous	17[a]	9.3	3.5	Donnet et al., 1990
Vacuum aspiration	6[b]	10.3	3.8	Blazar et al., 1980
Vaginal prostaglandin suppository	7	11.0	3.6	Blazar et al., 1980
Full-term live delivery	19[b]	37.4[c]	6.3	Gray et al., 1987

[a] Excluding one outlier with 15 weeks of anovulation
[b] Excluding two subjects who remained anovulatory for the entire period of follow-up
[c] Mean gestation based on the data of Hammes and Treloar (1970)

THE TOTAL IMPACT OF FETAL LOSS

We are now in a position to make a preliminary assessment of the overall fertility-reducing effect of fetal loss. We do this by estimating the expected time added to each birth interval by fetal death, following the logic of Figure 6.1. As before, q will denote the probability of fetal death per conception. In addition, n will be the total nonsusceptible period associated with a single fetal death, and w the expected fecund waiting time to next conception, irrespective of outcome.

The probability that no fetal loss occurs in a given birth interval is obviously $(1 - q)$. In this case, since the very first conception in the interval is fertile, the waiting time from first postpartum ovulation to next fertile conception is just w. If we assume successive pregnancy outcomes to be independent, exactly one fetal death will occur in the interval with probability $q(1 - q)$—the probability of an a-conception followed by an l-conception—and the waiting time to the next fertile conception will then be $w + n + w$ or $2w + n$. Similarly, exactly two deaths occur with probability $q^2(1 - q)$, entailing a waiting time of $3w + 2n$, and so on.

In general, the probability that exactly k fetal deaths will occur in a given birth interval is $q^k(1 - q)$. If k deaths do occur, then the total waiting time from first postpartum ovulation to next fertile conception is $(k + 1)w + kn$. When this time is weighted by its probability of occurrence and summed over all possible values of k, we have the expected waiting time to the next fertile conception in the presence of fetal loss:

$$\sum_{k=0}^{\infty} [(k + 1)w + kn]q^k(1 - q) = \frac{w + qn}{1 - q} \tag{6.14}$$

The expected or mean time *added* to each birth interval by fetal loss is then simply

$$\frac{w + qn}{(1 - q)} - w = \frac{(n + w)q}{1 - q} \tag{6.15}$$

We have already estimated n to be about 2.1 months. Although our discussion of fecundability, which determines the value of w, has been postponed until the following chapter, estimates of w generally fall between three and five months.[10] The results of combining these values of w with various levels of q are given in Table 6.9.

Clearly, at higher values of q, the time added by fetal loss is large—too large, in fact, to be plausible given what we know about the distribution of birth intervals in natural-fertility populations. The value of q

equal to 0.3 estimated from French and Bierman's data gives much more reasonable results: 2.2 months added to the interval when w equals 3, and 3.0 months when w equals 5. On balance, then, our best guess of the mean time added to each birth interval by fetal loss, at least for mid-reproductive age women, is about two to three months.

It should be recalled, however, that q may vary widely with maternal age. The higher values of q in Table 6.9 may not be unrealistic for women near the end of their reproductive careers. If so, then fetal loss may exert a powerful, even a dominant, influence on fertility for those women (Wood and Weinstein, 1990). We saw in Chapter 3 that birth intervals in natural-fertility populations typically grow longer just before the end of childbearing. The analysis here suggests that this effect may be at least partly attributable to an increasing risk of fetal loss (see Chapter 10).

For completeness, I should note that equation 6.14 gives the *mean* waiting time to the next fertile conception in the presence of fetal loss, but not the complete *distribution* of such waiting times. When we allow for heterogeneity in the risk of loss among women (Wood et al., 1992), the probability density function of the waiting time to the next fertile conception takes the general form

$$f(t) = \int_0^1 \left\{ \sum_{k=0}^{\infty} [\langle \zeta * \phi(t) \rangle_k * \zeta(t) q^k (1 - q)] \right\} \Psi(q) dq \qquad (6.16)$$

where q is the probability of fetal loss per conception, $\psi(q)$ is the PDF describing variation in the value of q among women, $\zeta(t)$ is the PDF of time to next conception regardless of outcome, $\phi(t)$ is the PDF for length

Table 6.9. Mean Time Added to Each Birth Interval by Fetal Loss, for Various Values of q (the total probability of loss per conception) and w (the expected fecund waiting time to conception)

q	Time Added (months)	
	w = 3 months	w = 5 months
0.1	0.6	0.8
0.3	2.2	3.0
0.5	5.1	7.1
0.7	11.9	16.6
0.9	45.9	63.9

of gestations ending in fetal loss, and $\zeta*\phi(t)$ is the convolution of $\zeta(t)$ and $\phi(t)$, i.e.,

$$\zeta * \phi(t) = \int_0^t \zeta(t - y)\phi(y)dy \qquad (6.17)$$

(The notation $\langle\zeta*\phi(t)\rangle_k$ denotes the k-fold convolution of ζ and ϕ.) Perhaps unsurprisingly, equation 6.16 does not have a simple analytical solution, but it can be solved numerically on a computer. This model can thus provide a basis for simulation studies of the fertility-reducing effects of fetal loss (see Chapter 12).

THE ETIOLOGY OF FETAL LOSS

At this point, it would be appropriate to turn to a discussion of inter-population variation in fetal loss. However, as should be all too clear by now, the current state of our understanding is such that it would be impossible to distinguish such variation from differential measurement error. It would scarcely be convincing to argue, for example, that the apparent differences among studies in Table 6.5 or 6.7 represent real differences among the populations involved rather than differences in the extent of selectivity or underreporting, not to mention sampling error. Recently, field studies have been initiated in Kenya and Bangladesh to estimate communitywide distributions of fetal loss rates using the new hCG assays, and studies of this sort should go a long way in the future toward solving the problems of selectivity and underreporting (see Leslie et al., 1993). As of this writing, however, those studies have yet to produce final results.

In the meantime, the question of interpopulation variation can be approached indirectly through a consideration of the etiology of fetal loss. If it can be shown that the risk of loss is affected by factors that are themselves known to vary among human societies, this would constitute fairly compelling evidence that intergroup differences in the level of loss are likely to exist. It was suggested above that the etiology of fetal loss varies by stage of pregnancy. According to our model, early fetal deaths are likely to involve some inherent defect in the zygote present at least as early as conception—and perhaps present in one of the gametes prior to conception—whereas late fetal deaths are more likely to be caused by some insult from outside the conceptus, from either the uterine or the external environment. This would seem to be the crux of the distinction between "abnormal" and "normal" conceptuses.

Cytological evidence provides strong support for this view. A large fraction of early fetal deaths involves some readily discernible chromosomal aberration or other major genetic defect, and the prevalence of such defects declines steadily with gestational age, at least after the seventh week of pregnancy (Table 6.10). This decline is consistent with the selection process hypothesized above, by which abnormal zygotes become a progressively smaller fraction of the total pool of surviving zygotes. The most common form of chromosomal aberration observed in early losses is autosomal trisomy, i.e., the presence within the fertilized egg of three copies of some particular nonsex chromosome instead of the normal two (Table 6.11). Monosomy of the X chromosome (Turner syndrome) is also fairly common in early losses.[11] Both these defects result from errors of meiosis during either oogenesis or spermatogenesis. Therefore, early losses are dominated by chromosomal anomalies present in the zygote from the moment of conception.[12] In contrast, euploid losses (abortuses not displaying an obvious chromosomal defect) are much more uniformly distributed across gestation (Figure 6.13).

There is little evidence that the frequency of chromosomal anomalies varies significantly among human populations, except perhaps in the special case of occupational exposure to high levels of toxic materials, such as paint solvents (Kline, 1986). Given that the frequencies of chromosomal aberrations in Table 6.11 are remarkably constant across studies, it may be reasonable to hypothesize that little interpopulation

Table 6.10. Percent of Fetal Deaths with Identifiable Chromosomal Aberrations by Gestational Age (*N* = number of abortuses karyotyped)

Gestational age (weeks)	London[a] %	London[a] N	New York City[b] %	New York City[b] N	Honolulu[c] %	Honolulu[c] N
≤7	—	0	26	91	37	82
8–11	50	121	53	594	54	641
12–15	44	376	46	632	50	620
16–19	21	168	24	266	40	209
20–23	10	145	12	248	44	55
24–27	0	83	14	91	46	28
Total	31	893	38	1,922	49	1,635

[a] Creasy et al., 1976
[b] Kline and Stein, 1987
[c] T. Hassold, reported in Kline et al. (1989)

Table 6.11. Frequency of Different Types of Chromosomal Aberrations in Recognized Fetal Deaths, Selected Studies (*N* = number of abortuses karyotyped)

Aberration	Paris		Hiroshima		Honolulu		London		New York City	
	%	N	%	N	%	N	%	N	%	N
Autosomal trisomy	54	495	57	135	50	404	50	143	49	389
Monosomy X	15	140	17	40	20	165	24	68	16	129
Triploidy	20	183	20	48	14	117	13	37	15	115
Tetraploidy	6	57	3	6	7	54	4	12	5	41
Rearrangement	4	35	1	3	5	43	3	9	3	27
Other[a]	1	11	2	5	3	24	6	18	11	88

[a] Includes mosaics, hypertriploidies, and other aberrations

Sources: Boué et al., 1975 (Paris); K. Ohama, reported in Kline et al., 1989 (Hiroshima); T. Hassold, reported in Kline et al., 1989 (Honolulu); Creasy et al., 1976 (London); Kline and Stein, 1987 (New York City)

variation in the risk of early fetal loss is likely to exist. Only further empirical research can settle this issue.

It is now fairly well established that the increase in risk of fetal loss with maternal age is attributable primarily to an increased incidence of chromosomal aberrations in the eggs of older women (Table 6.12).[13] The fact that there is at least one major biological risk factor for fetal loss that is known to change at later reproductive ages provides strong etiologic evidence that the apparent increase in risk with maternal age is not entirely an artifact of heterogeneity and selectivity. (The etiology underlying the apparent elevated risk of fetal death in teenage pregnancies, on the other hand, is more obscure, and it is more plausible on biological grounds to suggest that this elevation may be artifactual.) Moreover, if the incidence of chromosomal aberrations does not vary significantly across human populations, neither, presumably, should the age pattern of fetal loss unless there are other age-related risk factors that have yet to be identified.

Turning to late fetal losses, the hypothesis that such losses are more likely to be caused by environmental than endogenous factors is supported by observed historical trends in their incidence (Figure 6.14). In many western nations, the rate of loss after the twenty-eighth week of gestation has declined sharply in recent years. This decline parallels changes in neonatal and infant mortality rates, which in turn have been associated with improvements in nutrition, public hygiene, and both pre- and postnatal medical care. Although these secular trends do not prove that the latter factors are necessarily the *causes* of declining fetal death rates, they certainly implicate some aspect of the changing social environment. Moreover, the trend toward lower rates of late fetal loss

Table 6.12. Fetal Loss Rates of Chromosomal and Nonchromosomal Origin by Age of Mother, Based on 1,498 Karyotyped Abortuses

	Age of Mother (years)					
	<20	20–24	25–29	30–34	35–39	40+
Overall rate of fetal loss	0.207	0.162	0.192	0.261	0.366	0.500
Proportion of abortuses with detectable chromosomal aberrations	0.42	0.59	0.58	0.65	0.74	0.85
Decomposition of rate:						
Chromosomal origin	0.087	0.096	0.111	0.170	0.270	0.425
Nonchromosomal origin	0.120	0.066	0.081	0.091	0.096	0.075

From Leridon (1977); based on data of Boué et al. (1975)

stands in striking contrast to the invariance in rates of early loss (Figure 6.15).

Some of the environmental factors that have been identified in various studies as increasing the risk of late fetal loss are infections (e.g., amniotis), radiation, cigarette smoking, alcohol and caffeine consumption, exposure to toxic chemicals, and maternal nutritional status, including both undernutrition and obesity (Kline et al., 1989; Fenster et al., 1991). (In contrast, none of those factors has thus far been found to have a measurable effect on the risk of early fetal loss; see Wilcox et al., 1990.) That some aspect of the larger environment may be involved in late fetal loss is suggested by the finding that a woman's ethnicity and educational background affect her risk of loss (Shapiro and Bross, 1980). Although it is possible that the effect of ethnicity is caused by genetic differences among ethnic groups, it seems far more likely that ethnicity is important primarily as an indicator of socioeconomic status.

What, then, can we conclude about interpopulation variation in fetal loss? It is not known whether the incidence of chromosomal aberrations of the sort implicated in early fetal death varies among human populations; we can thus infer nothing at present about interpopulation variation in early loss. Variation in late fetal loss, however, is likely to be both ubiquitous and substantial. Whether such variation is great enough to induce demographically significant differences in fertility is uncertain. But at least we have developed the analytical machinery that will allow us to translate variation in the risk of loss into effects on birth-interval distributions once that variation has become better known.

NOTES

1. It has been pointed out that not all so-called intrauterine deaths take place in the body of the uterus; some occur in the oviducts or even in the peritoneal cavity. Hook and Porter (1980) therefore suggest using the term "intracorporeal death" in place of "intrauterine death." I reject this usage purely on esthetic grounds.

2. Because fetal loss by its nature involves not only a state (pregnancy) but also a transition from that state (fetal death), cross-sectional or current-status studies of it are not strictly possible.

3. This problem is likely to be most severe in natural-fertility populations characterized by prolonged breastfeeding. As suggested in Chapter 4, married women in such populations are unlikely to have much experience with regularly recurring menstrual cycles, in which case missed menses may be a poor indicator of pregnancy. In fact, as shown in Chapter 8, it is by no means unheard of for intensively nursing women to experience no menses whatsoever between pregnancies.

4. The sample was chosen in a way that almost guaranteed that it would be older than the total population of reproductively active married women, sim-

ply because each subject was required to have had at least one previous live birth.

5.   Recall that uncorrected gestational age, such as that shown in Figure 6.10, is really the time elapsed since the last menses. Thus, gestational ages 4–7 weeks represent only about two to five weeks of actual gestation.

6.   The later elevation in mortality rates probably reflects deaths associated with complications of delivery largely attributable to maternal factors rather than intrinsic to the conceptus; it is thus separate from the selection process described here.

7.   This model was developed independently by Boklage (1990) and myself (Wood, 1989). The version presented here corrects an error of parameterization made in my 1989 paper; I am grateful to Darryl Holman for bringing the error to my attention.

8.   In deriving these estimates, life-table probabilities of loss were converted into weekly hazard rates by assuming that $h(t) = -\{\ln [1 - q(t)]\}/4$. This conversion assumes that the hazard of loss is constant across each gestational-age interval in the life table; the denominator comes from the fact that this interval is four weeks long.

9.   In deriving this figure, it was necessary to correct for the fact that the period of exposure during the first two weeks of pregnancy is only half as long as the four-week periods for which the probabilities in the original French-Bierman life table were estimated.

10.   These estimates are themselves subject to error because of our inability to detect early pregnancies (see Chapter 7).

11.   Curiously, however, Y-monosomy is never observed. This almost certainly cannot be because zygotes without an X chromosome never occur, since there is no evidence that eggs lacking an X chromosome all die before fertilization—for if they did, how could individuals with Turner syndrome be produced? Rather, such zygotes must be so inviable that they are lost almost immediately after conception. This suggests that the prevalence of chromosomal aberrations among fetal losses occurring before the eighth week of gestation must be even higher than the figures given in Table 6.10.

12.   Recent evidence suggests that luteal phase defects are unimportant in causing early fetal losses (Baird et al., 1991).

13.   In view of the significant effect of paternal age found by Selvin and Garfinkel (1976), it is likely that chromosomal defects in the sperm of older men are also partly responsible for this increase (see Chapter 11).

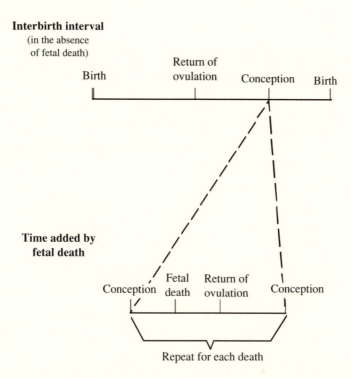

Interbirth interval
(in the absence
of fetal death)

**Time added by
fetal death**

Figure 6.1. The time interval between successive live births in the absence of fetal loss (*upper line*), and the time added to the interval by each fetal death (*lower line*). (Adapted from Bongaarts and Potter, 1983)

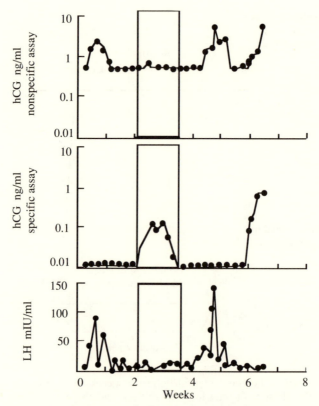

Figure 6.2. Daily urinary hormone concentrations in a
woman with a subclinical fetal loss followed by a clinically diagnosed pregnancy, illustrating the potential
cross-reactivity between human chorionic gonadotropin (hCG) and luteinizing hormone (LH). (*Shaded
bar* indicates menses.) In the *top panel*, hCG has been
assayed using the comparatively nonspecific radioimmunoassay of Wehmann et al. (1981); in the *middle
panel*, the much more specific immunoradiometric
assay of Armstrong et al. (1984) has been used. LH
(*bottom panel*) was assayed by a standard RIA system.
Had only the less-specific hCG assay been available, at
least two subclinical losses might have been diagnosed, both apparently corresponding with two successive LH surges. The one genuine loss, detected by
the more specific hCG assay, was actually missed by
the less specific assay, which is also less sensitive.
(Redrawn from Wilcox et al., 1993)

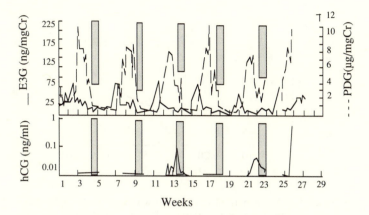

Figure 6.3. Daily urinary estrone-3-glucuronide *(solid line)*, pregnanediol *(broken line)*, and hCG *(bottom panel)* in a woman with two subclinical fetal losses followed by a clinically diagnosed pregnancy. *(Shaded bars* indicate menses.) Estrone–3–glucuronide and pregnanediol are urinary metabolites of estradiol and progesterone, respectively. (Redrawn from Wilcox et al., 1993)

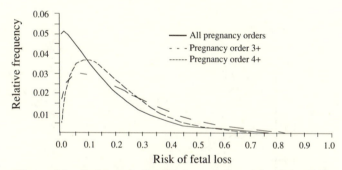

Figure 6.4. Estimated distribution of the risk of fetal loss among women, by pregnancy order. Based on beta distributions fit to data on 6,213 pregnancies ascertained retrospectively in a one-in-one-thousand sample of households in Australia, 1986–1987. (The fit of the beta model in each case was excellent.) The mean risk at all pregnancy orders was 0.144, with a variance of 0.014. The mean risk increases at higher pregnancy orders because the sample becomes selected for women who have experienced multiple pregnancy losses. (Redrawn from Santow and Bracher, 1989)

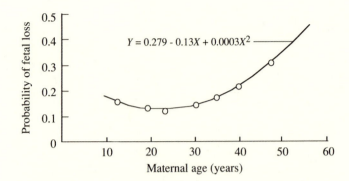

Figure 6.5. Age pattern of fetal loss in women from several populations in which induced abortion is not a competing risk (see Table 6.5). *Open circles* represent pooled data from nine populations, each adjusted to yield an overall loss across all ages of 150 per 1,000 conceptions. *Solid curve,* quadratic equation fit to these data by ordinary least squares ($r^2 = 0.998$). Since losses were ascertained retrospectively in all nine studies, the level of fetal loss is underestimated at all ages; only the shape of the age curve is of interest. (Redrawn from Wood and Weinstein, 1988)

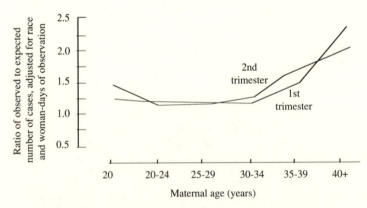

Figure 6.6. First and second trimester fetal losses by maternal age. Based on life table analysis of 1,507 fetal deaths during the first 28 weeks of gestation among pregnancies experienced by 32,182 women who were patients of a single group medical plan in northern California. (Redrawn from Harlap et al., 1980)

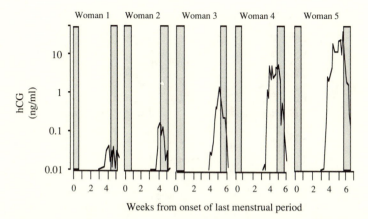

Figure 6.7. Daily urinary hCG concentrations in five women who experienced subclinical fetal losses. (*Shaded bars* indicate menses.) Although all these losses preceded a clinical diagnosis of pregnancy, they occurred at progressively later times following conception, attained higher levels of hCG before loss, and added more time to the apparent length of the menstrual cycle. (Redrawn from Wilcox et al., 1988)

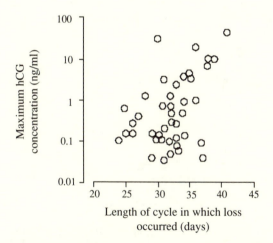

Figure 6.8. Relationship between maximum hCG excretion in 43 subclinical fetal losses and the apparent length of the menstrual cycle in which the loss occurred. Note that even though none of these pregnancies was diagnosed clinically, they added up to three weeks to the apparent length of the cycle. (Redrawn from Wilcox et al., 1988)

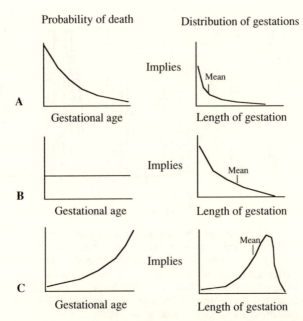

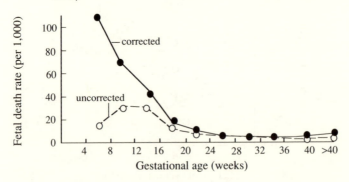

Figure 6.9. Relationship between the probability of fetal loss by gestational age *(left-hand panels)* and the distribution of gestations ending in fetal loss *(right-hand panels)* in three idealized cases. (Redrawn from Wood, 1989)

Figure 6.10. Estimates of the risk of fetal death by gestational age (weeks since onset of last menses), based on 3,083 diagnosed pregnancies, Hawaii, 1953-1956. The uncorrected rate is the fraction of all known fetal deaths that occur at a given gestational age. The corrected rate is the life-table estimate of the fetal death rate (see text). (Data from Bierman et al., 1965)

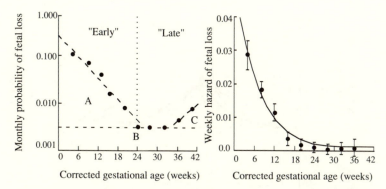

Figure 6.11. *(Left)* probability of fetal loss (on log scale) by corrected gestational age (time since onset of last menses minus two weeks). *Broken lines* divide the probablity of loss into three independent components: an early excess *(A)*, a constant fraction *(B)*, and a late excess *(C)* apparently reflecting complications of delivery. *Dotted line* demarcates "early" and "late" fetal losses based on the presence or absence of component *A*. (Data from French and Bierman, 1962.) *(Right)* fit of the two-point risk model of fetal loss (equation 6.10) to estimates of the weekly hazard of loss derived from the French-Bierman data: *closed circles*, empirical estimates ± one standard error; *solid line*, model-predicted hazards. Standard errors were computed by the method of Gehan (1969).

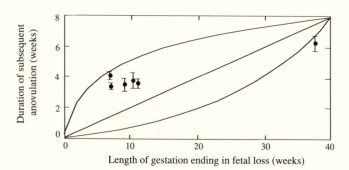

Figure 6.12. Relationship between the length of a pregnancy ending in fetal loss and the duration of residual infecundability following the loss. Smooth curves show various hypothetical relationships. Points are taken from the studies in Table 6.8 (mean observed duration of anovulation ± one standard error).

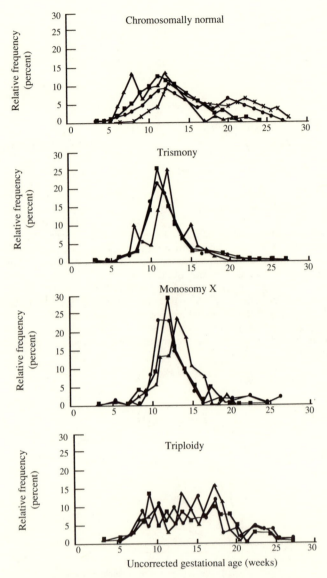

Figure 6.13. Percent distributions across gestation for chromosomally normal, trisomic, X-monosomic, and triploid fetal losses from four series of karotyped abortuses: ▲, Hiroshima; ■, Hawaii; ●, New York; ×, London. Numbers of losses of all types appear to be low during weeks 0-5 because comparatively few pregnancies have been diagnosed by that time. (Redrawn from Kline and Stein, 1987)

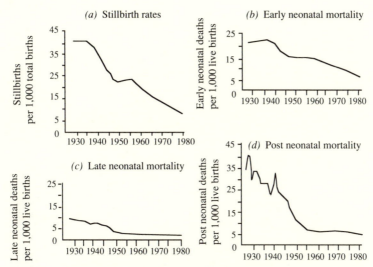

Figure 6.14. Changes in stillbirth rates (i.e., fetal loss after gestational age 28 weeks) and mortality rates at various stages of infancy, England and Wales, 1928–1981. (Redrawn from Macfarlane and Mugford, 1984)

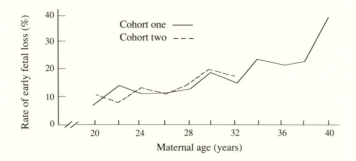

Figure 6.15. Rates of early fetal loss (before week 20 of gestation) by maternal age, based on self-reports from two cohorts of U.S. women: *Cohort one,* 2,070 women recruited in 1935–1944; *Cohort two,* 1375 women recruited in 1961–1970. At the time of recruitment all subjects were students at the University of Minnesota, and they were followed prospectively for up to 45 years. Sample consists of 2,408 pregnancies contributed by cohort one and 1,943 pregnancies contributed by cohort two. (Redrawn from Wilcox et al., 1981)

7

Fecundability and Coital Frequency

One of the most important components of each birth interval is one in which, at least on the surface, nothing much seems to happen: the *conception wait* or *fecund waiting time to conception*. During this interval, the woman is capable of conceiving and is exposed to regular, unprotected intercourse (i.e., intercourse in the absence of contraception), but she has not yet conceived. (Here we are referring to a single round of reproduction; she may well have conceived in the past.) At least one conception wait occurs during each birth interval, including the interval from marriage to the first birth and from the last birth to the onset of permanent sterility or menopause; moreover, each fetal death adds another conception wait to the interval in which it occurs (Figure 7.1). Depending upon the distribution of their lengths, these waiting times can add up to a substantial fraction of a woman's total reproductive life and may be a major factor limiting her fertility.

The waiting time to conception exists because of the probabilistic nature of conception itself. During any month of exposure to intercourse, conception is subject to many vicissitudes. Did ovulation take place? Did at least one act of intercourse fall within the fertile period? Did that act of intercourse actually result in fertilization? In other words, conception is by no means inevitable during any fixed period of exposure to unprotected intercourse, and even healthy, fecund couples who are actively trying to conceive may experience a lag of several months before succeeding.

To express the probabilistic element in this process, demographers speak of *fecundability*. This term was first used by the Italian demogra-

279

pher Corrado Gini in 1924, when he wrote "I call fecundability the probability that a married woman conceives during a single month, in the
absence of any Malthusian or neo-Malthusian practice intended to limit
procreation" (Gini, 1924).[1] More recent usage adds the stipulation that
the woman in question must be in a *susceptible state*—that is, she cannot
be experiencing any condition that renders her unable to conceive, such
as pregnancy, lactational infecundability, or pathological sterility.

Fecundability is of demographic significance because it determines
the distribution of fecund waiting times to conception. This seems intuitively obvious: the lower a couple's fecundability, the longer on average they will wait to conceive. Fecundability is also of interest from a
broader point of view because of the way it combines physiological
influences (e.g., ovarian function) and a fundamentally important
behavioral factor (i.e., sexual intercourse). Indeed, fecundability is one of
the principal points of interaction between the biological and behavioral
determinants of fertility.

It is convenient to distinguish several kinds of fecundability according to the outcome of the pregnancies involved. *Total fecundability* refers
to the monthly probability of any conception, regardless of outcome,
whereas *effective fecundability* refers to the monthly probability of a fertile conception, one that produces a live birth. Although these two forms
of fecundability are arguably the most fundamental from a theoretical
point of view, reference is often made to several other forms as well
(Leridon, 1977). The most important of these is *apparent fecundability,* or
the monthly probability of a conception that survives long enough to be
recognized by the woman and declared in an interview or registered in
a clinic. The definition of apparent fecundability involves some ambiguity, in that it is difficult to specify the exact set of conceptions to
which it refers (as noted in Chapter 6, it may be anywhere from two
weeks to four months after conception before a woman becomes aware
she is pregnant). Nonetheless, there is a compelling methodological reason for discussing apparent fecundability: it is currently impossible to
detect pregnancies that end in fetal loss during the first few days following conception (see Chapter 6). Thus, despite the theoretical importance of total fecundability, it is at present unobservable, and many
demographers simply choose to ignore it. The justification for not following their example here is the same as that offered in Chapter 6 for
not ignoring preimplantation fetal loss: it is useful to model the underlying processes as precisely as possible even if we are not yet in a position to measure them all. In the theoretical portions of this chapter, then,
the unmodified term *fecundability* will always refer to total fecundability, and other forms of fecundability will be designated by the appropriate qualifier, either *apparent* or *effective.*

FECUNDABILITY AND THE DISTRIBUTION
OF CONCEPTION WAITS

The relationship between fecundability and the distribution of conception waits depends upon the characteristics of the couples involved. The relationship is simplest to model in the so-called homogeneous case, in which all couples have the same fecundability and fecundability is constant over time. If, following Gini, we denote fecundability by p, then the probability that a randomly selected couple will conceive for the first time in Month 1 of exposure to intercourse while in a fecund state is simply p itself. The probability of conceiving in Month 2 is $(1 - p)p$, i.e., the probability of not conceiving in Month 1 times the conditional probability of conceiving in Month 2 given failure to conceive in Month 1. (That this conditional probability equals p implies that the failure to conceive in one month does not influence the probability of conceiving in subsequent months.) By extension, the length of the conception wait is a geometric random variable with parameter p and has the probability density function

$$f(t) = (1 - p)^{t-1}p \tag{7.1}$$

The mean or expectation of this PDF is

$$E(t) = \sum_{t=1}^{\infty} t(1 - p)^{t-1}p = \frac{1}{p} \tag{7.2}$$

and its variance is

$$\text{Var}(t) = (1 - p)/p^2 \tag{7.3}$$

In the homogeneous case, then, we have the remarkable result that the distribution of conception waits is entirely determined by the value of p. If we know fecundability, we know the distribution of waiting times.

The distribution of fecund waits implied by this simple model is shown in Figure 7.2, which summarizes the results of a Monte Carlo simulation of conception in 1,000 couples, each with fecundability equal to 0.2. (We set aside for the moment the question of whether this estimate of human fecundability is realistic.) As expected, the resulting distribution of conception waits takes on a reverse J-shape with a long upper tail and a single mode at Month 1 equal to $p \times 1,000$ or 200. The expected waiting time is $1/p$ or five months, very close to the mean "observed" in the simulation; the expected variance is 20 months. Despite the homogeneity of these couples, and despite the fact that most couples conceive within a few months of initiating exposure, a fairly large fraction of them must wait a year or more before conceiving. Indeed, the upper tail of the distribution never quite reaches zero but only approaches it asymptotically. Thus, there is always some small but nonzero chance that a couple will never conceive, no matter how long they try. Even in the homogeneous case, then, there is considerable variation in conception waits, all of which translates directly into variation in birth intervals.

What happens if we allow fecundability itself to vary? The answer depends upon what sort of variation we have in mind. To take the simplest case of "purely random" variation in p from month to month, suppose that the fecundability of a particular couple fluctuates at random, with values of p in different months varying independently of each other. Suppose, too, that the probability density function describing this variation in p is $g(p)$. That is, the probability that the couple's fecundability falls within the interval $[p, p + dp)$ during some particular month is equal to $g(p)dp$. Finally, suppose that $g(p)$ is the same across couples. Then each month, for any randomly selected couple,

$$E(p) = \int_0^1 pg(p)dp \tag{7.4}$$

and

$$f(t) = [E(1 - p)]^{t-1} E(p) \tag{7.5}$$

which is a geometric random variable analogous to equation 7.1. From this we immediately have

$$E(t) = [E(p)]^{-1} \tag{7.6}$$

and

$$\text{Var}\,(t) = \frac{E(1 - p)}{[E(p)]^2} \tag{7.7}$$

Thus, random variation in fecundability is identical to the homogeneous case, except that we need to think in terms of the mean fecundability $E(p)$ rather than some constant fecundability p.

The situation is more complicated when we allow for systematic heterogeneity in fecundability among couples. The mathematics for the most general case, which would allow for both persistent variation in fecundability among couples *and* random variation in fecundability within couples over time, have not yet been worked out, and even the more restricted case to be described here is mathematically daunting. Following the original exposition by Sheps (1964), we will assume that the fecundability of each couple is constant, but that fecundability varies systematically among couples according to the probability density function $h(p)$ with mean μ_p and variance $\sigma_p{}^2$. In this model, each couple draws its value of p at the outset from the distribution given by $h(p)$ and then is stuck with that value for the duration. In Month 1 of exposure, the proportion of couples conceiving is given by

$$f(1) = \int_0^1 p h(p) dp = E(p) = \mu_p \tag{7.8}$$

Thus, the probability of conceiving in Month 1 is just $E(p)$, as in the preceding case. However, for couples who did not conceive in Month 1, the conditional probability of conceiving in Month 2 is

$$f(2) = \int_0^1 p(1 - p) h(p) dp = \mu_p (1 - \mu_p) - \sigma_p{}^2 \tag{7.9}$$

Thus, unless $\sigma_p{}^2$ is zero, $f(2)$ is less than $E(1 - p)E(p)$, the value of $f(2)$ in both the homogeneous and random-variation cases.

In general, Sheps (1964) showed that the distribution of conception waits under systematic among-couple heterogeneity follows the *compound geometric distribution*

$$f(t) = \int_0^1 p(1 - p)^{t-1} h(p) dp \tag{7.10}$$

with mean

$$E(t) = 1/H_p \tag{7.11}$$

and variance

$$\text{Var}(t) = E(t)[1 - E(t)] + 2\text{Var}(1/p) \tag{7.12}$$

where H_p is the harmonic mean of p-values among couples.[2] The variance in equation 7.12 is greater than that expected in the homogeneous case by $2\text{Var}(1/p)$, a measure of how much variation in average waiting times to conception exists among couples.

What do these equations reveal about the effects of heterogeneous fecundability on the waiting-time distribution? If we compare the homogeneous and heterogeneous cases, assuming that p in the former equals $E(p)$ in the latter, we find some important differences (Figure 7.3). Both waiting-time distributions are still reverse J-shaped, and both still have a single mode at Month 1 equal either to p or to $E(p)$. But beginning in Month 2, the curve for the heterogeneous case drops off more rapidly, the difference between the two curves at this stage depending upon the variance in p in the heterogeneous case (see equation 7.9). Thus, we are more likely to observe short conception waits in the homogeneous case. Later, however, the two curves intersect, so the likelihood of observing long conception waits is greater in the heterogeneous case.

These differences can be explained in terms of a selection process that operates in the heterogeneous but not the homogeneous case. In the heterogeneous case, couples display a spectrum of fecundability values at the initiation of exposure. Those who conceive during the first month do not represent a random sample of all the couples initiating exposure; rather they include a disproportionate number of high fecundability couples—the higher the fecundability, the more likely the couple is to conceive during any month of exposure. Thus, high fecundability couples are preferentially removed from the initial pool, and the remaining couples become increasingly selected for low fecundability as time goes by. As a result of this selection process, the spectrum of variation in fecundability among the remaining couples becomes ever narrower even as the mean fecundability declines.

Whatever model we are considering, the probability that a couple conceives in month t rather than some other month can always be written $f(t) = A \times B$, where $A = $ Pr(conception in month $t|$ no conception in previous $t - 1$ months) and $B = $ Pr(no conception in previous $t - 1$ months). In the homogeneous case, the value of A never changes; it is just p. That is because couples who conceive late are identical to those who conceive early—they have exactly the same fecundability. And B always decreases by a constant multiplicative term $(1 - p)$ each month. Thus, in the homogeneous case $f(t)$ follows a negative geometric curve that drops off at a constant rate. In the heterogeneous case, however, the conditional probability in part A is always declining because of the selective removal of high fecundability couples, especially at early exposures: now couples who conceive early *are* different from those who conceive late. But the decline in part A tends to bottom out at higher exposures as the range of variation among remaining couples becomes more and more restricted. At the same time, the rate of decline in part B slackens as increments in the fraction of couples who have already conceived become smaller and smaller. Thus, in the heterogeneous case but not the homogeneous case, the rate of decline in $f(t)$ decelerates and approaches zero at higher exposures. In other words, when fecundability varies among couples $f(t)$ drops off rapidly at early exposures but very slowly at later ones.

The high upper tail of the waiting-time distribution in the heterogeneous case has a large effect on the mean conception wait. As shown by equations 7.2 and 7.11, the mean wait in the homogeneous case is simply $1/p$, while in the heterogeneous case it is $1/H_p$, where H_p is the harmonic mean of the fecundabilities. In general H_p is less than or equal to $E(p)$, and the two are equal only when σ_p^2 is zero. Thus, whenever there is couple-to-couple variation in fecundability ($\sigma_p^2 > 0$), the mean conception wait is longer than would otherwise be expected—and the greater the variation, the longer the mean wait. Not surprisingly, the variance in conception waits for the total sample is also greater in the heterogeneous case by an amount proportional to the among-couple variance in expected waits (compare equations 7.3 and 7.12). Thus, heterogeneity has the effect of inflating both the mean and variance of the waiting times to conception.

EMPIRICAL ESTIMATES OF FECUNDABILITY AND CONCEPTION WAITS

Which of these models corresponds most closely to reality? To answer this question, it would seem a simple matter to observe the waiting times to conception in a large sample of couples who are not using contraceptives, and then fit the various models to the data using the statistical methods described in Chapter 3 (since waiting times are duration variables, hazards analysis would seem to be the appropriate way to deal with them). Having ascertained which model fits the data best, it should then be possible to use that model along with the estimates of its parameters to infer the value of fecundability (or the mean and variance of fecundability, in the heterogeneous case). For several reasons, however, this straightforward approach is surprisingly difficult to implement:

(1) Since it is currently impossible to detect very early pregnancies, the waiting time to conception per se is unobservable, and pregnancies that end in fetal loss during the first few days of gestation are unlikely to be detected at all. For this reason, most empirical estimates pertain to apparent or effective fecundability rather than total fecundability.

(2) Data sets on the waiting time to conception almost always contain right-censored observations. All the models of fecundability suggest that the distribution of conception waits has a long upper tail that approaches zero asymptotically; thus, in a finite observation period we would expect there to be at least a few fecund couples who by chance did not conceive before observation was terminated. As discussed in Chapter 3, such right-censoring in itself poses no particular problem for hazards analysis. In the present case, however, there is a complication that makes the standard correction for right-censoring problematic: couples whose waiting times are censored may indeed be fecund, or they may be sterile and thus not exposed to the risk of conception at all. Sterile couples are in a nonsusceptible state and should therefore be excluded from the analysis, but in fact it is usually impossible to identify such couples on a priori grounds (see Chapter 10). When observations on sterile couples are included in the hazards analysis, as they almost

always are, the consequent overcorrection may result in a net underestimate of fecundability.

(3) Fecund waiting times to conception occur in all birth intervals, and in theory all intervals could supply data for estimating fecundability. It is often difficult, however, to determine exactly when ovulation resumes after a birth, especially where breastfeeding is prolonged and intensive. Thus, we are unlikely to know when exposure to the risk of conception actually begins in higher-order birth intervals. And even if we knew precisely when the first postpartum ovulation took place, prolonged breastfeeding is often associated with a variable period of postovulatory subfecundity that complicates estimation of fecundability (see Chapter 8).[3] To avoid these problems, most demographers restrict their analyses of fecundability to data on the first conception wait, that is, from marriage to the first conception. Unfortunately, this strategy introduces three further problems. First, since coital frequency tends to be at its highest during the period immediately following marriage (see below), estimates of fecundability from first conception waits are likely to be too high to represent the general population of fecund couples. Second, data on the first birth interval may include premaritally conceived pregnancies that appear to have *negative* conception waits, which is clearly absurd; however, eliminating such pregnancies (e.g., by removing all intervals from marriage to first birth that are less than about 8.0–8.5 months long) may introduce selectivity bias if couples who conceive before marriage are more fecund than those who do not. Third, if marriage itself is selective with respect to fecundability, as would be the case if premarital conception were a common reason for getting married, we will tend to overestimate fecundability in couples who marry at early ages and underestimate it in late-marrying couples.

(4) If fecundability is heterogeneous, its mean and variance cannot be estimated unless we specify its probability density function $h(p)$ beforehand. In other words, we need to know a great deal about couple-to-couple variation in fecundability before we can even determine whether such variation exists! The danger of circularity in such a situation is obvious. Following the suggestion of Henry (1961, 1964), most analyses assume that fecundability varies according to a beta distribution, such that

$$h(p) = \frac{p^{\alpha-1}(1 - p)^{\beta-1}}{B(\alpha,\beta)} \tag{7.13}$$

where $\alpha,\beta > 0$ and $B(\alpha,\beta)$ is the *beta function*, equal to $\int_0^1 p^{\alpha-1}(1 - p)^{\beta-1}dp$. This expression for $h(p)$ has two virtues: it constrains p to fall between zero and one, which as a probability it should, and it is extremely flexible. Majumdar and Sheps (1970) have worked out maximum likelihood estimators for α and β applicable to data on conception waits, and Suchindran and Lachenbruch (1975) have adapted them for use with right-censored observations.[4] More recently, Weinberg and Gladen (1986) have shown how the effects of covariates on fecundability can be incorporated into the analysis. Unfortunately, application of this model has produced results that are at best mixed: sometimes the beta model fits the data well, often it does not. Moreover, the beta distribution of fecundability becomes U-shaped at higher values of σ_p^2, which hardly seems to make biological sense (Wood and Weinstein, 1990).

Keeping all these problems in mind, we can turn to Figure 7.4, which summarizes data from five classic studies of conception waits in newly married U.S. couples: the Growth of the American Families Study (GAF) carried out in 1955 (Freedman et al., 1959; Weir and Weir, 1961), the Princeton study of Family Growth in Metropolitan America (FGMA) conducted in 1957 (Westoff et al., 1961), a survey conducted by P. K. Whelpton and C. V. Kiser in Indianapolis (IND) in 1941 and cited by Westoff et al. (1961), and two studies by G. W. Beebe at rural birth control clinics in Kentucky and West Virginia (KY-WV) and in North Carolina (NC) in the mid-1940s, cited by Tietze (1956). In each of these studies, the waiting time from marriage to first recognized conception was recorded for married couples who were not using contraception and who reported no marital disruption or spousal separation during the interval in question. The data are retrospective, elicited by interview at some variable time from two to fourteen years after the marriage of the couple. The period to which these data pertain (1927–1957) is probably early enough to limit the number of unreported induced abortions contaminating the results; in addition, "obviously subfecund gynecological patients" have been omitted (Westoff et al., 1961:51). Thus, these data can be taken to reflect the experience of healthy U.S. newlyweds who are not trying to avoid pregnancy. Samples sizes were 1,430 couples for the GAF study, 458 for FGMA, 1,783 for IND, 1,452 for KY-WV, and 481 for NC.

The distributions of conception waits in these studies are all reverse J-shaped (Figure 7.4, left panel), consistent with each of the models outlined above. More interesting, however, are the results of a life-table analysis of these data. The right-hand panel of Figure 7.4 shows the estimated monthly probabilities of conception, equal to the mean fecundability of couples who have not yet conceived. Under the homogeneous model, these probabilities should not change over time, reflecting both the constant fecundability of each couple and the absence of a selection process that renders the sample increasingly less fecund as time goes on. It is obvious that the data are not at all consistent with the homogeneous model. Far from being constant, the monthly probabilities drop sharply as the duration of exposure increases. This finding, which is remarkably consistent across samples, would seem to fit the model of heterogeneous fecundability quite well.[5] Assuming that the heterogeneous model of fecundability is correct, equation 7.8 shows that the relative frequency of conceptions occurring in Month 1 provides one possible estimate of the mean fecundability—or, more accurately in this case, the mean apparent fecundability. In Figure 7.4 (left panel) these frequencies all seem to cluster between about 0.25 and 0.35. Thus, we can conclude that the apparent fecundability of healthy U.S. couples during the first birth interval is, on average, approximately 0.30, but that its value varies substantially among couples.

Several studies have attempted to apply Henry's beta model of heterogeneous fecundability (equation 7.13) to observed distributions of conception waits, and some of their results are summarized in Table 7.1. Data from a variety of populations yield mean apparent fecundabilities of about 0.14 to 0.32, suggesting that the U.S. estimates just described may have been a bit on the high side (perhaps because subfecund gynecological patients were omitted). The variances in apparent fecundability estimated for the populations in Table 7.1 range from 0.006 to 0.027. Although these values may seem small, they have a remarkably large effect on the mean and variance of conception waits. As shown in the last two columns of Table 7.1, the mean conception wait in these studies is about 50 percent longer than would be expected in the homogeneous case, whereas the variance in waits is anywhere from four to seventy-two times larger than expected. However, as noted by Weinstein et al. (1993b), these inflated means and variances do not necessarily imply large effects of heterogeneity on the overall distribution of conception waits; they may simply reflect the disproportionate impact of the high upper tail of the heterogeneous distribution on the mean and variance.

Before we leave Table 7.1, it is worth mentioning that the fit of the beta model to these observations is far from uniformly good. As evaluated by standard goodness-of-fit tests, the correspondence of model and data ranges from good (e.g., the Hutterites) to fair (e.g., Taiwan) to terrible (the United States). Nor is there any profound theoretical reason for choosing the beta density function for fecundability over any other; the choice does not arise from a consideration of the processes underlying variation in fecundability, but is largely a matter of statistical convenience. Moreover, we have recently found that the beta model may not capture certain biologically plausible sources of heterogeneity in appar-

Table 7.1. Estimates of Apparent Fecundability, Mean Waiting Time to First Recognizable Conception, and Related Parameters Based on the Beta Model (equation 7.13)

Population	Mean fecund-ability	Variance in fecund-ability	Mean wait[a]	Variance in waits	E (t) ratio[b]	Var (t) ratio[b]
U.S.	0.14	0.006	10.0	287.5	1.4	7.0
Taiwan	0.16	0.006	8.2	139.3	1.3	4.4
Peru[c]	0.17	0.009	9.6	513.1	1.6	16.9
Crulai	0.18	0.009	8.1	208.0	1.4	8.2
Brazil	0.19	0.011	7.9	250.7	1.5	11.6
Tourouvre	0.21	0.014	7.7	423.1	1.6	23.6
Mexico[c]	0.21	0.014	7.4	310.8	1.6	18.1
Historical France	0.23	0.011	5.5	56.6	1.3	4.0
Tunis	0.25	0.020	6.7	864.3	1.7	72.0
Hutterites	0.27	0.014	4.8	44.9	1.3	4.5
Quebec	0.31	0.027	5.2	317.7	1.6	44.3

[a] In months.
[b] $E(t)$ ratio and $Var(t)$ ratio are ratios of the mean and variance in fecund waiting times, respectively, in the heterogeneous case versus the homogeneous case in which $p = E(p)$.
[c] Women age 25+ only.
Sources: Henripin, 1954b (reanalyzed by Bongaarts, 1975); Vincent, 1956 (reanalyzed by Leridon, 1977); Gautier and Henry, 1958 (reanalyzed by Bongaarts, 1975); Ganiage, 1960; Henry, 1964; Potter and Parker, 1964; Berquo et al., 1968 (reanalyzed by Leridon, 1977); Jain, 1969b; Charbonneau, 1970; Majumdar and Sheps, 1970; Balakrishnan, 1979.

ent fecundability (Wood and Weinstein, 1990). To date, however, no better distribution has been proposed.

A remarkably simple method for estimating fecundability has been developed by Bongaarts (1975). In deriving this method, Bongaarts has shown that the mean and variance in fecundability are likely to be highly correlated, suggesting in turn that the coefficient of variation in fecundability (c, equal to σ_p / μ_p) may be more or less invariant across populations. The figures in Table 7.1 support this conclusion: all the estimates of c fall between 0.44 and 0.57. Thus, Bongaarts suggests that fecundability in all human populations may be thought of as following a beta distribution with differing means and variances but with a coefficient of variation that always approximates 0.5.

Under the beta model, Leridon (1977:36) has shown that

$$\mu_p = \frac{1 + k}{E(t)} \tag{7.14}$$

where

$$k = \frac{c^2(1 - \mu_p)}{[1 - \mu_p - c^2(1 + \mu_p)]} \qquad\qquad (7.15)$$

Assuming that $c = 0.5$, the values of μ_p in Table 7.1 yield estimates of k that themselves fall close to 0.5. Thus, it should be possible to arrive at a fairly good estimate of μ_p simply by multiplying the inverse of the observed arithmetic mean conception wait by $(1 + k) \simeq 1.5$.[6]

Bongaarts and Potter (1983) have applied this method to data on the interval from marriage to the first birth derived from several studies of historical European populations (Table 7.2). Assuming that fetal loss adds about two months to the average birth interval, they arrive at estimates of mean fecundability ranging from 0.18 to 0.31. Given that these figures are supposedly estimates of *total* rather than apparent fecundability, and given that we have already seen estimates of apparent fecundability in the range of 0.14 to 0.32, it appears that this method may yield underestimates of μ_p—although it is not immediately obvious why it should.

In general, the difference between total and apparent fecundability depends upon two things, both of which may vary among populations as well as individuals: (1) the lag between conception and the first awareness of pregnancy, and (2) the probability of fetal loss prior to an awareness of pregnancy. Neither is easy to measure. In western women with ready access to medical facilities, the difference between total and apparent fecundability is largely a question of the frequency of subclinical fetal loss. As discussed in Chapter 6, recent attempts to estimate the level of subclinical loss by measuring the concentration of hCG in blood or urine have yielded widely varying results, probably reflecting differences in sample selectivity as well as assay sensitivity and specificity. The study of Wilcox et al. (1988) would appear to be the best to date with respect to such considerations. Out of 198 pregnancies detected in that study, mostly in women between the ages of 20 and 35, 43 were lost before a clinical diagnosis was made and before the woman herself became aware of the pregnancy. In other words, approximately 22 percent all pregnancies resulted in subclinical loss. These results suggest that estimates of apparent fecundability should be revised upward by at least a fifth in order to estimate total fecundability.

To summarize, estimates of apparent fecundability range from approximately 0.15 to 0.30. At present it is impossible to tell whether the variation in estimates from different studies is real or whether it reflects differences in methods and data quality. A conservative correction for subclinical fetal loss suggests that total fecundability may generally be about 20–25 percent greater than apparent fecundability. Most studies

Table 7.2. Observed and Estimated Mean Conception Waits (in months) and Total Fecundability in the First Birth Interval, Selected Historical Populations

Population	Average (and range) of mean intervals between marriage and first birth		Average (and range) of mean waiting times to first conception[a]		Average (and range) of estimated mean total fecundability in first birth interval[b]	
15 English parishes	19.0	(17.3–20.8)	8.0	(6.3–9.8)	0.22	(0.18–0.25)
13 German villages	17.9	(16.4–19.5)	6.9	(5.4–8.5)	0.23	(0.18–0.26)
2 Belgian communes	17.0	(16.7–17.3)	6.0	(5.7–6.3)	0.28	(0.28–0.29)
3 French communes	17.0	(16.3–17.3)	6.0	(5.3–6.3)	0.23	(0.18–0.31)

[a] Estimated from previous column by subtracting nine months for gestation and two months for fetal loss.
[b] Estimated assuming heterogeneous fecundability, such that mean fecundability = 1.5/(mean conception wait).
From Bongaarts and Potter (1983:29)

indicate significant couple-to-couple variation in fecundability, and we turn next to a consideration of one possible source of that heterogeneity.

FECUNDABILITY AND AGE

Despite the obstacles confronting the estimation of fecundability, most studies show a consistent effect of the female partner's age (Table 7.3, Figure 7.5). Fecundability is typically low in adolescence, but it rises rapidly to a peak during the early to mid-twenties. After age 25, fecundability declines steadily until reaching zero at menopause. Although it may be difficult to ascertain the exact level of fecundability at any age, the overall shape of the age curve in Figure 7.5 appears to be both common and more or less correct.[7]

There are many reasons fecundability might be expected to vary with age. In Chapter 4 we discussed some of the changes in ovarian function that occur during adolescent subfecundity and reproductive aging, including changes in the length of ovarian cycles, the likelihood of anovulatory cycles, and the prevalence of luteal insufficiency. In addition, since both Table 7.3 and Figure 7.5 pertain to apparent rather than total fecundability, changing levels of early fetal loss may also be involved.

Despite the likely contribution of all these physiological factors, James (1979) has argued that the decline in fecundability at higher ages is caused primarily by declining coital frequency. In order to assess this claim, it will be necessary to review studies of the relationships among age, coital frequency, and fecundability, a subject we take up below. For now, it will suffice to describe a recent study which seems to show quite convincingly that changing coital rates cannot be the sole cause of the decline in fecundability at later reproductive ages.

In a multicenter study coordinated by the Fédération Française des Centres d'Etude et de Conservation des Oeufs et du Sperme, cumulative pregnancy rates were analyzed in 2,193 French women who underwent artificial insemination with donor sperm (Fédération CECOS et al., 1982). The husbands of these women all appeared to be completely sterile as a result of azoospermia (a total absence of viable sperm in the semen), but there were no discernible reproductive impairments in the women themselves. The attraction of this study in relation to James's hypothesis is that "coital frequency" was essentially invariant with age: all subjects were inseminated at least once during the periovulatory segment of each cycle (as indicated by BBT and cervical mucus), and no one

Table 7.3. Mean Apparent Fecundability (± standard error) by Age of Wife, Selected Countries

Age at marriage (years)	Mexico		Costa Rica		Colombia		Peru		Taiwan	
	Fecund-ability	No. of women	Fecund-ability	No. of women	Fecund-ability	No. of women	Fecund-ability	No. of women	Fecund-ability	No. of women
14	.090 ± .003	169	.098 ± .006	62	.094 ± .004	94	.112 ± .005	104	.090 ± .004	69
15	.109 ± .003	245	.079 ± .004	81	.110 ± .005	140	.122 ± .005	125		
16	.145 ± .005	203	.163 ± .009	87	.144 ± .006	162	.112 ± .005	136	.093 ± .002	110
17	.145 ± .005	205	.184 ± .007	82	.203 ± .009	141	.135 ± .006	122	.128 ± .004	211
18	.151 ± .006	148	.110 ± .003	87	.186 ± .009	137	.158 ± .006	144	.121 ± .002	292
19	.180 ± .008	129	.165 ± .010	65	.220 ± .010	119	.143 ± .006	127	.151 ± .004	385
20	.183 ± .010	93	.179 ± .014	49	.139 ± .006	90	.135 ± .005	117	.180 ± .004	374
21	.216 ± .011	85	.107 ± .008	28	.136 ± .007	72	.115 ± .006	81	.209 ± .006	234
22	.165 ± .009	45	.172 ± .014	32	.194 ± .012	67	.119 ± .008	52	.226 ± .008	186
23	.253 ± .022	39	—†	18	.247 ± .020	40	.176 ± .013	42	.203 ± .007	125
24	.227 ± .024	26	—†	15	.240 ± .021	28	.235 ± .021	31	.276 ± .016	72
25+	.214 ± .014	65	.148 ± .011	46	.145 ± .008	87	.166 ± .009	83	.196 ± .013	132

† Sample too small (<25) for estimation of fecundability.
From Jain (1969a); Balakrishnan (1979)

was inseminated more often because of her age. Nonetheless, significant differences in cumulative pregnancy rates remained when the data were classified by the woman's age (Figure 7.6).[8] The cumulative rates were consistently lower in women ages 31 to 35 than in women below age 30, and sharply lower in women over 35. Since there was no attempt to match older women with older sperm donors, these differences are presumably attributable to age-related variation in the reproductive physiology of the women and not the men involved.

The results of the Fédération CECOS study have received widespread publicity, and they have even been used to argue that women who wish to have children should not postpone reproduction in order to pursue careers (DeCherney and Berkowitz, 1982). This recommendation has been sharply criticized for a variety of reasons (see Bongaarts, 1982a; Hendershot et al., 1982; Menken et al., 1986), two of which seem most salient here. First, the Fédération CECOS results do not appear to be representative of women who experience natural insemination. According to the 1976 National Survey of Family Growth, for example, noncontracepting U.S. couples who are not surgically sterile achieve 12-month cumulative pregnancy rates that are consistently higher than those attained in the Fédération CECOS study (Figure 7.6, arrows). Indeed, naturally inseminated women ages 35 to 39 do better than artificially inseminated women age 25 and less, despite the fact that they do not generally try to time intercourse to coincide with the fertile period. This fact strongly suggests either that storage of semen for artificial insemination may reduce its fertilizing capacity (most semen used in Fédération CECOS centers is frozen) or that the sample of women involved in the Fédération CECOS study may have been selected for subfecundity.[9] Second, even if the differences among age groups shown in Figure 7.6 were representative of naturally inseminated women, they translate into delays in the timing of births that may be measurable on a demographic scale but trivial in terms of an individual woman's reproductive life course or professional career. The median time to conception in Fédération CECOS women age 30 or less is about six cycles, whereas that of women ages 31 to 35 is seven cycles and that of women over 35 is eleven to twelve cycles. At worst, then, women over 35 may have to wait about six months longer on average to conceive than younger women, a difference that would scarcely seem to warrant a major change in a person's goals and aspirations.

These remarks are not criticisms of the original Fédération CECOS study, but of other writers' interpretations of it. Despite such problems of interpretation, the Fédération CECOS study is of genuine value in showing that fecundability is in fact affected by the woman's age even when coital frequency and her husband's age are controlled. Recently, the Fédération CECOS study was repeated in the Netherlands, with virtually identical results (van Noord-Zaadstra et al., 1991). As valuable as these studies are, however, they leave two questions unanswered: are age-related changes in female reproductive physiology sufficient *by themselves* to explain the observed changes in fecundability with age, and what particular aspects of female physiology are involved? Because of the difficulties in studying fecundability empirically, most attempts to answer these questions have been based on mathematical models of fecundability.

MODELS OF FECUNDABILITY

Earlier in this chapter, we dealt with models that took the value or distribution of fecundability for granted and derived its implications for conception waits. Here we want to develop models for a different purpose, namely, to identify the primary determinants of fecundability itself and to ascertain how they combine to produce particular values of fecundability. Such models allow us to perform artificial "experiments" on human fecundability that would be impossible in reality. For example, we can manipulate our models to hold coital frequency constant in order to isolate the effect of changing physiological processes on fecundability, or conversely we can hold the physiological processes constant while allowing coital frequency to vary.

The necessary components of any model of fecundability are easy to specify. For conception to take place in a given month, at least one ovulation must occur, at least one act of unprotected intercourse must fall within the fertile period associated with that ovulation, and insemination must actually result in conception. If we are modeling apparent rather than total fecundability, then the conceptus must also survive long enough for pregnancy to be recognized. Conception is probabilistic precisely because the probability of each of these occurrences is generally less than one.

On the physiological side, then, the important determinants of apparent fecundability are (1) the likelihood of ovulation per cycle; (2) the length of the cycle, which determines how many ovulations can occur in a month; (3) the length of the fertile period within each cycle; (4) the probability of conception from one act of unprotected intercourse in the fertile period (a probability that may vary with the day of the fertile period); and (5) the probability that the conception will produce a recognized pregnancy. The critical behavioral factors are (6) the frequency of intercourse and (7) the way in which acts of intercourse are distributed across the cycle relative to the fertile period. Clearly the wife's age may affect all the physiological factors listed here, while the husband's age may affect factors 3, 4, and 5.[10] Wife's age, husband's age, and marital duration, as well as a host of imponderables, are likely to affect the behavioral factors.

Before describing the various models that have been constructed, it will be helpful to develop some terminology. First, *coitus, coition,* and *insemination* will be used here as synonyms for unprotected intercourse. *Conception* will have its usual meaning of fertilization, but *impregnation* will refer strictly to pregnancies that survive long enough to be recognizable; thus, impregnation as defined here is roughly equivalent to implantation.

Next, following the pioneering work of Glass and Grebenik (1954), I define *restricted intercourse* as a pattern in which at most one insemination can occur in a single day, as opposed to *unrestricted intercourse,* which allows for multiple inseminations on the same day. As Lachenbruch (1967) has noted, these two cases can be thought of as sampling cycle days without replacement and with replacement, respectively. The term *uniform intercourse* will be used to describe a situation in which the probability of intercourse does not vary by cycle day, at least across the *intermenstruum* or the interval between menses. (Most models assume that intercourse does not occur during menses or, equivalently, that intercourse during menses is irrelevant since it is extremely unlikely to result in conception.) Similarly, *constant intercourse* refers to a case in which the probability of intercourse on a given day does not depend upon the time that has elapsed since the previous act of intercourse. Thus, uniform intercourse implies an absence of cycle day–dependence in the probability of intercourse, constant intercourse an absence of pure time-dependence.

In a useful review, Potter and Millman (1985) have suggested the following additional definitions:

p_o = the probability that a cycle is ovulatory,
p_h = the probability that at least one insemination occurs in a particular fertile period,
p_f = the conditional probability of conception given that at least one insemination has occurred in the fertile period, and
p_i = the conditional probability of impregnation given that conception has taken place.

If fecundability is defined in terms of a cycle rather than a month of exposure—and most of the models reviewed here are so defined—then total fecundability (p) is simply $p_o \times p_h \times p_f$, whereas apparent fecundability (p_a) is $p_o \times p_h \times p_f \times p_i$.

The distinction between a month and a cycle of exposure becomes critical at early and late reproductive ages when cycle length may depart substantially from a month (see Chapter 4). Only two of the models reviewed here make this distinction and thus permit investigation of the effects of varying cycle length. Throughout what follows, total and apparent fecundability per *cycle* will be denoted by p' and p_a', respectively, while those per *month* will be denoted by p and p_a. It is, of course, p and p_a that correspond to fecundability as originally defined by Gini (1924).

The earliest and still most influential attempt to model fecundability was by Glass and Grebenik (1954). Those authors assumed that insemination did not occur during menses, but that coitions were distributed across the intermenstruum as independent, identically distributed random events. In other words, they assumed both uniform and constant intercourse. They also assumed that cycles were always ovulatory. If M is the length of the intermenstruum, F the length of the fertile period, and n the total number of coitions that occur during the intermenstruum, then according to their model the probability that at least one coition will fall within the fertile period under unrestricted intercourse is

$$p_h = 1 - \left[\frac{M - F}{M}\right]^n \tag{7.16}$$

The same probability under restricted intercourse is

$$p_h = 1 - \left|\frac{\left(\dbinom{M - F}{n}\right)}{\dbinom{M}{n}}\right| \tag{7.17}$$

where $\dbinom{x}{y} = \frac{x!}{(x - y)! \, y!}$. The term $\dbinom{M - F}{n} / \dbinom{M}{n}$ in equation 7.17 is equivalent to the number of ways in which it is possible to distribute n coitions so they all fall outside the fertile period, divided by the number of ways to distribute n coitions across the intermenstruum as a whole. As Potter (1961) has noted, for a fixed value of n, the value of p_h is always greater in equation 7.17 than in equation 7.16. Glass and Grebenik go on to estimate total fecundability per cycle by assuming that both p_o and p_f are equal to one, whence $p' = p_h$.

This model has been used by Tietze (1960), Potter (1961), and many others either to estimate fecundability or, by taking fecundability as given,to estimate the length of the fertile period. Potter (1961) modified the model to allow for anovulatory cycles and subclinical fetal loss by setting $p_0 = 0.95$ and $p_i = 0.5$, values that seemed reasonable given the limited data then available. In a simulation model based on Glass and Grebenik's work, Lachenbruch (1967) added several additional refinements. First, he allowed the day of ovulation to vary within the cycle according to a distribution suggested by Farris (1956). Then he allowed the conditional probability of conception to vary by the day of the fertile period, following a distribution similar to that later estimated by Royston (1982) and described in Chapter 4. Finally, he investigated the effects of several nonuniform patterns of intercourse across the cycle, though always assuming no intercourse during menses.

Bongaarts (1983) considered a special instance of the Glass-Grebenik model in which $F = 2$ (i.e., a fertile period of two days). Following Potter (1961), Bongaarts assumed that p_0 was equal to 0.95, and he set $p_f = 0.95$ and $p_i = 0.5$ based on the work of A. T. Hertig cited in Chapter 6.[11] He also fixed the value of M at 26 days (corresponding to an average cycle length of 29 days minus three days of menstruation) so fecundability became strictly a function of coital frequency, which is what Bongaarts intended. It is simple to generalize Bongaarts's reparameterization of the Glass-Grebenik model to allow fertile periods of any duration, so the interaction of coital frequency and fertile period length can be investigated. With the values of p_0, p_f, p_i, and M chosen by Bongaarts, we have

$$p_a' = 0.45\left[1 - \prod_{j=0}^{F-1}\left(\frac{26 - n - j}{26 - j}\right)\right] \tag{7.18}$$

Figure 7.7 shows the values of p_a' generated by this equation for various coital rates and for fertile periods ranging from short ($F = 2$) to long ($F = 8$). Several features of this figure are worthy of remark. First and somewhat reassuringly, p_a' increases with coital frequency no matter how long the fertile period. Although the increase is roughly linear at lower coital rates, it becomes sharply curvilinear at higher rates as p_a' asymptotically approaches 0.45 (the value of $p_0 \times p_f \times p_i$). The curvilinearity is more marked for longer fertile periods; thus, when $F = 2$ the curve remains approximately linear over most of the range considered here, but when $F = 8$ it rapidly departs from linearity. All these features reflect the fact that we have set p_f, the probability of conception from one insemination in the fertile period, at a value close to one: for all practical purposes, we need to hit the fertile period only once for conception to occur. When the fertile period is as long as eight days, there is a very good chance that at least one insemination will fall within the fertile period even with only five acts of intercourse per cycle; moreover, increasing the number of coitions beyond that value has only a small effect on the likelihood of hitting the fertile period at least once. However, when the fertile period is only two days long, it takes far more coitions to guarantee at least one hit. Finally, it is apparent from Figure 7.7 that the more we increase F, the less and less difference it makes. Thus, it is important to know if the fertile period is as short as two days, but less critical to know where it falls in the range of four to eight days. The value of $F = 6$ suggested by Royston's (1982) analysis, cited in Chapter 4, would seem to rule out very low values of F, and perhaps that is all we need to know. The main features of Figure 7.7, especially the nonlinear relationship between fecundability and coital frequency and the differential effects of varying the length of the fertile period, have proven to be quite robust even in much more complicated models of fecundability. Thus, the Glass-Grebenik model is illuminating despite its simplicity, and because of it.

Potter and Millman (1986) consider a class of models similar to the Glass-Grebenik model but with an interesting twist. Laboratory studies by Guerrero and colleagues (Guerrero and Lanctot, 1970; Guerrero and Rojas, 1975) suggest that p_i may be affected by the amount of time gametes spend in the reproductive tract prior to fertilization and thus by the timing of insemination within the fertile period. Unfortunately, after investigating several different models Potter and Millman were unable to arrive at a single set of consistent parameter estimates describing this process of gametic aging.

In all versions of the Glass-Grebenik model, p_h is defined as the probability of *at least* one coition within the fertile period; one well-timed insemination is thus equivalent to a hundred. In formulating more realistic models, account must be taken of the fact that multiple coitions during a single fertile period do not contribute independently to p_f but in a sense compete with each other. That is, conception can occur on the kth insemination in the fertile period only if it did not occur on the previous $k - 1$ inseminations. If a total of K coitions occur in the same fertile period, and if the probability of conception does not vary across the fertile period but always equals a, then the total probability of conception from all coitions is

$$p_f = \sum_{i=1}^{K} (1 - a)^{i-1} a \tag{7.19}$$

If we allow the probability of conception to vary by day within the fertile period, with a_i being the conditional probability of conception given that insemination occurs on day i of the fertile period, then

$$p_f = 1 - \sum_{i=1}^{F} (1 - a_i)^{x(i)} \qquad (7.20)$$

where $x(i)$ equals one if coition occurs on day i and zero otherwise (Barrett and Marshall, 1969; Barrett, 1971, 1985; Schwartz et al., 1980).

As we have seen, Lachenbruch (1967) was the first to consider departures from uniform intercourse. Glasser and Lachenbruch (1968), on the other hand, were the first to allow for nonconstant intercourse, and in doing so they made a major advance in modeling fecundability. Glasser and Lachenbruch call the probability of coitus by day since last coition the *conditional risk of intercourse* (CRI), equivalent to the hazard rate $h(t)$ for intercourse given no intercourse on the previous $t - 1$ days. All the models considered above implicitly assume that CRI is constant, i.e., $h(t) = \kappa$. Glasser and Lachenbruch argue, with good reason, that this assumption is implausible and suggest that $h(t)$ is likely to increase with t. In other words, they argue that sexual intercourse becomes more probable the more time that has elapsed since the last intercourse—what might be called the "pressure cooker" model of sexual behavior.

Whatever form the CRI takes, the probability density function $f(t)$ for the length of time between successive coitions can always be found as

$$f(t) = h(t) \exp[-\int_0^t h(y)dy] \qquad (7.21)$$

In the case of constant intercourse $[h(t) = \kappa]$, this reduces to the negative exponential distribution

$$f(t) = \kappa e^{-\kappa t} \qquad (7.22)$$

Under this model,

$$E(t) = 1/\kappa \quad \text{and} \quad \text{Var}(t) = 1/\kappa^2 \qquad (7.23)$$

It can be shown that this model yields a Poisson distribution for the number of acts of intercourse in a fixed period of time, such as the fertile period (Cox, 1962).

Glasser and Lachenbruch suggest that a more realistic form for the CRI might be

$$h(t) = At^B, \qquad A > 0, \ B \geq 0 \qquad (7.24)$$

which allows for an increasing CRI. This equation has the virtue of flexibility. If, for example, $B = 0$, it reduces to the previous case of constant intercourse. If, however, $B = 1$, then $h(t)$ increases linearly with the time since last coitus, and if $B > 1$, $h(t)$ increases exponentially (Figure 7.8). If we reparameterize equation 7.24 so $B = \beta - 1$ and $A = \alpha\beta\beta$, we have

$$f(t) = \alpha\beta(\alpha t)^{\beta-1} \exp[-(\alpha t)^{\beta}] \tag{7.25}$$

This is a Weibull density function, which has a mean of

$$E(t) = (1/\alpha)\, \Gamma\left(\frac{\beta + 1}{\beta}\right) \tag{7.26}$$

and a variance of

$$\mathrm{Var}(t) = (1/\alpha^2)\, \{\Gamma\left(\frac{\beta + 2}{\beta}\right) - [\Gamma\left(\frac{\beta + 1}{\beta}\right)]^2\} \tag{7.27}$$

where the gamma function $\Gamma(z) = \int_0^\infty e^{-y} y^{z-1} dy$.

The advantage of assuming constant intercourse is that it produces a model with only one parameter. Thus, if we know the mean frequency of intercourse per week or month, we can specify the entire distribution of intervals between coitions. Glasser and Lachenbruch's model, in contrast, has two parameters, giving it greater flexibility but also making it more difficult to fit. However, if we are willing to fix B—for instance, at 1 so $h(t)$ increases linearly with time—then this model is as easy to fit as the other and is still probably more realistic.

Dobbins (1980) has recently pointed out that the pressure cooker model can have interesting implications for the distribution of coitus by cycle day. Using Glasser and Lachenbruch's Weibull model but with an additional assumption that little or no intercourse occurs during menses, he discovered that time dependence can induce an unexpected periodicity of intercourse by cycle day (Figure 7.9). What happens in Dobbins's model is this. Because of the low probability of intercourse during menstruation, long lags between coitions accumulate over the first few cycle days. Once menses is finished, the CRI is high for most couples; the burst of sexual activity is readily apparent in Figure 7.9 (bottom). Since the majority of couples will have had intercourse during this "burst," the CRI is rather uniformly low during the next several days. Long lags between coitus again accumulate, and as a consequence a smaller burst of intercourse occurs at midcycle, a kind of echo of the first. There is another echo late in the cycle, but it is so attenuated that it is almost unnoticeable.

Admittedly these results are dependent upon the particular values of $h(t)$ used. As Dobbins points out, the steeper the rise in $h(t)$, the greater the amplitude of the echoes and the slower the attenuation. But two features seem to be characteristic of all versions of the model: a peak in coital frequency immediately following menstruation, and a smaller peak at midcycle. Both peaks have been detected in empirical studies of sexual behavior, but they are usually considered to be unrelated (see below). Indeed, the secondary, midcycle peak is sometimes said to reflect the influence of ovarian hormones on female libido or receptiveness. The most significant contribution of Dobbins's work is the finding that the midcycle peak may be nothing more than a shadow of events occurring early in the cycle and have nothing to do with changes in hormones.

Trussell (1979) has used Glasser and Lachenbruch's Weibull function for intercourse in a simulation model of fecundability that included several additional refinements. First, fecundability was corrected to correspond to a month rather than a cycle of exposure (the first of the models considered thus far to make this correction), and the length of the cycle was allowed to vary according to a distribution based on the data of Bailey and Marshall (1970). The cycle day of ovulation was also allowed to vary as $(0.75C - 5)$, where C is total cycle length; this relation, also taken from Bailey and Marshall (1970), implies that most variation in cycle length is attributable to variation in the length of the follicular phase, which we know to be true (see Chapter 4). Finally, the model was run twice, first with a schedule of conception probabilities by fertile-period day used previously by Lachenbruch (1967) and then with another schedule based on the analysis of Barrett and Marshall (1969).

In a recent paper, we have adopted a different approach to modeling fecundability (Wood and Weinstein, 1988). For simplicity we revert to uniform but unrestricted intercourse with a constant CRI. If c is the event that conception occurs during one cycle of exposure to unprotected intercourse,

$$\Pr(c) = \sum_{n=0}^{\infty} \Pr(c \mid N = n)\Pr(N = n) \qquad (7.28)$$

where N is the number of acts of unprotected intercourse per cycle. This equation partitions the probability of conception into two independent components: $\Pr(N = n)$, which tells us something about the frequency of intercourse, and $\Pr(c \mid N = n)$, which tells us the probability of conception given a particular frequency of intercourse. Speaking somewhat loosely, we refer to these as the "behavioral" and "biological" components of fecundability, respectively, ignoring the likely contribution of biology (in the form of libido) to coital frequency. Consistent with the assumption of a constant CRI, we treat the behavioral component $\Pr(N = n)$ as a Poisson random variable and, in some versions of the model, allow it to change as a function of marital duration.

The biological component can be further partitioned as follows:

$$\Pr(c \mid N = n) = \sum_{i=1}^{n} \Pr(c \mid I = i)\Pr(I = i \mid o, N = n)\Pr(o) \qquad (7.29)$$

where o is the event that the cycle in question is ovulatory and I is the number of times that intercourse occurs during the fertile period of that cycle. Under the assumption that acts of intercourse are independent, $\Pr(I = i \mid o, N = n)$ becomes an example of the classic "occupancy problem" in probability theory: we randomly toss n pebbles into a series of boxes, each of a size proportional to the width of the fertile period, and then find the probability that exactly i of the pebbles fall into one particular box, the fertile period itself. The solution is

$$\Pr(I = i \mid o, N = n) = \binom{n}{i}(F/C)^i(1 - F/C)^{n-i}, \quad i = 0, \ldots, n \qquad (7.30)$$

where F and C are the length of the fertile period and the length of the cycle, respectively.

The final piece of the puzzle, $\Pr(c \mid I = i)$, can be found as follows. If $p_r(i)$ is the conditional probability that exactly r days with at least one act of intercourse occur in the fertile period given that $I = i$, then

$$p_r(i) = \frac{\binom{F}{r}\binom{i-1}{r-1}}{\binom{F+i-1}{i}}, \quad r = 1, \ldots, F \qquad (7.31)$$

If a is the probability that conception occurs from a single day of intercourse during the fertile period, then from equation 7.19

$$\Pr(c \mid I = i) = \sum_{r=1}^{F} p_r(i)[1 - (1 - a)^r], \quad i > 0 \qquad (7.32)$$

And since humans do after all reproduce sexually, $\Pr(c \mid I = 0) = 0$. All these pieces (equations 7.28–7.32) can be combined to find the probability of conception per cycle.

In our original paper (Wood and Weinstein, 1988) we use some tedious but straightforward algebra to show how the probability of conception per cycle can be converted into the probability of conception per month (i.e., total fecundability in the more usual sense). We also show that apparent fecundability is related to total fecundability as

$$p_a = \frac{p(1 - q_s)}{1 + pgq_s} \qquad (7.33)$$

where q_s is the probability of subclinical fetal loss and g is the duration of the nonsusceptible period associated with one subclinical loss. Assuming that subclinical losses add no additional time to the birth interval—and, as Figure 6.8 suggests, this assumption is almost but not quite true—then $g = 0$ and equation 7.33 reduces to $p_a = p(1 - q_s)$.

Finally, we add two further wrinkles that are of interest here. First, we allow all the biological factors, which have to do primarily with ovarian function, to vary with the female partner's age, using what we consider to be suitable empirical estimates for western women.[12] Second, we consider a special case that we call *maximal exposure*, in which intercourse occurs with probability one every day of the fertile period. This allows us to control for coital frequency, so any changes in predicted fecundability with the woman's age can be said to result from a pure age effect acting on female reproductive physiology. Maximal exposure is, in effect, our version of the Fédération CECOS study.

Figure 7.10 (top panel) shows the age patterns of total and apparent fecundability predicted by our model under maximal exposure. Despite the fact that coital frequency is held constant, both forms of fecundability change with the woman's age in response to changes in ovarian function and the risk of early fetal loss. Although total fecundability starts near zero, it increases rapidly and reaches a broad plateau between ages 20 and 40—in fact, it increases slightly to a peak at age 39, largely because ovarian cycles are growing shorter up to that age (see Chapter 4). After passing its peak, total fecundability plummets rapidly, reaching zero at the time of menopause. Because at least a few cycles are anovulatory at every age, and because insemination does not inevitably result in conception at any age, total fecundability never quite reaches one even with maximal exposure.

Because of the additional effect of early fetal loss, apparent fecundability follows a somewhat different age pattern under maximal exposure (Figure 7.10, top). Now the peak is at 25, after which age the curve drops off steadily as the risk of fetal loss increases. This difference suggests that any decline in fecundity between ages 30 and 40, such as that suggested by the Fédération CECOS study, is attributable not to a decline in the probability of conception per se, but to a decrease in the ability of the woman to carry a pregnancy to term.

Maximal exposure is, of course, unlikely to be the lot of most couples. It is, therefore, worth investigating a second case in which we hold the mean daily probability of intercourse constant over all ages at its expected value for a married woman age 25, as interpolated from the data of Kinsey et al. (1953). Although the overall shape of the resulting curves is similar to the shape of those produced by maximal exposure (Figure

7.10, bottom), fecundability is substantially lower. Instead of rising to a peak of 0.96, total fecundability now reaches only 0.62; apparent fecundability attains a maximum value of 0.42 instead of 0.68.

The most important conclusion to be drawn from these analyses is that fecundability varies considerably over the female reproductive life course even when coital rates are held constant. Apparent fecundability declines after age 25, and this decline is clearly attributable to early fetal loss, not to other aspects of ovarian function. When we remove the effects of fetal loss, the decline does not begin until about age 40.

Do these changes in female reproductive physiology fully account for the observed changes in apparent fecundability with age? A comparison of our model predictions with the values estimated by Bendel and Hua (1978), shown in Figure 7.5, suggests they do not. The model does well in predicting the increase in apparent fecundability up to age 20, but it predicts a drop-off at higher ages that is less rapid than that estimated from empirical data. Presumably there is an added effect of declining coital frequency after age 25, which suggests that James (1979) is partly (but only partly) correct in attributing the fall in fecundability at higher ages to decreasing marital coital rates. As noted above, this decrease in the frequency of intercourse may be an effect of the husband's age rather than or in addition to the wife's age. Moreover, since the fecundability values estimated by Bendel and Hua for ages 25+ were based on birth intervals of all orders, not just the first, marital duration may also have an effect on coital frequency. Only empirical analyses of coital patterns can untangle these effects to determine which is most important in explaining age-related changes in coital frequency.

EMPIRICAL PATTERNS OF INTERCOURSE

The review of mathematical models in the previous section has identified several aspects of sexual behavior that are potentially important for understanding and predicting fecundability: (1) Do marital coital rates change with age, and if so, is that change caused primarily by the wife's age, husband's age, or marital duration? (2) Are acts of intercourse randomly dispersed across the cycle, or does the likelihood of intercourse change with cycle day? (3) Is the conditional risk of intercourse constant, or does it increase with time since the previous act of intercourse? An additional question, suggested by our earlier discussion of the distribution of conception waits, is whether couple-to-couple variation in coital frequency is likely to be an important source of heterogeneous fecundability. This section reviews the empirical evidence relating to these issues; the next section uses the Wood-Weinstein model described above to assess the impact of the observed coital patterns on fecundability.

Before turning to the evidence, it is important to note some common problems with the validity and reliability of coital data. The most obvious of these are recall error, selectivity bias, and outright deception. Recall error is always a possibility in retrospective studies—that is, in studies based on coital histories rather than coital diaries—and it is typically more severe the longer the period of recall. This is shown by the frequent finding of inconsistencies when subjects are asked to recall how many acts of intercourse occurred during both the previous week and the previous month (Hornsby and Wilcox, 1989). Where practicable, prospective studies are always to be preferred.

In most cultural settings, sexual behavior is far from psychologically neutral, and the people willing to discuss it may not be representative of the wider population. Thus, the possibility of selectivity bias must always be kept in mind, particularly when dealing with preexisting coital diaries maintained by couples for their own edification.[13] Selectivity has also been cited as a special problem for the classic work of Kinsey and associates, since he recruited most of his subjects by cornering them at social occasions and requesting their sexual history—hardly a way to compile a random sample. Deception may also compromise the reliability of coital data. For example, in examining the findings of Kinsey and his colleagues (Figure 7.11a,b), are we really to believe that during the 1930s and 1940s the average unmarried teenage boy was having intercourse more often than once a week, or might some degree of braggadocio be involved? I see no reason to be as cynical as Malcolm Muggeridge, who once wrote, "Let me admit . . . that I never under any circumstances tell the truth about sexual experiences and proclivities, and assume that others are similarly constituted, and that therefore for me the data collected by Dr. Kinsey and other toilers in that vineyard [are] to put it mildly suspect" (Muggeridge, 1965). Still, caution is advisable.

In view of the potential problems, tests of the reliability and validity of coital data are always welcome. For example, if data are collected independently from both partners, their concordance can be checked. Even more ingeniously, Udry and Morris (1967) collected daily urine samples along with coital diaries in order to test the correlation of sexual behavior and hormone levels; they found that sperm fragments could be detected in the wife's urine for about 48 hours after the last insemination, thus providing an independent check on the validity of at least some of the diaries.

The Effect of Age and Marital Duration

It has long been known that coital frequency in married couples tends to decline as the husband and wife grow older (Figure 7.11a,b). The decline is parallel in the two sexes. Since the ages of both partners increase at exactly the same rate, it is difficult to determine which is more important. And of course, barring dissolution of the marriage, marital duration also increases at the same rate. Nor is it possible to rule out a cohort effect: perhaps younger couples have sex more often not because they are younger or because their marriages are more recent, but because they have grown up with more relaxed sexual mores.

When we look at nonmarital rather than marital coital rates (Figure 7.11a,b), the decline with age is much less apparent and may not occur at all. This finding suggests that age may be less important than marital duration. It also suggests that biological explanations of declining coital frequency—attributing it, for example, to a decrease in female libido associated with declining testosterone levels (Persky et al., 1978)—may be incorrect, since such biological factors should affect marital and non-marital coital rates equally.

A recent review by Udry et al. (1982) shows similar declines in marital coital frequency by age in a variety of cultural settings (Table 7.4). Multivariate analysis of these data (Table 7.5) suggests that, although the effect of both husband's and wife's ages are significant in some situations, the effect of wife's age is almost always larger and more often significant (see also Udry and Morris, 1978). Moreover, the range of variation in regression coefficients is greater for wife's age than for husband's, implying that the effect of the former may explain more inter-population variation in coital frequency than the latter.

When the same data are arrayed by year of survey, as in Table 7.4, the coital rates for any one age group are almost always higher in later years. This suggests a possible cohort effect, although not one large enough to account fully for the apparent effect of age. Despite the differences between years, however, once age has been taken into account the apparent similarity of coital rates across disparate populations is surprising.

The apparent effect of marital duration on coital rates takes a somewhat different form (Figure 7.11c). Instead of a consistent, more or less linear decline, the change in coital frequency in U.S. couples is most rapid during the first few years of marriage and then levels off for several years, with perhaps a secondary sharp decline at very high durations. The same general pattern has been observed in rural Bangladesh (Ruzicka and Bhatia, 1982) and other countries (Frank et al., 1992).

Table 7.4. Mean Monthly Coital Rate by Age of Respondent for Married Women, Selected Countries

Age of woman (years)	Thailand 1969			Japan 1975			Belgium 1975			United States[a] NFS 1965			NFS 1970			FPEP 1974		
	Mean	SD	No.	Mean	SD	No.	Mean	SD	No.	Mean	SD	No.	Mean	SD	No.	Mean	SD	No.
<25	8.0	7.3	173	10.5	5.0	110	12.5	6.5	911	9.2	6.7	706	10.6	8.7	1,093	11.6	9.1	565
25–34	6.1	5.6	377	8.2	4.7	390	10.4	5.2	1,545	7.2	4.9	1,344	8.9	6.8	1,935	9.2	7.7	781
35–44	5.8	6.9	245	6.6	4.4	117	8.2	8.2	1,531	5.5	4.0	1,462	6.5	5.1	1,532	5.9	5.8	287
Total	6.4	6.5	795	8.3	4.8	617	10.3	5.5	3,987	6.9	5.2	3,512	8.5	7.0	4,560	9.5	8.2	1,633

[a]NFS, National Fertility Survey; FPEP, Family Planning Evaluation Project.
From Udry et al. (1982)

Table 7.5. Multiple Regression of Monthly Coital Rate
on Spouses' Ages, Selected Countries

Country	N	Wife's Age $\hat{\beta}$	F	Husband's Age $\hat{\beta}$	F
Thailand	795	−0.09	2.87	−0.04	0.65
Japan	617	−0.19	7.30*	−0.11	2.27
Belgium	3,987	−0.31	70.21*	−0.05	2.14
United States					
NFS 1965	3,512	−0.14	16.50*	−0.16	21.14*
NFS 1970	4,560	−0.16	31.84*	−0.08	7.85*
FPEP 1974	1,633	−0.20	32.25*	−0.08	4.89*

$\hat{\beta}$, estimated standardized regression coefficient.
* $P < 0.05$.
From Udry et al. (1982)

To date, only four studies have attempted to assess the effects of age and marital duration simultaneously, all of them using data on U.S. couples (Trussell and Westoff, 1980; Udry, 1980; Jasso, 1985; Kahn and Udry, 1986). In a stepwise multiple regression analysis of four-year panel data collected from 513 married couples during the 1970s, Udry (1980) found a significant effect of both the husband's and wife's ages on coital frequency whenever age alone was entered into the analysis; when marital duration was also entered, however, both age effects vanished and only the duration effect remained. James (1983) has interpreted Udry's results as indicating that marital duration is the most important factor underlying the decline in coital frequency over the life course, and that wife's age appears to be more significant than husband's age in many analyses merely because it is more highly correlated with marital duration (this occurs because the variance in wife's age at marriage is less than that of husband's age in most populations).

In a more elaborate regression analysis designed to control for period and cohort effects, Jasso (1985) found significant negative effects of both husband's age and marital duration but a significant *positive* effect of wife's age. Although it is not yet clear how to reconcile these findings with those of the earlier studies, Kahn and Udry (1986) point out that Jasso's results may be an artifact of model misspecification. Moreover, they note that the apparently significant effect of wife's age vanishes after the removal of only eight outliers from a total of 2,063 observations. Nonetheless, it is intriguing that Jasso's results are consistent with the finding that general sexual responsiveness peaks at about age 17 in males but not until the mid-forties in females (Kaplan, 1986).

To summarize, a steady decline in marital coital frequency over the life course has been observed in all societies studied thus far. Although the relative contributions of spouses' ages and of marital duration have yet to be resolved, marital duration seems to have the most consistent effect. This, of course, is not to suggest that the decline with age is unreal, but merely that age per se may not be its primary cause. The next section will return to the question of whether the declines in coital rate documented above are sufficient by themselves to explain the observed decline in fecundability with age, or whether age-related changes in reproductive physiology must also be involved.

The Probability of Intercourse by Cycle Day

It is well-documented that the likelihood of coitus in U.S. couples is markedly lower during menstruation than in any other part of the cycle (see Morris and Udry, 1976, 1983). Although comparable data on sexual behavior are lacking for most societies, cultural proscriptions against menstrual intercourse are widespread (Róheim, 1933; Ford, 1945; Davenport, 1965). For the intermenstruum, however, evidence of cycle-day dependence in the probability of intercourse is, to say the least, equivocal.

Extrapolating from estrous behavior in nonhuman primates, several researchers have suggested that the probability of intercourse in humans should be elevated at midcycle, coinciding with the fertile period (Adams et al., 1978; Daly and Wilson, 1983). This elevation, in their view, is likely to be caused directly or indirectly by the hormonal changes known to be occurring in the female at midcycle, and it is presumably selectively advantageous because it maximizes the likelihood of conception. From a purely demographic point of view, such an elevation would obviously be of great significance since it would mean that fecundability is actually much higher than coital rates estimated for the cycle as a whole would imply.

In one of the best treatments of this problem, James (1971) has collated prospective coital data from several sources, yielding information on 385 cycles experienced by 140 married women (mostly educated, middle-class white women in the United States and Great Britain). Since changes in female sexual behavior over the cycle (if they exist) may be caused by fluctuating hormone levels, James excluded all women using oral contraceptives on the grounds that they interfere with normal hormonal patterns. For obvious reasons, he also excluded women using the rhythm method. Finally, he only considered days that were "available" for intercourse—specifically, days on which menstruation did not occur and on which neither spouse was ill or absent.

When the remaining data are broken down by total cycle length (Figure 7.12), they provide at best ambiguous evidence for a midcycle rise in coital frequency. According to a series of likelihood ratio tests, none of the patterns in Figure 7.12 departs significantly at the 0.05 level from a uniform distribution of coitus over the intermenstruum. Only the curve for 29-day cycles approaches the borderline of significance ($0.10 > P > 0.05$), and this is primarily because of an elevation in coital rates immediately following menses rather than at midcycle.

Nonetheless, there are suggestive (if not significant) midcycle rises in coital frequency in cycles of all lengths except 30 days. Even if those rises were significant, however, they would not necessarily demonstrate that hormonal changes over the female cycle affect sexual behavior. There are at least two alternative explanations. First, some of the couples may be trying to conceive and therefore deliberately concentrating intercourse at about the time they expect ovulation to occur (the exclusion of couples using certain forms of contraception makes this possibility more likely). Second, in accordance with the model of Dobbins (1980) discussed above (Figure 7.9), the midcycle rise may be nothing more than an attenuated echo of a larger rise following menstruation.

The data in Figure 7.12 enable us to distinguish among these alternative hypotheses to at least some extent. As shown in Chapter 4, most variation in cycle length is attributable to variation in the length of the follicular phase. Thus, if the midcycle rise in coital rates is caused by hormonal changes associated with ovulation, its day of occurrence should be later in longer cycles. Clearly this is not the case. The rise generally falls on day 14 to 16, with no obvious relationship to total cycle length. However, since couples cannot know at the onset of a cycle how long it will be, the data are consistent with a deliberate attempt to time intercourse so that it falls within the presumed fertile period. They are also consistent with Dobbins's time-dependent model of coital frequency.

Adams et al. (1978) suggest that the failure to discern a clear-cut rise in coital frequency at midcycle derives from the fact that male- and female-initiated intercourse are not differentiated. If the probability of intercourse really is sensitive to hormonal changes occurring in the female and not the male, fluctuations in coital rate are more likely to be observed if we confine our attention to those acts of intercourse that she initiates. And indeed, when all forms of female-initiated sexual behavior are considered, and when cycle day is counted backwards from the onset of the next menses in order to provide a somewhat more reliable indicator of the probable time of ovulation, Adams et al. (1978) find increases in the frequency of both autosexual and heterosexual behavior that seem to correspond to the fertile period (Figure 7.13).[14] However, whereas the rise in autosexual behavior represents a significant departure from uniformity, that for heterosexual behavior does not. Once again, the evidence for elevated coital rates during the fertile period is equivocal.

It is apparent that part of the problem in these studies is the lack of a reliable method for detecting the fertile period. In an attempt to overcome this difficulty, Hedricks et al. (1987) collected coital diaries and daily urine samples simultaneously from 25 married, noncontracepting women in urban Zimbabwe. After assaying the urines for LH, they were able to analyze changes in coital frequency relative to the preovulatory surge in LH secretion (Figure 7.14). As in the previous study, there are suggestive changes in sexual behavior over the cycle, with an apparent peak in coital rate corresponding to the day on which the LH surge begins. Once again, however, the departure from uniformity is not significant (one significant difference at the 0.05 level in 20 comparisons is no more than would be expected by chance). In addition, since the subjects of this study were all affluent, educated women who had chosen not to use contraception, we cannot rule out the possibility that some of them were deliberately trying to conceive. To muddle things even further, a recent reanalysis of these data correcting the urinary hormone levels for creatinine concentration (a rough adjustment for the diluteness of the urine specimen) has made the association of intercourse with the LH surge less clear-cut than it originally seemed, although it has also suggested a relationship between the probability of intercourse and urinary *estrogen* excretion (Hedricks, 1994).

In sum, there are tantalizing hints of midcycle changes in coital frequency. But even if these changes are real, the available samples may all be too small to reject the null hypothesis of uniform intercourse over the intermenstruum. Moreover, despite the best efforts of researchers in this area, apparent flaws in study design often make it difficult to distinguish among competing hypotheses. For now, the cautious conclusion must be that increases in coital frequency corresponding to the fertile period remain a possibility but are yet to be demonstrated convincingly in humans.

The Probability of Intercourse by Time Since Last Intercourse

Unlike the probability of intercourse as a function of cycle day, the question of time dependence in general has received almost no attention. This is true despite its importance in the modeling efforts of Glasser and Lachenbruch (1968) and Dobbins (1980). In an analysis of coital diaries maintained by 241 married couples, Barrett (1970) found that the probability of intercourse on a given day was usually lower if intercourse had occurred on the preceding day; however, substantial numbers of couples behaved differently—either their probability of intercourse was unaffected by the previous day's experience, or it actually *increased* when intercourse had taken place previously. At any rate, this study did not address the question of time dependence in the CRI spanning periods longer than a single day.

In a reanalysis of one of the coital diaries discussed above, a reanalysis inspired by Dobbins's model, James (1980) estimated the hazard function $h(t)$ for intercourse over days one to six to be 0.37, 0.39, 0.51, 0.76, 0.80, and 1.00, respectively. (He also found that the probability of intercourse was affected not only by the interval since last coitus, but also by the interval before that.) If such a pattern proves to be common, it would provide strong support for Dobbins's explanation of the increase in coital frequency following menstruation, as well as at midcycle.

As predicted by Dobbins's model, apparent periodicities in coital rates within a single cycle have been observed in many studies (see, for example, Figure 7.15). Although such periodicities are not absolute proof that Dobbins's model is correct, they provide prima facie support for it. Given the available evidence, I suggest that this model is at least as plausible an explanation for changes in coital rates over the cycle as are hormonal changes.

Differences in Sexual Behavior as a Possible Source of Heterogeneous Fecundability

Couple-to-couple variation in coital frequency has been well documented: Figure 7.16 shows the usual pattern seen with prospective data. Unfortunately, since each subject is represented in this figure by an average of more than four cycles, variation within couples is confounded with variation among couples. In addition, since couples of widely differing ages are included, the figure mixes variation associated with age—and doubtless many other factors as well. However, even when wife's age and within-couple variability are controlled, the amount of couple-to-couple variation in coital rates remains considerable (Figure 7.17).

As James (1981) has pointed out, if a given couple's coital rate is more or less constant over a short period of time (say, a week), and if coital rates are distributed among couples as a gamma random variable, then the distribution of acts of intercourse per unit of time among couples should follow a negative binomial distribution. In fact, the fit of this distribution to recall data is rather poor (Figure 7.17, dotted lines). The principal cause of the poor fit seems to be that more couples than expected are having *no* intercourse, which distorts the rest of the expected distribution. When a *truncated* negative binomial distribution is fit to the data after couples with zero coitions are removed, the goodness-of-fit improves considerably. Thus, James suggests that the couples in Figure 7.17 are a combination of two distinct subgroups, one of which follows the negative binomial model, the other consisting of deliberate abstainers or cases of spousal separation. Interestingly, the negative binomial model provides a much better fit to prospective data, such as those collected by McCance and his colleagues during the 1930s (Figure 7.16). This finding suggests that abstainers may be less likely than nonabstainers to volunteer for a prospective study of sexual behavior.

All the available evidence suggests that variation among couples in coital rates is substantial. What this implies for the distribution of fecundability is not yet clear; the answer seems to depend in large part on what model of fecundability is used to draw the implication (see James, 1981). In the following section, I use our own model (Wood and Weinstein, 1988) to assess the impact of heterogeneous coital rates, as well as other coital patterns, on fecundability.

ASSESSING THE IMPACT OF OBSERVED COITAL PATTERNS

What effects do the patterns of variation in coital rates described above have on fecundability? Are those effects larger or smaller than the effects of variation in reproductive physiology? In view of the problems involved in the direct estimation of fecundability, these questions have proven extremely difficult to answer based purely on empirical analysis, except in special (and perhaps unrepresentative) cases such as the Fédération CECOS study. In this section, I explore these issues using our model of age-specific fecundability. In particular, I pose three questions: (1) Does the decline in coital rate with age or marital duration explain the observed decline in fecundability with age, or are physiological factors also important? (2) Does observed couple-to-couple variation in coital rates account for a substantial fraction of the estimated heterogeneity in fecundability or, again, must physiological factors be invoked? (3) Finally, if physiological factors do appear to be important, which ones are most important?

Application of the model has already shown that fecundability varies considerably over the female reproductive life course even when coital frequency is held constant (Figure 7.10). Although these results suggest that age-related changes in physiology do exert important effects on age patterns of fecundability, they do not tell us whether such effects are more or less important than the effects of changing coital rates. Figure 7.18 (left panel) compares the effects of physiology and coital behavior more directly. Declining coital frequency is clearly of some importance in explaining the drop in fecundability between ages 25 and 35, but physiological factors become increasingly important at later ages. Another way to show the same thing is to calculate the fraction of the decline in apparent fecundability after age 25 attributable to physiological changes in the female partner (Figure 7.18, right panel). This fraction increases almost linearly between the ages of 30 and 45, and it reaches 50 percent by age 35. Before this age, most of the decline must be caused by changes in coital rate, but after age 35 the decline is predominantly biological. Our model thus supports the conclusions (but not necessarily other authors' interpretations) of the Fédération CECOS study. The model further suggests that the biological portion of the decline principally reflects an increasing risk of early fetal loss, at least until age 40 when increasing cycle length and anovulatory cycles begin to play an important role.

Of course, this absolute separation of coital behavior and physiology is artificial—which is precisely why we resort to models in order to accomplish it. In reality, coitus and physiology vary together by age and may interact in interesting ways to produce the observed age patterns of fecundability. This possibility is strongly supported by the model results summarized in Figure 7.19, which shows effective fecundability as a function simultaneously of coital rate and physiological age. Clearly, the impact of coital rate is itself dependent upon the physiological characteristics of the woman involved. The effects of coital frequency are almost nonexistent at younger and older reproductive ages, when extreme physiological characteristics tend to predominate. During the most important segment of the reproductive life course, from about ages 20 to 35, fecundability is quite sensitive to variation in coital rate at *low* rates but comparatively insensitive at higher rates, reflecting the nonlinear relationship between fecundability and coital rate discussed above. The most striking feature evident in Figure 7.19 is the broad plateau during the prime reproductive years, suggesting that effective fecundability is fairly invariant over a wide range of coital frequencies at those ages.

If coital patterns play a secondary role in explaining age-specific variation in mean fecundability, do they still explain a major fraction of the heterogeneity in fecundability among couples at any one age? Our model suggests not. We used the model to analyze data on coital rates in two populations, Taiwan and the United States, and obtained a rather surprising result: the observed levels of intrapopulation variation in coital rates generated variances in fecundability that were consistently much smaller than those that have been estimated for these same populations (Weinstein et al., 1993a). Using observed levels of heterogeneity in coital frequency to model apparent fecundability induces at most a 15 percent increase in the mean waiting time to conception relative to the homogeneous case—only a fraction of the increase reported in most empirical studies (see Table 7.1). Indeed, an increase in the mean wait of only 25 percent was induced by inflating the observed variance in coital rates in the United States by 150 percent; for Taiwan, which has comparatively low coital frequencies, the observed variance in coital rates had to be *doubled* in order to increase the mean wait by 25 percent. If these findings turn out to be generally true, then we must conclude that couple-to-couple variation in coital frequency is not an important source of heterogeneous fecundability. Heterogeneous fecundability must be attributable primarily to physiological causes.

But which physiological causes? Too little is known at present about individual-level variation in reproductive biology to be certain, but our model suggests some preliminary answers. Table 7.6 summarizes the results of sensitivity tests of the model, in which each physiological component is allowed to vary through its known range of variation while all the other components are held constant (for the technical details of these sensitivity tests, see Wood and Weinstein, 1988). Large changes in apparent fecundability are induced by varying the probabilities of early fetal loss and of anovulatory cycles, whereas smaller but still substantial changes are induced by varying the length of ovarian cycles. Interestingly, the effects of both the length of the fertile period and the probability of conception from a single insemination in the fertile period depend upon coital frequency. Under maximal exposure, those effects are small, whereas under the more "natural" coital rates typically observed within marriage they are much larger. This result makes sense: because our model assumes that intercourse follows a Poisson distribution across the cycle, there is a good chance that several days without intercourse can occur in succession under anything less than maximal exposure. Thus, given either a short fertile period or a low probability of conception per insemination, marital coital frequencies entail a low probability of conception. The probability of conception is less sensitive to changes in these parameters under maximal exposure because there are always multiple inseminations during each fertile period. These two parameters are, in fact, the principal sources of the inter-

Table 7.6. Sensitivity Tests of a Model of Fecundability: Changes in Apparent Fecundability Induced by Variation in Several Aspects of Female Reproductive Physiology

	Response of Apparent Fecundability[a]	
Physiological factor	*Maximal exposure*	*Marital exposure[b]*
Early fetal loss	+++	+++
Frequency of anovulatory cycles	++	+++
Length of ovarian cycle	+	+
Length of fertile period	±	++
Probability of conception[c]	±	++

[a] Percent change induced by varying each of the model's parameters between its highest and lowest known values for human populations: ±, 0–14.9%; +, 15–29.9%; ++, 30–44.9%; +++, 45–59.9%.
[b] Coital rate set at value expected for marital intercourse at wife's age 25 years.
[c] Probability of conception from a single insemination within the fertile period.
Modified from Wood and Weinstein (1988)

action between coital behavior and reproductive physiology noted above.

In sum, the sensitivity tests suggest that fetal loss, anovulatory cycles, the length of the fertile period, and the probability of conception per insemination within the fertile period are all potentially important sources of heterogeneous fecundability. Although it is premature to say which of these factors are actually important, recent applications of our model suggest that observed levels of variation in the risk of fetal loss may account for a large fraction of the heterogeneity in effective fecundability (Wood and Weinstein, 1990). The contribution of other physiological factors is still unknown.

NOTES

1. By "Malthusian practice" Gini presumably meant sexual abstinence, whereas "neo-Malthusian practice" refers to any other form of contraception.

2. In a sample of n couples, each with fecundability p_j, the harmonic mean fecundability is $H_p = n[\Sigma_j(1/p_j)]^{-1}$. This should be contrasted with the arithmetic mean $\bar{p} = \Sigma_j p_j / n$. The expectation $E(p)$ corresponds to (or, more accurately, is estimated by) the arithmetic mean, not the harmonic mean.

3. Goldman et al. (1985) suggest estimating probabilities of conception at various times postpartum in higher-order intervals without attempting to distinguish ovulatory and anovulatory women. At any given stage of the postpar-

tum period, however, the sample is likely to contain a complex mixture of fully fecund women, subfecund women who have resumed menses but not full ovarian function, and women who are completely infecund because of breastfeeding, making it extremely difficult to interpret such probabilities.

4.   Once α and β have been estimated, we can compute $f(1) = \alpha/(\alpha + \beta)$ and, for $t > 1$, $f(t) = \alpha\beta(\beta + 1) \cdots (\beta + t - 1)/[(\alpha + \beta)(\alpha + \beta + 1) \cdots (\alpha + \beta + t - 1)]$. We can also find $\mu_p = \alpha/(\alpha + \beta)$ and $\sigma_p^2 = \alpha\beta/[(\alpha + \beta)^2(\alpha + \beta + 1)]$. The mean and variance of the distribution of conception waits can be estimated as $E(t) = (\alpha + \beta - 1)/(\alpha - 1)$ if $\alpha > 1$, and $\mathrm{Var}(t) = \alpha\beta(\alpha + \beta - 1)/[(\alpha - 1)^2(\alpha - 2)]$ if $\alpha > 2$. For other parameters estimable from α and β, see Majumdar and Sheps (1970).

5.   There is an alternative explanation. It may be that what we are witnessing is not a selection process based on couple-to-couple heterogeneity, but rather an actual decrease over time in the fecundability of each couple who has not yet conceived. Such a decrease could happen if, for example, coital frequencies decline rapidly during the first few months of marriage. Evidence reviewed below suggests that such a decline in coital frequency may in fact occur in many societies, but it is probably not sufficient by itself to explain the drop in probabilities shown in Figure 7.4.

6.   The obvious advantage of this method is its simplicity. It has one major disadvantage, however, in that it provides no way to assess either goodness-of-fit or the reliability of the estimate of μ_p that it yields. Nor does it allow direct estimation of the variance in fecundability, thus making it impossible to test the assumption that $c = 0.5$.

7.   As discussed in Chapter 11, it is by no means clear that this pattern is strictly an effect of the female partner's age since most studies do not control for any possible effect of husband's age. However, since most estimates of fecundability are based strictly on the first birth interval, marital duration is usually not an important confounder despite its high correlation with spouses' ages. This is true of all the estimates in Table 7.3, as well as the values in Figure 7.5 up to age 25. After that age, Figure 7.5 shows fecundability values estimated indirectly from birth intervals of all orders and thus confounds not only the husband's and wife's ages but also marital duration.

8.   The heterogeneity chi-square for all four curves in Figure 7.6 is highly significant ($P < 0.01$). Controlled comparisons show that ages ≤ 25 and 26–30 do not differ significantly from each other, but that both ages differ from the two older groups ($P < 0.05$ in comparison with ages 31–35, $P < 0.001$ in comparison with ages 36+).

9.   If, for example, much of the azoospermia in the husbands of these women was attributable to gonorrhea, which can cause blockage of the epididymis, vas deferens, or seminal vesicles, then the women themselves may have been at an elevated risk of contracting this disease and thus may have been infertile owing to tubal occlusion.

10.   Recall from Chapter 4 that the length of the fertile period is determined primarily by the life span of the gametes, sperm as well as egg. Characteristics of the sperm and semen may also affect the probability of conception from an insemination in the fertile period and the likelihood of subclinical fetal loss (see Chapter 11).

11.   This estimate of p_f is almost certainly too high (cf. Royston, 1982). Compare the value of 0.64 estimated by Rolfe (1982) using EPF, a figure that is likely to be inflated by false positive diagnoses of pregnancies. On the other hand, Bongaarts's value of p_i is probably too low. According to the study of subclini-

cal fetal loss done by Wilcox et al. (1988), based on a sample far larger than Hertig's, p_i would seem to be about 0.8. On balance, however, these discrepancies probably make little difference: since p_f and p_i occur in Bongaarts's model only as products, their errors tend to cancel out.

12. For the age pattern of variation in total cycle length and in the onset of menarche and menopause, we use the data of Treloar and colleagues (Treloar et al., 1967; Treloar, 1974). Estimates of the age-specific probability that a cycle is ovulatory are taken from Vollman (1977). The length of the fertile period and the probability of conception per insemination in the fertile period are derived from the results of Royston (1982). We assume that the probability of subclinical fetal loss for women age 20–24 is 0.3; this figure is slightly higher than the estimate of Wilcox et al. (1988), reflecting the fact that they used a somewhat conservative criterion for identifying a loss and that they missed very early losses that preceded the excretion of detectable levels of urinary hCG. The risk of loss is allowed to vary with woman's age as shown in Figure 6.5.

13. An instructive example is related by James (1971), who attempted to place a letter requesting that couples who had maintained coital diaries contact him. Several newspapers and journals, including the *British Medical Journal* and *Lancet*, refused to publish the letter. It was accepted by only two, one of which was *Penthouse*. One may be excused for suspecting that, with regard to sexual behavior, the readers of *Penthouse* may not represent a perfectly random sample of all possible respondents. Of course, neither may the readers of *Lancet*.

14. Since all the women in this study were using some form of contraception (exclusive of the pill or the rhythm method), the midcycle rise in coitus is unlikely to reflect deliberate attempts to conceive. At any rate, such attempts would not explain the midcycle increase in autosexual behavior. It should also be mentioned that the midcycle increase in both auto- and heterosexual behaviors is obliterated in subjects using the combined steroid pill, consistent with a hormonal effect on female sexuality during the normal cycle.

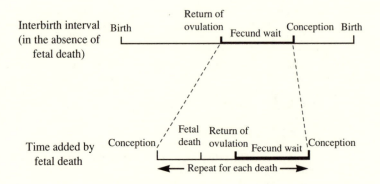

Figure 7.1. The completed birth interval, showing the occurrence of fecund waiting times to conception. (Adapted from Bongaarts and Potter, 1983)

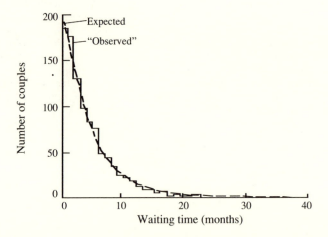

Figure 7.2. The distribution of fecund waiting times to conception. *"Observed"*, waiting times from a Monte Carlo simulation of 1,000 couples, all with fecundability equal to 0.2 (mean wait = 4.996 months). *Expected*, distribution generated by equation 7.1 (mean wait = 5 months). (Redrawn from Wood, 1989)

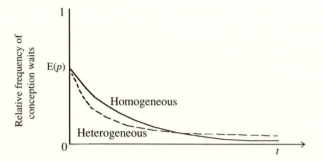

Figure 7.3. Schematic representation of the distribution of conception waits under homogeneous fecundability *(solid line)* and heterogeneous fecundability *(broken line)*, when *p* in the homogenous case equals E(*p*) in the heterogeneous case. For simplicity, each distribution is drawn as if it were continuous. (Redrawn from Wood, 1989)

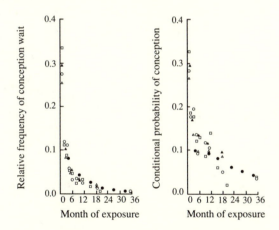

Figure 7.4. The distribution of conception waits *(left)* and the monthly probability of conception *(right)* estimated from data on the first birth interval in U.S. couples. The data, which all pertain to recognized conceptions, come from five studies: the Growth of American Families survey *(closed circles)*, the Family Growth in Metropolitan America survey *(open circles)*, a survey conducted in Indianapolis *(open squares)*, and studies of rural family planning clinics in Kentucky and West Virginia *(closed triangles)* and in North Carolina *(open triangles)*. (Data from Tietze, 1956; Freedman et al., 1959; Weir and Weir, 1961; Westoff et al., 1961)

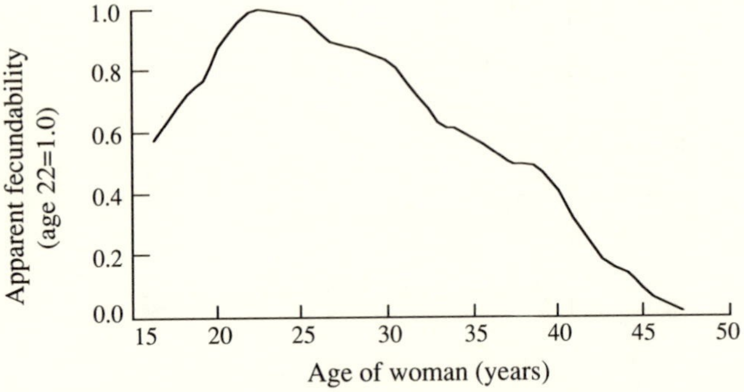

Figure 7.5. The age curve of apparent fecundability (scaled so the maximum value is one). Based on Bendel and Hua's (1978) reanalysis of data from Taiwan (Jain, 1969a) and the Hutterites (Sheps, 1965). Estimates for ages 16–24, based on the Taiwanese data, pertain strictly to the first birth interval. For ages 25+, estimates are derived by an indirect method from data on birth intervals of all orders in the Hutterites.

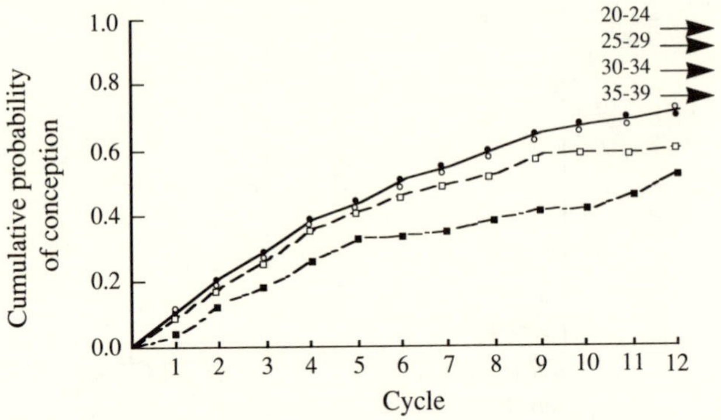

Figure 7.6. Monthly cumulative pregnancy rates following artificial insemination with donor sperm in four age-groups of French women: ≤ 25 years *(closed circles)*, 26–30 *(open circles)*, 31–35 *(open boxes)*, and > 36 *(closed boxes)*. *Arrows*, 12-month cumulative pregnancy rates in naturally inseminated U.S. couples who were continuously married but not surgically sterile or using contraceptives during the 12-month exposure period; numbers indicate the ages (in years) of the female partner. (French data from Fédération CECOS et al., 1982; U.S. data from Mosher and Pratt, 1982)

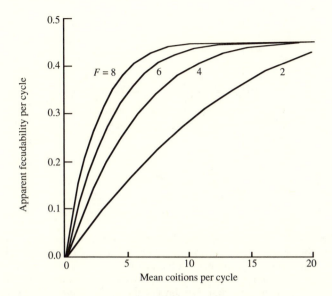

Figure 7.7. The relationship between coital frequency and apparent fecundability per cycle for several values of F, the length of the fertile period. The curves are based on the model of Glass and Grebenik (1954) as reparameterized by Bongaarts (1983), assuming $p_o = 0.95$, $p_f = 0.95$, $p_i = 0.5$, and $M = 26$ days.

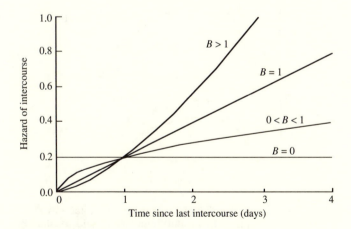

Figure 7.8. Different forms of the conditional risk of intercourse obtained by varying the parameter B in equation 7.24. In all cases, $A = 0.2$.

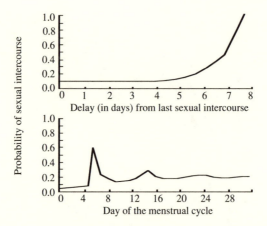

Figure 7.9. A model of time-dependent sexual intercourse. *(Top)* the hazard function for intercourse used in the model. This function was combined with an assumption that the likelihood of intercourse is much lower during menstruation, so $h(t) = 0.01t$ during the first five days of the cycle. *(Bottom)* the probability of intercourse by cycle day induced by this model. (Redrawn from Dobbins, 1980)

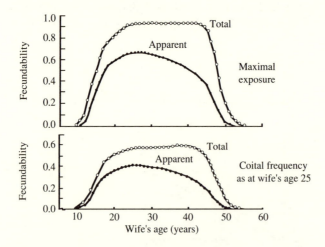

Figure 7.10. Model predictions of age-specific total and apparent fecundability when coital frequency is held constant but the parameters of female reproductive physiology are allowed to vary. *(Top)* maximal exposure or daily intercourse. *(Bottom)* coital frequency held constant at the level observed in married U.S. women age 25, interpolated from Kinsey et al. (1953). (Redrawn from Weinstein et al., 1993a)

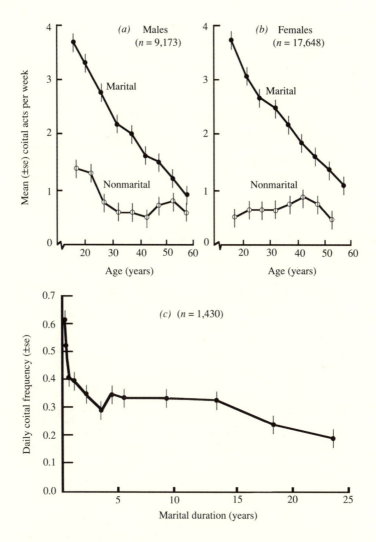

Figure 7.11. Changes in coital frequency in U.S. couples by *(a)* age of male partner, *(b)* age of female partner, and *(c)* marital duration. (Data from Kinsey et al., 1948, 1953; James, 1983)

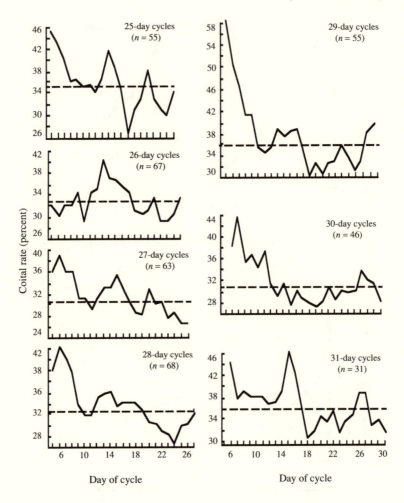

Figure 7.12. Coital rates across the intermenstruum in British and U.S. couples, broken down by cycle length and expressed as percentage of available days (smoothed by taking three-day moving averages). *Broken lines* represent average daily coital rates for all the cycles of each length. Sample sizes *(n)* refer to the number of cycles observed at each length. (Redrawn from James, 1971; original data from McCance et al., 1937; Kinsey et al., 1953; Udry and Morris, 1968; James, 1971)

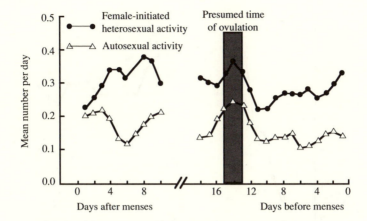

Figure 7.13. Female-initiated sexual behavior by cycle day, as recorded daily by 23 U.S. women over 109 cycles. (Data were smoothed by taking three-day moving averages.) None of the subjects was using oral contraceptives or the rhythm method, but all were using some other form of contraception. (Redrawn from Adams et al., 1978)

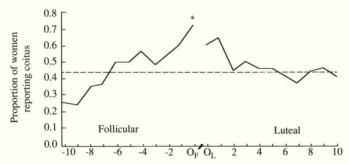

Figure 7.14. Distribution of coitus relative to the LH surge in 25 women of Harare, Zimbabwe, who were not pregnant or lactating and who were using no contraceptives. Data are pooled across several cycles from each subject. Data from the follicular phase are organized around the day of onset of the LH surge (O$_F$), whereas luteal phase data are organized around the day of peak LH secretion (O$_L$). The *broken line* represents the mean coital rate (0.44 ± 0.06) across all cycle days; the *asterisk* (*) denotes a value significantly different from the mean ($P < 0.05$). (Redrawn from Hedricks et al., 1987)

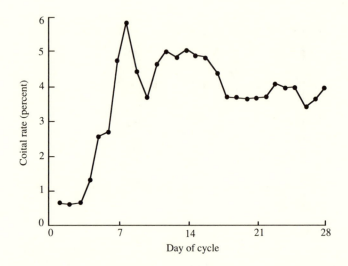

Figure 7.15. Daily coital rate expressed as the percentage of all coitions occurring on a given cycle day. Data are from 1,123 coitions recorded prospectively by 52 women (McCance et al., 1937). Compare Figure 7.9. (Redrawn from James, 1971)

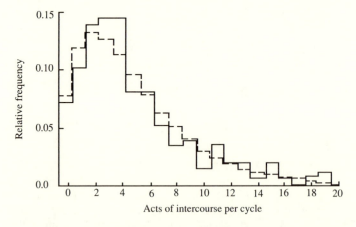

Figure 7.16. Variation in the number of coitions per cycle, based on 237 cycles monitored prospectively in 56 women (McCance et al., 1937). The mean for all cycles was 4.8 coitions, with a variance of 14.9. *Solid line,* observed frequencies; *dotted line,* a negative binomial distribution fit to these data ($\chi^2 = 9.73$, d.f. $= 11$, $P > 0.5$). (Redrawn from James, 1981)

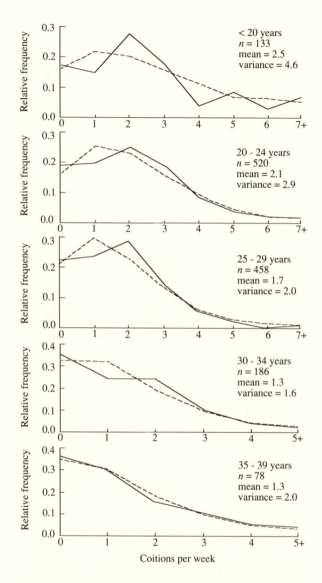

Figure 7.17. Variation in coital rate per week, classified by age of the female partner. Based on recall of previous week's experience by 1,375 women (one week per woman) (Cartwright, 1976). *Solid lines,* observed frequencies; *broken lines,* negative binomial distributions fit to the data. Note the decline in mean coital rate with age. (Redrawn from James, 1981)

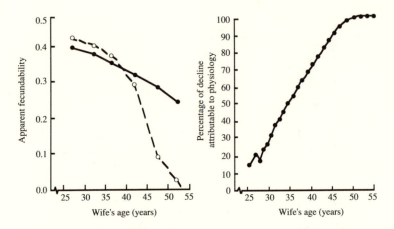

Figure 7.18. (*Left*) comparison of two sets of model-predicted age-specific apparent fecundabilities: parameters of reproductive physiology held constant at wife's age 25 while coitus varies by wife's age (*solid line*), and coitus held constant at wife's age 25 while reproductive physiology varies by wife's age (*broken line*). Marital coital rates are interpolated from Figure 7.11b. (*Right*) the fraction of the age-related decline in apparent fecundability relative to age 25 attributable to changes in female reproductive physiology. (Redrawn from Weinstein et al., 1993a)

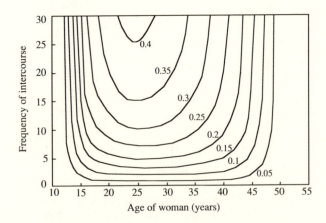

Figure 7.19. Contour map representation of model-predicted values of effective fecundability by frequency of intercourse and physiological age. Contour lines are drawn at increments in effective fecundability equal to 0.05, from zero to 0.40. (Redrawn from Weinstein et al., 1990)

8

Breastfeeding and Postpartum Infecundability

By far the most variable segment of the birth interval is the time from parturition to the first postpartum ovulation, often referred to as the period of *postpartum infecundability* because the woman is incapable of conceiving again as long as it lasts. Numerous studies indicate that this period varies widely, both within and among human populations: some women ovulate within one or two months of childbirth, others may wait two years or longer. It is now well-established that the length of post-partum infecundability largely reflects patterns of breastfeeding. Indeed, *lactational infecundability* is used as a virtual synonym for postpartum infecundability.

The postpartum return of ovulation is impossible to detect unless plasma, urine, or saliva for assaying ovarian steroids is collected from women on a frequent basis (preferably daily) over the course of the postpartum period, or unless follicular growth and the collapse of the follicle at ovulation are monitored by daily ultrasonography. As discussed below, this sort of prospective study has now been done in enough different settings to verify that breastfeeding can indeed cause substantial delays in ovulation. However, most demographic studies have had to rely on more easily detected signs of returning fecundity, principally the first postpartum menses. In reviewing these studies it is important to bear in mind that the periods of postpartum amenorrhea and anovulation, although often similar, are not coterminous; sometimes ovulation precedes the first menses, sometimes it follows it (see below). Furthermore, light, sporadic bleeding (*lochia*) can occur during the immediate postpartum period as high levels of estradiol and progesterone are cleared from the maternal circulation and the endometrium recovers from the trauma of birth. In studies of amenorrhea, it is important to distinguish lochia from genuine menses so the period of amenorrhea is not underestimated.

THE PHYSIOLOGY OF LACTATION

Although later sections of this chapter will emphasize the contraceptive effects of breastfeeding, the primary function of lactation is, of course, nutritive. Numerous studies indicate that in most natural-fertility settings, this nutritive function is of critical importance: a short subsequent birth interval, typically associated with truncated breastfeeding, can significantly increase the risk of death during early childhood (DaVanzo et al., 1983; Hobcraft et al., 1983; Knodel and Hermalin, 1984; Palloni and Millman, 1986; Palloni and Tienda, 1986; Thapa et al., 1988). In rural Egypt, to cite one example, it has been estimated that children who are breastfed for fewer than three months are about 30 percent less likely to survive to age two years than those who are breastfed twelve months or longer (Janowitz et al., 1981). Under such circumstances, lactation is as vital to successful reproduction as gametogenesis, fertilization, and pregnancy since a failure of milk production will often result in early infant death.

Morphology and Development of the Breast

Milk is secreted by the mammary glands, which are made up of masses of glandular tissue arrayed in lobes, with fibrous connective tissue linking the lobes and adipose tissue deposited between them (Figure 8.1). There are about 15–20 lobes per breast, and each lobe consists of several tiny *lobules* interconnected by ducts and blood vessels. When fully developed, the lobules consist of clusters of rounded *alveoli* (sing. *alveolus*), which are the actual site of milk synthesis. The alveoli open into the smallest branches of the milk-collecting ducts, which unite to form larger *lactiferous ducts;* each lactiferous duct drains one lobe of the mammary gland. The lactiferous ducts converge toward the *areola*, beneath which they form *lactiferous sinuses* that serve as milk reservoirs. Each sinus opens directly onto the surface of the nipple.

The immature, prepubescent mammary gland consists predominantly of lactiferous ducts with few, if any, alveoli present and only minor ductal branching. During puberty, as the concentrations of estrogen, growth hormone, and adrenal corticosteroids (steroids produced by the cortex of the adrenal gland) rise, the lactiferous ducts sprout branches and form small, spheroidal masses of tissue that later develop into true alveoli. As regular ovarian cycles are established and the breast is exposed to successive waves of estrogen and progesterone, alveolar and ductal proliferation continues. Before pregnancy, however, the secretory and ductal tissue is limited in volume relative to the connective tissue and fat laid down in the breast during puberty.

The hormonal changes of pregnancy promote rapid growth of the mammary gland. During early pregnancy, under the influence of estrogen and progesterone, and perhaps also prolactin and placental lactogen, the ductal-lobular-alveolar system grows vigorously. Prominent lobules form, the interior spaces (*lumina*) of the alveoli become dilated, and the cells making up the alveoli differentiate into their mature, secretory form. By the end of the fourth month of pregnancy, the breast is fully prepared for lactation, but active milk synthesis and secretion are postponed until after childbirth. It is only after parturition that the breast becomes fully functional.

In sum, the full course of hormonal changes that occur during pregnancy is critical to support *mammogenesis,* or the establishment of a glandular morphology capable of producing large quantities of milk. However, actual *lactogenesis,* or establishment of an actively secreting mammary gland, cannot occur until after parturition. The relevant hormonal changes of pregnancy and the *puerperium* (the immediate postpartum period) are summarized in Figure 8.2. During pregnancy, the rising levels of estrogen and progesterone produced mainly by the growing placenta are essential to mammogenesis, as (albeit less clearly) are the rising levels of placental lactogen (also produced by the placenta) and prolactin (secreted mostly by the maternal anterior pituitary but also by the placenta). The possible effects of prolactin aside, then, mammogenesis is caused primarily by hormones secreted into the maternal bloodstream by embryonic and fetal tissues—the developing infant in effect prepares its mother to feed it. At the same time, it now appears that high concentrations of progesterone actually *inhibit* lactogenesis; it is only after progesterone is withdrawn at parturition with the expulsion of the placenta (Figure 8.2) that the breast begins to synthesize and secrete milk. Thus, the hormones of pregnancy simultaneously prepare the breast for lactation and postpone the initiation of lactation until it is needed after childbirth. Prolactin, which drops after parturition but remains above prepregnancy levels, is essential for *galactopoiesis* or the maintenance of milk secretion.

The hollow alveolus of the mature, secretory breast is lined with a single layer of epithelial cells (Figure 8.3). These cells are the site of milk synthesis and secretion during lactation, and they carry a rich supply of receptors for steroids, placental lactogen, and prolactin. Each alveolus is contained within a capsule of basement membrane, embedded within which are streaks of contractile *myoepithelial cells*. These cells, which share functional characteristics with smooth muscle cells, are important for squeezing milk from the alveolar lumen into the lactiferous duct during milk ejection (see below).

The protein fraction of milk, made up largely of lactalbumin and lactoglobulin, is synthesized within the Golgi apparatus of the epithelial cells and then packaged into vacuoles, which move to the adluminal surface of the cell to be released into the lumen by exocytosis. Milk fat is synthesized within the endoplasmic reticulum of the same cells; after synthesis, membrane-bound lipid droplets move toward the surface of the cell oriented toward the lumen, increasing in size as they do so. The droplet pushes against the outer cell membrane, causing it to bulge into the lumen. Eventually the cell membrane constricts behind the lipid droplet and pinches off to release the membrane-bound droplet into the alveolar lumen. All these cellular processes occur only in the presence of high levels of prolactin in the maternal bloodstream.

Milk Composition

The composition of milk varies greatly over the course of lactation. During the first few days of the puerperium, the mammary glands secrete a small amount of a thick, yellowish, sticky fluid known as *colostrum*. Then a transitional period of about 2–3 weeks follows, resulting in the production of *mature milk*. Compared with mature milk, colostrum is high in protein and low in both fat and milk sugar (*lactose*); it is also rich in maternal immunoglobulins, especially secretory IgA and IgG. Colostrum appears to play an important role in establishing the normal bacterial flora of the infant's digestive tract, as well as providing epidermal growth factors needed for the final maturation of the infant gut.

The maternal immunoglobulins secreted into breast milk play a critical role in protecting the newborn from infectious disease. This protection is especially important during the puerperium, since the infant's own immune system is immature and has yet to be exposed to the full range of pathogens present in its environment. Indeed, it is likely that a large portion of the protective effect of lactation against disease and death during the first few months of life is immunological rather than purely nutritional in nature. It has long been known that certain segments of the maternal gut are surrounded by specialized masses of lymphoid tissue known as *Peyer's patches*. Recent studies have found that this tissue, in effect, monitors the foreign antigens to which the mother is exposed; antigen taken up through the Peyer's patches stimulates proliferation of lymphocytes, which enter the mother's bloodstream and "home in" on the breast. There, these lymphocytes lodge and produce the secretory IgA found in breast milk. By this system of so-called passive immunity, the infant is provided with high levels of antibody specific to the pathogens in its environment (which it normally shares with its mother) at a time when its own capacity for active immunological protection is low.

Mature human milk, like that of other primates, is remarkably dilute, being especially low in fat and protein (although high in lactose) when compared with other mammalian species (Patton and Jensen, 1976). These differences have important implications for the frequency of suckling. As noted by Findlay (1974), "species nursing more or less continuously give a relatively dilute milk, low in fat and protein composition, while those species which nurse seldom (like the rabbit which nurses only once a day, and many aquatic mammals) have richer milk." These facts suggest that suckling frequency has been comparatively high throughout much of primate evolution, which as we shall see has important implications for both galactopoiesis and the contraceptive effects of breastfeeding.

Prolactin and the Maintenance of Lactation

In humans, unlike most other mammals, galactopoiesis is entirely dependent upon concentrations of the peptide hormone prolactin (PRL) in the maternal circulation. PRL appears to act directly on the alveolar cells to stimulate the synthesis and secretion of milk proteins, lactose, and lipids. Prolactin is synthesized and secreted by a specialized line of cells known as the *lactotrophs* located in the anterior pituitary gland. During pregnancy, the pituitary doubles in size, a change that appears to be attributable almost entirely to an increase in the size and number of lactotrophs (Goluboff and Ezrin, 1969; Thorner and Login, 1981). This overgrowth of the lactotrophs enables the mother to secrete large, transitory pulses of PRL in response to suckling (Figure 8.4). These pulses cannot possibly represent elevated de novo synthesis of PRL, for the rapid response of PRL to suckling does not allow enough time for any significant production of new mRNA or for the processing of mature PRL polypeptide; rather, the pulses involve the release of preexisting PRL stores contained within special secretory granules in the cytoplasm of the lactotrophs.

This reflex release of PRL from the anterior lobe of the pituitary is induced by the *suckling stimulus,* or more precisely by mechanical stimulation of the nipple and surrounding tissues (Table 8.1).[1] Since this response involves both hormonal and neuronal elements, it is an example of a *neuroendocrine reflex.* The afferent limb of this reflex arc consists of neural pathways that convey sensory information from the nipple to the spinal cord, and from there to the hypothalamus (Figure 8.5). This information acts upon the hypothalamus, and hence the anterior pituitary, in at least two ways. First, it suppresses the activity of specialized hypothalamic neurons that normally secrete the catecholamine *dopamine* into the portal circulatory system linking the hypothalamus and anterior pituitary. Dopamine acts as a powerful *prolactin-inhibiting factor.*[2] Thus, suppression of the so-called *dopaminergic neurons* in the hypothalamus leads to a transitory elevation in prolactin secretion by the pituitary lactotrophs. In addition, recent evidence suggests that suckling induces a different set of hypothalamic neurons to secrete *vasoactive intestinal polypeptide* (VIP) into the portal system. VIP is an extremely potent *releaser* of PRL, and receptors for VIP have been found on the lactotrophs. These two mechanisms for inducing a burst of prolactin secretion appear to act in concert, since suppression of dopamine actually increases the sensitivity of the lactotrophs to stimulation by VIP.

Prolactin, carried by the bloodstream to the mammary alveoli, represents the efferent limb of this neuroendocrine reflex (Figure 8.5). Although the amount of PRL secreted during any one pulse is affected by the strength and duration of nipple stimulation, the total concentration of PRL in the maternal bloodstream is largely determined by *suckling frequency* or, equivalently, by the time interval between suckling episodes. As shown in Figure 8.4, PRL is cleared from the maternal circulation fairly quickly after the end of the suckling stimulus (plasma

Table 8.1. Serum Prolactin Concentrations (mean ± s.e., in ng/ml) in Nonlactating Women before and after Manual Stimulation

Stimulus site	N	Baseline	Immediately after stimulation	15 minutes after stimulation
Breast and nipple	8	5.1 ± 2.0	108.1 ± 14.3	37.4 ± 3.0
Breast alone	8	5.8 ± 0.8	60.8 ± 4.3	22.6 ± 2.7
Sternum	4	7.9 ± 1.8	7.1 ± 2.3	9.3 ± 1.4
Forearm	4	9.5 ± 2.0	5.5 ± 0.3	5.8 ± 1.5

From Kolodny et al. (1972)

PRL has a half-life of approximately two hours). If another suckling episode does not intervene, PRL approaches the concentration observed in nonlactating women about 4–6 hours after nipple stimulation. Thus, the plasma concentration of PRL is highly correlated with suckling frequency, and persistent elevations in plasma PRL are maintained in the maternal circulation only if nursing occurs more often than once every few hours. Plasma concentrations of PRL during lactation appear to be more important than any other factor in determining the amount of milk secreted. The fact that nipple stimulation *during* an episode of breastfeeding induces PRL release that *subsequently* induces further milk secretion has led Johnson and Everitt (1988:319) to remark that "the baby actually orders its next meal during its current meal."

The Milk Ejection Reflex

A final element in the neuroendocrine control of lactation is the *milk ejection reflex* (MER) or the *milk let-down response*. Milk ejection has to do, not with producing milk, but with making it available to the nursing child. Central to the MER is a reflex contraction of the "basket" of myoepithelial cells surrounding each alveolus (Figure 8.3). This contraction squeezes milk from the alveolar lumen into the lactiferous duct, which then drains into the lactiferous sinus. Since each sinus opens directly onto the nipple, the MER makes milk available for removal by the "suckling" infant—something of a misnomer since the infant removes milk by a stripping action rather than through negative pressure generated by sucking.

The suckling stimulus is the most important, although not the exclusive, inducer of milk ejection. As diagrammed in Figure 8.6, stimulation of the nipple sends sensory information to the hypothalamus via the spinal cord and brain stem. There, this information stimulates specialized neurons that arise within the hypothalamus but terminate in the *posterior* pituitary gland to synthesize and release the peptide hormone *oxytocin* into the maternal bloodstream.[3] Upon reaching the mammary gland, oxytocin induces contraction of the myoepithelial cells and expulsion of milk from the alveolar lumina. The consequent build-up of intramammary pressure may actually cause milk to spurt from the nipple.

Although mechanical stimulation of the nipple is the most potent stimulus of oxytocin release and milk ejection, the MER can be conditioned to occur in response to other stimuli, such as a baby's cry (Figure 8.7). Thus, this neuroendocrine reflex differs from others in that it is partly a learned response. Note that this sort of conditioning does not occur in the case of prolactin release, for which stimulation of the area around the nipple appears to be the only effective stimulus.

This section has emphasized the central role of the suckling stimulus both in sustaining lactation and in making milk available to the nursing infant. As later sections will detail, the suckling stimulus is also important for the maintenance of lactational infecundability. In the complete absence of suckling, the mammary gland rapidly regresses, and lactation typically ceases by the second or third week postpartum.

THE CONTRACEPTIVE EFFECT OF BREASTFEEDING

We turn now to a detailed consideration of the role of breastfeeding in birth spacing. From a functional point of view, this role is likely to be as important to infant survival as are the nutritive and immunological contributions of breast milk. As noted above, the high concentrations of progesterone found in the maternal circulation during the advanced stages of pregnancy shut down active milk synthesis and secretion. Thus, a subsequent pregnancy, if it occurs too soon, can interfere with the nursing of the previous child; indeed, pregnancy has been found to be the most frequently cited reason for weaning (aside from death of the nursing child) in a variety of natural-fertility settings. By delaying the next pregnancy, breastfeeding actually extends the period over which the child can enjoy its nutritional and immunological benefits.

Many studies have confirmed the contraceptive effects of breastfeeding. One of the earliest to do so was a study of 368 women in rural Rwanda who were not using artificial contraceptives and who were monitored prospectively for up to four years postpartum (Bonte and van Balen, 1969). A subgroup of 209 women had children who survived and breastfed until the next pregnancy was detected. The infants of another 50 women died during the first week of life, thus providing a nonnursing control group. Figure 8.8 presents the results of a life-table analysis of the time to the next diagnosed pregnancy in the nursing and nonnursing women. As can be seen, nursing entailed a marked delay in the next conception: the estimated mean time to conception among nonnursing women was 8.4 months and among nursing women 19.6 months, a difference of almost a year. Thus, at least in this study, breastfeeding appears to have acted as a powerful birth-spacing mechanism.

Broadly similar results have now been obtained in many other studies conducted in a wide variety of cultural settings (e.g., Biswis, 1963; Potter et al., 1965a; Jain et al., 1970; Berman et al., 1972; Chen et al., 1974; Delvoye and Robyn, 1980; Huffman et al., 1980; Díaz et al., 1982;

Habicht et al., 1985; Simpson-Hebert and Makil, 1985; Wood et al., 1985b; Zablan, 1985; Elias et al., 1986; Stern et al., 1986; Santow, 1987; Goldman et al., 1987; Rivera et al., 1988; Rosetta, 1989; Vitzthum, 1989; Gray et al., 1990; Lewis et al., 1991; Tracer, 1991). To pick one example, Jones (1988a, 1988b, 1989) has analyzed data on the postpartum experience of 382 women living in rural Java. Each subject was interviewed monthly about her current menstruation status and her "average" daily breastfeeding pattern over the previous month. Although this type of recall data on breastfeeding doubtless has many limitations, this study still produced striking results. When women were classified into four groups on the basis of nursing intensity (more precisely, how much time was spent nursing during each 24-hour period), low-intensity women resumed menses significantly earlier than high-intensity nursers (Figure 8.9), with a difference in the median times to first menses of almost nine months between the lowest- and highest-intensity groups (Table 8.2).

The self-reports of menstrual bleeding used in the Javanese study are subject to various errors, including the problem of distinguishing true menses from lochia. Recently, several studies have tracked changes in ovarian steroids (assayed in serum, urine, or saliva) during the puerperium as a more direct way to monitor the resumption of ovarian function (e.g., Duchen and McNeilly, 1980; Konner and Worthman, 1980; Díaz et al., 1982; Brown et al., 1985; Wood et al., 1985b; Gray et al., 1988; Lewis et al., 1991). One of the longest average durations of ovarian quiescence estimated in this fashion was found in a study of Gainj women from highland Papua New Guinea (Wood et al., 1985b), where breastfeeding appears to delay follicular development and ovulation by a median of more than 20 months (Figure 8.10). Although extreme, this value is not entirely surprising since highland New Guinea (along with Nepal and a few other regions) appears to be characterized by unusually prolonged and intensive breastfeeding practices (Wood, 1992). In addition, individual-level observations of anovulation lasting two years or more are by no means uncommon whenever women nurse frequently and over a very long period, even in the contemporary United States (Stern et al., 1986).

Some studies have estimated the contraceptive use-effectiveness of breastfeeding as if it were a form of artificial birth control. In India, for example, Sehgal and Singh (1966) found that lactational amenorrhea affords contraceptive protection that compares quite favorably with that provided by barrier methods, such as condoms and diaphragms, and is only slightly less effective than oral contraceptives and intrauterine devices in the United States (Table 8.3). Note, however, that lactational amenorrhea does not provide perfect protection. Since a "failure rate" of about nine pregnancies per 100 woman-years of exposure was observed

Table 8.2. Breastfeeding Intensity and the Return to Menses Postpartum in 382 Rural Javanese Women[a]

Breastfeeding intensity	Minutes per bout	Bouts per day	Bouts per night	Total bouts/ 24-hr	Minutes per day	Minutes per night	Total minutes per 24-hr	Median duration amenorrhea (months)[b]
(1) Low	5.12	4.88	3.35	8.23	24.70	17.11	41.81	11.47
N = 41	(0.09)	(0.16)	(0.20)	(0.24)	(0.82)	(1.07)	(1.39)	(0.91)
(2) Medium-low	9.15	4.66	2.65	7.31	42.31	24.02	66.33	14.01
N = 81	(0.27)	(0.08)	(0.04)	(0.10)	(1.43)	(0.70)	(2.02)	(1.07)
(3) Medium-high	8.53	5.14	3.71	8.85	42.41	31.23	73.64	17.65
N = 199	(0.15)	(0.08)	(0.04)	(0.10)	(0.73)	(0.58)	(1.20)	(0.78)
(4) High	8.11	7.23	4.09	11.32	56.64	32.54	89.19	20.35
N = 61	(0.24)	(0.16)	(0.10)	(0.24)	(1.79)	(1.02)	(2.67)	(0.85)
Group differences:[c]	1-2	1-4	1-2	1-2	1-2	1-2	1-2	1-2‡
	1-3	2-3	1-3	1-4	1-3	1-3	1-3	1-3‡
	1-4	2-4	1-4	2-3	1-4	1-4	1-4	1-4‡
	2-3	3-4	2-3	2-4	2-4	2-3	2-3	2-3‡
			2-4	3-4	3-4	2-4	2-4	2-4‡
			3-4				3-4	3-4‡

[a] All breastfeeding variables are expressed as means; time to first postpartum menses is expressed as a median; standard errors are in parentheses. Only group differences that are significant at the α = 0.05 level or better are shown.

[b] P < 0.001, generalized Wilcoxon test for homogeneity among groups.

[c] Group differences for all but last column, P < 0.05, ANOVA, Scheffé's multiple comparison test.

† P < 0.05, generalized Wilcoxon test for paired comparisons.

‡ P < 0.001, generalized Wilcoxon test for paired comparisons.

From Jones (1989)

Table 8.3. Pregnancy Rates (pregnancies per 100 woman-years of exposure) during Different Methods of Contraceptive Protection

	Pregnancy Rate	
Form of protection	India	United States
Lactational amenorrhea	8.9	–
Lactational menstruation	42.7	–
Oral contraceptive	–	4–10
Condom + spermicidal agent	15.0	5
IUD	–	5
Condom	13.8	10
Diaphragm + spermicide	9.6	17
Coitus interruptus	16.8	20–25
Rhythm	38.5	21
Douche	40.8	–
No method	132.2	90

From Sehgal and Singh (1966); Hatcher et al. (1976); Simpson-Hebert and Huffman (1981)

during amenorrhea, it is obvious that at least some women ovulate before their first postpartum menses. As we shall see, this is not an uncommon occurrence.

As noted by Bongaarts and Potter (1983), the contraceptive effects of breastfeeding can also be discerned at the aggregate level, in that populations with prolonged *average* breastfeeding durations also have significantly longer *average* durations of postpartum amenorrhea. Figure 8.11 (left panel), which shows this relationship, is widely cited as evidence for the contraceptive effect of breastfeeding; it also forms the basis for Bongaarts's decomposition of the proximate effects of breastfeeding on the total natural marital fertility rate (see Chapter 3). However, as Habicht et al. (1985) have pointed out, it is strange to use the *total* length of breastfeeding as an explanatory variable when amenorrhea typically ends well before the complete cessation of breastfeeding. As shown in the right panel of Figure 8.11, the average durations of amenorrhea fall below a line with a slope of one for all populations except those with extremely short durations of breastfeeding.[4] In other words, the resumption of menses typically precedes complete weaning. The regression in Figure 8.11 thus predicts the occurrence of an event (menses) using a *later* event (weaning), as if future events can influence the past. It is therefore difficult to interpret Bongaarts and Potter's regression model in terms of any mechanisms that make physiological sense. Nonetheless, the relationship between the duration of breastfeeding and the duration of amenorrhea is striking and is almost certainly telling us something. It would be explicable if (1) suckling *intensity* (e.g., some measure of suckling frequency or the total duration of suckling during any 24-hour period) were important in the maintenance of amenorrhea, and (2) suckling intensity *before* the first menses were positively correlated with the total duration of breastfeeding (i.e., populations that nurse for a long time also nurse more intensively throughout the whole period of breastfeeding). As detailed in the following section, the first of these assumptions is clearly true; later sections provide evidence that the second assumption may also be correct.

Another way in which the aggregate-level effects of breastfeeding can be detected is through the effects of early childhood death on the length of the subsequent birth interval. If a child dies before weaning, breastfeeding obviously ends, and the mother is likely to resume ovulating earlier than she otherwise would have; she thus becomes exposed to the risk of conception, and her next child is likely to be born earlier than if the first child had lived. Numerous studies of natural-fertility populations indicate that infant death occurring more than nine months before the next birth can reduce the birth interval by as much as 35 percent (Table 8.4).[5] It should be obvious from Table 8.4, however, that this effect of early death varies considerably in magnitude across populations. As discussed by Knodel (1988), this variation can be shown to be consistent

Table 8.4. Intervals between Successive Births According to the Survival of the Infant Born at the Beginning of the Interval, Selected Natural-fertility Populations

| | | Mean Birth Interval (Months) | | |
Community	Date of marriage	Following infant death	Following infant survival	% reduction in mean interval by infant death
Mommlingen	1840–1880	19.4	30.0	35.3
Crulai	1674–1742	20.7	29.6	30.1
Cocos–Keeling Islands	1888–1947	20.6	28.6	28.0
Geneva (Bourgeoisie)	1600–1649	19.8	27.4	27.7
Sachsén (miners)	pre-1880	20.5	26.9	23.8
French Canada	1700–1729	19.4	25.0	22.4
Le Mesnil–Beaumont	1740–1799	22.8	27.7	17.7
Bas-Quercy	1700–1792	26.2	31.8	17.6
Geneva (Bourgeoisie)	pre-1600	25.9	30.8	15.9
Schönberg	1840–1890	20.0	22.0	9.1
Anhausen	1840–1890	19.7	19.9	1.0

Compiled by Masnick (1979) from the following sources: Geissler (1885), Henripin (1954b), Henry (1956, 1958), Gautier and Henry (1958), Smith (1960), Ganiage (1963), Valmary (1965), Knodel (1968).

with what is known about breastfeeding behavior in the tabulated populations: the longer and more intensive the breastfeeding, the larger the impact of infant death. And this observation highlights another important point: both breastfeeding behavior and the fertility-reducing effect of breastfeeding vary enormously among natural-fertility populations.

THE NEUROENDOCRINE CONTROL OF LACTATIONAL INFECUNDABILITY

After childbirth, estrogen and progesterone are cleared from the maternal circulation in about two to four days (West and McNeilly, 1979), and hCG is cleared after another 10–12 days (Cameron and Baird, 1988). In the absence of breastfeeding, plasma levels of FSH and LH gradually increase over a 30-day period, leading to a resumption of menstruation at an average of about 1.5–2.0 months postpartum (Bonnar et al., 1975; Howie and McNeilly, 1982). Fully normal ovulation and luteal function are typically reestablished in the nonnursing mother by the second cycle postpartum (Howie et al., 1982; Poindexter et al., 1983). In the nursing mother, in contrast, resumption of normal ovarian function can be delayed by many months. Over the past 10–15 years, it has been a major goal of reproductive biologists to elucidate the mechanisms through which breastfeeding exercises this powerful contraceptive effect.

The Centrality of the Suckling Stimulus

It is now clear that the duration of lactational infecundability is critically dependent upon suckling intensity, although it is not yet certain which particular aspect of suckling behavior—suckling frequency, duration of the suckling episode, or the total duration of suckling over a 24-hour period—is most critical. Individual examples of the relationship between suckling intensity and the maintenance of lactational infecundability are shown in Figures 8.12 and 8.13, taken from the work of Howie and McNeilly (1982) in Edinburgh. Figure 8.12 shows a woman who maintains a fairly high suckling frequency up to the time at which she introduces solid food into the infant's diet at about 22 weeks postpartum. As her suckling intensity declines in subsequent weeks, we begin to see evidence of some follicular development as reflected in urinary estrogen levels. After breastfeeding is suspended at week 40, enough follicular development occurs to cause menses at week 48. Note, however, that this first menses is not preceded by ovulation, which does not occur until week 51–52 (as indicated by the concentration of pregnanediol). Thus, this woman's first "menses" is apparently estrogenwithdrawal bleeding.

A slightly different pattern is shown in Figure 8.13. In this woman, suckling delays resumption of ovarian function, but now several rounds of incomplete follicular development and menses occur before the first definitive ovulation at week 52.[6] Note in both cases that resumption of any form of ovarian activity is postponed until after the introduction of supplementary solid food, after which time suckling intensity declines rapidly.

The aggregate data from this study are shown in Figure 8.14, centered on the time of introduction of supplementary food. As suckling intensity declines, follicular development begins, followed about four weeks later on average by ovulation.

Figure 8.15, which is particularly informative about the role of suckling, contrasts the breastfeeding patterns of early- and late-ovulating women in the Edinburgh study. Clearly, late-ovulating women experience more suckling episodes per day, a longer total duration of breastfeeding, and fewer supplemental feeds than do early ovulators.

The relationship between suckling frequency and the delay of follicular development and ovulation has now been confirmed in numerous studies in which ovarian function has been monitored by plasma or urinary steroids or by ultrasound (Duchen and McNeilly, 1980; Konner and Worthman, 1980; Anderson and Schioler, 1982; Díaz et al., 1982; Brown et al., 1985; Gross and Eastman, 1985; Hennart et al., 1985; Wood et al., 1985b; Glasier et al., 1986; Stern et al., 1986; Rivera et al., 1988; Gray et al., 1993). However, these studies also suggest that there may be wide variation in the suckling intensity at the point where individual women resume ovarian function. This fact has made it difficult to specify a minimum suckling intensity necessary to make breastfeeding a reliable contraceptive. In the Edinburgh study, it was reported that ovulation occurred when suckling frequency had fallen to less than six times per day and suckling duration to less than 60 minutes per day (Howie et al., 1982). More recently, in a study of U.S. women, 80 minutes of suckling per day, in association with a minimum of six suckling episodes, were reported as highly predictive for the maintenance of lactational amenorrhea (Stern et al., 1986). Among the !Kung of Botswana and the Gainj of Papua New Guinea, a pattern of very frequent suckling episodes (40–50/day) of very short duration (approximately three minutes each) was found to suppress ovulation for almost two years (Wood et al., 1985b; Stern et al., 1986). It may be that a pattern of few suckling episodes per day, each of fairly long duration, has an effect similar to one of many episodes of short duration. It also appears that the total daily suckling duration may become more important as suckling frequency declines (McNeilly et al., 1985).

In sum, although the contraceptive importance of suckling intensity, broadly defined, is now well-established, much remains to be learned about how specific aspects of suckling behavior affect the resumption of ovarian activity. Studies of this problem will be greatly facilitated by discovering which particular hormonal consequences of suckling are critical to lactational infecundability (see below). In addition, as noted by McNeilly (1993), "the pattern of suckling in relation to amenorrhea must be established [separately] for each society"—and, one might add, quite possibly for each woman.

Sequence of Changes in the Return of Full Fecundity

As should be obvious from Figures 8.12 and 8.13, the resumption of ovarian function during the postpartum period is not a simple "off/on" process. Rather, it involves a gradual set of continuous changes, from the earliest signs of follicular development, through eventual ovulation, to the reestablishment of normal luteal function. The first postpartum menses may fall almost anywhere in this sequence of changes: in some cases it precedes the first ovulation (as in Figures 8.12 and 8.13) but in other cases it may follow the first ovulation (Figure 8.16). A classic study by Pérez et al. (1971) of 200 Chilean women showed that the likelihood that ovulation precedes the first menses depends upon the intensity of suckling at the time of menses and how many months have elapsed since parturition: the lower the suckling frequency and the later the menses, the higher the probability that ovulation will already have occurred (Table 8.5). Subsequent studies have verified these results (Gross, 1981; Brown et al., 1985; Gray et al., 1990). If ovulation does precede menses, it is entirely possible for the next conception to occur without any intervening menstruation (Figure 8.17).

For this reason, the first postpartum menses cannot be considered an absolutely reliable indicator of the precise time at which fecundity returns. In addition, fecundity does not resume all at once, for breast-feeding has been shown to have "residual" contraceptive effects *after* the first menses. These residual effects were first detected in purely demographic studies (Sehgal and Singh, 1966; Jain et al., 1979) and only later confirmed by physiological research (Howie and McNeilly, 1982). For example, Sehgal and Singh (1966) estimated that the pregnancy rate during lactation was about 0.09 per woman-year of exposure before the first menses, but about 0.43 after the first menses (Table 8.3). Although the

Table 8.5. The Probability That the First Postpartum Menstruation
Is Preceded by Ovulation, Estimated from a Sample of 200
Chilean Women Whose Ovulatory Status Was Determined
by Endometrial Biopsy

	Breastfeeding Status[a]		
Time since parturition (months)	full	partial	none
0	0.00	0.00	0.00
1–2	0.00	0.29	0.83
> 2	0.58	0.79	0.93

[a] Full breastfeeding is defined as breastfeeding in the absence of any
supplemental foods other than water. Partial breastfeeding is breastfeeding
with supplementation.
From Pérez et al. (1971)

latter "failure" rate may seem high in comparison to the former, it is
only about a third of the 1.32 pregnancies per woman-year observed in
the same study when neither lactation nor any other form of contraception was practiced. Thus, suckling continues to provide *partial* contraceptive protection after the first menses. Physiological studies have
shown that this partial protection involves at least three phenomena: (1)
long cycles following the first menses, (2) short and otherwise inadequate luteal phases, and (3) a high probability that the first few cycles
will be anovulatory (Pérez, 1981; Brown et al., 1985). Most of these residual effects of breastfeeding are attenuated fairly rapidly over the first
two or three cycles postpartum, but others may persist until the fifth
cycle or beyond as long as suckling continues at a fairly high rate
(Brown et al., 1985).

The Role of Prolactin

Although the importance of the suckling stimulus is now well-established, the actual neuroendocrine mechanisms linking suckling and
ovarian quiescence during lactation are poorly understood—in fact, the
mechanisms appear murkier today than they did a decade ago. By about
1980, it seemed clear to most researchers that the primary hormone
involved in lactational infecundability was prolactin, even though it was
unknown at what level—hypothalamus, pituitary, or ovary—PRL exerted its contraceptive effects. More recently, it has been questioned
whether PRL has any direct involvement at all in lactational infecundability. This uncertainty about the role of PRL has left us with no very

clear understanding of the neuroendocrine control of lactational infecundability.

In earlier years, prolactin was considered to be a causative agent in lactational infecundability for two principal reasons. The first of these was purely correlational: in numerous studies, plasma PRL was observed to be substantially elevated in nursing mothers throughout the period of ovarian inactivity, and to be decreasing by about the time that ovarian function was reestablished (Howie and McNeilly, 1982; Gross and Eastman, 1985; Wood et al., 1985b; Stern et al., 1986). In Figures 8.12–8.14 and 8.16, for example, ovarian function resumes at about the time that plasma prolactin is dropping to the level observed in nonnursing women (broken line). This association may be spurious since PRL levels can be thought of, in effect, simply as proxy measures for suckling frequency. Moreover, the plasma concentration of PRL at which ovarian activity resumes appears to vary widely among women, with some women experiencing ovulatory cycles while PRL levels are still four to five times higher than those observed in normally cycling, nonlactating women (Howie et al., 1982; Hennart et al., 1985; Stern et al., 1986).

The second reason many researchers believed that prolactin must play an active role in ovarian suppression was clinical: pathological hyperprolactinemia, associated, for example, with tumors of the lactotrophs, often involves a variety of menstrual disorders ranging from irregular cycles to luteal insufficiency to complete amenorrhea (Lehtovirta et al., 1979; Molitch and Reichlin, 1980). Treatment with the drug bromocriptine, which mimics the prolactin-inhibiting effects of dopamine, simultaneously causes a regression in the lactotrophic tumors, suppression of PRL, and reestablishment of normal cycles (Mornex et al., 1978; Lehtovirta et al., 1979; Yuen et al., 1982; Bassetti et al., 1984). However, while the therapeutic effects of bromocriptine on ovarian function may be mediated partly by PRL, there is evidence that the administration of bromocriptine induces follicular development even in women with normal levels of PRL (Seppälä et al., 1976). In addition, recent evidence from Papua New Guinea suggests that women can maintain regular cycles even in the face of PRL levels characteristic of clinical hyperprolactinemia (P. Johnson et al., 1987; Holman et al., 1992). Therefore, the functional link between menstrual dysfunction and hyperprolactinemia per se is uncertain.

Complementing these correlational and clinical studies, McNatty and colleagues (McNatty et al., 1974, 1976; McNatty, 1977, 1979) performed a series of in vitro experiments on the effects of PRL on steroidogenesis in cultured human granulosa cells. Those experiments indicated that high concentrations of PRL in the culture medium were associated with a reduction in the number of viable granulosa cells and a reduced ability of each cell to produce estradiol. In addition, high levels of PRL inhibited the secretion of progesterone in a dose-dependent manner (McNatty et al., 1974). Unfortunately, McNatty and his coworkers were unable to demonstrate the existence of PRL receptors on the granulosa cells, a crucial step in establishing a causal linkage between a hormone and its putative target. In addition, other researchers have questioned whether their experiments realistically mimicked in vivo physiological conditions, among other criticisms (for a review, see McNeilly et al., 1982). Nonetheless, their results remain the strongest evidence available that prolactin may have some direct involvement in lactational infecundability. Combined with other studies indicating that PRL has little if any measurable inhibitory effect on pituitary or hypothalamic function in humans (McNeilly, 1993), these experiments also imply that any contraceptive effects of PRL must be exerted primarily and perhaps exclusively on the ovary, e.g., through inhibition of follicular development and steroidogenesis.

Despite this evidence, the contraceptive role of PRL was called into question by an influential study of the estradiol-induced LH surge by Schallenberger et al. (1981). These investigators worked with ten female rhesus monkeys who had spontaneously given birth to healthy, full-term infants. Beginning on the day of parturition, the investigators periodically administered enough bromocriptine to the mothers to obliterate any measurable prolactin during the study period. Six mothers were then allowed to suckle their infants ad libitum, whereas four were separated on the day of birth from their infants. The capacity of a high concentration of estradiol in the maternal circulation to elicit an LH surge was assessed in both the suckled and nonsuckled females by an injection of estradiol benzoate on the thirtieth day postpartum and at 30-day intervals thereafter for six months. The results of this experiment are shown in Figure 8.18.

In all six suckled females, estradiol failed to induce an LH response on the first attempt at 30 days postpartum. At 60 days, half did not respond, and at 90 days all but one responded with an LH surge (Figure 8.18, top panel). All the suckled females responded at 120 days postpartum. In sharp contrast to the observations in suckled animals given bromocriptine, all the identically treated nonsuckled females responded to the very first and all subsequent estradiol stimuli with an LH surge (Figure 8.18, bottom panel). Thus, *even in the absence of prolactin,* suckling delayed the resumption of a normal gonadotropin response to estradiol by 1–3 months.

Several influential authors have interpreted this experiment as showing that PRL is of no importance whatsoever to lactational infecundability. For several reasons, such an extreme conclusion may not be warranted. First, there is an inherent problem of study design. By abolishing prolactin, the bromocriptine treatment also resulted in complete cessation of lactation in all the experimental animals; consequently, the infants, even those that were allowed to suckle, were fed a commercial formula twice a day. Although "the volume of formula was adjusted to prevent satiation in an attempt to ensure maximum suckling activity" (Schallenberger et al., 1981), it is difficult to believe that the nursing mothers experienced anything like a normal suckling regime. More important, the duration of suppression of the LH surge by suckling in the bromocriptine-treated mothers was considerably shorter than that observed in otherwise similarly handled females who are not treated with bromocriptine. According to a study by Plant et al. (1980), the sorts of LH discharges observed at 2–3 months postpartum in bromocriptine-treated monkeys are not observed in non-bromocriptine-treated monkeys until 7–10 months postpartum (Figure 8.19). This may simply reflect the fact that the bromocriptine-treated mothers were not lactating and therefore their bottle-fed infants may not have been nursing as frequently. But, as Schallenberger et al. (1981) are careful to note, these findings "do not exclude the possibility that the more prolonged inhibition of gonadotropin release in untreated lactating monkeys may be due to a synergism between the effect of suckling per se and elevated serum prolactin concentrations." Finally, the experiment of Schallenberger et al. (1981) addresses only lactational infecundability arising from suppression of the LH response to estradiol stimulus; it does not exclude the possibility that PRL exerts an effect at some other level, for example, on follicular development.

This experiment is nonetheless of great importance in showing unequivocally that prolactin cannot be the *sole* cause of lactational infecundability, if it is a cause at all, for suckling-induced suppression of the hypothalamo-pituitary-ovarian axis occurred even in the absence of measurable prolactin. Even if prolactin eventually does turn out to be involved, there must be a second "arm" of the system linking the suckling stimulus and ovarian quiescence.

Suppression of Pulsatile GnRH Secretion

An important clue about the nature of this second arm was provided by a study carried out by Glasier et al. (1986) during the mid-1980s. In research conducted in Edinburgh, those investigators administered pulsatile GnRH at 6–7 weeks postpartum to four nursing mothers ages 22–33 years whose ovaries showed no hormonal signs of activity at the beginning of the study. The dose of GnRH administered in this experiment (0.1 μg/kg/90 min) was known from other studies to induce ovulation in pathologically anovulatory, nonbreastfeeding women. Despite the fact that suckling frequency did not change significantly during the study period and that plasma PRL levels did not decline, follicular development was induced in all four subjects, and all except one of them subsequently ovulated (Figure 8.20). Moreover, when pulsatile infusion of GnRH was suspended, the still-breastfeeding mothers all returned to a state of ovarian quiescence. Thus, in breastfeeding mothers the pituitary is able to respond to artificially administered GnRH by releasing LH, and the ovary is able to respond to gonadotropin stimulation by initiating follicular development. This study strongly suggests, therefore, that whatever the primary cause of lactational infecundability may be, it must be acting somehow to suppress GnRH release. In other words, the contraceptive effects of suckling appear to operate in large part at the level of the hypothalamus—precisely the level at which prolactin does *not* appear to exert a major influence.

However, this experiment does not necessarily exclude other effects of the suckling stimulus, especially since one out of four experimental subjects failed to ovulate, and the other three displayed abnormal luteal development following ovulation. Thus, additional mechanisms may be acting on the ovary to suppress luteal function, consistent with the "residual" contraceptive effects of breastfeeding noted above after the first postpartum ovulation. In light of the experiments by McNatty et al. (1974, 1976) suggesting that PRL may block the normal development and secretory activity of the granulosa cells and corpus luteum, there may still be some scope for a contraceptive effect of PRL at the level of the ovary.

Nonetheless, the idea that the hypothalamus is the primary locus of lactational infecundability is supported by other experiments in which nursing women have been given amounts of estradiol that would be sufficient to induce a midcycle LH surge in normally cycling women. It will be recalled from Chapter 4 that estradiol secreted by the developing follicle has important feedback effects on pulsatile GnRH secretion during the follicular phase. During the early follicular phase, the feedback effect is negative, and increasing estradiol reduces pulsatile secretion of GnRH; after a certain threshold concentration of estradiol is exceeded, however, the feedback becomes positive, and an LH surge is triggered. Several studies (Baird et al., 1979; Glass et al., 1981; Vermer and Rolland, 1982) have shown that administration of amounts of estradiol well over this regulatory threshold does not induce a positive feedback release of LH in nursing women. Indeed, not only is positive feedback suppressed during breastfeeding, but the negative feedback effect of estradiol appears to be enhanced (Baird et al., 1979).

This enhancement may involve the endogenous opiate β-endorphin. As discussed in Chapter 4, elevated β-endorphin levels tend to suppress pulsatile GnRH secretion, and this action appears to be a major mediator of the negative feedback effects of progesterone during the luteal phase of the normal ovarian cycle (Crowley et al., 1985; Lincoln et al., 1985). It has been known for some time that β-endorphin activity within the hypothalamus increases in response to the suckling stimulus in rats, sows, and ewes (Srinathsinghji and Martini, 1984; Mattioli et al., 1986; Gordon et al., 1987), but evidence for such a response in humans is more recent. In a study of eight healthy women ages 24–36 years who had given birth to normal babies and who were studied on the third or fourth day postpartum, Franceschini et al. (1989) were able to show a transient elevation in plasma β-endorphin peaking about 20 minutes after the initiation of suckling (each suckling episode in this study lasted 20–30 minutes) and ending perhaps 1–2 hours later (Figure 8.21).[7] Now that evidence exists for a suckling-induced β-endorphin response in humans, this endogenous opiate must be considered a prime candidate for the mechanism by which the suckling stimulus suppresses pulsatile GnRH secretion. However, the evidence pertaining to the role of β-endorphin in lactational infecundability is still somewhat circumstantial, and its importance in humans cannot be considered well-established.

If β-endorphin proves to be the primary agent linking the suckling stimulus and suppression of GnRH release, its dynamics become very interesting. The data summarized in Figure 8.21 suggest that the suckling-induced β-endorphin and PRL pulses in plasma are dynamically similar: their peaks occur at about the same time, and the levels of both decline in roughly parallel fashion. Further analysis of these data suggests that the clearance rates of PRL and β-endorphin from the maternal circulation are also roughly similar, the half-life of β-endorphin in plasma appearing to be about 30 percent higher than that of PRL. Thus, PRL should be a good proxy measure for the effects of β-endorphin, even if PRL is in no way directly involved in lactational infecundability. This possibility, which may explain why PRL has long appeared to be involved in lactational infecundability, has an important practical implication: since the plasma concentration of PRL is generally more than an order of magnitude higher than that of β-endorphin, PRL is easier to assay in the peripheral circulation; indeed, methods have recently been developed for assaying PRL from finger-sticks (Worthman et al., 1991). Consequently, PRL may turn out to be a useful biomarker for the activity of β-endorphin.

It appears that suppression of LH secretion may be of greater importance to lactational infecundability than suppression of FSH. Indeed, several studies suggest that plasma FSH increases to levels characteristic of the nonlactating early follicular phase by about the fourth week postpartum, even in the presence of intensive breastfeeding, but LH remains low as long as suckling is sufficiently frequent (Delvoye et al., 1978; Duchen and McNeilly, 1980; Schallenberger et al., 1981; Glasier et al., 1983; Gross and Eastman, 1985). In the study by Glasier et al. (1983), mean basal levels of LH did not increase significantly until normal ovulatory cycles had resumed. However, in several of the women in that study, basal LH was quite high from the immediate postpartum period on—sometimes even higher than in cycling women. What tended to distinguish cases of lactational infecundability from cycling controls was a suppression of *pulsatile* LH secretion (Glasier et al., 1984): during about 70–80 percent of the period of suckling-induced ovarian inactivity, LH pulses were significantly lower in frequency and amplitude. McNeilly (1993) interprets these findings as suggesting that breastfeeding in effect stalls the cycle in the early follicular phase, perhaps by rendering the GnRH-secreting neurons hypersensitive to estradiol negative feedback (precisely the effect that β-endorphin might be expected to have). It is important to emphasize that some follicular development begins well before the first postpartum ovulation (see Figures 8.12 and 8.13) but as long as suckling frequency remains high, follicular development stops after only a modest amount of estradiol has been secreted. However, once ovulation has resumed, breastfeeding continues to exert some inhibitory effect on follicular development and luteal function. Could these residual effects of breastfeeding be evidence that prolactin does indeed have some effect on ovarian function even after pulsatile GnRH release has been reestablished?

Figure 8.22 summarizes the view of the neuroendocrine control of lactational infecundability developed in the preceding paragraphs. A prominent feature of this figure is the "two arm" mechanism hypothesized above: β-endorphin suppresses pulsatile GnRH release in the hypothalamus, while prolactin suppresses follicular development and steroidogenesis in the ovary. Admittedly, several parts of this figure remain controversial, especially the direct involvement of prolactin in the suppression of ovarian function. But what is beyond question is the pivotal role played by the suckling stimulus in the maintenance of lactational infecundability. As Short (1984a) has remarked, breastfeeding is a "push-button" contraceptive, and its effectiveness depends upon how often the button is pushed.

How Long Can Suppression of Ovarian Function Be Sustained?

The studies reviewed above suggest that breastfeeding can delay the resumption of normal ovarian activity by anywhere from a few months to more than two years. But is there a physiological upper limit to lactational infecundability, or can it be maintained indefinitely, at least in principle? A final answer to this question will have to await a better understanding of the neuroendocrine mechanisms involved. However, if prolactin does play an important part, then recent studies suggest that lactational infecundability may be indefinitely extendable.

As noted earlier, pregnancy entails a massive increase in the number and size of pituitary lactotrophs. In the absence of suckling, the lactotrophs regress rapidly during the puerperium, so plasma PRL drops to prepregnancy levels by about 30 days postpartum (Bonnar et al., 1975). Suckling itself may actually retard this regression. In a study of 21 nursing mothers in Papua New Guinea, for example, we found that once the effects of suckling frequency were controlled statistically, time since parturition had no independent effect on PRL secretion (Wood et al., 1985b). In other words, there appeared to be no inherent tendency for the lactotrophs to regress over more than three years postpartum. Thus, it appears that pituitary regression can be delayed and plasma PRL maintained at high levels indefinitely *as long as suckling frequency remains sufficiently high.*

An interesting demographic analysis of survey data has lent support to this idea. Habicht et al. (1985) applied the model summarized in Figure 8.23 to retrospective data collected from 5,312 women in rural Malaysia. Under their model, the authors assume that menstrual bleeding resumes when the plasma level (H) of some maternal hormone—and they are careful not to specify what that hormone may be—drops below a threshold value V. This event can occur during full (unsupplemented) breastfeeding (represented by the line with slope f), during partial (supplemented) breastfeeding (lines with slope p), or when no breastfeeding at all is occurring (lines with slope n). Habicht et al. (1985) used nonlinear least-squares to estimate f/n and p/n from data on the timing of first postpartum menses, nursing status at the time of first menses (full, partial, or none), and nursing history before the first menses (e.g., did full precede partial?). Perhaps the single most remarkable result to emerge from fitting this model to the Malaysian data was that the estimated value of *f/n* was significantly negative; under the quite reasonable assumption that the value of *n* must itself be negative, this result implies that *f* was actually *positive*. Thus, as long as the initial hormone level (*I*) was greater than *V* and full nursing was maintained, *H* could never fall below *V*; in other words, amenorrhea could be maintained forever if full nursing were continued.[8] Incidentally, these findings reinforce one implication of the Edinburgh study (Figure 8.15), namely, that the presence or absence of supplemental feeding is a major determinant of suckling frequency.

In reality, suckling frequency almost always declines during the postpartum period. Nutritional studies indicate that unsupplemented lactation can supply a growing infant's nutritional requirements for only about the first 4–6 months postpartum; after that time, some degree of supplementation is necessary for normal growth, and typically the required amount of supplementation increases steadily as the child grows older. Thus, in most situations suckling frequency tends to decline more or less monotonically with time since parturition. As a consequence, the duration of lactational infecundability rarely extends beyond about two years. There is, however, one human institution within which this normal decline in suckling frequency does not occur, namely, wet-nursing. Fildes (1988) has summarized historical evidence on wet-nursing practices in France and England during the seventeenth and eighteenth centuries, indicating that at least some wet-nurses were rendered permanently amenorrheic and sterile. Although we cannot know whether such women would have cycled normally had they not been wet-nurses, these historical data may provide the strongest evidence that lactational infecundability can be extended indefinitely if suckling frequency remains high.

The Resumption of Menses Following Complete Weaning

What happens when nursing is suspended altogether? One assumption underlying the model of Habicht et al. (1985) is that all the slopes marked *n* in Figure 8.23 are equal, i.e., that the time to resumption of ovulation (or menses) once breastfeeding has been discontinued is the same, no matter when during the postpartum period discontinuation occurs. Moreover, their statistical analysis provides an indirect estimate of about 1–2 months for the time from discontinuation to first menses. To my knowledge, the study of Jones (1989) in Java discussed earlier in this chapter is the only one that has addressed this issue directly. Jones was able to compare the menstrual experience of three groups of women: (1) women who never breastfed, (2) women who completely weaned their children while still amenorrheic, and (3) women whose children died while they were still amenorrheic. As shown in Figure 8.24, the distributions of times from birth (group 1), weaning (group 2), and infant death (group 3) to first postpartum menses are similar, with medians ranging from about 2 to 3 months. Moreover, there was no significant difference in the time to first menses between women who weaned after 14 months postpartum (equivalent to the Javanese "year") and those who weaned before 14 months ($P > 0.14$). Combining these results with those of Habicht et al. (1985), we can conclude that menses resumes in about two months in the absence of suckling, and that this figure holds true no matter when during the puerperium suckling is suspended.

Age and Lactational Infecundability

One final aspect of lactational infecundability that has received very little attention from physiologists is the possible effect of a woman's age on the duration of infecundability. Since there is some evidence for changes in the hypothalamo-pituitary-ovarian axis at later reproductive ages (see Chapter 9), it seems reasonable to ask whether those changes are reflected in patterns of postpartum infecundability.

Several demographic studies have suggested that older women may experience longer durations of lactational amenorrhea (Table 8.6). Since none of those studies controlled for suckling frequency, Corsini (1979) has argued quite plausibly that this apparent age effect is really a cohort phenomenon reflecting changes in breastfeeding behavior. In many developing-world settings, older mothers might be expected to conform to a more traditional, less "westernized" regime of intensive nursing in comparison with younger cohorts. And the fact that the age effect is vir-

Table 8.6. Mean Duration of Amenorrhea by Age of Woman, Selected Studies

Age (years)	Mean Duration of Amenorrhea (Months)				
	Bombay	Punjab	Taiwan	Bangladesh	Boston
< 20	4.0	7.4	9.1		4.3
20–24	5.0	9.4		16.4	
25–29	5.6	10.7	10.2		4.5
30–34	6.2	12.3	11.2	23.2	
35+	6.3	13.1	–	–	4.6

Compiled by Corsini (1979) from Potter et al. (1965a), Salber et al. (1966), Jain (1967), Jain and Sun (1972), Chen et al. (1974).

Table 8.7. Estimated Coefficients (β̂) for Proportional Hazards Model of the Time to First Postpartum Menses, with Breastfeeding Variables as Covariates, 382 Rural Javanese Women[a]

	Model 1		Model 2		Model 3		Model 4	
Covariate	β̂(s.e.)	exp(β̂)	β̂(s.e.)	exp(β̂)	β̂(s.e.)	exp(β̂)	β̂(s.e.)	exp(β̂)
(1) Nursing bout								
≤ 6 min long	0.93(0.17)‡	2.54	0.87(0.33)†	2.39	0.95(0.17)‡	2.59	0.85(0.17)‡	2.35
(2) ≤ 6 feeds/day	0.34(0.16)*	1.41	0.39(0.19)*	1.48	0.43(0.17)*	1.54	0.46(0.17)‡	1.58
(3) ≤ 3 feeds/night	0.34(0.14)*	1.41	0.98(0.41)*	2.66	0.96(0.41)*	2.61	0.83(0.41)*	2.30
(1) × (2)			0.23(0.40)	1.26				
(1) × (3)			-0.25(0.37)	0.78				
(2) × (3)			-0.66(0.44)	0.52	-0.68(0.43)	0.51	-0.64(0.44)	0.52
Age 15–24 years							0.76(0.20)‡	2.13
25–34 years							0.50(0.19)‡	1.65
Log likelihood	-1352.686		-1351.297		-1351.606		-1343.315	
Model χ²	46.78‡		51.53‡		48.59‡		64.02‡	
Degrees of freedom	3		6		4		6	

[a] Covariates are treated as dummy variables; excluded categories are: > 6 min for length of nursing bout, >6 for number of feeds/night, >3 for number of feeds/day, 35–49 years for age.

* $P < 0.05$
† $P < 0.01$
‡ $P < 0.001$
From Jones (1988b)

tually absent in Boston women, who are presumably not undergoing the same secular trend, would appear to support that idea. However, without direct statistical controls for suckling intensity, this suggestion could not be tested.

Once again, Jones's study in Java appears to provide an answer to this question (Jones, 1988b). This author performed a series of proportional hazards analyses of the time to menses under several models (summarized in Table 8.7). When he included age effects as well as several measures of suckling intensity (Model 4) he found that women ages 15–24 years had an estimated hazard of first menses at any given time postpartum that was more than twice as high as that of 35–49 year olds, while the estimated hazard experienced by 25–34 year olds was about 65 percent higher than that of older women.[9] Both differences are highly significant ($P < 0.01$). Therefore, in the only study to date that has simultaneously measured suckling intensity and maternal age, age has been found to have a substantial effect on the duration of lactational amenorrhea. Older women appear to experience longer durations of lactational infecundability, at least in this one instance, and this age effect cannot be explained away as an artifact of changing suckling intensity. At present, there is no well-established physiological reason why this result ought to be true; we must wait for the reproductive physiologists to provide us with an explanation.

DURATIONS OF LACTATIONAL INFECUNDABILITY: EMPIRICAL DISTRIBUTIONS AND MODELS

The fertility-reducing effects of breastfeeding depend upon the amount of time added to the birth interval by lactational infecundability—or more accurately, upon the *distribution* of times added. Unfortunately, good data on the distribution of infecundability are difficult to obtain. A number of well-known defects potentially contaminate such data, depending largely on the type of data-collection system used to generate them (for discussions, see Lesthaeghe and Page, 1980; Bracher and Santow, 1981; Ford and Kim, 1987). In the case of recall data, rounding errors often result in massive "heaping" of observations at preferred durations, typically 3, 6, 12, 18, and 24 months (Figure 8.25). This problem is usually more severe the longer the period over which recall is elicited. Current-status data (i.e., data from surveys in which each nursing subject is asked at a known time postpartum whether she has resumed menstruating) are largely free of heaping but tend to be biased: as shown by Bracher and Santow (1981), such data overestimate the duration of amenorrhea because women who discontinue breastfeeding early and thus resume menstruation early are underrepresented. By far the best data to use are observations collected prospectively, especially if large numbers of women are involved and interviews are repeated at short intervals. Unfortunately, such surveys tend to be costly. To date, only three large prospective studies of lactational amenorrhea (and none of anovulation) have been conducted, one in India (Potter et al., 1965a), one in Bangladesh (Chen et al., 1974), and one in Guatemala (Delgado et al., 1978). All three studies were in rural areas that, at the time the data were collected, could be regarded as approximating conditions of natural fertility.

Figure 8.26 shows the distributions of lactational amenorrhea from these three studies. These distributions differ not only in average length but also in shape: in some instances (e.g., Bangladesh) the distribution is distinctly bimodal, presumably reflecting the presence of women who do not breastfeed or do so for only a short period, as well as those who breastfeed for a considerable period; in others (e.g., Guatemala), the postpartum menstrual experience of women appears to be somewhat more homogeneous, producing a unimodal distribution of lactational amenorrhea. Can a single model be constructed that accounts for this observed range of variation?

Over the years, several attempts have been made to model the distribution of lactational infecundability. To smooth out the problem of heaping on multiples of six months, for example, Lesthaeghe and Page (1980) developed a model based on the logit transform, similar in structure to the well-known two-parameter logit model of mortality developed by Brass (1975). After using a three-point moving average to attenuate heaping, Lesthaeghe and Page transform the smoothed monthly proportions of women still amenorrheic, $P(t)$, into logits using the formula

$$Y(t) = \text{logit } P(t) = \frac{1}{2} \ln\left[\frac{P(t)}{1 - P(t)}\right] \tag{8.1}$$

This transformation linearizes $P(t)$, stretching it out between $+ \infty$ and $- \infty$. Next, they use ordinary least squares to regress $Y(t)$ on a set of standard logits, $Y_s(t)$, generated from the prospective data collected by Delgado et al. (1978) in Guatemala:

$$\hat{Y}(t) = \hat{\alpha} + \hat{\beta} Y_s(t) \tag{8.2}$$

Finally, the model-estimated survival function for amenorrhea is found by the reverse transform

$$\hat{S}(t) = 1 - \frac{\exp[2\hat{Y}(t)]}{1 + \exp[2\hat{Y}(t)]} \tag{8.3}$$

This procedure smooths the observed distribution while ensuring that the model-predicted survival function falls within the interval (0,1). As shown in Figure 8.27, this model largely achieves its goal of eliminating heaping; in addition, it is able to generate both unimodal and bimodal distributions. Unfortunately, however, the model tends to underestimate the proportion of women who experience very short durations of amenorrhea.

In one of the earliest modeling efforts in this area, Barrett (1969) specified the length of postpartum amenorrhea as

$$A = 1 + x_1 + x_2 \tag{8.4}$$

where A equals the number of months the woman remains amenorrheic, and the x_i follow identical and independent geometric distributions. This model gives rise to a modified Pascal distribution,

$$f(t) = \binom{t}{1}(1 - \theta)^{t-1}\theta^2, \quad t = 1, 2, \ldots, \theta > 0 \tag{8.5}$$

where θ is a parameter to be estimated. Barrett's model has the virtue of producing conditional monthly probabilities of resuming menses that increase with the time elapsed since childbirth; this rise is consistent with the known effects of suckling since, as discussed above, suckling intensity tends to decline with time postpartum. However, as Potter and Kobrin (1981) have pointed out, the model does not fit the available data on lactational amenorrhea particularly well, especially in producing excessively high variances for a given mean duration of amenorrhea. To overcome this problem, Potter and Kobrin (1981) suggest the following generalization of Barrett's model:

$$A = C + x_1 + x_2 + \cdots + x_k \tag{8.6}$$

where C and k are positive integers. The resulting modified Pascal distribution now takes the form

$$f(t + C) = \frac{t + k - 1}{k - 1}(1 - \theta)^t\theta^k \tag{8.7}$$

Although this model provides a somewhat better fit to the data than does that of Barrett (1969), neither model allows for the kind of bimodal distribution that is sometimes observed empirically. To allow for bimodality, Potter and Kobrin (1981) propose a mixed geometric–negative binomial distribution, with the geometric representing the experience of women who resume menses soon after parturition. The chief advantage of this distribution is that it permits modeling of a heterogeneous population made up of two subgroups of lactating mothers, one of which nurses intensively and the other of which nurses infrequently, if at all. Unfortunately, although this extension does allow for bimodality, it now tends to *overestimate* the fraction of women who resume menses immediately after childbirth (Figure 8.28).

More recently, Ford and Kim (1987) have suggested a different mixture model for the same two subgroups of women:

$$f(t) = qf_1(t) + (1 - q)f_2(t), \quad 0 \le q \le 1 \tag{8.8}$$

where q is a mixing parameter, and $f_1(t)$ and $f_2(t)$ are independent, nonidentical Weibull distributions with parameters ϕ_1, θ_1 and ϕ_2, θ_2, respectively. Figure 8.29 shows the fit of this model to prospective data on the resumption of menses. Clearly, this model provides the most satisfactory fit of any of the models considered thus far.

The models of Lesthaeghe and Page, Barrett, Potter and Kobrin, and Ford and Kim (equations 8.1–8.8) are good examples of *empirical models*, whose principal justification is that they have been shown to fit at least some observed distributions. None of them, however, is an *etiologic model*; that is, none is derived from existing knowledge of the mechanisms that underlie the contraceptive effects of breastfeeding. An excellent example of an etiologic model of lactational infecundability is that of Ginsberg (1973). Ginsberg assumed that the most important variable predicting the duration of anovulation is suckling frequency, a quite reasonable assumption as it turns out. For a fixed suckling frequency he further assumed that anovulation follows a gamma PDF with rate parameter λ. (The assumption of a gamma PDF is the most arbitrary part of the model.) The parameter λ was then made to depend upon the intensity of nursing, assuming the value λ_1 under full nursing (nursing in the absence of supplementation), λ_2 in the case of partial nursing (nursing with supplementation), and λ_3 if weaning has occurred. Based upon what we know about the biology of lactational anovulation, we would expect that $\lambda_1 < \lambda_2 < \lambda_3$, although these relationships are not a necessary outcome of the model. Finally, Ginsberg assumed that, over the postpartum period, full nursing always precedes partial nursing, which always precedes weaning. Under these assumptions, he derived a modified gamma PDF for the duration of infecundability:

$$f(t) = \lambda(t)e^{-\tau(t)}\tau(t)^{k-1}/(k - 1)!, \quad k \ge 1 \tag{8.9}$$

where k is a constant to be estimated, $\tau(t) = \int_0^t \lambda(y)dy$, and $\lambda(t)$ equals λ_1 up to the introduction of supplemental foods, λ_2 up to the time of weaning, and λ_3 from that time on. Ginsberg also derived MLEs for this model, although not for censored observations. Figure 8.30 shows the excellent fit of this model to the data of Pérez et al. (1971), verifying the prediction that $\hat{\lambda}_1 < \hat{\lambda}_2 < \hat{\lambda}_3$. In the absence of good information on the physiological effects of suckling (and such information *was* largely absent in the early 1970s when Ginsberg developed his model), this analysis provided strong support for a causal role of suckling intensity in the maintenance of lactational amenorrhea.

In the only other modeling effort to include any explicit effects of breastfeeding behavior, Billewicz (1979) suggested the following survival function for postpartum amenorrhea:

$$S(t) = 1 - \frac{1}{\sqrt{2\pi}}[1 - B(t)^2 K(t)] \int_0^t e^{\frac{-z^2}{2}} dz, \quad t > 0.9 \tag{8.10}$$

where $B(t)$ is the proportion of women who are still breastfeeding at t months postpartum, $z = \alpha \ln(t - 0.9) - \beta$, $K(t) = (\gamma - t)/\delta$ for $t \le \gamma$ and zero otherwise, and α, β, γ, and δ are constants to be estimated. Billewicz fit this model numerically to published cross-sectional data by a series of successive approximations, with results shown in Figure 8.31. As can be seen, this model produces mixed results: sometimes the fit is good (e.g., urban India) but at other times it is quite poor (e.g., Sweden or rural Ethiopia). In view of these mixed results and the rather ad hoc method used by Billewicz to fit his model, the model of Ginsberg (1973) is to be preferred.

Efforts such as these are fundamentally important to our goal of developing dynamic models of birth-spacing processes. Taken together, the particular models reviewed here provide three additional insights. First, to fit empirical distributions well it is necessary to allow for heterogeneity in breastfeeding behavior within each population; thus, the mixture models of Potter and Kobrin and of Ford and Kim perform significantly better than does the model of Barrett, which in effect treats women as homogeneous. Second, it is generally possible to achieve a better fit when the model incorporates data on breastfeeding *behavior*, as in the models of Ginsberg and Billewicz. Third, and perhaps most important, these analyses suggest that there are important commonalities in patterns of amenorrhea across widely differing human populations, commonalities that reflect the common physiological mechanisms and behaviors underlying the empirical patterns. Although, in detail, the distributions of lactational infecundability may vary widely across populations, the variation is apparently constrained, perhaps in a way we can hope to understand.

SUCKLING PATTERNS AND LACTATIONAL INFECUNDABILITY

In view of the fundamental importance of suckling intensity—perhaps the only point in our understanding of lactational infecundability that is universally accepted—it may seem surprising that few empirical studies of breastfeeding behavior have attempted to measure suckling frequency in natural-fertility settings on the basis of anything other than subject recall (Whitehead et al., 1978; Huffman et al., 1980; Konner and Worthman, 1980; Wood et al., 1985b; Vitzthum, 1989; Panter-Brick, 1991; Tracer, 1991; Worthman et al., 1993). As mentioned earlier, recall data on suckling frequency are unlikely to be reliable; at best, they may provide an indication of "average" rates of breastfeeding over the period of recall. By far the most useful observations of suckling intensity come from prospective studies in which a trained observer monitors a mother-offspring pair directly, timing suckling episodes with a stopwatch. As might be imagined, such studies are enormously labor-intensive and time-consuming, and few have involved more than a handful of subjects.

In comparing the prospective studies of suckling behavior that have been done to date, several methodological difficulties should be borne in mind. Most of these problems are statistical in nature and are most easily introduced at relevant points during the following discussion of specific studies. But there is also a more general problem of definition: there is no consensus among researchers about what constitutes a single *suckling episode*, the basic unit of observation. In many recall studies, especially among women in industrialized societies, a single "feed" is defined as the time interval over which the child is presented with the breast, even if the child does not suckle continuously throughout that interval. Such a definition is well-suited for the breastfeeding patterns that characterize most western societies, in which the mother and child remove themselves to a separate room at a specially scheduled time, and the mother bares her breast at the beginning of the feed and then finally covers it at the end. Such a pattern often includes an element of "ritualistic" behavior: the mother may change and powder the baby at the beginning of the feed, for example, and she may attach emotional significance to the special time she gets to spend alone with her infant.[10]

This definition of a breastfeeding episode is, however, of little use in natural-fertility settings. In most traditional societies, breastfeeding is unembarrassing and freely conducted in public. Often, especially in tropical settings, the mother's breasts are bare throughout much of the day, and no great effort is needed to make the breast available to the child. Breastfeeding may be more or less "on demand"; that is, the mother may suckle the child whenever it cries in a way that suggests hunger—or even at other times, in which case suckling may be as much for comfort and pacification as for nutrition. (In terms of lactational infecundability, this distinction is largely irrelevant since nipple stimulation and its hormonal consequences occur in either case.) Such a pattern of breastfeeding is typified not by a few, discrete, regularly scheduled "feeds," but rather by sporadic suckling throughout the day and night.

In our research we have adopted an extreme definition of a single suckling episode: two episodes are distinguished if the nipple visibly leaves the child's mouth for at least five seconds. Compared with definitions used in other studies (e.g., feeds must be separated by at least 30 minutes), our criterion will lead to the appearance of a very high suckling frequency. But we would argue that our definition makes sense given the current state of knowledge. By measuring suckling behavior with a fine-grained scale, we can subsequently look for higher-order temporal clustering of episodes that may provide a more natural definition of a feed for the society under investigation. With a fine-grained measurement scale we can always aggregate observations upward, but with a coarse-grained scale it is impossible later to disaggregate the data onto a finer scale. Our definition appears to be the one most widely adopted in recent anthropological work (Vitzthum, 1989; Tracer, 1991; Worthman et al., 1993). It is important to emphasize, however, that whatever the definition of a suckling episode chosen, it makes a difference, because the choice will largely determine the results of any analyses done on the resulting data. This fact must be kept in mind when results from studies based upon differing definitions are compared.

One of the first prospective studies of suckling behavior in a natural-fertility setting was conducted by Konner and Worthman (1980) among the !Kung San of Botswana. These investigators monitored each of 17 mother-infant pairs for six hours (in two-hour periods spread over separate days), recording the precise timing of a variety of behaviors, including suckling.[11] The infants involved ranged in age from about 3 to 32 months. Four of their sets of observations are summarized in Figure

8.32, which should convey something of the richness of these data. Konner and Worthman's study was the first to document what appears to be true on-demand feeding: the child is given the breast almost every time it frets or cries, and the overall pace of suckling is largely determined by the child's sleeping patterns (Figure 8.32). Moreover, suckling is clearly not scheduled at regular intervals; rather it occurs frequently and sporadically and is often of very short duration.

Pooling their observations, Konner and Worthman (1980) estimate that suckling episodes among the !Kung are separated by only 13.2 ± 1.3 minutes on average, and that each episode lasts an average of only 1.9 ± 0.2 minutes (means ± standard errors). This is unquestionably a pattern of very frequent and short suckling episodes. There are, however, two reasons to suspect that their estimate of the mean interepisode interval may be biased downward by a substantial amount. First, prospective observations on suckling are subject to right-censoring. Since Konner and Worthman's estimate is based strictly on completed interepisode intervals, it ignores the open interval between the last nursing episode observed and the end of the observation period. The probability that a randomly selected interval is open rather than closed depends upon its length; that is, short intervals are more likely to be completed within the observation period than are long intervals. Thus, the completed intervals that have been recorded are not an unbiased sample of all intervals begun during the observation period. The way to circumvent this censoring problem is to perform a hazards analysis on the data (see Chapter 3).

The second bias is one that we have called *disproportionate weighting* of the observations (Wood et al., 1985b). This problem frequently affects estimates calculated from pooled data on different mother-offspring pairs. Such pooling places disproportionate weight on shorter interepisode intervals, since pairs with frequent episodes contribute more intervals to the observation pool than pairs with more widely spaced episodes. This problem can be avoided quite simply by calculating the mean interval separately for each mother-offspring pair, thus weighting all pairs equally. If a single global mean is desired, it is best found as the mean of the mean intervals across mother-offspring pairs.

An additional difficulty in interpreting the !Kung data has to do with the age composition of Konner and Worthman's sample. If suckling frequency changes with the child's age, as it very commonly seems to do, then a particular mean frequency will in part reflect the age composition of the sample from which it is estimated. Unfortunately, Konner and Worthman do not provide details of the age composition of their sample, making it difficult to compare their observations with other samples.

Figure 8.33 summarizes data from a study of another natural-fertility population, the Gainj of highland Papua New Guinea (Wood et al., 1985b). In this study, 21 mother-offspring pairs were monitored continuously from approximately 8 AM to 4 PM, corresponding roughly to a normal gardening day. Inspection of the full set of data shown in Figure 8.33 gives the impression that suckling episodes tend to be clustered in time. This impression is confirmed by a hazards analysis of the interval between suckling episodes (Figure 8.34, top panel). Compared with a situation in which the probability of initiating suckling is constant across time (the solid curves in Figure 8.34), more short interepisode intervals (< 5 min), fewer intervals of intermediate length, and more very long intervals (> 160 min) are observed in the pooled data set than expected. It might be thought that this pattern is simply the result of pooling observations involving offspring of many different ages; for example, the very short intervals might be contributed by newborns and the very long ones by older children. But in fact, essentially the same pattern is observed in individual children at each age (Figure 8.34, bottom panels). Regardless of the child's age, suckling is not distributed homogeneously across time.

To avoid the problems of censoring bias and disproportionate weighting, the data from this study were analyzed using hazards methods as summarized in Figure 8.35. The rate of suckling appears to decline from about two episodes per hour in very young infants to about one per hour in three-year-olds; however, according to a proportional hazards analysis of these data using the child's age as a covariate, this decline is not statistically significant ($P > 0.40$). The finding that suckling frequency declines slowly if at all as the child ages may go a long way toward explaining the unusually prolonged period of lactational anovulation that has been estimated for this same population (see Figure 8.10).

However, the apparent slow decline in suckling intensity inferred for the Gainj may partly be an artifact of selectivity arising from the cross-sectional design of the study (i.e., the fact that age patterns are not inferred by following each mother-offspring pair as the offspring grows older, but reconstructed indirectly by pooling observations on many offspring of differing ages). In the nature of things, suckling frequency can be observed only in mothers who are still nursing at a given time postpartum. If early- and late-weaning mothers have nursing patterns that differ from each other, the sample will become increasingly selected to reflect the experience of late weaners as time goes on. If, for example, early-weaning mothers nurse comparatively infrequently, the sample will be selected toward higher suckling frequencies at longer durations postpartum. The effect of this selectivity on the apparent time trend in suckling frequency is illustrated in Figure 8.36. Clearly, the rate of decline in the *mean* suckling rate of women who are still nursing is lower than the rate of decline actually experienced by the women who contribute the observations. The only obvious way to eliminate this problem is to monitor each mother-infant pair over the entire course of lactation, obtaining frequent repeat measures on each pair and then analyzing the rate of change in suckling intensity on a pair-by-pair basis. Equally obviously, such a study is likely to be extremely costly.

Largely because of the sorts of methodological difficulties reviewed above, no very clear understanding of variation in suckling behavior has emerged from a purely empirical approach to the problem. A more productive approach, I would argue, would be to develop a *theory* of suckling behavior—that is, an understanding of the causal mechanisms that drive suckling frequency—an *ecology of breastfeeding*, as it were. Although such a theory is not yet available, a few obvious generalizations can be made. For example, tropical populations whose food staples include starchy root crops such as cassava, sweet potato, or yams, which are high in bulk but low in fats and proteins, may lack appropriate supplemental foods on which to wean infants and may therefore tend to breastfeed for comparatively long periods of time. In contrast, pastoral peoples with access to reliable sources of nonhuman milk (e.g., from cattle, camels, goats, sheep, or horses) may tend to wean their children at comparatively early ages. But such crude generalizations are of little help in making detailed predictions about breastfeeding behavior, and they are essentially useless for understanding intrapopulation variation in suckling.

Konner and Worthman's analysis of the complementarity of suckling and infant sleep (Figure 8.32) suggests one approach to the development of a more useful ecology of breastfeeding: it may be important to understand the set of *competing demands* for the mother and infant's attention, time, and energies that may distract from breastfeeding. Turning this logic around, we need to investigate the *opportunity costs* of breastfeeding, as well as its more tangible physiological costs for the lactating mother.

The observations on suckling collected among the Gainj appear to reflect such competing demands. As mentioned before, suckling episodes are not distributed uniformly with respect to time among the Gainj. If they *were* so distributed, the number of episodes occurring each hour would follow a Poisson distribution, and such is not the case. However, it is possible to fit the observations well with a mixture of *two* Poisson distributions (Figure 8.37), suggesting that there may be two categories of time during the day; in one suckling frequency is low (about once every two hours) and in the other it is high (about five episodes per hour), but the time spans within each category are homogeneous. One interpretation of these results is that at certain times (the high-frequency hours) there are few if any constraints on suckling; these times presumably reflect true "on-demand" feeding, the pace of which is set largely by the child. However, at other times (the low-frequency hours) competing demands intervene to lower the suckling rate markedly. It is unclear what these demands might be, although they may reflect the child's sleeping patterns or the mother's gardening activities (cf. Panter-Brick, 1992). It is striking that the "constrained" hours represent about 80 percent of the total observation time, with only 20 percent of the time given over to true on-demand feeding.

One of the most remarkable studies of the competing demands for a nursing mother's time has been carried out in Matlab, a rural district in southern Bangladesh (Huffman et al., 1980). In this study, suckling patterns were observed in 216 mother-offspring pairs in which the offspring were 17–25 months old and the mothers still amenorrheic at the time observation was initiated. Perhaps the most striking finding of this study was a marked seasonality in suckling intensity (Figure 8.38), with a distinct trough from November to February and a peak in June and July.[12] Elsewhere, this pattern has been shown to account for a substantial proportion of the seasonal variation in fertility observed in Matlab, although by no means all of it (Becker et al., 1986; McElrath, 1992). The trough in suckling behavior in November–February corresponds closely to the period during which women are most intensively involved in subsistence activities, especially harvesting and processing rice. Apparently, agricultural production and breastfeeding are in some sense in competition for a nursing mother's time. The June–July peak in suckling intensity corresponds approximately to the period when agricultural produce is least abundant and when few appropriate supplemental foods are available for the nursing child (Chen et al., 1974).

Another issue addressed in the Matlab study was the reason for terminating breastfeeding. As detailed in Table 8.8, the only important reason for termination of nursing during the first year of life mentioned by informants was infant death. After the first 18 months, a new pregnancy was the most frequently cited reason, consistent with the likely im-pact of pregnancy on galactopoiesis described above. On average, however, nursing continued until about the sixth or seventh month of the next pregnancy (Table 8.9), suggesting that steroid concentrations in the pregnant mother's circulation may not be high enough to interfere with continued milk production until the third trimester. And Tables 8.8 and 8.9 underscore another important point: despite the contraceptive effect of breastfeeding, a large fraction of pregnancies in natural-fertility populations begin before the previous child has been completely weaned.

Table 8.8. Proportion of Women Terminating Breastfeeding by Infant's Age
and Reason for Stopping, Matlab, Bangladesh

	Reason for Terminating Breastfeeding					
Child's age *(completed months)*	*Child* *death*	*Next* *pregnancy*	*Insufficient* *milk*	*Other*[a]	*All*	*N*
0–6	0.97	0.00	0.01	0.02	1.00	256
7–12	0.80	0.09	0.02	0.10	1.00	93
13–18	0.57	0.27	0.02	0.14	1.00	56
19–24	0.15	0.53	0.14	0.18	1.00	88
25–30	0.09	0.43	0.25	0.23	1.00	64
All	0.67	0.18	0.06	0.10	1.00	557[b]

[a] Includes infant becoming too old or stopping willingly, infant sick, mother divorced or separated from child.
[b] Excludes five cases in which timing of cessation of breastfeeding was unknown.
From Huffman et al. (1980)

Table 8.9. Life-table Estimates of the Probability
of Breastfeeding during Pregnancy by Number
of Months Pregnant, Matlab, Bangladesh

No. of *months* *pregnant*	*No. of* *women*	*Probability* *of still* *breastfeeding*
1	295	1.00
2	266	0.98
3	228	0.96
4	170	0.88
5	110	0.78
6	60	0.64
7	23	0.45
8	8	0.31
9	0	0.19

From Huffman et al. (1980)

IS NOCTURNAL BREASTFEEDING
OF SPECIAL IMPORTANCE?

It has been known for some years that the secretion of prolactin displays a marked circadian rhythm in nonlactating individuals (Sassin et al., 1972). Recently, such circadian variation has also been observed in the suckling-induced PRL pulse (Gross et al., 1979; Liu and Park, 1988; Díaz et al., 1989). Specifically, the PRL response to suckling is approximately 4–6 times greater at 12–4 AM than at 8 AM (Figure 8.39). Moreover, the nocturnal PRL surge is significantly greater at any given time postpartum in women who have remained amenorrheic than in those who have resumed menses (Figure 8.40). These observations have led several authors to suggest that nighttime nursing may be of special importance in the maintenance of lactational infecundability. This suggestion is supported by data from Java, indicating that a comparatively small number of nighttime feeds (three or more) has an effect on the duration of amenorrhea that is as large as or larger than that of a higher number (six or more) of daytime feeds (Table 8.10). If this finding proves to be general, behavioral data on nocturnal suckling patterns will become of great interest.

Unfortunately, such data are difficult to collect under field conditions without recourse to invasive techniques that are likely to disrupt normal feeding patterns. It seems unlikely, however, that daytime and nighttime suckling rates should be identical. In many cultures, the nursing child always sleeps with its mother, and the breast is exposed to the child throughout the night. In our own research in Papua New Guinea, we have observed suckling taking place while the mother slept, and I am convinced that we have observed it even when the *child* was asleep.

In general, then, nighttime feeds are likely to be more frequent than daytime feeds. But the fact is we do not know—except for one isolated

Table 8.10. Life-table Estimates of the Time to First Postpartum Menses, 382 Rural Javanese Women Classified by Three Breastfeeding Variables

| Breastfeeding variable | N | Months to First Menses | | Test statistic[a] |
		Mean	Median	
Duration of feed				
≤ 6 min	55	12.2	12.0	32.77‡
> 6 min	327	18.1	17.4	
No. feeds/day				
≤ 6	299	16.6	15.6	8.44*
> 6	83	19.2	19.2	
No. feeds/night				
≤ 3	106	15.1	13.3	13.60‡
> 3	276	18.0	18.0	

[a] Generalized Wilcoxon test for difference between groups (d.f. = 1,
* $P < 0.05$, ‡ $P < 0.001$).
From Jones (1988b)

report from rural Kenya (van Steenbergen et al., 1981). In that study, suckling was more frequent at night than during the day, except at very early ages, and the decline in suckling frequency with time since parturition was somewhat less for nighttime suckling than for daytime suckling (Figure 8.41). Unfortunately, it is difficult to assess these results for three reasons. First, the report gives no indication of how nighttime suckling frequency was actually measured, much less of whether the measurement technique was reliable. Second, infants were test-weighed after each feed, day or night, to determine the volume of milk consumed; it is difficult to believe that this procedure did not disrupt normal sleeping patterns and thus, perhaps, nighttime feeding itself. Third, only mean suckling frequencies have been published, not standard deviations or variances; consequently, it is impossible to assess the difference between daytime and nighttime nursing statistically. Although the patterns reported in this study were more or less what might be expected on a priori grounds, it would be a major step forward in our understanding of the ecology of breastfeeding and the reproductive consequences of suckling if we could develop a simple, noninvasive method for measuring nighttime feeds.

CONCLUSIONS

It is now well-established that breastfeeding can exert a powerful contraceptive effect, potentially delaying the return of full fecundity during the postpartum period by two years or more. It is also clear that the *magnitude* of the effect actually realized varies considerably both within and among human populations. Recent research has shown that this variation reflects differences in suckling behavior, making the postpartum infecundable period, like the fecund waiting time to conception, a major point of interaction between the biological and behavioral determinants of fertility. The frequency of suckling in turn reflects many factors, including women's work patterns and the availability of weaning foods. While the importance of the suckling stimulus in maintaining lactational infecundability is now beyond dispute, much remains to be learned about why so much variation in suckling behavior actually exists. In addition, several major questions persist about the actual physiological mechanisms linking the suckling stimulus and the suppression of ovarian activity during the postpartum period. Still, there is no longer any doubt that lactational fecundability is one of the most important factors limiting reproduction in most natural-fertility populations.

NOTES

1.   Evidence suggests that there is a sharp increase in nipple sensitivity to tactile stimulation during the puerperium (Robinson and Short, 1977; Drife and Baynham, 1988).

2.   In contrast to other pituitary hormones such as LH and FSH, PRL is secreted in large amounts even when the portal circulation is severed so that releasing hormones cannot reach the pituitary. Unlike the gonadotropins, therefore, PRL secretion is regulated primarily by *inhibiting factors* and only secondarily by releasing factors. Dopamine, which is a potent PRL-inhibiting factor, binds to special receptors on the plasma membrane of the lactotroph and is taken into the cell. It has been suggested that dopamine acts within the lactotroph to increase lysosomal degradation of PRL, making less of the hormone available for release (Johnson and Everitt, 1988).

3.   There is a fundamental difference between the anterior and posterior lobes of the pituitary gland. As detailed in Chapter 4, the anterior lobe consists mainly of glandular tissue that releases its products in response to neuroendocrines like GnRH secreted into the portal circulation by specialized neurons located entirely within the hypothalamus. The posterior lobe, in contrast, consists of neural tissue; moreover, the neurons that make up this tissue actually originate in the hypothalamus, where their nuclei and much of their axons lie. The axons of these neurons run down the pituitary stalk and terminate within the posterior lobe. Thus, the hormones secreted by the posterior lobe of the pituitary are actually neurohormones, and the portal circulation is not normally involved in stimulating or inhibiting the secretion of these hormones.

4.   This latter fact is easily accounted for by noting that 1–2 months of amenorrhea occur even in the absence of nursing (Howie and McNeilly, 1982).

5.   It is important to restrict this analysis to infant deaths occurring more than nine months before the next birth, i.e., before the likely time of the next fertile conception. As mentioned earlier in this chapter, the next pregnancy may interfere with lactation and thus place the previous child at an elevated risk of death; thus, an infant death occurring within nine months of the next birth may be an *effect* of a short birth interval rather than its cause. This is a classic example of what epidemiologists call a *simultaneity problem* in which causation occurs simultaneously in both directions: early death may *cause* a shorter birth intervals (by disrupting nursing) or may be an *effect* of a shorter interval (through pregnancy interfering with lactation).

6.   Although earlier evidence of possible ovulation is seen at week 36, 44, and 48, the peak levels of pregnanediol excreted at those times do not exceed the authors' minimal criteria for ovulation. Even if ovulation had occurred at those earlier times, the associated luteal phases were very far from having been adequate to support a pregnancy.

7.   As Franceschini et al. (1989) stress, there is no clear evidence from their study that the elevation in β-endorphin observed in plasma is strictly or even primarily of hypothalamic origin.

8.   Although Habicht et al. (1985) are somewhat skeptical about whether the true value of f is really positive or merely equal to zero, their qualitative conclusion would be the same in either case. They also point out that, because of variation around the regression line, individual women can have negative values of f and thus resume menses while fully breastfeeding; in fact, about 2.3

percent of the women in their sample did precisely that. And even if real, the positive slope of f is an aggregate characteristic and may not reflect what is happening in individual women. The positive value of f could, for example, reflect selectivity if women who introduce supplemental foods very late in the postpartum period also experience very frequent suckling during the time leading up to supplementation.

9.   The magnitude of these effects is given by the $\exp(\hat{\beta})$ values in Table 8.7. In the proportional hazards model, the outcome variable on which each covariate is exerting its effect is the hazard of a first menses. For a given covariate, a positive value of β or, equivalently, a value of $\exp(\beta)$ greater than one indicates that the covariate increases the hazard and thus *decreases* the time to first postpartum menses.

10.   I owe this observation to Anna Glasier.

11.   Unfortunately, these authors do not specify their definition of a suckling episode; however, it is clear from their data that it must have been extremely fine-grained.

12.   The overall downward trend in Figure 8.38 is presumably an effect of time since parturition. Since the study was restricted to women who had not yet resumed menses at 17–25 months postpartum, the general level of suckling observed in this sample is likely to be substantially higher than among all women in Matlab with children ages 17–25 months. It is unlikely, however, that this upward bias in mean suckling intensity affects the apparent seasonality of suckling.

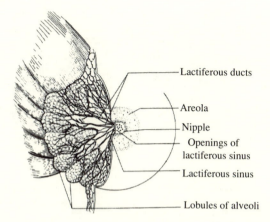

Figure 8.1. Anatomy of the right breast of a pregnant woman. Note the radial arrangement of lobules. When fully developed, these lobules consist of clusters of rounded alveoli that open into the smallest branches of the milk-collecting ducts. These, in turn, unite to form larger lactiferous ducts, each draining a lobe of the mammary gland. The lactiferous ducts converge toward the areola, beneath which they form lactiferous sinuses that serve as small milk reservoirs. After narrowing in diameter, each lactiferous sinus runs separately up through the nipple to open directly onto its surface. (Redrawn from Johnson and Everitt, 1988)

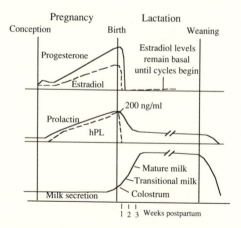

Figure 8.2. The sequence of hormonal changes in the maternal circulation during pregnancy and the puerperium that underlies the onset of lactation. Withdrawal of progesterone removes a block to prolactin-induced milk secretion by the mammary gland. (Redrawn from Johnson and Everitt, 1988)

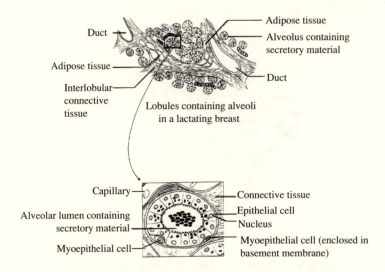

Figure 8.3. Microscopic structure of lobules *(above)* in a lactating mammary gland, and a higher-power view of an alveolus *(inset)*. Note the rich vascular supply from which the secretory epithelial cells draw precursors used in the synthesis of milk and prolactin for support of galactopoiesis. The myoepithelial cells situated between the basement membrane and the epithelial cells form a contractile "basket" around each alveolus. (Redrawn from Johnson and Everitt, 1988)

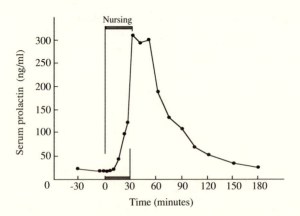

Figure 8.4. The suckling-induced prolactin pulse in a mother at 22 days postpartum. Before the beginning of the nursing episode shown here, 3-4 hours had elapsed since the previous episode. (Redrawn from Noel et al., 1974)

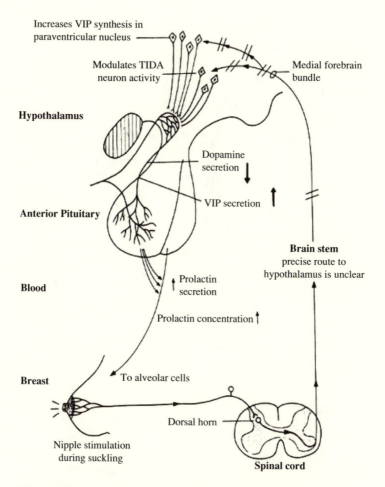

Figure 8.5. Neuroendocrine pathways involved in the suckling-induced reflex release of prolactin. The exact route taken by sensory information between brain stem and hypothalamus is unknown. Within the hypothalamus the activity of dopaminergic neurons (neurons secreting the catecholamine dopamine) is suppressed by this sensory input, while the secretory activity of neurons containing vasoactive intestinal polypeptide (VIP) is increased. Since dopamine is a prolactin-inhibiting factor and VIP is a prolactin releaser, both effects are probably important in driving the suckling-induced prolactin pulse. TIDA neuron, tubero-infundibular dopamine neuron. (Redrawn from Johnson and Everitt, 1988)

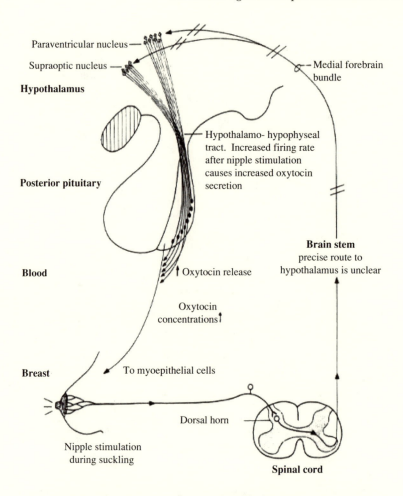

Figure 8.6. Neuroendocrine pathways in the suckling-induced reflex
release of oxytocin. The precise route between the brain stem and
hypothalamus is unknown but probably involves the medial
forebrain bundle. (Redrawn from Johnson and Everitt, 1988)

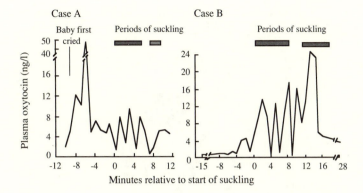

Figure 8.7. Plasma oxytocin profiles in a woman during suckling. In Case A (representing the most typical pattern) there
was a release of oxytocin before the baby was applied to the
breast, presumably reflecting a conditioned release triggered by the sound of the baby crying. Case B shows an
example in which oxytocin release occurred only after the
onset of suckling. The marked fluctuations in oxytocin levels seen in both examples probably reflect the pulsatile pattern of hormone release. (Redrawn from McNeilly et al.,
1983)

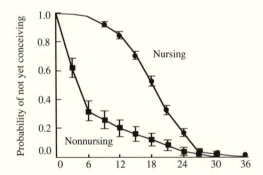

Figure 8.8. Kaplan-Meier estimates (± s.e.) of the survival
function for time to next diagnosed pregnancy in
mothers who nursed throughout the interval *(circles)*
or who terminated nursing because of neonatal death
(squares) from a sample of 259 women in rural Rwanda who were not using artificial contraceptives. The
background characteristics of nursing and nonnursing
mothers were otherwise similar. (Data from Bonte and
van Balen, 1969)

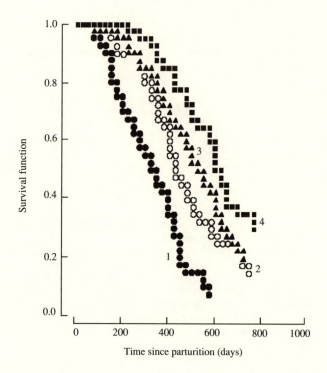

Figure 8.9. Breastfeeding and postpartum amenorrhea in 382 rural Javanese women monitored prospectively: Kaplan-Meier estimates of the proportion of women still amenorrheic by time postpartum (in days) for: *(1)* low-intensity breastfeeding women, *(2)* medium-low-intensity breastfeeding women, *(3)* medium-high-intensity breastfeeding women, and *(4)* high-intensity breastfeeding women. (For a characterization of the suckling patterns of these four groups of women, see Table 8.2.) Each woman was interviewed monthly about current menstruation status and "average" daily breastfeeding patterns over the previous month. (Redrawn from Jones, 1989)

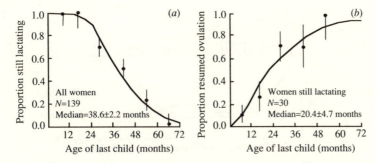

Figure 8.10. Duration of lactation *(a)* and resumption of ovarian-
function *(b)* during lactation by age of child, Gainj women
whose last child is still living. Resumption of ovarian function
was diagnosed from serum estradiol and progesterone levels.
The data in *(a)* were derived from a populationwide survey,
whereas *(b)* is based on a smaller, more intensive study.
Smooth curves and estimated medians are taken from inverse-
variance-weighted regressions fit to the probit transforms of the
data. (Redrawn from Wood et al., 1985b)

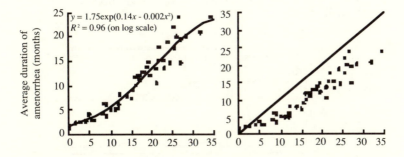

Figure 8.11. *(Left)* regression of average (mean or median) duration of
postpartum amenorrhea on average (mean or median) duration of
breastfeeding, 47 populations and subpopulations. *(Right)* the
same data as in the left panel, plotted relative to a regression line
with a slope of 1. Note that all groups fall below this line except
those with very short durations of breastfeeding and amenorrhea,
which suggests that amenorrhea usually ends well before the
complete termination of breastfeeding. In other words, an earlier
event (resumption of menses) is being predicted by a later event
(weaning) in this regression. (Redrawn from Bongaarts and Potter,
1983)

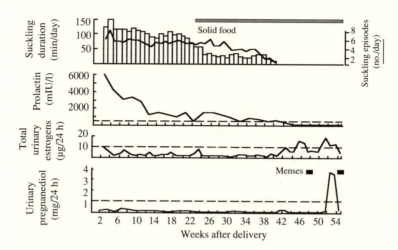

Figure 8.12. Plasma prolactin concentrations and return of postpartum ovarian activity in relation to breastfeeding patterns in a mother from Edinburgh. The broken line for prolactin indicates the upper level observed in normal, nonpregnant, nonnursing women. The broken lines for estrogen and pregnanediol indicate the minimal levels required for a diagnosis, respectively, of follicular development and ovulation. (Redrawn from Howie and McNeilly, 1982)

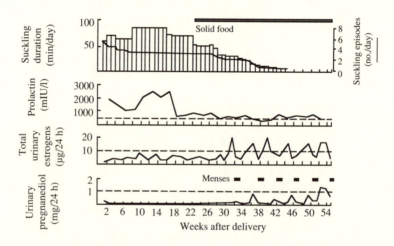

Figure 8.13. Plasma prolactin concentrations and return of postpartum ovarian activity in relation to breastfeeding patterns in a mother from Edinburgh. Conventions as in Figure 8.12. Note the series of anovulatory cycles that occurs before the first definitive ovulation at about 52 weeks. (Redawn from Howie and McNeilly, 1982)

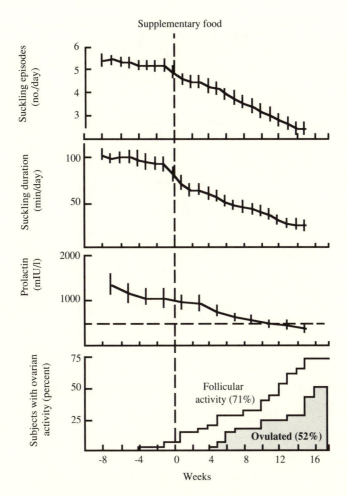

Figure 8.14. Mean suckling rate, suckling duration, basal prolactin, and ovarian activity in Edinburgh mothers (*N* = 27) before and after the introduction of supplemental feeding. Values are mean ± s.e. *Broken line,* upper limit of prolactin for normal, nonlactating, nonpregnant women. (Redrawn from Howie and McNeilly, 1982)

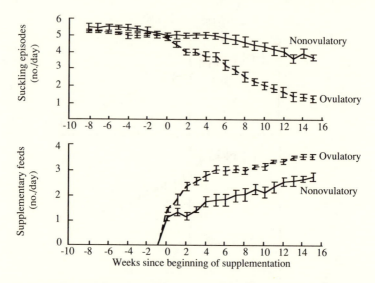

Figure 8.15. Comparison of infant feeding patterns in mothers ovulating within 16 weeks postpartum (*N* = 14) and mothers suppressing ovulation after the introduction of supplemental feeds
(*N* = 13). Values are mean ± s.e. (Redrawn from Howie and
McNeilly, 1982)

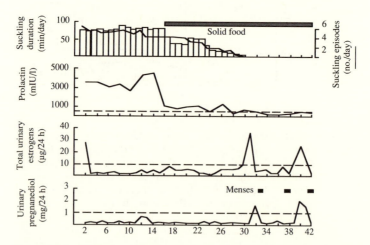

Figure 8.16. Plasma prolactin concentrations and return of postpartum ovarian activity in relation to breastfeeding patterns
in a mother from Edinburgh. Conventions as in Figure 8.12.
Note that this woman ovulates before her first postpartum
menses. (Redrawn from Howie and McNeilly, 1982)

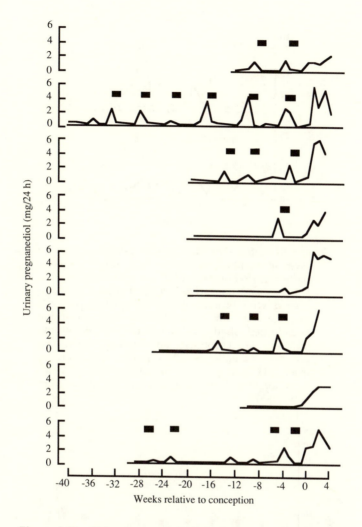

Figure 8.17. Urinary pregnanediol profiles in breastfeeding Edinburgh mothers during the weeks leading up to the first postpartum conception, showing the variability in menstrual and ovulatory experience. Dark bars indicate menses. Note that the second and fourth women from the bottom conceived on the first postpartum ovulation and experienced no menses before conception. (Redrawn from McNeilly, 1993)

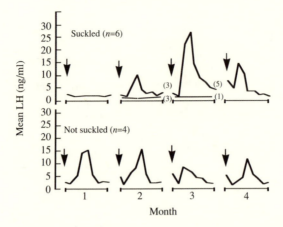

Figure 8.18. Effects of injections of estradiol benzoate given at monthly intervals (*arrows*) during the postpartum period on serum LH concentrations in suckled (*top*) and nonsuckled (*bottom*) rhesus monkeys treated with bromocriptine to suppress prolactin secretion. The time in months following parturition is indicated on the *x*-axis. In the suckled group, numbers of responders and nonresponders are given in parentheses. Data points represent the mean LH concentrations in samples taken twice daily for four days. (Redrawn from Schallenberger et al., 1981)

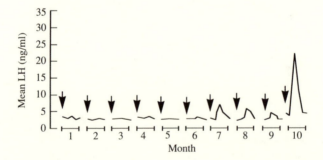

Figure 8.19. Effects of injections of estradiol benzoate given at monthly intervals (*arrows*) on serum LH concentrations in naturally suckled, non-bromocriptine-treated rhesus monkeys (months 1-8, N = 6; month 9, N = 5; month 10, N = 4). The sample size declines during the later months of the study because animals were eliminated from the experiment once the ability of estrogen to elicit LH surges had been established. (Redrawn from Plant et al., 1980)

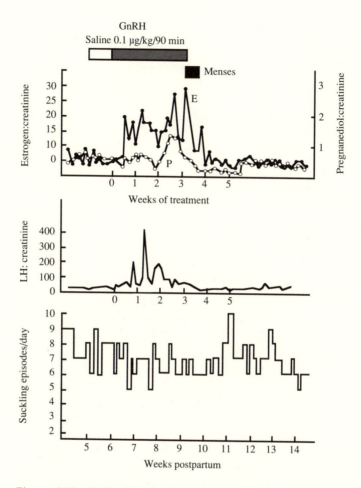

Figure 8.20. Follicular development (indicated by urinary estrogen:creatinine ratio), gonadotropin surge (indicated by LH:creatinine ratio), and ovulation (indicated by pregnanediol:creatinine ratio) induced in a nursing woman at six weeks postpartum by the pulsatile infusion of GnRH (standardization by creatinine concentration corrects for variation in the diluteness of the urine specimen). In the top panel, estrogen is indicated by closed circles and pregnanediol by open circles. Note the complete quiescence of the ovaries both before and after GnRH treatment. Full breastfeeding continued throughout the study at the rate indicated in the bottom panel. (Redrawn from Glasier et al., 1986)

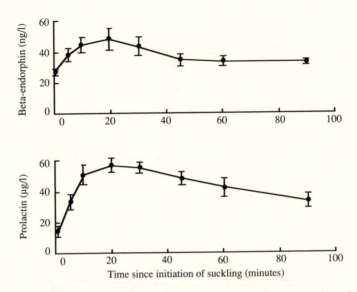

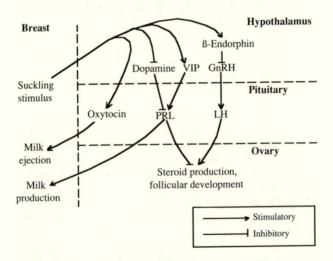

Figure 8.21. Plasma concentrations of β-endorphin *(top)* and prolactin *(bottom)* by time since the initiation of suckling in eight women (mean ± s.e.). (Redrawn from Franceschini et al., 1989)

Figure 8.22. Schematic representation of the cascade of hormonal effects initiated by the suckling stimulus. VIP, vasoactive intestinal polypeptide; GnRH, gonadotropin-releasing hormone; PRL, prolactin; LH, luteinizing hormone.

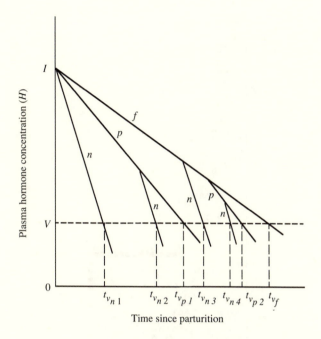

Figure 8.23. Theoretical model of the relationships among breastfeeding, plasma hormone levels, and the duration of postpartum amenorrhea. H, basal level of hormone (perhaps β-endorphin or prolactin) that suppresses menses; I, initial level of hormone at parturition; V, hormonal threshold below which menses resumes; n, slope of decline in H when no breastfeeding is occurring; p, slope of decline in H during partial (supplemented) breastfeeding; f, slope of decline in H during full (unsupplemented) breastfeeding; t_v, duration of postpartum amenorrhea. (Redrawn from Habicht et al., 1985)

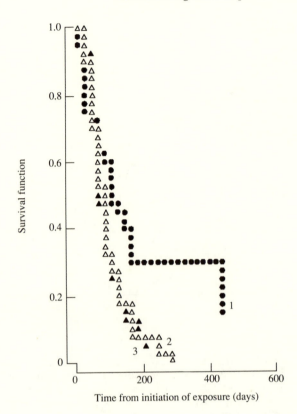

Figure 8.24. Kaplan-Meier estimates of the survival function for time to resumption of menses in three groups of rural Javanese women: (1) women who were still amenorrheic at the time of complete weaning, counted from the time of weaning; (2) women who never breastfed, counted from the time of parturition; and (3) women who were still amenorrheic when their nursing child died, counted from the time of infant death. The sample sizes for the three groups are 44, 30, and 38, respectively. The median times to menses for the three groups are 3.4, 2.6, and 2.1 months, respectively, but the differences are not significant ($P > 0.35$). (Redrawn from Jones, 1989)

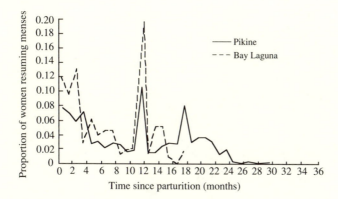

Figure 8.25. "Heaping" on 12 months in retrospectively reported data on resumption of menses during the postpartum period in two populations, Pikine (Senegal) and Bay Laguna (Philippines). Sample sizes are 304 for Pikine and 165 for Bay Laguna. Note that some heaping is also evident at three months in Bay Laguna and at 18 months in Pikine. (Redrawn from Lesthaeghe and Page, 1980; based on data from Osteria, 1973; Delaine, n.d.)

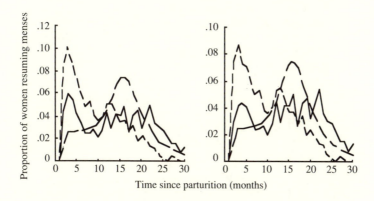

Figure 8.26. Distributions of reported durations of postpartum amenorrhea from three large-scale prospective studies: Matlab, Bangladesh (—), Narangwal, India (- - -), and four rural communities in eastern Guatemala (– –). (*Left*) all cases; (*right*) excluding cases in which the nursing child died before weaning. Sample sizes are 1,852, 2,093, and 621, respectively. Because of poor quality, data from Guatemala have been smoothed by the logit method of Lesthaeghe and Page (1980). (Redrawn from Ford and Kim, 1987; based on data from Potter et al., 1965a; Chen et al., 1974; Delgado et al., 1978)

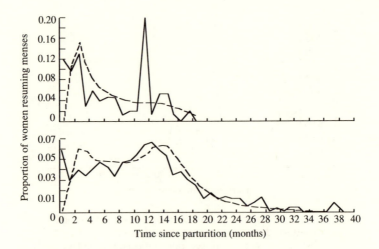

Figure 8.27. Fit of the Lesthaeghe-Page logit model to data on postpartum amenorrhea from Bay Laguna, Philippines (*top*), and Khanna, India (*bottom*). (*Solid line*) data on the resumption of menses; (*broken line*) fitted logit model. Data from Bay Laguna are retrospective ($N = 165$), whereas those from Khanna are prospective ($N = 1390$). The estimated parameters of the logit model are $\alpha = -0.957$ and $\beta = 0.902$ for Bay Laguna, $\alpha = -0.463$ and $\beta = 0.982$ for Khanna. Note that in both applications, the logit model substantially underestimates the proportion of women resuming menses during the first month postpartum. (Redrawn from Lesthaeghe and Page, 1980)

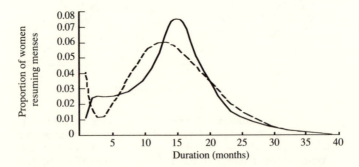

Figure 8.28. Fit of the Potter-Kobrin mixed geometric-negative binomial model (*broken line*) to prospective data (*solid line*) on the duration of postpartum amenorrhea from rural Guatemala ($N = 621$). Data have been smoothed by the logit method. (Redrawn from Potter and Kobrin, 1981; based on data from Delgado et al., 1978)

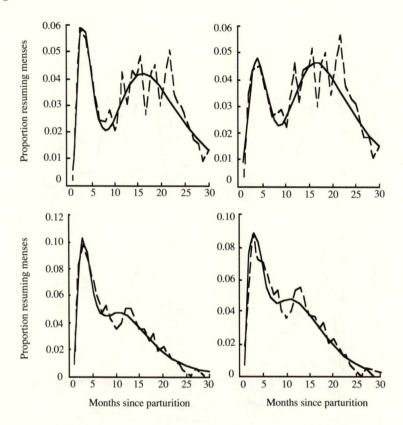

Figure 8.29. Fit of the Ford-Kim mixture model (*solid lines*) to prospective data on the duration of postpartum amenorrhea (*broken lines*) in Matlab, Bangladesh (*top*), and Narangwal, India (*bottom*). (*Left*) all cases; (*right*) excluding cases in which the nursing child died before weaning. Sample sizes are 1,852 for Matlab and 2,093 for Narangwal. (Redrawn from Ford and Kim, 1987; based on data from Potter et al., 1965a; Chen et al., 1974)

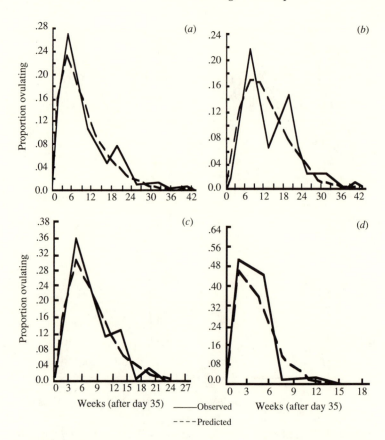

Figure 8.30. Fit of the Ginsberg model to prospective data on the time to first postpartum ovulation (as determined by endometrial biopsy) in 165 Chilean women. (*a*) All cases. (*b*) Women who were fully nursing (i.e., without supplementation) at day 35 postpartum (*N* = 73). (*c*) Women who were partially nursing (i.e., with supplementation) at day 35 postpartum (*N* = 61). (*d*) Women who were not nursing at day 35 postpartum (*N* = 31). The fit is excellent in all four applications. (Redrawn from Ginsberg, 1973; based on data from Pérez et al., 1971)

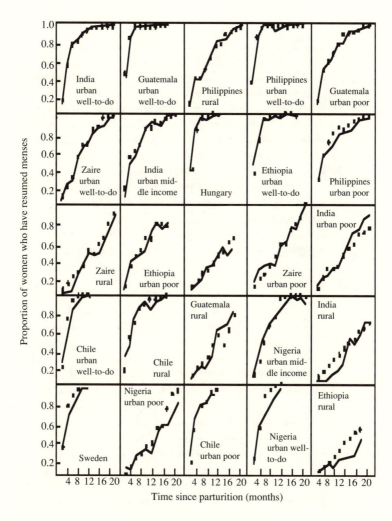

Figure 8.31. Fit of the Billewicz model (*points*) to data on the duration of postpartum amenorrhea (*lines*), selected populations. (Redrawn from Billewicz, 1979)

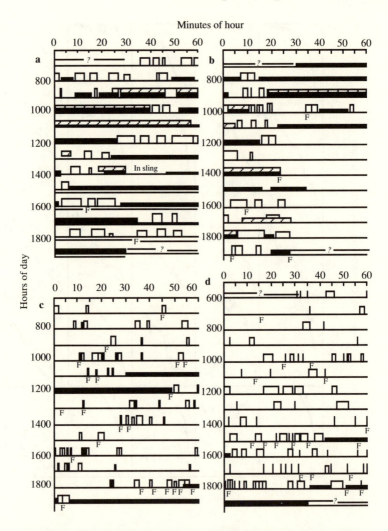

Figure 8.32. Four 13-hour (dawn to dusk) continuous observations of nursing !Kung infants, Botswana. (*a, b*) Newborn boy at 3 and 14 days, respectively. (*c*) 52-week-old girl. (*d*) 79-week-old boy. Open bars and tall vertical lines, nursing; closed bars, sleep; F, fretting or crying. Slashed lines represent the time held by mother, recorded for newborn only. (Redrawn from Konner and Worthman, 1980)

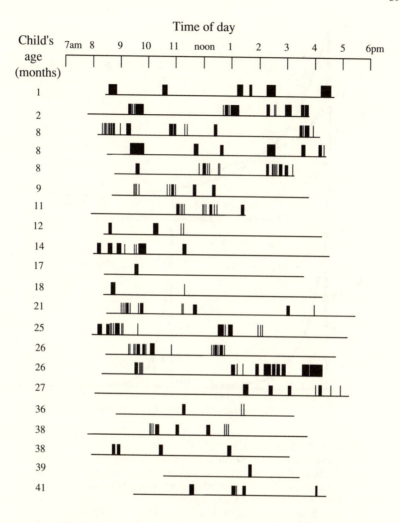

Figure 8.33. Continuous observations of 21 nursing Gainj mother-off-spring pairs, Papua New Guinea, by age of the nursing child. Each horizontal line represents one pair, and its length is proportional to the total period of observation for that pair. Vertical lines and heavy bars represent nursing episodes. Episodes were recorded to the nearest second, and two episodes were distinguished if the nipple left the child's mouth for at least five seconds. A total of 217 suckling episodes was observed during 159.4 hours of observation.

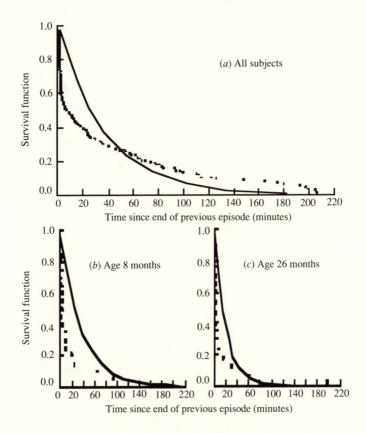

Figure 8.34. Kaplan-Meier estimates (*solid squares*) of the survival function for the time between suckling episodes, with
the best fitting exponential survival function superimposed
(*solid lines*), 21 Gainj mother-offspring pairs. (*a*) All cases
combined (217 suckling episodes). (*b*) A single case in
which the nursing child was 8 months old (20 episodes). (*c*)
A single case in which the nursing child was 26 months old
(18 episodes). In all cases, including those not shown here,
the Kaplan-Meier estimates consistently depart from the
exponential curve in the direction illustrated, suggesting
that suckling episodes tend to be clustered in time.

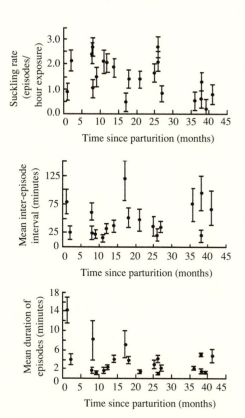

Figure 8.35. Summary of hazards analyses of the timing of suckling in 21 nursing Gainj mother-offspring pairs, by age of child. Each point represents a
single mother-offspring pair. (*Top*) mean (± s.e.) suckling rate. Exposure
periods were calculated by subtracting the total time spent suckling from
the total observation period for each mother-offspring pair; standard errors
were estimated by the method of Bartholomew (1963). (*Middle*) mean (±
s.e.) time interval between suckling episodes, estimated by the method of
Kaplan and Meier (1958) to correct for right-censoring. The correction for
censoring increased the estimated mean interval by about 17.4 percent on
average over the mean computed from completed intervals alone. (*Bottom*)
mean (± s.e.) duration of suckling episodes. Possible effects of the nursing
child's age on suckling rate and interepisode intervals were tested using a
proportional hazards model; no significant effect was found (*P* > 0.4). Nor
were there significant differences among pairs in suckling rate or
interepisode interval (logrank test statistic [Mantel version] = 23.48, d.f. =
20, *P* > 0.2). Possible age effects on duration of the suckling episode were
estimated by regressing the natural log of all the episode durations on time
since parturition (*r*² = 0.003); the log transform was used in this regression
to normalize and linearize the data, as well as to correct for heteroscedas-
ticity. No significant age effect was found (*t* = 0.793, d.f. = 1, *P* > 0.4). However, there was significant intersubject heterogeneity in episode duration (*F*
= 4.69, d.f. = 20, 196, *P* < 0.001). Mean episode duration across all subjects
was 3.0 ± 0.3 min. (Reanalysis of data reported by Wood et al., 1985b)

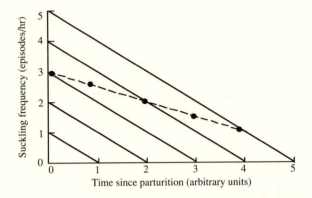

Figure 8.36. The effect of selectivity bias on the apparent mean suckling rate estimated from cross-sectional data on mother-offspring pairs at varying times since parturition. (*Solid lines*) true changes in suckling frequency over time in five mother-offspring pairs. (*Solid circles*) mean suckling frequencies computed from observations on those mother-offspring pairs who are still nursing. Because pairs with low suckling frequencies are progressively selected out of the sample as time goes on, the apparent decline in mean suckling rate is less than the actual decline experienced by each pair.

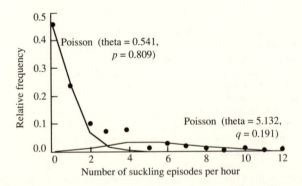

Figure 8.37. Fit of a double-Poisson mixture model (*solid curves*) to observations on the number of suckling episodes per hour (*solid circles*), 21 Gainj mother-offspring pairs (153 completed hours of observation). *Theta*, estimated parameter of each of two Poisson distributions; p, the proportion of hours falling within distribution 1; q (= 1 - p), the number of hours falling within distribution 2. The fit of the model is adequate (χ^2= 16.2, d.f. = 9, 0.10 > P > 0.05).

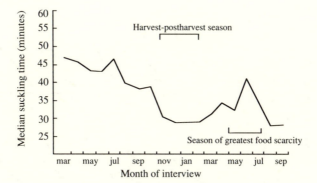

Figure 8.38. Median suckling time (total time spent suckling during a continuous 8-hour observation period) in 216 mother-offspring pairs in which the offspring were 17–25 months old and the mothers still amenorrheic at the time observation was initiated, Matlab, Bangladesh. Each pair was observed once a month over the entire 18-month study period unless the mother conceived again or the child was completely weaned. (Redrawn from Huffman et al., 1980)

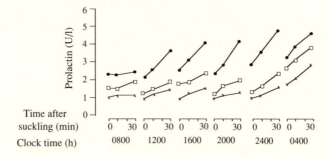

Figure 8.39. Plasma prolactin concentrations (geometric mean) before and 10 and 30 min after the initiation of suckling at six different clock times in three sampling periods: 3–4 months postpartum (●), 6–7 months postpartum (□), and 9–11 months postpartum (×). Subjects were 10 normal Chilean women ages 20–30 years who were fully breast-feeding and amenorrheic at three months postpartum and whose infants had a normal growth rate. (One woman whose child had been weaned was excluded in the third sampling period.) Plasma PRL levels increased significantly after suckling ($P < 0.05$) at all hours except 8 AM and decreased with the time since parturition ($P < 0.05$). (Redrawn from Díaz et al., 1989)

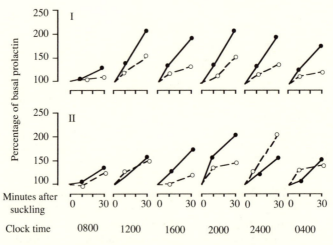

Figure 8.40.   Percent increase in plasma prolactin concentrations
(geometric mean) 10 and 30 min after the initiation of suckling at six different clock times in five women who were
amenorrheic (●) and five women who were cycling (○) on
day 180 postpartum.   Subjects are as described in Figure
8.39. (*Top*) 3-4 months postpartum (sampling period I); (*bottom*) 6-7 months postpartum (sampling period II). The percent increase was significantly higher in amenorrheic
women (*P* < 0.05) at all hours in sampling period I and at
4 PM and 8 PM in sampling period II than in women who
resumed cycling. (Redrawn from Díaz et al., 1989)

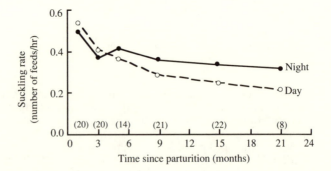

Figure 8.41.   Daytime (approx. 8 AM to 8 PM) and nighttime
(approx. 8 PM to 8 AM) suckling frequency by child's age,
85 nursing Akamba mother-offspring pairs, rural Kenya.
The sample size at each age is indicated in parentheses.
(Redrawn from van Steenbergen et al., 1981)

9

Menarche and Menopause

The outer limits of the female reproductive span are set by *menarche,* the first menstrual bleeding experienced by a woman, and by *menopause,* her last menstrual bleed. It seems intuitively obvious that the ages of menarche and menopause can influence a woman's reproductive output: the earlier the menarche or the later the menopause, the longer the period of time over which a woman can potentially reproduce. The reality, however, is somewhat more complicated. Marriage, if it occurs well after menarche (and if a woman's age at marriage is uncorrelated with her age at menarche), can dampen or erase menarche's effects on fertility. To complicate things further, menarche does not signal an abrupt transition from a completely nonfecund state to one of full fecundity, nor menopause the opposite transition. Menarche is followed—and menopause preceded—by a period of subfecundity, the length of which may itself be a major determinant of the age pattern of natural fertility.

In terms of measurement, menarche is a comparatively unambiguous event, marked in many societies by ritual and ceremony. Menopause is less certain: at a given moment, how can a woman know that her most recent menses was her *final* one? The standard criterion used in clinical research is that a woman is deemed postmenopausal if she has experienced at least 12 months since her last menses in the absence of a known pregnancy. But this criterion may be somewhat unspecific since cycles become longer and more variable during the period immediately preceding menopause: a perimenopausal woman (i.e., one near the age of menopause) may not have menstruated for many months, perhaps even a year, and yet still not have undergone menopause.[1] When assessment of menopause is based on self-reports of bleeding experience, cohort studies are generally more reliable than cross-sectional or retrospective studies, since the latter tend to overestimate the proportion of women who are menopausal at a given age.

But what causes menarche and menopause? Recall from Chapter 4 that normal ovarian cycles are maintained by a complex regulatory system involving the hypothalamus, pituitary, and ovaries. Central to this system is the GnRH pulse generator, located in the hypothalamus; this neural complex controls pulsatile secretion of the pituitary gonadotropins LH and FSH, which in turn act upon the ovary to initiate follicular development and induce ovulation. Although the GnRH pulse generator has its own endogenous rhythm, its baseline, frequency, and amplitude can all be modified by changes in the circulating concentration of ovarian steroids—estradiol during the follicular phase and progesterone during the luteal phase. Crucial to maintaining regular cycles is the negative feedback of estradiol on the pulse generator during the early stages of the follicular phase, followed by an abrupt change to positive feedback just before the midcycle surge in LH and FSH. As detailed below, menarche (or more correctly, puberty) involves the switching on of this regulatory system, and menopause its switching off. However, the two switches operate in fundamentally different ways. The critical event in puberty is activation of the GnRH pulse generator by the central nervous system (CNS). In menopause, the critical event is loss of steroid feedback control, occurring principally at the level of the ovary.

THE BIOLOGY OF MENARCHE

Endocrine Changes during Puberty

Throughout most of prepubertal life, LH secretion is low and shows no clear sign of pulsatility (Figure 9.1). But in the earliest stages of puberty LH secretion increases, primarily during sleep, and begins to follow a pulsatile pattern. The nighttime elevation in LH secretion represents the initial activation of the GnRH pulse generator. At this stage, the ovary appears not to be involved in modulating the pulse generator; rather, activation seems to originate entirely from within the CNS (Plant, 1988). By mid to late puberty, pulsatile LH secretion is clearly occurring (Figure 9.1); indeed, secretion of LH is at a comparatively high level because steroid regulation has not yet been established (compare the bottom two panels of Figure 9.1). By mid-puberty, circulating levels of LH are high enough to induce follicular development in the ovary.

Recall that primordial follicles (and the primordial oocytes they contain) have been present in the ovary since before birth. It appears that the follicles are capable of maturation much earlier than the normal age of menarche. Studies of precocious puberty indicate that follicular development can begin412 even in 3–4-year-old girls if high levels of GnRH, LH, and FSH are present for pathological reasons, for example, as a result of hypothalamic tumors (Styne and Grumbach, 1991). Similarly, the normal course of pubertal development can be induced in girls suffering from delayed puberty by episodic administration of GnRH (Stanhope et al., 1988). Thus, the critical process in puberty is not maturation of the ovary or pituitary gland but activation of the GnRH pulse generator. This is shown clearly by experiments on immature rhesus monkeys (Wildt et al., 1980), in which artificial infusion of GnRH in the right pulsatile pattern can induce follicular development, an LH surge, and even ovulation (Figure 9.2). The "immature" pituitary and ovary are thus perfectly capable of responding in adult fashion if presented with adult patterns of GnRH secretion.[2]

Because LH secretion is initially low and increases only gradually during puberty, it is not surprising that full follicular development, followed by ovulation and formation of a normal corpus luteum, does not occur immediately after the GnRH pulse generator is first activated. On the contrary, much of puberty is marked by successive waves of partial follicular development ending in atresia (Figure 9.3). The earliest of these waves does not even go far enough to produce menstruation. As discussed in Chapter 4, the endometrium of the uterus in cycling adult women initially proliferates during the follicular phase under the influence of rising concentrations of estradiol produced by the developing follicle, and it continues to proliferate under the influence of progesterone during the luteal phase; it is this blood-rich endometrium, sloughed off at the end of the cycle, that constitutes menstrual bleeding. Before menarche, follicles begin to develop but do not produce enough estradiol to cause significant endometrial proliferation. Thus, no detectable menses occurs. One such wave of incomplete follicular development can be observed in the case shown in Figure 9.3. Moreover, menarche itself does not typically signal the start of ovulation. The girl described in Figure 9.3 actually undergoes 11 rounds of menses before her first definite ovulation (as indicated by the elevation of pregnanediol in her urine) more than a year after menarche.[3] The bleeding associated with these anovulatory cycles represents estrogen-withdrawal bleeding rather than full-scale menses. Her first several menstrual cycles are also long and quite variable in length, probably because some of them actually contain two or more rounds of partial follicular development. The pattern shown in Figure 9.3, characterized by long, variable, and anovulatory cycles, combined with ovulatory cycles having inadequate luteal-phase rises in progesterone, appears to be typical of the perimenarcheal transition in all women.

Menarche as an Aspect of Puberty

The initiation of ovarian cycles is only one of many morphological changes that occur during puberty (Figure 9.4). Indeed, menarche and the first ovulation occur fairly late in the sequence of pubertal events. In particular, initiation of ovarian function falls near the end of the adolescent growth spurt, the elevated somatic growth rate that occurs in western girls between about ages 10 and 14. Although the growth spurt is later in boys, spermatogenesis generally begins *before* the male growth spurt; thus, although most pubertal changes occur later in boys than in girls, fecundity begins earlier.

Although there is considerable heterogeneity among girls in the age at which puberty begins and the rate at which pubertal changes occur, menarche tends to occur at the same stage of puberty in all girls. Thus, the age at menarche is associated more strongly with measures of developmental age than with strict chronological age (Figure 9.5). This observation has given rise to the hypothesis that the central nervous system somehow monitors morphological development and turns on ovarian function only when the female is physically mature enough to support a pregnancy and subsequent lactation (Foster et al., 1985). A more extreme form of this hypothesis identifies body fat as the critical component for supporting reproduction and posits a sharp threshold in the ratio of body fat to lean body mass below which ovarian function is switched off and above which it is switched on (Frisch and Revelle, 1970). This idea gains some plausibility from the recent finding that progesterone synthesis is in part dependent upon the availability of cholesterol bound to low-density lipoprotein, which is partly a reflection of dietary fat intake (Carr et al., 1981). However, studies indicate that skeletal development is a far better predictor of the age at menarche than is body fat (Ellison, 1982), and numerous researchers have criticized the idea of a sharp threshold effect of body fat on both statistical and substantive grounds (Johnston et al., 1975; Trussell, 1978).

The close relationship between skeletal age and the age at menarche is largely attributable to the fact that the same underlying neuroendocrine changes initiate both ovarian function and the adolescent growth spurt (Plant, 1988; Stanhope, 1989). The trophic effects of growth hormone and insulinlike growth factor I on cell proliferation and differentiation in the growth plates of the long bones are greatly enhanced by the increased concentration of gonadal steroids that marks puberty. Thus, skeletal maturation and the switching on of the reproductive system are two distinct outcomes of a single set of changes in the hypothalamo-pituitary-gonadal axis. Since pubescent girls also respond to increases in ovarian estradiol by laying down subcutaneous fat, it is not surprising that a correlation between body fat and the age at menarche exists. This correlation does not, however, indicate an effect of fat on sexual maturation but rather the reverse: it is activation of ovarian steroidogenesis that leads to an increase in body fat, not the other way around.

Even through the critical body fat hypothesis has been rejected, it is clear that poorly nourished and/or more slowly growing girls experience later menses than well-nourished, rapidly growing ones (Chowdhury et al., 1977; Foster et al., 1986; Riley et al., 1993). In some general sense, then, it is at least plausible that female fecundity may be adjusted to begin at a sufficiently advanced stage of physical development to support reproduction. From this point of view, it is not surprising that fecundity begins earlier in males than in females, since spermatogenesis is far less costly in metabolic terms than are pregnancy and lactation.

Adolescent Subfecundity

Over thirty-five years ago, the anthropologist Ashley Montagu (1957) put forward the concept of "adolescent sterility" to explain an ethnographic observation that had been made in certain traditional Pacific and Southeast Asian societies, namely that unmarried teenage girls often engaged in regular sexual activity without becoming pregnant. He suggested that the reason for their failure to conceive was physiological: adolescent girls are sterile despite having undergone menarche. Modern studies have largely confirmed this idea, except we now know that perimenarcheal girls are not literally sterile (which implies a complete inability to produce a live birth) but merely subfecund. From a biometrical point of view, the existence of adolescent subfecundity increases the importance of variation in the age at menarche, for it partly counteracts the effect of marriage in decoupling menarche from subsequent reproductive events.

Some of the earliest direct evidence for the causes of adolescent subfecundity is shown in Figure 9.6, which summarizes the characteristics of 3,275 cycles experienced by 481 German women ages 12–50 (Döring,

1969). Timing ovulation according to the midcycle rise in BBT, it appears that the first several years after menarche are characterized by a high fraction of anovulatory cycles and cycles with short hyperthermic phases (the hyperthermic phase is the portion of the cycle between the midcycle rise in BBT and the next menses; it thus approximates the luteal phase). Although it is now known that the BBT method consistently underestimates the frequency of ovulation (Johansson et al., 1972), this bias has been assumed to affect all ages equally.[4] Under this assumption, the frequency of anovulatory cycles is higher at perimenarcheal ages than at mid-reproductive ages, even if it is not as high as Figure 9.6 suggests. This conclusion has now been substantiated using more reliable methods for detecting ovulation (Apter et al., 1978; Lipson and Ellison, 1992).

A second cause of adolescent subfecundity has to do with cycle length (WHO, 1986b). As shown in Table 9.1, both the mean and variance in cycle length decline consistently over the first 36 cycles following menarche—the mean by about 40 percent, the variance by fully 96 percent. Most of the extreme variance immediately following menarche is attributable not to short cycles, but rather to a sizable fraction of very long cycles (Table 9.1). As noted earlier, these apparently long cycles

Table 9.1.   Characteristics of the First 36 Cycles Following Menarche in 298 British Girls

Cycle number	No. of girls	No. of cycles	Mean length	Variance in length	% short cycles (<17 days)	% long cycles (>57 days)
1	263	263	49.5	948.6	2.3	27.4
2	263	263	39.8	470.9	1.9	14.8
3	263	263	35.4	309.8	4.9	10.6
4	263	263	34.8	256.0	3.8	7.2
5	263	263	32.5	171.6	3.0	4.6
6	263	263	32.6	139.2	3.0	6.1
7–9	292	876	33.5	201.6	2.3	5.7
10–12	292	876	32.4	132.2	1.7	3.7
13–18	257	1542	31.4	79.2	2.1	2.2
19–24	209	1254	30.8	72.2	1.9	1.6
25–30	157	942	30.4[a]	62.4[a]	1.4	1.6
31–36	116	696	29.5	37.2	1.6	0.6

[a] Excluding one case with an interval of 208 days; otherwise mean = 30.7 and variance = 96.0.

From Billewicz et al. (1980)

may actually represent two or more waves of follicular development unaccompanied by either ovulation or menses. Whatever their cause, long cycles reduce fecundity by lowering the number of fertile periods expected to occur during any fixed period of exposure to unprotected intercourse.

Another large-scale study of cycle characteristics, again based on the BBT method, suggests that perimenarcheal cycles are long principally because of long follicular phases (Figure 9.7). Indeed, according to this study perimenarcheal cycles have luteal phases that are slightly shorter than those of cycles at mid-reproductive ages, consistent with the findings in Figure 9.6. It might be objected that, since these studies used the error-prone BBT method, they are unreliable. However, a recent study that timed ovulation according to the day of maximum LH secretion (Lenton et al., 1984a, 1984b) has confirmed the general pattern suggested by earlier research (Table 9.2).

In sum, adolescent subfecundity appears to involve three factors: an elevated risk that a given cycle will be anovulatory, a distinct subfraction of very long cycles, and cycles with short luteal phases, which probably entail an elevated risk of fetal loss should conception occur. The *duration* of adolescent subfecundity is less well characterized and probably varies considerably for each girl, but it appears to be between one and two years on average.

Table 9.2.   Age-related Variation in the Length of the Follicular Phase and the Fraction of Short Luteal Phases in 226 British and 73 Swedish Women, in whom Ovulation Was Timed According to the Day of Maximum LH Secretion

Age (years)	No. of cycles	Length of follicular phase (geometric mean)	Percent of luteal phases ≤ 10 days
18–24	42	14.2*	12.0*
25–29	125	12.9	2.4
30–34	91	13.1	4.4
35–39	35	12.1	0.0
40–44	18	10.4**	5.6
45–50	14	10.5**	14.2

* $P < 0.05$ when compared with 25–29 year age group
** $P < 0.01$ when compared with 25–29 year age group
From Lenton et al. (1984a, 1984b)

Age at Menarche and the Duration
of Adolescent Subfecundity

A recent prospective study of more than 200 Finnish schools girls ages
7–17 years has suggested that the duration of adolescent subfecundity
(or at least one aspect of adolescent subfecundity) is positively correlated with the age at menarche (Apter and Vihko, 1983; Vihko and Apter,
1984). In this study, girls who reached menarche late also experienced a
longer period before 50 percent of their cycles were ovulatory (Figure
9.8). This finding, if confirmed, is potentially important because it
implies that variation in the age at menarche may have a larger effect
on the onset of full fecundity than has heretofore been suspected. However, this result was not replicated in a large-scale study of 681 U.S. girls
(MacMahon et al., 1982). And a hazards analysis of prospective data on
more than 1,500 Bangladeshi girls ages 10–20 years suggested the opposite result: the time from menarche to the first fertile conception was significantly *shorter* for girls with late menarche compared to early maturers, even when age at marriage was controlled (Foster et al., 1986). In
part, this latter finding may reflect what has been found in other studies, that early menarche is associated with an elevated risk of fetal loss
during the first few pregnancies (Leistøl, 1980; Casagrande et al., 1982;
Wyshak, 1983).

On balance, the evidence for a relationship between the age at menarche and the duration of adolescent subfecundity is mixed. Moreover,
such evidence as does exist addresses the relationship strictly within,
not between, populations. Thus, even if Apter and Vihko's finding turns
out to be generally true, it would not necessarily mean that *populations*
with late average ages at menarche were also characterized by prolonged adolescent subfecundity. To date, no direct evidence for an interpopulation relationship exists. A recent endocrinological study in highland Papua New Guinea suggests that one particular population with a
late median age at menarche (18.6 ± 0.7 years) also has an unusually
long first birth interval, which may partly reflect prolonged adolescent
subfecundity (Wood, 1992). However, the reasons for the long birth
interval are unknown and may include low coital frequency as well as
(or perhaps rather than) any physiological factors.

Menarche and Subsequent Reproductive Events

Even if its duration does not vary by age at menarche, adolescent subfecundity still acts to link menarche and subsequent reproductive
events. In western developed countries this linkage has been largely but

not completely broken by two factors, early menarche and late marriage. Marriage now occurs so much later than menarche that adolescent sub-fecundity has negligible effects on marital fertility. To some extent, the same was true of historical Europe, which as a region was characterized by unusually late average ages at marriage (see Chapter 11); however, the available evidence suggests that menarche was also later than at present (see below). Even in the contemporary western world, however, there remains a small but consistent effect of age at menarche on the ages at marriage and first conception (Table 9.3).[5]

In most natural-fertility settings, it appears that menarche and marriage are much more closely linked. Ethnographic evidence suggests that menarche is considered a prerequisite for marriage in many parts of the world, although premenarcheal marriage was not uncommon for certain periods and social classes in South and East Asia. In general, marriage typically occurs within a few years of menarche. In Bangladesh, for example, age at marriage is closely tied to the age at menarche for both Hindu and Muslim women (Chowdhury et al., 1977). Similarly, among the Gainj of highland Papua New Guinea, a young woman is considered marriageable when her breasts have reached a stage of development that normally falls about a year after menarche (Wood et al., 1985a); the median age at marriage among the Gainj is estimated to be about two years later than the median age at menarche (Wood, 1992).

In traditional societies like the Gainj, the age at menarche would be expected to have a strong effect on the timing of subsequent reproductive events, and such seems to be the case. Udry and Cliquet (1982) have

Table 9.3. Mean Age at First Marriage and First Conception (for women ever pregnant) by Age at Menarche, in Two Cohorts of U.S. University Students

Age at menarche	1935–1939 Recruitment cohort (N = 679)		1961–1965 Recruitment cohort (N = 615)	
	Age at first marriage	Age at first conception	Age at first marriage	Age at first conception
<12	25.0	26.6	22.9	25.4
12	24.8	26.7	23.0	25.5
13	25.2	27.3	23.2	25.7
14	25.2	27.3	23.1	25.7
15+	26.3	28.1	23.9	26.1
All ages	25.2	27.1	23.1	25.6

From Sandler et al. (1984)

recently collated data from several prospective studies in which ages at menarche, first sexual intercourse, first marriage, and first birth were recorded for the same respondents (Figure 9.9). In industrialized countries such as Belgium and the United States, characterized by early menarche and late marriage, the age at menarche has only a weak—but nonetheless consistent—effect on the timing of later events. In developing countries such as Pakistan and Malaysia, where traditional practices still hold sway in most rural areas and where marriage follows more closely upon menarche, the impact of menarche on later events is much more pronounced. In Pakistan, for example, a year's difference in the age at menarche translates into almost a full year's difference in a woman's age at her first birth (Figure 9.9). As a general point, then, the effects of menarche and adolescent subfecundity on later fertility are largely mediated by the relationship between the ages at menarche and marriage. Those effects might therefore be expected to vary widely across different cultural settings.

THE BIOLOGY OF MENOPAUSE

We turn now from the onset of reproductive function to its termination at menopause. The salient feature of menopause is a removal of estradiol and progesterone feedback control of the GnRH pulse generator. The distinctive hormonal profile associated with this process is shown in Figure 9.10. The woman in this example experienced her last menses at age 48; it is clear from her plasma progesterone levels that the cycle preceding her final menses was anovulatory. Following the last menses is an interval of one or two weeks during which some estradiol is present, perhaps reflecting one last round of partial follicular development. After this interval, the levels of both estradiol and progesterone (if they are present at all) are well below the assay's limits of detection, and they will remain that low for the rest of the woman's life. Coinciding with the decline in ovarian steroids is a marked elevation in LH and FSH, indicating that the hypothalamo-pituitary axis has been freed from negative feedback by ovarian steroids. From this point on, at least until very late in her life, this woman's gonadotropin levels will be high and quite variable, and she will experience ephemeral spikes of LH and FSH coinciding with the "hot flashes" of menopause (Casper and Yen, 1981). As noted earlier, the combination of no detectable ovarian steroids and high gonadotropins is the unmistakable endocrine signature of menopause.

What accounts for the loss of feedback regulation by the ovarian steroids? The answer, baldly stated, is that the ovary runs out of eggs. More precisely, the remaining pool of viable primordial follicles (i.e., those not yet lost to atresia or ovulation) apparently becomes too depleted to produce the amount of estradiol necessary to support further follicular development and feedback control of gonadotropin section (Nelson and Felicio, 1987). This concept has been around for several decades, but it was long rejected because of autopsy estimates of the number of primordial follicles present in the ovaries of women at various ages (Figure 9.11). Based on a linear regression fit to the logarithm of these estimates (the straight line in Figure 9.11, right panel), it appears that something like 5,000 viable follicles are still present at about age 50: too many, it would seem, to support the hypothesis that menopause is caused by follicular depletion. It appears, however, that the relationship between the log of the number of remaining follicles and age may not be linear, and indeed a quadratic regression fits the data better (curved line in Figure 9.11, right panel). According to this curve, far fewer than 1,000 primordial follicles, and perhaps as few as 100–200 or less, are still present at menopause. It should be noted, however, that the apparent curvilinearity in these data is largely attributable to the two lowest data points. Before age 40 the relationship with age does in fact appear to be log-linear, implying a constant rate of follicular loss; there may, however, be some acceleration in loss at about age 40.

Until recently, these autopsy data were the only evidence available for follicular depletion. The special nature of this sample makes selectivity bias a concern, since women who undergo autopsy may not be representative of living women. In addition, nothing is known about the cycling status of these women at the time of their deaths, i.e., who was menopausal and who still cycling. Consequently, the relationship between follicular depletion and cycling status remained uncertain.

Recent evidence provides more direct support for the follicle depletion hypothesis. In a study of 17 women of known cycling status who underwent elective hysterectomy and oophorectomy at ages 45 to 55, Richardson et al. (1987) observed follicle reserves ranging from 1,000 to 2,500 in normally cycling women, to a few hundred in perimenopausal women, to virtually none in postmenopausal women (Figure 9.12). There is some confounding by age in Figure 9.12 since regularly cycling women in this study were slightly younger than perimenopausal women, who were in turn slightly younger than postmenopausal women. However, an analysis of covariance designed to adjust for these age differences still indicates a strong relationship between a woman's cycling status and her number of remaining follicles (Table 9.4).

Table 9.4.   Analysis of Covariance of the Effects of Cycling Status (whether a woman is cycling normally, experiencing the irregular menses of the perimenopausal transition, or postmenopausal) and Age on the Number of Remaining Primordial Follicles in One Ovary[a]

Source	Sum of squares	DF	Mean square	F	P
Cycling status	5,608,198.06	2	2,804,099.03	9.628	0.003
Age	216,989.29	1	216,989.29	0.745	0.404
Error	3,785,996.39	13	291,230.49		
$R^2 = 0.64$					

[a] The sample consists of 17 women ages 45–55 undergoing elective hysterectomy and oophorectomy. Cycling status was determined by a menstrual history confirmed by plasma hormone levels. Women were classified as postmenopausal if they had experienced at least one year since the last menses. One ovary chosen at random from each woman was serially section for determination of follicle number. In the analysis, cycling status is treated as a class variable, whereas age is treated as a continuous covariate.
Data from Richardson et al. (1987)

These data, coming as they do from oophorectomies, may still be unrepresentative of normal women. However, the oophorectomies were elective, and none of the subjects had any prior conditions known to affect follicle numbers or follicular loss. In addition, Richardson et al. (1987) summarize evidence indicating that ovaries obtained through autopsy and oophorectomy yield similar estimates of the number of remaining follicles at each age. Each of these two data sources may be biased, but they are very unlikely to be biased in the same way.

This study provides the clearest evidence to date of a causal relationship between follicular depletion and the onset of menopause. In addition, when the results of this study are combined with the earlier autopsy data, they provide powerful evidence that the rate of follicular depletion is indeed accelerated after age 40, so that essentially no follicles are left at menopause (Figure 9.13). As a recent review has put it, "There is little doubt that the ovarian exhaustion of follicles is the pacemaker of reproductive senescence in women" (vom Saal and Finch, 1988).

Follicular depletion may not, however, be the whole story. Other evidence suggests that there may also be some deterioration in hypothalamo-pituitary function with age. For example, in some perimenopausal women, "postmenopausal" levels of LH and FSH have been observed in the presence of measurable estradiol (Metcalf et al., 1981), suggesting failure of estradiol feedback at the level of the hypothalamus or pituitary rather than the ovary. However, such a hormonal profile is not a regular feature of the perimenopausal period (Metcalf et al., 1981, 1982). In addition, animal experiments rule out any important role of hypothalamo-pituitary failure in reproductive senescence. For example, when ovaries from young hamsters are transplanted into old hamsters, they continue to cycle normally (Krohn, 1962; Blaha, 1964). On balance, then, it appears that follicular depletion is indeed the key process underlying menopause.

Menopause is thus the endpoint of a process that begins before birth, whereby an initial reserve of some seven million primordial follicles (and the oocytes they contain) is gradually diminished, principally through atresia (see Chapter 4). Entry into the menopausal transition is apparently not caused by an absolute exhaustion of this reserve; rather, the remaining pool of follicles appears to fall below some threshold number necessary to sustain efficient estradiol regulation of gonadotropin secretion. Acceleration of follicular loss during perimenopausal life may well correspond to the crossing of that threshold. Either the threshold or the amount of estradiol produced by the remaining pool of follicles near the threshold appears to be subject to considerable random variation. As a result, women's cycles tend to become extremely variable in length just before menopause (Treloar et al., 1967). As shown in Figure 9.14 this variability is primarily attributable to a fraction of extremely long cycles; minimum cycle length is much less variable. These long perimenopausal cycles, which contribute to declining fecundity near the end of the reproductive span, may reflect the presence of incomplete follicular development without menstrual bleeding, just as occurs during the perimenarcheal period.

These biological considerations suggest a simple model for the etiology of menopause (Figure 9.15, top left panel). Follicles are lost, primarily through atresia, at a more or less constant rate throughout prereproductive life and most of reproductive life. Eventually the number of remaining follicles crosses a threshold (broken line in Figure 9.15), below which the basal level of circulating estradiol is too low to maintain regular cycles. Once this threshold has been passed, cycles become extremely irregular and the rate of follicular loss increases until menopause (complete follicular depletion) occurs. Under this simple model, there are likely to be only three important sources of heterogeneity among women (Figure 9.15): differences in the size of the initial pool of follicles, differences in the rate of follicular loss, and differences in the level of the threshold. All three sources of heterogeneity are biologically plausible, and it is known that at least the first two are important in determining differences among inbred strains of mice (Jones and

Krohn, 1961). If parametric hazards models of this process could be developed, it might be possible to estimate levels of variation in these three factors from data on variation in the age at menopause.

Far less is known about the physiological and environmental correlates of menopause than about those of menarche, in part reflecting the greater difficulty in correctly ascertaining the age at menopause in large-scale surveys. However, nutritional status has been implicated in the timing of menopause, just as it has for menarche. Studies in the United States and Bangladesh have shown that leaner women reach menopause earlier than heavier women (MacMahon and Worcester, 1966; Karim et al., 1985). In addition, an aggregate-level comparison of two populations in Papua New Guinea that differ in nutritional status suggests that chronic malnutrition may depress the age of menopause (Scragg, 1973). All these studies, however, leave a number of important potential confounding variables (e.g., SES, ethnic affiliation, marital status, parity) uncontrolled. Large-scale multivariate studies that attempt to control such confounders have generally not confirmed the effect of nutritional status on the age at menopause (van Keep et al., 1979; McKinlay et al., 1985). Indeed, the *only* covariate that has been found to affect the age at menopause significantly in these studies is long-term cigarette smoking, which for reasons that are entirely obscure lowers the age of menopause by about 1.7 years on average (McKinlay et al., 1985).

DIFFERENTIALS AND TRENDS IN MENARCHE AND MENOPAUSE

We turn now to a survey of the available biometric evidence on variability in the ages at menarche and menopause, considering both variation among women of particular populations and variation among populations. We also need to consider temporal variation to determine if *secular trends* or consistent, long-term changes have occurred in the timing of menarche and menopause. Such trends, if they exist, may reveal something about the underlying causes of variation in the ages at menarche and menopause.

Individual-level Variation

Arguably the single best data set on the timing of menarche and menopause, and on the relationship between the two at the individual level, was compiled by Alan Treloar and his colleagues (and their scientific heirs) among predominantly white, middle-class female students at the University of Minnesota (Treloar, 1974; Treloar et al., 1967). The first cohort of subjects was recruited into this study during the 1930s, and a second cohort was initiated during the 1960s. (We have already encountered a number of these subjects in Table 9.3 and in earlier chapters.) More recently, the daughters and granddaughters of the original cohort have been added to the sample, providing records on a total of more than 4,000 women. From all these women, detailed menstrual and reproductive diaries have been collected on an ongoing, prospective basis.

Figure 9.16 summarizes data on a subset of 324 of these woman who, by the mid-1970s, had been in the study long enough to provide reports of the age at both menarche and menopause (menopause was defined as a year or more of no menses in the absence of a known pregnancy). Because this was a cohort study, the data on menopause are expected to be fairly reliable. The mean age at menarche in these women was 13.6 years, with a variance of 1.9; the distribution of ages at menarche is slightly right-skewed and approximates a lognormal distribution (i.e., the log of age at menarche is approximately normal). The distribution of ages at menopause is more symmetrical, with a mean of 49.5 years. Clearly, the age at menopause is much more variable than that at menarche: the variance in the age at menopause is 9.1, almost five times greater than that of menarche.

Even more striking is the relationship—or rather, lack of relationship—between the two ages (Figure 9.17, top panel). Age at menopause was not significantly correlated with age at menarche in this sample. As a consequence of this absence of correlation and of the much greater variance in ages at menopause, variation in the total length of menstruating life (age at menopause minus age at menarche) is attributable primarily to variation in the age at menopause (Figure 9.17, bottom panels).

The absence of any correlation between the ages at menarche and menopause may seem surprising. A negative association might have been expected if healthy, well-nourished girls tended to have both early menarche and late menopause. Conversely, a positive association might have been anticipated if menarche and menopause were both linked to the same underlying developmental processes. But in fact, given what we know about the etiology of menarche and menopause it should be no surprise that the two are uncorrelated—and also that the age at menopause is much more variable than the age at menarche. We have seen that the physiological mechanisms driving menarche and menopause are completely different, one involving CNS activation of the GnRH pulse generator and the other, depletion of ovarian follicles.

Thus, there is no obvious biological reason to expect that the two should be correlated at all. In addition, the age at menarche is tightly linked with overall physical maturation, which, though variable to some degree, is constrained to fall within rather narrow limits (except in rare conditions such as precocious puberty). Menopause, in contrast, is the final endpoint of a process of follicular depletion that begins before birth and continues throughout premenopausal life, a period of some 45–50 years. Variation in the rate at which follicles are lost during any phase of this period might be expected to produce variation in the age at menopause. It is no wonder, then, that menopause is more variable than menarche and thus plays a larger role in determining the length of an individual woman's reproductive span.

It is important to bear in mind, however, that the absence of an individual-level correlation within any given population does not necessarily mean that there is no correlation in mean ages at menarche and menopause *across* populations. Nor is interpopulation variation in the mean age at menopause necessarily the most important determinant of interpopulation variation in the mean length of the reproductive span.

The Secular Trend in Age at Menarche

It is widely believed that the average age at menarche has declined among western women during the past century and a half. The most famous illustration of this decline—indeed, one of the most frequently reproduced figures in this whole area of study—is reproduced yet again as Figure 9.18. On the basis of these data, Tanner (1962) has concluded that "age at menarche has been getting earlier by some 4 months per decade [i.e., declining linearly] in Western Europe over the period 1830–1960." Although some researchers have argued that this secular trend is a reflection of improving nutrition, other factors, such as changing prevalences of infectious disease and decreased consanguinity, have also been cited. But many of the pre-1900 estimates in Figure 9.18 have been criticized for being based on anecdotal evidence or on small, highly select samples (Bullough, 1981)—and indeed, all the estimates before 1945 were drawn from Finland and northern Scandinavia, which may not be representative of Europe as a whole. More recent compilations indicate far more variability both within and among regions of Europe (Figure 9.19). An overall decline may have occurred, but it apparently occurred at widely varying rates and, at least locally, was probably reversible. In addition there appears to have been substantial heterogeneity by socioeconomic class (Figure 9.20), which is left uncontrolled in most analyses of the secular trend.

The best evidence for a secular decline in the age at menarche comes from the University of Minnesota cohort study initiated by Treloar and his colleagues (Figure 9.21). This sample has the advantage of being well-defined and comparatively homogeneous, representing predominantly middle-class subjects of northern European (mostly German and Scandinavian) background. In this sample, a significant decline in the mean age at menarche occurred between the 1900–1910 and 1920–1930 birth cohorts (the pre-1900 birth cohort was too small to yield much statistical power). By 1920, however, the decline appears to have stopped at about 12.8–13.0 years (Figure 9.21).[6] This finding suggests that there may be a biological minimum below which the age at menarche will not fall, perhaps reflecting the close relationship between menarche and overall physical maturation.

Another series of data that are pertinent here is summarized in Figure 9.22. Included are the results of four large surveys of well-defined populations in various parts of England from the mid-nineteenth to the mid-twentieth century. There may well be some selectivity in these surveys, especially those restricted to school girls and hospital patients, and there are important differences in study design; the 1949 curve is particularly worrisome because it involves retrospective reports of the age at menarche collected some 45–70 years after the event in question. Nonetheless, these curves provide what is probably the best possible comparison of ages at menarche spanning more than a century. It appears that most of the change occurred sometime between 1890 and 1949, spanning the period of maximum change in the Minnesota cohort study (Figure 9.21). The rate of change had clearly decelerated by 1949–1960. In addition, despite marked changes in the *median* age at menarche, little if any change occurred between 1845 and 1962 in the *origin* of the curves, i.e., in the youngest observed ages at menarche. This finding reinforces the idea of a biological minimum age at menarche. By 1962, the age of onset curve appears to be approaching a step function close to the presumed minimum.

None of these studies has directly addressed the effect of changing nutritional status on the secular trend in menarche. Although it seems plausible to speculate that nutrition may be one factor involved, this hypothesis has never been subjected to a critical test.

It has also been speculated that a secular *increase* has occurred in the average age of menopause in European populations over the same period we have been discussing. Recent analyses make it clear, however, that no such trend has occurred (Gray, 1976; McKinlay et al., 1985), strengthening the conclusion that environmental factors play little role in determining the age of menopause.

Table 9.5. Estimates of the Median Age at Menarche (in years) in Contemporary
Populations

Population	Period	Median age at menarche	Source
EUROPE			
Italy, Carrara	1968	12.6	Eveleth and Tanner, 1976
Greece, Athens	1979	12.6	Dacou-Voutetakis et al., 1983
Spain, Madrid (high SES)	1968	12.8	Eveleth and Tanner, 1976
Hungary, Budapest	1959	12.8	Eveleth and Tanner, 1976
England, Warwick	1981	12.9	Dann and Roberts, 1984
Italy, Naples (low SES)	1969–1970	12.9	Carfagna et al., 1972
Denmark, Copenhagen	1982–1983	13.0	Helm and Helm, 1984
England, London	1966	13.0	Eveleth and Tanner, 1976
Belgium	1965	13.0	Eveleth and Tanner, 1976
USSR, Moscow	1970	13.0	Eveleth and Tanner, 1976
Yugoslavia, Lipik	1972	13.0	Eveleth and Tanner, 1976
Hungary	1965	13.1	Eveleth and Tanner, 1976
Norway, Oslo	1970	13.2	Eveleth and Tanner, 1976
France, Paris	1966–1968	13.2	Eveleth and Tanner, 1976
Hungary, Szeged	1958–1959	13.2	Eveleth and Tanner, 1976
Finland, Helsinki	1965–1969	13.2	Helm and Helm, 1984
Romania, Urban	1963–1966	13.3	Eveleth and Tanner, 1976
Sweden, Stockholm	1980s	13.3	WHO, 1986a
Netherlands	1965	13.4	Eveleth and Tanner, 1976
England, Northeast	1967	13.4	Eveleth and Tanner, 1976
Czechoslovakia, Bratislava	1960–1962	13.4	Eveleth and Tanner, 1976
Yugoslavia, Gypsies	1961–1966	13.6	Eveleth and Tanner, 1976
Romania, rural	1963–1966	14.2	Eveleth and Tanner, 1976
USSR, Kirghiz (700 m elev.)	1968	14.4	Eveleth and Tanner, 1976
USSR, Kirghiz (2500 m elev.)	1968	15.2	Eveleth and Tanner, 1976
NORTH AMERICA			
United States, blacks	1960–1970	12.5	Eveleth and Tanner, 1976
Mexico, Xochimilco	1966	12.8	Eveleth and Tanner, 1976
United States, Massachusetts	1965+	12.8	Zabin et al., 1986
United States, whites	1960–1970	12.8	Eveleth and Tanner, 1976
United States, California	1950s	12.8	Zabin et al., 1986
Cuba, whites	1973	12.9	Eveleth and Tanner, 1976
Cuba, mixed race	1973	13.0	Eveleth and Tanner, 1976
Canada, Montreal	1969–1970	13.1	Eveleth and Tanner, 1976
U.S., California (Japanese)	1971	13.2	Eveleth and Tanner, 1976
Guatemala, Guatemala City	1965	13.3	Eveleth and Tanner, 1976
U.S., Alaskan Eskimos	1968	13.8	Eveleth and Tanner, 1976
SOUTH AMERICA			
Chile, Santiago (mid SES)	1971	12.3	Eveleth and Tanner, 1976
Venezuela, Carabobo	1978	12.7	Farid-Coupal et al., 1981
Brazil, Japanese	1965	12.9	Eveleth and Tanner, 1976
Chile, Santiago (high SES)	1971	13.0	Eveleth and Tanner, 1976
AUSTRALIA AND NEW ZEALAND			
New Zealand, Maori	1970	13.0	Gray, 1979
Australia, Sydney	1970	13.0	Eveleth and Tanner, 1976
New Zealand	1970	13.0	Eveleth and Tanner, 1976
Australia, inland town	1970	13.2	Eveleth and Tanner, 1976

Table 9.5. (continued)

Population	Period	Median age at menarche	Source
AFRICA			
Seychelles	1978	13.2	Grainger, 1980
Uganda, Baganda (high SES)	1959–1962	13.4	Eveleth and Tanner, 1976
Nigeria, Ife	1980s	13.8	WHO, 1986a
Senegal, Dakar	1970	14.6	Eveleth and Tanner, 1976
Ethiopia, rural highlands	1970s	14.7	Pawson, 1976
Tanzania, Nyakyusa	1969	14.9	Eveleth and Tanner, 1976
Egypt, Nubians	1966	15.2	Eveleth and Tanner, 1976
Rwanda, Tutsi	1957–1958	16.5	Eveleth and Tanner, 1976
Botswana, Dobe !Kung	1963–1969	16.6	Howell, 1979
Rwanda, Hutu	1957–1958	17.0	Eveleth and Tanner, 1976
NEAR EAST AND ASIA			
Turkey, Istanbul (high SES)	1965	12.4	Eveleth and Tanner, 1976
Singapore (mid SES)	1968	12.7	Eveleth and Tanner, 1976
Hong Kong	1980s	12.8	WHO, 1986a
India, Madras (urban)	1960	12.8	Eveleth and Tanner, 1976
Singapore (low SES)	1968	13.0	Eveleth and Tanner, 1976
India, Assam (urban)	1957	13.2	Eveleth and Tanner, 1976
Burma, urban	1957	13.2	Eveleth and Tanner, 1976
Turkey, Istanbul (low SES)	1965	13.2	Eveleth and Tanner, 1976
India, Kerela (urban)	1960	13.2	Madhavan, 1965
Israel, Tel Aviv	1959–1960	13.3	Eveleth and Tanner, 1976
Iran, urban	1963	13.3	Eveleth and Tanner, 1976
Tunisia, Tunis (high SES)	1970	13.4	Eveleth and Tanner, 1976
Sri Lanka, Colombo	1980s	13.5	WHO, 1986a
Iraq, Baghdad (high SES)	1969	13.6	Eveleth and Tanner, 1976
Malaysia	1976–1977	13.7	Ann et al., 1983
India, all urban	1956–1965	13.7	Eveleth and Tanner, 1976
Sri Lanka, Jaffna	1981	13.8	Prakash and Pathmanathan, 1984
Iraq, Baghdad (low SES)	1969	14.0	Eveleth and Tanner, 1976
Tunisia, Tunis (low SES)	1970	14.0	Eveleth and Tanner, 1976
India, Ladakh	1980s	14.1	Malik and Hauspie, 1986
India, Madras (rural)	1960	14.2	Eveleth and Tanner, 1976
Sri Lanka, Peradeniya	1980s	14.4	WHO, 1986a
India, Kerela (rural)	1960	14.4	Madhavan, 1965
India, all rural	1956–1965	14.4	Eveleth and Tanner, 1976
Bangladesh, Matlab (Muslims)	1976	15.8	Chowdhury et al., 1977
Tibet	1970s	16.1	Pawson, 1976
Nepal, rural Tibetans	1981	16.2	Beall, 1983
Philippines, Agta foragers	1980–1982	17.1	Goodman et al., 1985
Nepal, Sherpas	1970s	18.1	Pawson, 1976
PACIFIC ISLANDS			
Yap	1946–1948	14.5	Hunt and Newcomer, 1984
Papua New Guinea, Megiar	1967	15.5	Eveleth and Tanner, 1976
Papua New Guinea, Kaiapit	1967	15.6	Eveleth and Tanner, 1976
Papua New Guinea, Karkar	1971	15.6	Eveleth and Tanner, 1976
Papua New Guinea, Lufa	1971	16.5	Eveleth and Tanner, 1976
Papua New Guinea, Chimbu	1965	17.5	Eveleth and Tanner, 1976
Papua New Guinea, Bundi	1967	18.0	Eveleth and Tanner, 1976
Papua New Guinea, Lumi	1967	18.4	Eveleth and Tanner, 1976
Papua New Guinea, Gainj	1983	18.6	Wood, 1992

*Interpopulation Variation in Age at Menarche
and Menopause*

When studying variation in the ages at menarche and menopause among human populations, it should be borne in mind that different study designs can create the appearance of different age patterns where none exist. Unquestionably, prospective or cohort studies are the most reliable but also the most costly and time-consuming. Cross-sectional studies, which are far more common, can be of two sorts: retrospective or current status. In retrospective cross-sectional studies, women are asked whether they have reached menarche or menopause and, if they have, at what age. Such data may be biased by the fact that only women who have already experienced the event in question can report an age, i.e., the data may be right-censored. In addition, recall error may be a problem, especially if the recalled event was in the distant past. Several studies comparing the actual (known) age at menarche with self-reports of age many years later have shown that recall errors can be quite large, though usually not biased in any systematic way (Livson and McNeill, 1962; Damon et al., 1969). The current status research design avoids recall error. Under this design, one simply ascertains whether each woman at a given age has or has not experienced the event of interest, and then uses logit, probit, or rankit analysis to estimate means or medians (Finney, 1971). However, this approach requires us to assume that the ages at onset of menarche and menopause have not changed appreciably in the past several decades.

All study designs are faced with the problem that a woman may not know at a given time whether she is in fact postmenopausal. It is usually necessary, therefore, to use some more or less arbitrary waiting time criterion for the assessment of menopause—that is, if a woman is to be deemed postmenopausal, a certain minimum period must have elapsed before the interview during which she experienced no menstruation or pregnancy. As already noted, the most commonly used waiting time is 12 months, but some studies may use other diagnostic criteria. This difference may itself be a source of apparent variation among studies: the longer the waiting time used, the lower the fraction of women who will appear postmenopausal at any given age. This potential source of heterogeneity must be considered when assessing interpopulation variation in the age at menopause.

Bearing these caveats in mind, what sort of variation do we observe worldwide in age patterns of menarche and menopause? Table 9.5 summarizes estimates of the median age at menarche in contemporary populations, broken down by region. There appears to be considerable variation both within and between regions, with estimates ranging from 12.3

Table 9.6. Estimates of the Median Age at Menopause (in years) in Contemporary Populations

Population	Period	Median age at menopause	Study design
United States, whites	1981–1982	51.4	current status
Netherlands	1969	51.4	current status
England	1965	50.8	current status
New Zealand	1967	50.7	current status
South Africa, whites	1971	50.4	current status
Scotland	1970	50.1	current status
United States, whites	1966	50.0	current status
United States, whites	1934–1974	49.8	cohort
South Africa, blacks	1971	49.7	current status
United States, blacks	1966	49.3	current status
Dobe !Kung	1963–1973	49.1	current status
Yap	1946–1948	48.6	current status
South Africa, blacks	1960	48.1	retrospective
Papua New Guinea	1973		current status
well nourished		47.3	
poorly nourished		43.6	
Nepal, Tibetans	1981	46.8	current status
Papua New Guinea, Gainj	1978	46.2	current status*
India, Punjab	1966	44.0	mixed cohort, current status
Bangladesh	1975–1978	43.6	cohort
Central Africa	1970s	43.0	retrospective

* Based on hormonal assessment; otherwise based on self-report.
From Gray (1976), Beall (1983), Hunt and Newcomer (1984), Karim et al. (1985), McKinlay et al. (1985), Wood et al. (1985a), Rahman and Menken (1993).

to 18.6 years. In general, developed countries in Europe, North America, and the Pacific Rim have the lowest medians, as do high SES groups in developing countries. Within countries, there are consistent differences between SES groups, between ethnic groups like blacks and whites in the United States and South Africa (which are also different SES groups to a large extent), and even between groups at different altitudes. It seems plausible that all these differences reflect underlying differentials in nutritional status to at least some degree, although other factors may also be implicated. The highest known median ages at menarche tend to be found at high altitudes in Nepal and Tibet or at much lower altitudes in highland Papua New Guinea (Lufa, Chimbu, Bundi, Lumi, Gainj). Although chronic undernutrition has been suggested as a cause of late menarche in each of these areas, it is far from clear that these groups are significantly less well nourished than most rural populations in other parts of the developing world.

Data on age patterns of menopause are less common, for reasons that have already been discussed. From Table 9.6, it appears that women of western European origin living in developed countries have the latest ages at menopause, with medians clustering near age 50. Women from other parts of the world—or black women living in white-dominated societies—have lower and much more variable medians, ranging from about 43 to 50. Again, it seems plausible that nutritional differences play some role in these differentials, but the evidence does not permit us to exclude possible genetic differences or methodological problems arising from age misstatement, recall bias, or differing cultural definitions of menopause.

If the figures in Tables 9.5 and 9.6 can be taken as representative of worldwide variation, it appears that menopause is somewhat more variable across populations in age at onset than is menarche. The range in medians spans about 8.4 years for menopause and 6.3 years for menarche, and the variance in medians calculated from the tabulated estimates is 7.68 for menopause and 2.25 for menarche. On this basis, it would appear that variation in menopause is likely to be a more important cause of interpopulation variation in the length of the reproductive span than is variation in menarche, similar to the findings on intrapopulation variation. However, since the tabulated groups are far from being a random sample of all contemporary populations, it is uncertain how seriously we should take this conclusion.

NOTES

1. Endocrine assays may offer a solution to this problem. As discussed below, menopause is marked by a fairly unambiguous hormonal signal: any time we observe a woman with no detectable ovarian steroids but very high concentrations of gonadotropins, even on a single blood draw or urine specimen, we can be reasonably certain the woman is postmenopausal. To date, however, there have been few applications of this approach to comparative research.

2. It is unclear how activation of the GnRH pulse generator occurs. Most researchers believe that *adrenarche* (initiation of steroid secretion, especially of androgens, by the adrenal cortex) plays a critical role since precocious puberty is often associated with adrenal hyperplasia (Pescovitz et al., 1984). In addition, experiments on rodents have shown an age-dependent effect of exogenously administered opiates such as β-endorphin on gonadotropin secretion (Wilkinson and Bhanot, 1982), leading to the speculation that changes in endogenous opiate activity within the CNS are necessary for activation of the GnRH pulse generator and initiation of pubertal development (Kelch et al., 1983).

3. Note that there may have been two or three ovulations just before the first definitive one; if so, the associated corpora lutea did not produce enough pro-

gesterone to support a pregnancy. The first definitive ovulation is followed by another cycle with an inadequate luteal phase.

4. This assumption may not be correct. The midcycle rise in BBT partly reflects the initiation of progesterone secretion by the corpus luteum, and luteal adequacy is now known to vary with age (see Chapter 4). Thus, there may be a greater likelihood of missing ovulation in a perimenarcheal or perimenopausal subject using the BBT method than in a mid-reproductive age subject.

5. The apparent effect of age at menarche on age at first conception is probably attributable to the former's effect on age at marriage, since there is no consistent relationship between age at menarche and the interval from marriage to first conception.

6. Other studies have also suggested that the decline in the age at menarche has slackened or stopped, e.g., in Denmark at 13.2 years (Helm and Helm, 1984), in England at 12.9 years (Dann and Roberts, 1984), and in Japan at 12.4 years (Hoshi and Kouchi, 1981). Another study conducted in the United States suggests that the decline has ended at 12.8 years (Zacharias et al., 1976), consistent with the findings of the Minnesota cohort study.

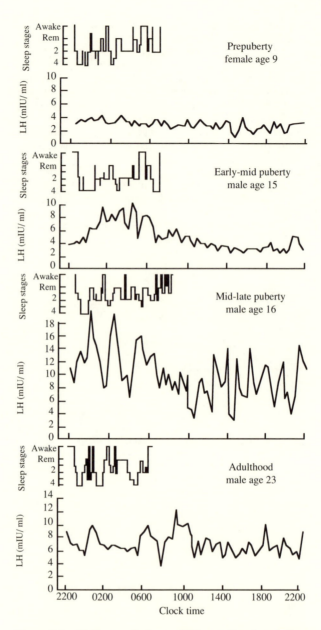

Figure 9.1. Patterns of plasma LH secretion in relation to sleep stages, from prepuberty to early adulthood. Observed patterns in males and females are essentially identical. (Redrawn from Weitzman et al., 1975)

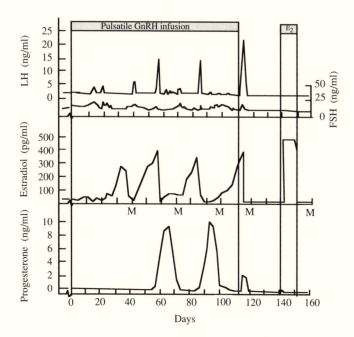

Figure 9.2. Induction of ovulatory cycles in an immature female rhesus monkey by the pulsatile infusion of GnRH (1 mg/min for six minutes once every hour). The animal shown is about 12 months old, which is approximately 16 months before the normal age at menarche and 20 months before ovulatory cycles are normally observed. The period of GnRH infusion is shown by the horizontal bar (Days 0–110); LH, FSH, estradiol, and progesterone were undetectable before GnRH infusion. Note that the first estradiol peak (ca. Day 30) did not elicit a full LH response. However, subsequent estradiol peaks induced LH surges at 28-day intervals, corpus luteum formation (progesterone peaks), and eventual menses (M). Cessation of GnRH infusion immediately before the LH surge ca. Day 112 was followed by prompt re-entry into a non-cyclic, prepubertal state. Subcutaneous implantation of an estradiol-containing capsule between 140 and 150 days (as indicated by bar) to produce high levels of estradiol in plasma *did not* induce an LH response in the absence of exogenous GnRH, confirming that the central site of action in the induction of puberty is the GnRH pulse generator, not the ovary. (Redrawn from Wildt et al., 1980)

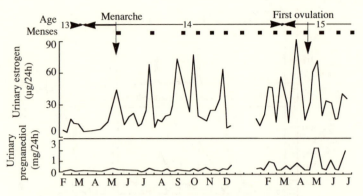

Figure 9.3. Urinary estrogen and pregnanediol excretion in a girl undergoing puberty between the ages of 13 and 15. Note that waves of estrogen excretion precede menarche, and that the first definitive ovulation (indicated by pregnanediol excretion > 2mg/24h) occurs about a year after menarche. (Redrawn from Brown et al., 1978)

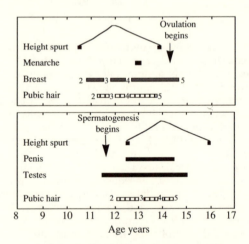

Figure 9.4. Diagram of the sequence of physical changes occurring during puberty in girls (*top*) and boys (*bottom*). The "average" western child is represented. Note that while pubertal changes are normally accelerated in girls compared with boys, production of mature gametes begins at an earlier stage of puberty and hence at a younger age in boys. (Adapted from Marshall, 1970; Marshall and Tanner, 1970)

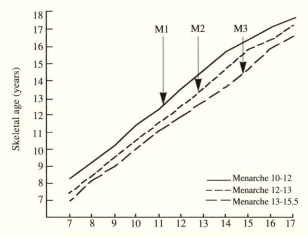

Figure 9.5. Relationship of age at menarche to skeletal maturity. Skeletal development ages (Todd Standards) for early, average, and late menarche groups of girls, from age 7 to maturity. M1, M2, M3, mean age at menarche for each group. Mixed longitudinal data. (Redrawn from Tanner, 1962)

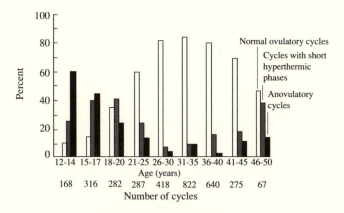

Figure 9.6. Anovulatory cycles, cycles with short hyperthermic phases, and normal ovulatory cycles in girls and women ages 12 to 50 years. Cycle quality was evaluated by the BBT method: anovulatory cycles are those with no midcycle elevation in BBT, whereas cycles with short hyperthermic phases are those with < 10 days between the midcycle rise in BBT and the onset of next menses. The latter are assumed to involve short luteal phases. Both types of ovarian dysfunction appear to be more frequent at perimenarcheal and perimenopausal ages. (Redrawn from Döring, 1969)

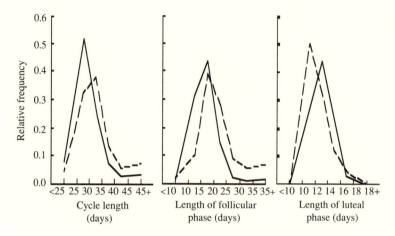

Figure 9.7. Observed distributions of menstrual cycle length (*left*), length of the follicular phase (*middle*), and length of the luteal phase (*right*) in Japanese women ages 13–19 years (*broken lines*, N = 380 cycles) and 20–24 years (*solid lines*, N = 1,350 cycles). Phases of the cycle were distinguished by the midcycle rise in BBT. (Data from Matsumoto et al., 1962)

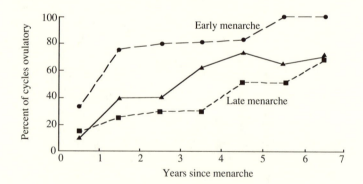

Figure 9.8. Percent of cycles ovulatory by time since menarche in Finnish girls who, at the time of menarche, were less than 12 years old (*circles*, 4–8 subjects at each time point), 12 years old (*triangles*, 6–14 subjects), and 13 years old or older (*squares*, 11–24 subjects). Numbers of subjects vary because of loss to follow-up. Ovulation was assessed by serum progesterone concentration > 2 ng/ml. Both age at menarche and years since menarche have significant effects on the percent of cycles ovulatory ($P < 0.001$ for both). (Redrawn from Apter and Vihko, 1983)

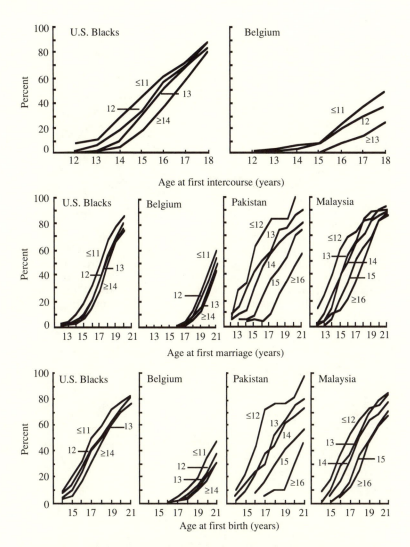

Figure 9.9. The relationship between age at menarche (in years, numbers within each panel) and age at subsequent reproductive events: first intercourse (*top*), first marriage (*middle*), and first birth (*bottom*). Sample sizes: 1,350–1,477 (U.S. Blacks), 2,170–3,357 (Belgium), 185 (Pakistan), and 549–552 (Malaysia, excluding Chinese Malaysians). Sample sizes vary according to which subsequent event is being analyzed because of differential loss to follow-up. (Redrawn from Udry and Cliquet, 1982)

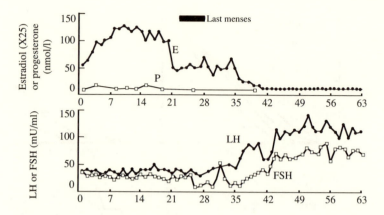

Figure 9.10. Changes in the plasma concentrations of ovarian steroids (*top*) and gonadotropins (*bottom*) during the menopausal transition in a woman age 48 years. Note that the fall in estradiol concentration results in the last menses and is followed by a marked rise in the concentrations of LH and FSH. E, estradiol-17β; P, progesterone; LH, luteinizing hormone; FSH, follicle-stimulating hormone. (Redrawn from van Look et al., 1977)

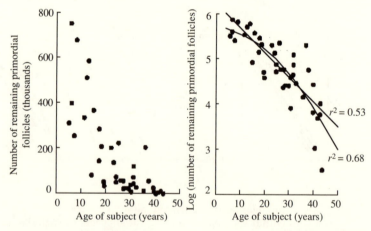

Figure 9.11. Age-related changes in the number of primordial follicles present in both ovaries, based on cross-sectional autopsy data: linear plot (*left*) and semi-log plot (*right*). A quadratic regression (*curved line*) fits the semi-log data significantly better than a linear regression (*straight line*), suggesting near-depletion of follicles by the time of menopause. (Data from Block, 1952)

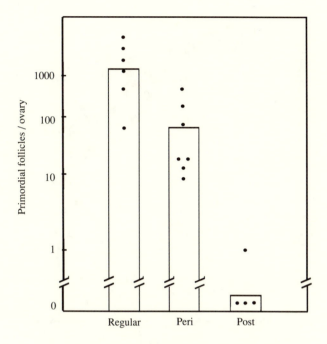

Figure 9.12. The relationship between primordial follicle number per ovary and menstrual status in 17
U.S. women 45–55 years of age who underwent
elective hysterectomies and oophorectomies. Cycling status was determined by a menstrual history confirmed by plasma hormone levels. The
effect of menstrual status on follicle number is
highly significant ($P < 0.001$ by ANOVA). All
pairwise contrasts between groups are also significant at the $\alpha = 0.05$ level or better. *Regular,*
women cycling regularly during the 12 months
before surgery; *Peri,* women of perimenopausal
ages, as indicated by irregular cycles with or
without hot flashes; *Post,* postmenopausal
women, or women who had experienced no
menses for at least 12 months before surgery.
(Redrawn from Richardson et al., 1987)

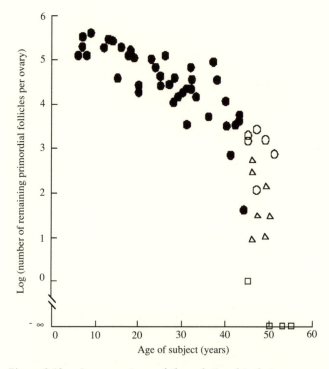

Figure 9.13. A comparison of the relationship between age and primordial follicle number per ovary in Block's (1952) autopsy study of 43 girls and women ages 6–44 years (*closed circles*), with that in the Richardson et al. (1987) study of oophorectomy data from 17 women ages 45–55 years (*open symbols*). Since only one ovary was examined in the study of Richardson et al., a single ovary was selected at random from each subject in Block's study. Insofar as these two data sets are comparable, follicular depletion appears to accelerate during the years preceding menopause. *Open symbols*, women from Richardson et al.'s study: ○, women who were experiencing regular menses; △, perimenopausal women; □, postmenopausal women.

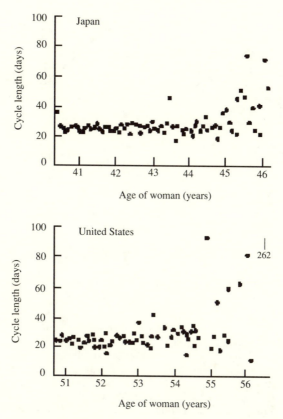

Figure 9.14. Patterns of cycle-length variation during the last six years of menstrual life in two typical cases: (*top*) a Japanese woman experiencing menopause at age 46, and (*bottom*) a U.S. woman experiencing menopause at age 56. (Data from Matsumoto et al., 1962; Treloar et al., 1967)

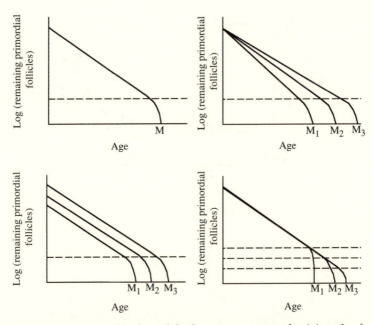

Figure 9.15. A graphical model of menopause, emphasizing the fact that menopause is the final, discrete outcome of a continuous, decades-long process of follicular depletion (*upper left*). It is assumed that follicular loss occurs at a constant rate until a threshold is reached (*broken line*). Below that threshold, the remaining pool of follicles secretes too little estradiol early in the follicular phase to maintain regular cycles. Accelerated follicular depletion is hypothesized to result from the loss of efficient estradiol feedback control. Menopause *(M)* occurs when the pool of remaining follicles is more or less completely depleted. Under this model, variation among women in the age of menopause can result from heterogeneity in the rate of follicular loss (*upper right*), heterogeneity in the initial number of follicles (*lower left*), or heterogeneity in the level of the regulatory threshold (*lower right*). If acceleration in follicular depletion occurs during perimenopausal life, additional variation could result from heterogeneity in the rate of acceleration.

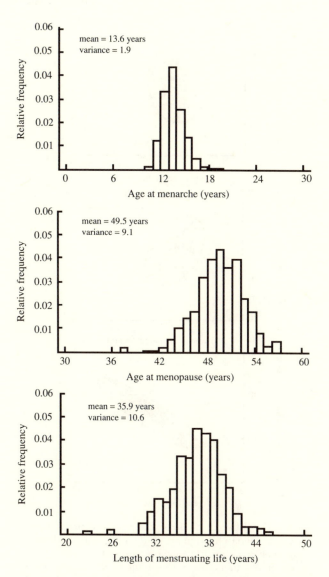

Figure 9.16. The distributions of age at menarche (*top*), age at menopause (*middle*), and length of menstruating life (*bottom*) in 324 U.S. women born in 1910–1925 and monitored prospectively from about age 20. Length of menstruating life is the difference between age at menopause and age at menarche. (Redrawn from Treloar, 1974)

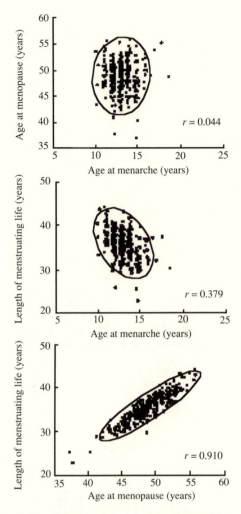

Figure 9.17. Jittered plots of the relationships among age at menarche, age at menopause, and the length of menstruating life in the same sample of women as shown in Figure 9.16. *Ellipses*, 95 percent joint confidence limits. The correlation between ages at menarche and menopause is not significant ($P > 0.05$). The effect of age at menarche on length of menstruating life is weak ($r^2 = 0.144$), whereas that of age at menopause is strong ($r^2 = 0.828$). Variation in the length of the reproductive span is thus determined principally by the age at menopause. (Redrawn from Treloar, 1974)

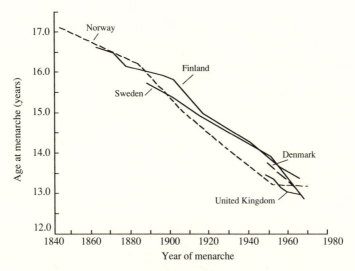

Figure 9.18. The purported trend in age at menarche in Europe. (Redrawn from Eveleth and Tanner, 1976)

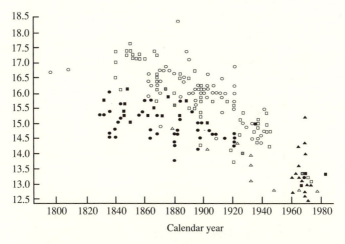

Figure 9.19. Estimates of the mean or median age at menarche by calendar year from 1790 to 1980. Symbols refer to England (■); France (●); Germany (○); The Netherlands (▨); Scandinavia (□); the United States (△); and Belgium, Czechoslovakia, Hungary, Italy, Poland (rural), Romania (urban and rural), Russia, Spain, and Switzerland (all denoted ▲). Twenty-seven points are identical and do not appear in the figure. (Redrawn from Wyshak and Frisch, 1982)

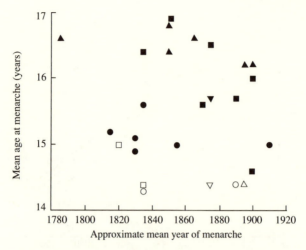

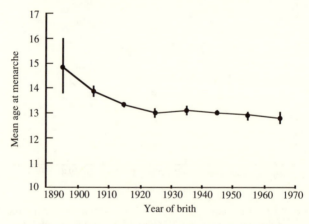

Figure 9.20. Estimates of the mean age at menarche in Europe from the late eighteenth to early nineteenth centuries, broken down by region and socioeconomic class. Solid symbols represent working-class samples, open symbols middle-class samples. Regions: Great Britain (*circles*), Scandinavia (*squares*), Germany (*triangles*), and Russia (*inverted triangles*). (Estimates from Tanner, 1981)

Figure 9.21. Mean age at menarche (± 95 percent confidence interval) by year of birth, 3,839 U.S. university students. Menarche was recorded retrospectively, but usually within 10 years of its occurrence. The mean age at menarche differs significantly among birth cohorts ($P < 0.001$). (Redrawn from Sandler et al., 1984)

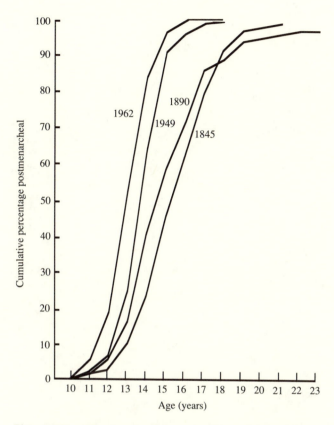

Figure 9.22. Changes in the age pattern of menarche in
England from four surveys: 1845 (retrospective data
from maternity hospital patients in Birmingham,
1830–1845, N = 623); 1890 (retrospective data from a
survey of women ages 60–85 in Sheffield, 1989–1991,
N = 237); 1949 (cross-sectional data on school girls ages
10–17 in Oxford city and county, N = 1,338); 1962 (retrospective data on girls ages 15–19 in West Riding,
Yorkshire, N = 685). Note that the curve for the 1962
sample may be shifted slightly to the left by right-
censoring. (Redrawn from Brown, 1966)

10

The Onset of Permanent Sterility

Although menopause marks the absolute upper limit of the reproductive span, childbearing typically ends several years before menopause, even in the absence of contraception. Table 10.1 summarizes estimates of the average age of women at their last birth in several natural-fertility settings. Perusal of this table will reveal a striking degree of agreement among those averages: almost all fall within a narrow range from about 38 to 41 years. There are, moreover, reasonable explanations for the few that fall outside that range. For example, in the Punjab villages included in Table 10.1 there is apparently some use of contraception, including surgical sterilization (Wyon and Gordon, 1971:162). Similarly, at the time they were first studied the !Kung San had a high prevalence of pathological sterility, apparently associated with gonorrhea (Howell, 1979). Either situation is expected to result in early termination of reproduction.

It is noteworthy that when the data can be disaggregated appropriately, there is a small but consistent effect of the wife's age at marriage on her age at last birth (Table 10.1). A naive interpretation of this finding is that there is some degree of cryptic birth control in all these populations, so that couples who marry early and thus attain their desired family size early also terminate their reproduction at an early age. But in fact this effect seems to be entirely attributable to data truncation. Couples who marry at age 30, for example, cannot have a last birth before age 30, at least not within marriage. Thus, their mean age at last legitimate birth must be later than that of couples who marry, say, at age 20.

Table 10.1. Estimates of the Mean Age of Women at Last Birth in Natural-fertility Populations

Population	Mean Age at Last Birth (Years)
GERMAN VILLAGES (PRE–1850)	
Grafenhausen	39.7
Oshselbronn	39.2
Werdum	39.4
Three Bavarian villages	40.6
Four Waldeck villages	40.7
All villages:	
Age at marriage: 20–24	39.2
25–29	40.3
30–34	41.0
NORTH AMERICAN COMMUNITIES	
Canada (seventeenth century)	40.1
(eighteenth century)	41.0
Bois-Vert	40.7
St. Jean-de-Cherbourg	38.6
Hingham	39.1
Quakers	39.6
Hutterites	40.9
Bris	38.7
Mormons (1800–1869)	40.5
FRENCH PARISHES	
Crulai	40.0
Tourouvre-au-Perche	40.3
25 parishes in Paris Basin (sixteenth–seventeenth centuries)	39.5
ENGLAND	
England (mid-nineteenth century)	41.7
16 English parishes (1600–1799):	
Age at marriage: 20–24	38.2
25–29	39.7
30–34	40.9
SWEDEN (1841–1900)	
Age at marriage: 20–24	38.9
25–29	39.7
NONEUROPEAN COMMUNITIES	
Timling (Nepal)	39.3
Ghana (rural)	38.8
Matlab (Bangladesh)	38.8
11 Punjab villages (may include some contraception)	37.0
!Kung San (Botswana)	34.4

From Gautier and Henry (1958); Hyrenius (1958); Potter et al. (1965c); Charbonneau (1970, 1979); Frisch (1978); Knodel (1978b, 1988); Howell (1979); Mineau et al. (1979); Bongaarts and Potter (1983); Gaisie (1984); Wilson (1984); Karim et al. (1985); Fricke (1986)

Allowing for this truncation effect, the mean ages at last birth in natural-fertility populations appear to cluster near 40, thus preceding the average age at menopause by something like five to ten years (compare Table 9.6).[1] How can we explain this lag between a woman's last birth and the onset of menopause, especially if we assume that she is not deliberately trying to terminate childbearing? In part, the lag between last birth and menopause can be explained by some degree of postpartum infecundability, as well as the normal waiting time to the next fertile conception. As we have seen, this waiting time is determined by effective fecundability, which in turn reflects the monthly probability of conception (which is declining at perimenopausal ages) and the risk of fetal loss (which is increasing at the same ages). A fairly long lag between the last birth and menopause is therefore to be expected.

But does this explanation fully account for the observed lag? Many demographers believe that it does not. They argue that as part of the normal aging process, women enter a state of absolute and irreversible *sterility*, or physiological inability to produce a live birth, some time before menopause. Of course, temporary sterility can also occur, for example, in association with breastfeeding; but what we are concerned with here is a state of *permanent sterility* that demographers believe to be a universal concomitant of increasing age.[2]

Recently it has been suggested by Wilson et al. (1988) that the onset of permanent sterility is the dominant process determining the typical age pattern of natural marital fertility (Figure 10.1). This suggestion is important for it implies that permanent sterility, not menopause, is the ultimate factor limiting a woman's reproductive span—which in turn implies that the effective reproductive span may be considerably shorter than suggested by data on the age at menopause. To evaluate this claim it will be necessary, after clearing up some definitional problems and reviewing what is known about the biology of sterility, to discuss methods for estimating the prevalence of sterility from demographic data. As should quickly become apparent, estimating age at onset of permanent sterility is one of the most problematic undertakings in the entire field of fertility analysis. Indeed, near the end of the chapter we shall ask whether the demographic approach to the onset of permanent sterility has any biological meaning at all.

PROBLEMS OF DEFINITION

One source of confusion about sterility is that the word can mean various things: it can be temporary or permanent, partial or complete, primary or secondary. Several conditions can induce a temporary inability to conceive that is readily reversible; obvious instances include lactational anovulation and inadequate luteal function associated, for example, with anorexia nervosa or endurance exercise (Pirke et al., 1989). Many of the same conditions can also induce partial sterility or subfecundity, as can such disorders as malaria, endometriosis, polycystic ovarian disease, and hypothyroidism. Since all these conditions lengthen the expected waiting time to a conception rather than rendering a woman permanently unable to conceive, it may be better to think of them as acting upon fertility via lowered fecundability rather than via sterility per se. Accordingly, I will use the term *sterility* to refer strictly to sterility that is both *complete* and *permanent*. For consistency, the term *subfecundity* will be used throughout what follows in lieu of "partial sterility" and *infecundability* for "temporary sterility."

The distinction between *primary sterility* and *secondary sterility* is defined in at least two different ways which are by no means equivalent. For some authors the distinction depends upon whether the subject has ever been *capable* of reproducing rather than whether she has in fact reproduced. By this definition, menopause usually causes secondary sterility, regardless of the woman's prior reproductive experience or lack thereof. Although this distinction makes some biological sense—for example, when differentiating congenital causes of sterility from those beginning later in life—it is associated with an insurmountable measurement problem: how do we determine if someone was capable of reproducing in the absence of a history of reproduction? Because of this difficulty, many demographers use a different definition: primary sterility is sterility that begins before any offspring have been produced, whereas secondary sterility occurs after the birth of at least one child. From a biological perspective, this distinction is useful because of the possibility of *puerperal sepsis:* parturition itself may entail some risk of infection, which may in turn result in pelvic inflammatory disease, scarring of the fallopian tubes, and subsequent sterility (McFalls and McFalls, 1984). For convenience, I adopt this second definition of primary and secondary sterility: primary sterility precedes the birth of any live-born children, secondary sterility begins after at least one such birth.

The proper antonym for sterility is *fecundity*, not *fertility* as is often encountered in the medical literature. *Infertility* is the nonproduction of a live birth, and it may or may not be a sign of sterility. A nun, for example, may be perfectly fecund and yet infertile. And indeed, so may a regularly cohabiting couple if they happen by chance or by choice not to produce a live birth. Even ignoring contraception, there are several reasons a fecund couple might not go on to produce a(nother) child:

they may have low coital frequency, low fecundability, or an elevated risk of fetal loss, any of which would cause them to have a long waiting time to their next fertile conception but would not render them absolutely sterile. In addition, they may not be subfecund at all but may simply fall in the long upper tail of the distribution of normal waiting times to the next fertile conception.

This distinction is worth repeating: sterility is the actual physiological inability to produce a live birth, infertility the mere absence of such a birth. Infertility may result from sterility, but it need not do so. Unfortunately, demographic estimators of the prevalence of sterility and its progression with age are forced to use data on infertility, not sterility per se. As a consequence, they often overestimate the prevalence of sterility to at least some extent.[3] Thus, the "onset of permanent sterility" as defined by demographers should be considered a purely demographic phenomenon, one that has to do with the age at last childbearing but that need not correspond to any real physiological transition.

THE BIOLOGY OF STERILITY

The various causes of genuine, permanent sterility to which women are potentially subject can be classified into three categories: (1) congenital and genetic, (2) infectious, and (3) senescent. The first category (which refers to congenital and genetic defects in the woman, not her offspring) includes several chromosomal abnormalities. For example, Turner syndrome (45,X0 karyotype) is characterized by gonadal dysgenesis and rapid follicular depletion: in effect, females with this condition undergo the menopause *before* entering puberty (Jacobs, 1982). Testicular feminization (androgen resistance syndrome) is a condition in which a person who is chromosomally male has a superficially normal female phenotype (Griffin and Wilson, 1989). Most cases of this syndrome appear to be caused by an androgen-receptor defect, rendering target tissues insensitive to masculinization (Pinsky and Kaufmann, 1987). Individuals with testicular feminization exhibit complete failure of gonadal function and abnormal development of the reproductive tract. Cases of karyotypically normal gonadal dysgenesis are also known (Feinman, 1991), and these may be associated with mutations in genes responsible for ovarian maturation or maintenance (e.g., the genes that code for LH receptors). Other causes of congenital sterility include congenital adrenal hyperplasia, the immotile cilia syndromes, endometriosis of genetic origin, and congenital malformations of the uterus and cervix (Simpson, 1985).

From a demographic perspective, the salient characteristic of all the congenital and genetic syndromes is that they are individually rare and uncommon even in the aggregate. Although good prevalence estimates are unavailable for most of these syndromes, it is extremely unlikely that collectively they affect as many as 5 percent of all women in most populations (Simpson, 1985). The expected incidence of these conditions varies by etiology. Primary sterility of genetic origin, for example, must be maintained by mutation-selection balance since the genetic fitness of individuals affected by this condition is necessarily zero. For dominant mutations the expected phenotypic incidence at birth is approximately 2μ, where μ is the mutation rate per locus per generation; for recessive mutations the expected incidence is μ (Ewens, 1979:13). For most human genes μ has been estimated to fall between 10^{-5} and 10^{-6} (Gelehrter and Collins, 1990:54). The incidence of Turner syndrome among newborns, which reflects the frequency of meiotic nondisjunction in their parents, is approximately one in 7,500 (Jacobs, 1982); incidence rates for other chromosomal aberrations are roughly similar (Weatherall, 1991:24). The incidence of nonheritable congenital defects causing sterility is unknown but probably quite variable depending upon environmental exposures to teratogens. However, the incidence of *all* congenital malformations, the great majority of which have no known direct effect on fecundity, is only about 20 per 1,000 live births (Carter, 1982). Thus, although congenital and genetic syndromes cannot be ignored as causes of primary sterility, they are not expected to have a large effect at the population level. Moreover, by their nature they cannot account for any *new* cases of adult-onset sterility. Thus, although they may contribute to the prevalence of preexisting sterility (i.e., sterility already present at the time of sexual maturation), they can have nothing whatsoever to do with the incidence of new sterility.

Female sterility of infectious origin involves blockage of the fallopian tubes by scar tissue, or tubal scarring short of occlusion but severe enough to inhibit transport of ova by normal ciliatory movement. Where they occur, such tubal lesions are almost always sequelae of pelvic inflammatory disease (PID). Ignoring causes of PID that are of recent origin, such as IUD use, dilatation and curettage, or hysterosalpingography, PID is caused primarily by gonorrhea, chlamydia, and other sexually transmitted infections, puerperal sepsis, or far less commonly, genital tuberculosis (McFalls and McFalls, 1984; Muir and Belsey, 1980).[4] Where they are practiced, female circumcision and traditional methods of inducing abortion may also result in PID (McFalls and McFalls, 1984). Although good epidemiological data are scarce, sexually transmitted diseases (STDs) that cause PID appear to be uncommon in most natural-fertility settings, except in such regions as Central Africa and some of the

Pacific islands (Muir and Belsey, 1980; Frank, 1983). The fact that these latter regions are atypical is underscored by their unusual epidemiological patterns, which differ markedly from what is observed in most natural-fertility populations (Frank, 1983; Cates et al., 1985). For example, couples who report infertility problems in sub-Saharan Africa are significantly younger on average than in other regions, and they are more likely to have experienced infertility of long duration and to have a history of STD (Cates et al., 1985). And even in sub-Saharan Africa high prevalences of pathological sterility seem to be a fairly recent phenomenon, perhaps less than a century old, and are apparently the result of increasing contact with the larger world (Muir and Belsey, 1980; Caldwell and Caldwell, 1983; Frank, 1983). In most natural-fertility settings, there is no obvious reason to expect the prevalence of pelvic inflammatory disease to be high at any premenopausal age or to increase dramatically at later ages—except perhaps as a result of puerperal sepsis.

Although several authors have emphasized the potential importance of puerperal sepsis as a cause of secondary sterility (Gray, 1979; McFalls and McFalls, 1984), no study has ever measured its actual demographic impact in any natural-fertility setting. The elevated risk of PID associated with childbirth presumably reflects the dissolution of the cervical mucus plug and opening of the cervix and uterine body to the external environment, as well as any tearing of maternal tissue that may occur during delivery; the use of nonsterile medical instruments may also be a factor in some settings. The magnitude of the risk is essentially unknown but probably quite variable. Although Curtis and Huffman (1950) and Hellman et al. (1971) have argued that puerperal sepsis rarely results in tubal occlusion, Weström (1975) estimates that some degree of tubal scarring occurs in about 25 percent of Swedish women suffering from pelvic infection resulting from childbirth, induced abortion, and dilatation and curettage (presumably the percentage would be higher in the less aseptic conditions found in much of the developing world). Nor are the pathogens involved in puerperal sepsis well characterized. Where gonorrhea is widespread, *Neisseria gonorrhoeae* may be an important infectious agent; however, only about 10 percent of all cases of gonorrhea-related PID in sub-Saharan Africa appear to be attributable to childbirth (Grech et al., 1973). Another possible pathogen is β-hemolytic streptococcus (McFalls and McFalls, 1984), especially where nonsterile medical instruments are used. But in most cases, the pathogen is simply unknown. Indeed, the safest general conclusion about the importance of puerperal sepsis as a cause of secondary sterility in most natural-fertility populations is that it too is unknown.

Senescent sterility, or age-related loss of reproductive function, is of course expected to be universal in women who survive long enough. As detailed in previous chapters, reproductive aging in women appears to be dominated by three fairly well-characterized processes: (1) an increasing cumulative incidence of chromosomal aberrations and germinal mutations resulting in an elevated risk of fetal death (Kline and Stein, 1987), (2) progressive depletion of primordial follicles in the ovary, resulting in less effective estradiol feedback control of the hypothalamo-pituitary-ovarian axis and ultimately in menopause (Richardson et al., 1987), and to a lesser extent (3) reduced responsiveness of the pituitary and hypothalamus to feedback by ovarian steroids (Metcalf et al., 1981, 1982).[5] In theory, these various age-related physiological changes should be manifested demographically as a decline in fecundability, an increase in the risk of pregnancy loss, and a final cessation of reproduction at the menopause. Only the last, the actual onset of menopause, can be considered a cause of complete, irreversible sterility as distinguished from partial subfecundity. In other words, there is no obvious biological reason to expect any aspect of normal aging, short of the menopause itself, to render a woman permanently sterile.

In sum, biological arguments would seem to suggest that subfecundity may be fairly widespread in humans, especially at later reproductive ages; that secondary sterility may result at least occasionally from childbirth; but that *primary* sterility is likely to be uncommon in women who have not yet reached menopause, except where STDs attain unusually high prevalences. In general, epidemiological studies tend to support this conclusion (Barad, 1991). In one large-scale study in Great Britain, for example, the frequency of involuntary childlessness was estimated to be 4.5 ± 0.9 percent in women born in 1935 and 3.3 ± 0.7 percent in women born in 1950 (G. Johnson et al., 1987). Similarly, in an intensive epidemiological study of a recently contacted tribal population in Papua New Guinea, we found that only 1.1 ± 1.1 percent of ever-married women were still childless at age 45 (Wood et al., 1985a). Demographic studies, in contrast, often yield much higher estimates of the level of primary sterility. Several analyses of data from natural-fertility settings have suggested comparatively high prevalences of primary sterility (8–10 percent) among newlywed couples married before wife's age 25, followed by an initially slow but accelerating increase in prevalences through the late twenties and thirties, reaching 50 percent or more by wife's age 45 (Vincent, 1950; Henry, 1965; Trussell and Wilson, 1985; Barrett, 1986; Menken et al., 1986). But if the known biology and epidemiology of sterility suggest that such high prevalences are unrealistic, then the estimated prevalences may reflect biases and uncertainties in the demographic data and analyses. I discuss some of these problems later in this chapter.

ANALYTICAL ISSUES

Before we review the standard demographic estimators of the prevalence of sterility, several frequently recurring analytical issues should be addressed. The chief of these has already been mentioned. Demographic data provide information on the occurrence or nonoccurrence of live births; they never tell us anything more directly about the ability or inability to produce a live birth. Thus, demographic analyses of the prevalence and age pattern of permanent sterility are forced to estimate patterns of sterility by some means from observations on infertility. As will become clear, this unavoidable fact is the source of numerous difficulties of inference and interpretation.

Most of the estimators to be discussed in this chapter are designed to be used strictly with data from natural-fertility populations. This restriction is necessary if physiological sterility is not to be confused with deliberate birth control, including surgical sterilization. It is not acceptable to limit attention to couples who have never used contraceptives, or who have terminated use, in controlled-fertility settings. Couples who do not practice contraception where it is otherwise widespread are likely to represent a highly select sample of all couples, and in particular are likely to be selected for at least a self-assessment of subfecundity or sterility (Leridon, 1977).

In most studies, attention is further restricted to infertility within marriage; otherwise, an absence of exposure to regular intercourse could be mistaken for physiological sterility. This restriction may itself introduce biases, because illness can simultaneously delay marriage and cause sterility, and because childlessness is often a cause of divorce (Larsen and Menken, 1989). Moreover, intact marriages are not necessarily characterized by continuous cohabitation and regular sexual relations. Demographic estimators can often do little to correct for abstinence or spousal separation.

Some investigators have used arbitrary waiting-time criteria for determining whether a woman is sterile (e.g., Westoff and Pebley, 1981; Vaessen, 1984; Mosher, 1985). For example, Westoff and Pebley (1981) consider sterile all currently married, nonpregnant women who have not used contraceptives for at least five years since their last birth. By this criterion, all women who have been married less than five years are deemed fecund, as are all women who have used any form of contraception in the past five years, though both groups may well include some sterile women. In addition, all married women who did *not* conceive during five years when contraceptives were not used are deemed sterile, perhaps including fecund women who happened to fall in the upper tail of the distribution of waiting times to conception.

To reduce the arbitrariness of such waiting-time criteria, many demographic estimators of sterility use data from completed maternity histories, including the widely used methods of Vincent (1950) and Henry (1961, 1965). According to these estimators, any woman who survives in an intact marriage until age 50 without reproducing is considered to have been sterile from the time of marriage. Although this approach has its merits, it does not fully remove the arbitrariness of using a fixed waiting-time criterion: a woman may not reproduce before age 50 for many reasons other than biological sterility, especially if she marries at a comparatively late age. In fact, each age group of women has, in effect, its own unique arbitrary waiting-time criterion: a woman who marries at age 20 is considered sterile if she does not reproduce after 30 years, whereas one who marries at 45 is deemed sterile after only five years. More seriously, the need for completed maternity histories effectively limits use of these methods to historical demography (Trussell and Wilson, 1985).

By modifying Vincent's method for use with incomplete maternity histories, Larsen and Menken (1989) have greatly increased the scope of sterility estimation. Even better is the approach of Heckman and Walker (1990) and Larsen and Vaupel (1993), who have estimated fecundability and sterility simultaneously using multistate hazards models. In this approach, right-censored observations at any duration of exposure contribute information about both the fraction sterile and the upper tail of the waiting-time distribution among the nonsterile, thus removing the need for any arbitrary waiting-time criterion. Later in this chapter, I describe a similar method developed by our research group (Wood et al., 1994a).

ESTIMATORS OF PERMANENT STERILITY

The Parity Progression Method

The earliest attempt to estimate the age-specific prevalence of permanent sterility was by Vincent (1950). He estimated the prevalence of primary sterility among women age a as $[1 - \alpha_0(a)]$, where $\alpha_0(a)$ is the zero-order parity progression ratio (i.e., the proportion of women who go on to have at least one child) among women marrying for the first time at age a. Using historical data from France, rural Quebec, England, and Wales, Vincent obtained the estimates given in Table 10.2. According to this method, the prevalence of primary sterility is low at early ages but then increases almost exponentially, reaching about 70 percent by age 45. More recently, Menken and Larsen (1986) applied Vincent's method to several other populations, all of them of European origin (Figure 10.2); they found essentially the same pattern, although their estimates tend to run somewhat higher than those in Table 10.2.

There are several potential problems associated with this method. First and perhaps most obvious, marriage may not represent the initiation of sexual activity, and a woman who had a previous child out of wedlock may be suffering from secondary sterility at the time of marriage rather than from primary sterility. Second, as Larsen and Menken (1989) point out, the method is only useful if entry into marriage is spread over a wide age range, so there are sufficiently large samples of newlywed women at each age. Such a pattern is often found in historical Europe (Flinn, 1981), but not in many non-European cases. Third, marriage must not be selective with respect to fecundity—i.e., there should be no tendency for more fecund women to marry early, as would be the case if premarital pregnancy were a major reason for marriage. If marriage is selective for fecundity, the apparent effect of age on sterility is likely to be exaggerated.

Vincent's method for primary sterility can be extended to assess secondary sterility as well. Thus, the proportion of couples who become sterile after having n children can be estimated as $(1 - \alpha_n)$, where α_n is the nth-order parity progression ratio as defined in Chapter 2. The age to which this estimate applies lies somewhere between the mean ages at the nth and $(n + 1)$th births in the population being studied (for this reason, Vincent's method for estimating secondary sterility should not be used in cross-population comparisons because women of different populations attain a given parity at different mean ages). Taking data from the same populations as previously, Vincent (1950) compared $(1 - \alpha_n)$ for various orders of n with $[1 - \alpha_0(a)]$ at the appropriate ages at marriage to assess the additional effect of parturition on sterility above that of aging alone. He found, first, that each parity progression does entail an elevated risk of sterility but a very small one, amounting to only about 2 to 3 percent. He also found that the increase in the proportion of couples sterile at each parity progression was not affected by parity itself, at least up to parity five.[6] In other words, an elevated risk of sterility (albeit a small one) is associated with each live birth , but the magnitude of the elevation is independent of the number of children already born. Henry (1953a, 1953b) found similar effects of parturition in Norway but not Japan, whereas Holmberg (1970) detected similar effects in Sweden. These results strongly suggest that the additional risk of sterility associated with puerperal sepsis must generally be small. Moreover,

Table 10.2. Percentage of Sterile Couples among Newlyweds:
Historical Data from Four Groups of European Origin

	Age of wife at marriage					
	20	25	30	35	40	45
Percent of couples sterile	4	6	10	16	33	69

From Vincent (1950)

this additional risk may not really be an effect of parturition per se, but rather of marital duration operating, for example, on coital frequency.

The Method of Henry (1961, 1965)

Two slightly more complicated estimators of the age-specific prevalence of sterility (and here we lump both primary and secondary sterility) were developed by Henry (1961, 1965). Both estimators are most easily explained with reference to Figure 10.3. Consider an age interval $[a, a + k)$ for which we wish to estimate the prevalence of sterility; following Trussell and Wilson (1985), we call this the "index interval." With respect to this interval, Henry distinguishes three categories of married women: category 0, women who do not give birth after the beginning of the index interval (regardless of whether they had one earlier); category 1, women who have at least one child during the index interval but none thereafter; and category 2, women who have at least one child during the index interval *and* at least one other child after the interval. Let N_i be the number of women in the ith category ($i = 0, 1, 2$) and B_i be the number of births during the index interval to the N_i women in the ith category (obviously B_0 will equal zero). Finally, write $N = N_0 + N_1 + N_2$ and $B = B_0 + B_1 + B_2 = B_1 + B_2$. The overall marital fertility rate for the index interval is then

$$r(a) = B/N \tag{10.1}$$

and the marital fertility rate conditioned on subsequent fertility is

$$r'(a) = B_2/N_2 \tag{10.2}$$

To derive his first estimator of the prevalence of sterility within the index interval, Henry assumes that all subsequent infertility reflects sterility, so that $r(a) = r'(a) \cdot p_f(a)$, where $p_f(a)$ is the fraction of couples in the index interval who are fecund, i.e., nonsterile. Under this assumption, the prevalence of sterility is

$$p_s(a) = 1 - p_f(a) = 1 - r(a)/r'(a) = 1 - (BN_2)/(B_2N) \tag{10.3}$$

Henry's second estimator is a simplification of this expression based on the further assumption that fertility is the same for all women who had at least one live birth in the index interval, irrespective of whether they are subsequently fertile. In other words, $B_1/N_1 = B_2/N_2$, which implies $B_1/B_2 = N_1/N_2$. Hence,

$$\frac{r(a)}{r'(a)} = \frac{N_2}{N}\frac{B}{B_2} = \frac{N_2}{N}\left(\frac{B_1 + B_2}{B_2}\right) = \frac{N_2}{N}\left(1 + \frac{N_1}{N_2}\right)$$
$$= \frac{N_2}{N}\left(\frac{N_1 + N_2}{N_2}\right) = \frac{N_1 + N_2}{N} = \frac{N - N_0}{N} = 1 - \frac{N_0}{N}$$

(10.4)

Thus, under this latter assumption (but *only* under this assumption), $p_s(a)$ reduces to N_0/N, which is Henry's second estimator of sterility. Note that Vincent's estimator of primary sterility is the same as this simplified expression whenever the latter is calculated for the age group in which marriage occurs (Trussell and Wilson, 1985). Thus, in a table in which estimates using Henry's second estimator are cross-classified by both age and age at marriage, the entries along the principal diagonal are equivalent to Vincent's estimates. Henry's method makes explicit the fact that Vincent's estimator is based on an assumption that Vincent himself failed to recognize, namely, that the current fertility of women who later become sterile is the same as that of women who do not. In effect, Vincent's estimator assumes that sterility is not preceded by sub-fecundity.

Henry (1961) applied his second method to data from several natural-fertility populations (Table 10.3). The mean prevalences for his European populations are remarkably similar to the estimates derived by Vincent (compare Table 10.2). But this similarity amounts to an inconsistency: Henry's estimates include both primary and secondary sterility, whereas Vincent's include only primary sterility; thus, Henry's esti-

Table 10.3. Age-specific Prevalences (percentage) of Sterility among Married Women in Several Natural-fertility Populations

Population	Woman's age				
	20	25	30	35	40
England (mid-nineteenth century)	3	7	12	19	32
Canada	3	4	6	10	22
Geneva (wives of men born before 1650)	3	8	13	21	39
Europeans of Tunis	3	7	10	16	28
Crulai (Normandy)	2	3	7	15	36
Rural Japan (pre–World War II)	4	10	19	33	53
Average of European populations	3	6	10	16	31

From Henry (1961)

mates should be consistently higher than Vincent's. All of Henry's European estimates are quite similar to each other, but his only non-European case, rural Japan, yields substantially higher estimates. It is unknown whether such high levels of sterility are characteristic of many non-European populations, whether there was widespread pathological sterility or deliberate birth control in prewar rural Japan, or whether there are problems of data quality in the Japanese case.

Calibrating the Age Curve of Sterility

If infertility were synonymous with sterility, then all these estimators would be related somehow to sterility. But they would still leave open the question: sterility *when*? That is, how are the age-specific prevalences based on these estimators related to the actual age at *onset* of sterility? A women who marries at age *a* and subsequently fails to reproduce may already have been sterile at the time of marriage, or she may become sterile many months later after a normal fecund wait; she may even become pregnant, lose the pregnancy to fetal loss, and only later become sterile. Similarly, a woman may have her last child at age *a* and be infertile thereafter, but it is unlikely that she became permanently sterile the very moment she had her last birth—unless, of course, sterility was *caused* by childbirth. Thus, the age for which a given prevalence of sterility is estimated may or may not be the age to which it actually pertains. And the relationship between the two will change with age as the expected duration of waiting times to the next fertile conception changes.

Larsen and Menken (1989) call the putative age for which an age-specific prevalence of sterility is calculated the "calculation age" and the true age to which it pertains the "reference age." Discovering the relationship between the two has been referred to as "calibrating the age curve of sterility" (Trussell and Wilson, 1985). Henry (1965) was the first to recognize the need for calibration. Although he advocated using the midpoint of the calculation age interval as the reference age, he realized that the lag between the two ages would increase at later maternal ages because of declining fecundability. Leridon (1977) suggested that the true reference age is $a + w$, where w is the expected waiting time to the next fertile conception at age a; he then used independent empirical estimates of w to calibrate both Vincent and Henry's age curves of sterility. This approach has a fundamental flaw, to which we return below.

The problem of calibration cannot be solved by data analysis alone, for data are effectively silent about the reference age. Consequently, recent analyses have turned to *microsimulation* as a method of calibration (Barrett, 1972, 1986; Trussell and Wilson, 1985; Larsen and Menken,

1989). In this approach, artificial maternity histories are generated for a collection of women by Monte Carlo simulation on a computer. The simulation takes as input known age schedules of marriage, fecundability, lactational infecundability, and so forth, including a known schedule of age-specific risks of becoming permanently sterile. Each woman is allowed to experience these age-specific probabilities for each year during the reproductive span, and the final output consists of completed maternity histories for all the women in the simulation. Usually several thousand maternity histories are generated to limit stochastic variation.

The advantage of microsimulation in the present context is that the underlying age curve of permanent sterility is known, having been specified by the programmer. Thus, we can know the precise age at which exactly $p_s(a)$ women are sterile. The disadvantage is that we must draw the age schedule of sterility from some source, and that source is likely to be one of the sets of estimates we are trying to calibrate. The way around this dilemma is to run simulations with many different age curves of sterility and find estimators for which the relationship between calculation and reference ages is robust to the details of the underlying curves.

The microsimulations of both Trussell and Wilson (1985) and Barrett (1986) identified Henry's second estimator as the most robust of the ones reviewed here. The reference ages for this estimator yielded by these two studies are given in Table 10.4. After age 35, both studies show that the reference age is consistently later than the midpoint of the calculation age interval. In Trussell and Wilson's simulations, the lag between the calculation and reference ages increases steadily with age; in Barrett's simulation, the relationship is U-shaped, with a long lag at age 25–29, a comparatively short lag at 30–34, and a steadily increasing lag thereafter. The reason for this discrepancy is unclear.

Figure 10.4 shows the proportion of couples sterile by wife's (calibrated) age, estimated by Trussell and Wilson (1985) for 16 rural English parishes in the mid-sixteenth to early nineteenth centuries. The different lines show the breakdowns by wife's age at marriage. Thus, the solid line, for which wife's age and wife's age at marriage are the same, corresponds to Vincent's estimator of primary sterility. When calibrated by microsimulation, the curves for both primary and secondary sterility (broken lines) corresponding to different ages at marriage are consistently staggered; in other words, the proportion of couples sterile by a given age increases with marital duration. Presumably this increase reflects the effect of puerperal sepsis since the longer a woman has been married, the more children she is likely to have produced. It may also reflect an increase in the cumulative incidence of STDs with increasing length of exposure to intercourse, as well as any decline in coital fre-

Table 10.4. Reference Ages to which Henry's Second Estimator Applies, for Wife's Age at Marriage 20–24

Simulation study	Calculation age				
	25–29	30–34	35–39	40–44	45–49
Trussell and Wilson (1985)	25.6	31.5	37.8	43.1	47.6
Barrett (1986)	27.9	30.2	37.8	43.6	48.6

quency associated with marital duration. Whatever its cause, this staggering of curves was not apparent before the estimates were calibrated by microsimulation (Trussell and Wilson, 1985).

In my opinion, calibration of the age-at-onset of sterility is a problem precisely because earlier analyses failed to make a clear distinction between the proportion of women already sterile at age *a* (the *prevalence* of sterility) and the number of new cases of sterility between *a* and *a* + Δ*a* (the *incidence* of sterility). Were those two parameters specified separately in a model and then estimated simultaneously, they could be cumulated to provide an interpretable age-at-onset curve, whereupon this difficulty would evaporate.

Problems with the Standard Demographic Estimators

As already noted, the standard demographic estimators all equate sterility and infertility and thereby fail to distinguish true sterility from long waiting times to the next fertile conception. That is, they confound sterility and effective fecundability. Most demographers who have developed and used the standard estimators recognize this problem (Henry, 1965; Trussell and Wilson, 1985; Barrett, 1986; Larsen and Menken, 1989). Indeed, the suggestion by Leridon (1977) that the reference age for sterility be set equal to the calculation age plus the mean waiting time to a fertile conception is a direct response to the problem. But this "solution" only compounds the difficulty, for he combines estimates of the mean conception wait derived from right-censored observations (which include an unknown fraction of sterile couples) with estimates of sterility derived from observations of infertility (which include an unknown fraction of fecund couples with right-censored waiting times).

Clearly, the confounding of sterility and effective fecundability can cause overestimation of the true prevalence of permanent sterility at any one age. Can it also account for the apparent increase in the cumulative incidence of sterility by wife's age? It can, because of two biases. The first bias is caused by a differential truncation of observations that occurs whenever we use data on completed fertility, i.e., data on couples whose marriages remain intact until the wife reaches 50. Since newlyweds who marry when the wife is, say, twenty have 30 years in which to produce a live birth before age 50, whereas those who marry at wife's age 40 have only ten, the prevalence of sterility at age 40 will inevitably *appear* to be higher than at age 20 even if no real difference exists. The second bias, which exacerbates the first, reflects the fact that the expected waiting time to the next fertile conception increases with age, even in couples who are not yet permanently sterile (see Chapter 7). The simultaneous effect of both biases on the estimated age-specific prevalence of sterility is shown in Figure 10.5. Clearly, differential truncation and differences in waiting-time distributions together may account for a substantial fraction of the apparent increase in the cumulative incidence of sterility with increasing age. The apparently universal age pattern of permanent sterility thought to characterize natural-fertility populations may thus be nothing more than an artifact.

An Alternative Approach

The obvious way to avoid the biases associated with differential truncation and differing waiting-time distributions is to estimate effective fecundability and sterility simultaneously for different age groups of women while truncating the observations of all women at the same upper limit. Recently, several new methods have been proposed for doing just that (Heckman and Walker, 1990; Larsen and Vaupel, 1993; Wood et al., 1994a). Since these methods are all closely related and broadly similar, I will concentrate here on our own approach (Wood et al., 1994a). In our analyses, we use data only on the first birth interval within a woman's first marriage; hence, our method estimates primary but not secondary sterility.

The model used in our analyses is outlined in Figure 10.6. We assume that, at any point during observation, women must be in one of three distinct, nonoverlapping states: fecund, sterile, or pregnant. At the beginning of exposure, assumed to coincide with marriage, a fraction s of women are in the sterile state, $1 - s$ are fecund, and no women are pregnant. The fraction pregnant gradually increases over time as women are exposed to the risk of conception. Because the data we use allow us to recognize only pregnancies that end in live birth, any pregnancies that terminate in fetal loss are effectively submerged in the fecund state. In view of the nonsusceptible period associated with each lost pregnancy (i.e., the truncated gestation ending in loss, plus any period of post-loss infecundability), "fecund" is something of a misnomer for this state; nonetheless, we retain the term for convenience.

Only a limited number of transitions among states are permissible. Because we deal strictly with the first pregnancy following a woman's initial entry into exposure, pregnancy is an absorbing state and back-transitions from it are impossible. By definition, transitions cannot occur from the sterile to the pregnant state. Moreover, since we are interested solely in permanent sterility, sterility is also treated as an absorbing state. Thus, only two transitions are included in the model, both originating in the fecund state: fecund couples may become pregnant, or they may become sterile. Because we carefully distinguish the prevalence of sterility at the outset of exposure from the incidence of new sterility, there is no need to calibrate the age at onset of sterility.

Fecundability, defined for present purposes as the *hazard* of a transition from the fecund to the pregnant state,[7] is assumed to be constant over time for any one couple and equal to $\lambda_c e^z$, where λ_c is the component of fecundability common to all fecund couples, and z is the component specific to the individual couple. We assume that z is normally distributed with mean zero and variance σ^2 to be estimated—in other words, we allow fecundability to vary among couples. This feature of the model is important for sterility estimation precisely because heterogeneous fecundability tends to elevate the upper tail of the distribution of conception waits (see Chapter 7). If we did not take this effect into account, we would be inclined to regard couples with long, right-censored observations as sterile when in fact they may just be subfecund. Consistent with the present definition of fecundability as a hazard, the proportional-hazards specification $\lambda_c e^z$ constrains fecundability to fall within $(0, +\infty)$ when z varies between $-\infty$ and $+\infty$. Under this model, fecundability is lognormally distributed among nonsterile women, with a mean of $\lambda_c e^{\frac{1}{2}\sigma^2}$ and a variance of $\lambda_c^2 e^{\sigma^2}(e^{\sigma^2} - 1)$. In interpreting the results of our analyses, it is important to remember that mean fecundability is partly determined by the value of σ^2. When $\sigma^2 = 0$, the mean reduces to λ_c and the variance to zero.

The instantaneous hazard of becoming sterile is treated as a constant equal to λ_s; thus, the risk of becoming sterile during marriage is assumed to be the same in all women. From a biological perspective, this is perhaps the most obvious shortcoming of the current model. From a statistical point of view, however, it is probably unavoidable since allowing λ_s to vary among women would risk over-parameterizing an already parameter-rich model. In most natural-fertility settings, the single most important source of variation in the risk of primary sterility is undoubtedly female age, especially in situations where STDs are uncommon and hence reproductive senescence and menopause are the principal causes of such sterility. (Since we are not trying to estimate secondary sterility, puerperal sepsis is irrelevant.) The impact of assuming an invariant risk of sterility can be minimized by applying the model in piecewise fashion to separate age groups of women. This solution has the advantage of showing how the prevalence (s) and incidence (λ_s) of sterility change with age, as well as how fecundability itself changes. In our analyses, we group exposures into five-year intervals, thus truncating the observations of *all* women at a maximum of five years regardless of their age at marriage.

Elsewhere we describe maximum likelihood methods for estimating this model from data sets on the waiting time to first fertile conception that include right-censored observations (Wood et al., 1994a). For illustration, Table 10.5 applies the model to data on rural Sri Lankan women interviewed in 1987 as part of the worldwide Demographic and Health Survey (see Fisher, 1988, for details of data collection and Arnold, 1991, for an assessment of data quality).[8] As can be seen, the sample sizes rapidly attenuate; this unavoidable fact, which reflects the marriage patterns of this population, limits how much we can infer about women in the later segments of reproductive life. The small samples at later ages compared with those at earlier ages also raise concerns about potential selectivity bias associated with late ages at marriage. These problems should be borne in mind when interpreting the results for women ages 30 years and older.

Each panel of Table 10.5 corresponds to a single age-group of women. The first two lines within each panel represent the full model, in which all four parameters are free to take on their maximum likelihood values; subsequent lines show a series of reduced models in which one parameter at a time is fixed at zero.[9] Table 10.5 also gives the mean and variance of the distribution of effective fecundability among nonsterile couples, as well as the log likelihood at the maximum for each model. The latter number can be used in likelihood ratio tests of nested models (see Chapter 3).

Leaving sterility aside for the moment, there are several interesting findings about fecundability in Table 10.5. First, since likelihood ratio tests consistently reject the reduced models in which $\sigma = 0$, heterogeneity in fecundability appears to be significant at every age examined ($P < 0.001$). Second, the estimated value of λ_c, the shared component of fecundability, declines monotonically with age, consistent with the studies of artificial insemination with donor sperm summarized in Chapter 4 (Fédération CECOS et al., 1982; van Noord-Zaadstra et al., 1991). Mean fecundability, in contrast, shows no clear decline with age under the full model and may even increase at later ages. This pattern is largely attrib-

Table 10.5. Estimated Parameter Values for Full and Nested Models Fitted to Data on the Waiting Time from Marriage to First Fertile Conception, Sri Lanka, 1987 Demographic and Health Survey

Woman's age (years)	Sample size	Model	Parameter estimate (standard error)				Estimated effective fecundability of fecund couples		Log likelihood
			λ_c	σ	s	λ_s	mean	variance	
20–24	1,398	Full	0.11766 (0.00829)	1.00936 (0.09715)	0.00021 (0.00103)	<0.00001 (0.00017)	0.19554	0.06545	−3763.645
		$\sigma = 0$	0.10019 (0.00294)	0.00000	0.01891 (0.02477)	0.00039 (0.00017)	0.10019	0.00000	−3848.703‡
		$s = 0$	0.11701 (0.00791)	1.01842 (0.00005)	0.00000	0.00002 (0.00017)	0.19625	0.06775	−3764.884
		$\lambda_s = 0$	0.11766 (0.00799)	1.00936 (0.09188)	0.00021 (0.00048)	0.00000	0.19554	0.06545	−3763.645
25–29	655	Full	0.10186 (0.01062)	0.95534 (0.16636)	0.01090 (0.01768)	<0.00001 (0.00132)	0.16057	0.03741	−1733.984
		$\sigma = 0$	0.08357 (0.00389)	0.00000	0.03437 (0.03679)	0.00001 (0.00132)	0.08357	0.00000	−1775.122‡

		$s = 0$	0.09952 (0.00982)	0.99410 (0.00058)	0.00000 –	0.00095 (0.00132)	0.16290	0.04336	-1736.980*
		$\lambda_s = 0$	0.10186 (0.01026)	0.95534 (0.15227)	0.01090 (0.00830)	0.00000 –	0.16057	0.03741	-1733.984
30–34	235	Full	0.06995 (0.01530)	1.36077 (0.28568)	0.00128 (0.00566)	<0.00001 (0.00029)	0.17581	0.14769	-604.481
		$\sigma = 0$	0.04776 (0.00446)	0.00000 –	0.02867 (0.06830)	<0.00001 (0.00029)	0.04776	0.00000	-633.158‡
		$s = 0$	0.06802 (0.01427)	1.41101 (0.00013)	0.00000 –	0.00007 (0.00029)	0.18316	0.18488	-605.287
		$\lambda_s = 0$	0.06995 (0.01448)	1.36077 (0.24953)	0.00128 (0.00249)	0.00000 –	0.17581	0.14769	-604.481
35–39	72	Full	0.02772 (0.02468)	2.05725 (1.43952)	0.06313 (0.24781)	<0.00001 (0.00365)	0.22408	1.60645	-100.965
		$\sigma = 0$	0.02441 (0.00830)	0.00000 –	0.19057 (0.20936)	0.00013 (0.00365)	0.02441	0.00000	-106.454‡
		$s = 0$	0.01728 (0.01633)	2.31453 (0.00191)	0.00000 –	0.00097 (0.00365)	0.24010	3.49741	-101.685
		$\lambda_s = 0$	0.02714 (0.02334)	2.07262 (1.29398)	0.06059 (0.13229)	0.00000 –	0.22621	1.70295	-100.965

* $P < 0.05$ against full model
‡ $P < 0.001$ against full model
From Wood et al. (1994a)

utable to a marked increase in σ at later ages, which tends to drag mean fecundability upward even as λ_c declines.[10] In other words, the age curve of mean fecundability is at least as much a reflection of the couple-to-couple variation in fecundability as it is of the underlying, "common" component of fecundability shared by all couples.

Turning to sterility, there are several noteworthy features of Table 10.5. First, λ_s is effectively zero at all ages in Sri Lanka. Thus, although there is surely some nonzero risk of becoming sterile at each age, it is much too small to detect even with large samples. Similarly, s is small at all ages—although in one instance where a comparatively large estimate is combined with an adequate sample size (ages 25–29), the value of s does differ significantly from zero. Note that the estimates of s, even when nonsignificant, tend to increase (albeit inconsistently) with age, rising from about 0.0002 at ages 20–24 to 0.063 at 35–39. Even though this result suggests that the prevalence of primary sterility may indeed rise to some extent across the twenties and thirties, the magnitude of the increase estimated here is very much smaller than that suggested by earlier demographic analyses (compare Tables 10.2, 10.3, and Figure 10.2). Granted, those analyses all involve data from other populations, which may partly explain their different patterns of sterility. But I suggest that the differences mostly reflect the untoward effects of differential truncation and varying waiting-time distributions in the earlier analyses.

According to our analysis, then, there is little evidence that sterility is important at any of the ages we examine. The prevalence of preexisting sterility (s) almost never differs significantly from zero, and the incidence of new sterility (λ_c) is always negligibly small. Admittedly, the estimates at higher ages have large standard errors, and for those ages we simply do not have the statistical power to reject a reduced model with no sterility. Still, the biological evidence summarized earlier in this chapter would not lead us to expect there to be much primary sterility at ages well before the menopause, and our analyses appear to be consistent with that evidence. This finding is especially compelling because we have allowed for heterogeneous fecundability: since a failure to distinguish subfecund and sterile couples would have biased our estimates of sterility upward, it is unlikely that the true prevalence of primary sterility is much higher than that estimated here. I conclude, therefore, that primary sterility is probably not an important determinant of natural fertility, at least before age 40, in countries like Sri Lanka where STDs are uncommon. Its importance after age 40 can be clarified only by further analysis.

Since an increasing risk of sterility is almost certain to be more important near the end of reproductive life, it is unfortunate that the current version of our model is applicable only to the first birth interval. In the

Sri Lankan case this limitation has created problems of interpretation even for women in their thirties, by which age small samples and possible selectivity have tended to obscure the results. I suggest, therefore, that the most crucial extension of the current model is to adapt it for application to higher-order birth intervals (for discussions of the analytical problems involved, see Larsen and Vaupel, 1993; Wood et al., 1994a).

Another Hazards Model of Sterility

Pittenger (1973) has developed a much simpler parametric hazards model for the onset of permanent sterility. Strictly speaking, this is not a method for estimating age-specific prevalences of sterility; rather, the model is fit to a preexisting set of estimates, e.g., estimates derived by either Vincent or Henry's method. The model is useful, however, in underscoring the commonalities and differences in age patterns of sterility across populations.

According to Pittenger's model, the hazard for onset of sterility at age a is

$$h(a) = \begin{cases} h(a_0)\phi^{a-a_0}, & a \geq a_0 \\ 0, & a < a_0 \end{cases} \tag{10.5}$$

where a_0 is the earliest age at which permanent sterility can occur, $h(a_0)$ is a constant representing the hazard of sterility at age a_0, and ϕ is the annual rate of increase in the hazard. This model assumes, in effect, that some fraction of couples are sterile from the very outset of reproductive life, and that the remaining couples experience a nonlinearly increasing hazard of becoming sterile after that time, as they might if sterility were caused by an accumulating incidence of insults (e.g., tubal scarring). The proportion of couples sterile at age a is equal to the complement of the survival function associated with equation 10.5:

$$p_s(a) = 1 - \exp\left[\frac{h(a_0)(1 - \phi^{a-a_0})}{\ln \phi}\right], \quad a \geq a_0 \tag{10.6}$$

This model provides a good fit to data from a wide variety of natural-fertility populations (Pittenger, 1973; Trussell and Wilson, 1985; Wood et al., 1985a). When fitted to the age-calibrated prevalences of primary sterility estimated by Trussell and Wilson (1985) for 16 English parishes (the solid line in Figure 10.4), the resulting estimates of a_0, $h(a_0)$, and ϕ are 5.67, 0.00043, and 1.14345, respectively (since $\hat{a}_0$ falls well before the beginning of the reproductive span, these parameter values should probably not be interpreted too literally). Except for a rather high starting level, most of the resulting curve falls squarely within the observed range of variation in the age at onset of menopause (Figure 10.7). One interpretation of these results is that there is indeed a small fraction of women who are sterile from the very outset of their reproductive lives, and that primary sterility does increase slightly with age during the twenties and early thirties, but that most of the apparent age-related increase in sterility is attributable to nothing more or less than the onset of menopause.

STERILITY AND THE AGE PATTERN OF NATURAL FERTILITY

We are now in a position to reevaluate the claim by Wilson et al. (1988) mentioned at the beginning of this chapter. Recall their suggestion that the onset of permanent sterility as defined demographically is the single most important factor determining the age pattern of marital fertility observed in most natural-fertility populations. What Wilson and his colleagues are really claiming is that the curve of $\alpha_0(a)$ values calculated for age at marriage (i.e., the complement of Vincent's estimator of primary sterility) corresponds closely to the curve of age-specific natural marital fertility rates standardized so that the rate for 20–24-year-old women equals one (see Figure 10.1). Figure 10.8 summarizes additional data on this point. This figure differs from Figure 10.1 in three important respects: (1) the $r(a)$ and $\alpha_0(a)$ curves are estimated from data on the *same* population, (2) the reference ages to which the $\alpha_0(a)$ values pertain have been calibrated following the microsimulation study of Trussell and Wilson (1985), and (3) the values of $\alpha_0(a)$ have been adjusted upward so that the value corresponding to 20–24-year-olds equals one. This latter adjustment is necessary to match the adjustment in $r(a)$ values, since an $r(20)$ of one is a fertility rate that necessarily excludes sterile women.

Now the match between $r(a)$ and $\alpha_0(a)$ schedules is much less close than it previously appeared to be. In particular, marital fertility declines more rapidly than does the proportion of nonsterile couples. Add to this the fact that the $\alpha_0(a)$ values in Figure 10.8 equate infertility with sterility and thereby underestimate the fraction of nonsterile couples at each age, and it becomes evident that something over and above permanent sterility is going on.

If Wilson and his colleagues were correct in believing that natural fertility rates decline with age solely because of the onset of permanent sterility, then the birth-spacing patterns of women known *not* to be sterile would be unaffected by age under a regime of natural fertility. But such is not the case. As shown in Chapter 3, birth intervals grow longer near the end of the reproductive span, even among women who go on to have at least one more child. Figure 10.9 shows another example from a natural-fertility setting, this time with the correct translation so that the direct effect of age is visible (right panel). Clearly, birth intervals grow consistently longer with age, at least when disaggregated by completed family size. And since these are all *completed* birth intervals, they necessarily exclude any couples who are permanently sterile. Therefore, the decline in natural fertility with age is not simply a matter of some couples being lost to permanent sterility. It is also a birth-spacing phenomenon.

I suggest that we should recognize a pattern of *senescent subfecundity* among nonsterile women at later reproductive ages, comparable to adolescent subfecundity at the beginning of the reproductive span. In previous chapters, we have identified the basic elements of senescent subfecundity: long and variable cycles, a higher fraction of anovulatory cycles, and an elevated risk of fetal loss, all of which should lower effective fecundability and thus lengthen birth intervals, but none of which will necessarily render a woman sterile. Unfortunately, the impact of these physiological changes on fertility is difficult to assess from real data because of the confounding effects of declining coital rates with age or marital duration. We can, however, evaluate their impact using the mathematical models developed in Chapters 6 and 7.

Recall from equation 6.13 that the expected waiting time to the first fertile conception is equal to

$$(w + qn) / (1 - q) \tag{10.7}$$

where w is the mean waiting time to first conception (regardless of outcome), q is the probability of fetal loss per conception, and n is the duration of the nonsusceptible period associated with each fetal loss. Assuming homogeneity, then w is simply equal to $1/p$, the inverse of total fecundability. Since we know (or think we know) quite a bit about age-related changes in q and p, we can pose the following question: how do senescent changes in q and p affect completed birth intervals when we hold coital frequency, as well as other birth interval components, constant? In particular, suppose that fetal loss changes according to the age curve that was shown back in Figure 6.5, adjusted upward so the risk of loss at maternal ages 20–24 equals 0.3 (the value estimated by Wilcox et al., 1988). Further suppose that total fecundability follows the curve in the lower panel of Figure 7.10. That curve was generated by the Wood-Weinstein model of fecundability described in Chapter 7, allowing ovarian function to change with age according to the patterns observed in normal western women while holding coital frequency constant at the value observed in married U.S. women age 25 (for details, see Weinstein et al., 1990). Finally, suppose that n is equal to two months at all ages, and that lactational infecundability always lasts 16 months, near the middle of the range of values tabulated by Bongaarts and Potter (1983). What, then, do completed birth intervals look like according to equation 10.7?

The answer is shown in Figure 10.10. By allowing only fetal loss and the physiological determinants of total fecundability to vary with age, we get something similar to the age pattern of birth spacing that is actually observed in natural-fertility populations. Moreover, almost all the predicted change in birth spacing is attributable to fetal loss; only after age 45 does fecundability have much impact. Thus, for most women in most natural-fertility populations, the so-called onset of permanent sterility may simply be a situation in which the risk of fetal loss is so high—and later, fecundability so low—that there is no longer enough time before menopause for another fertile conception to occur. There may be no true, absolute sterility before menopause except in the occasional pathological case.

If this is true, then the fact that the average age of women at their last birth clusters near 40 in most natural-fertility populations still needs explaining. It is important to bear in mind that an *average* of 40 is entirely consistent with a sizable fraction of last births occurring well after that age, and at least a few women may actually give birth immediately before they become menopausal. But leaving that point aside, imagine an "average" woman who produces her last child at age 40. Since there is a small but apparently real negative effect of maternal age on the hazard of first postpartum menses (see Chapter 8), she is likely to experience a somewhat longer duration of lactational infecundability than most younger women—so let us suppose that she resumes ovulating at age 42 or 43. By that age her fecundability is starting its rapid decline so she might expect to wait as many as eight or nine months until her next pregnancy, which she has a very high risk of losing. If she loses two or three pregnancies in succession, with an increasingly long fecund wait separating each one, she is likely to be 45 or 46 years old and distinctly perimenopausal without having yet brought another pregnancy to term. When we recall from Chapter 9 that the average age at menopause is several years earlier in the developing world than in industrialized countries, 40 is perhaps what we should expect as an average age at last birth in traditional societies even if women do not become absolutely sterile until they reach menopause.

NOTES

1.   Because estimation of the mean age at last birth requires *completed* reproductive histories, Table 10.1 is heavily dominated by family reconstitution studies using historical European data. It may yet turn out that the supposedly "universal" invariance of mean ages at last birth applies mainly to Europe before the modern decline in fertility.

2.   Most studies of the age pattern of permanent sterility are framed in terms of the woman's age, obscuring the fact that sterility can be attributable to a defect in the male partner or in both partners simultaneously. The present chapter deals exclusively with female sterility; the male contribution will be discussed in Chapter 11.

3.   Incidentally, modern medical practice does no better. The standard clinical criterion for diagnosing pathological sterility is the failure of a couple to conceive after trying to do so for 12 months (Speroff et al., 1986). This criterion does not distinguish between couples with genuine sterility and those who are merely in the upper tail of the distribution of waiting times to conception. Recent studies suggest that this diagnostic criterion may entail a risk of false positive diagnosis of sterility as high as 80 percent or more (Trussell and Wilson, 1985; Menken et al., 1986; Marchbanks et al., 1989).

4.   The effect of genital tuberculosis on fecundity is difficult to assess since genital TB is often asymptomatic. However, pathological evidence suggests that genital involvement occurs in only about 4–12 percent of all cases of TB (Medlar, 1955); moreover, only about 50 percent of all women with genital TB become sterile (Ojo et al., 1971). Thus, even in populations with extraordinarily high prevalences of tuberculosis, genital TB is unlikely to cause sterility in more than about 5 percent of women.

5.   Until recently, it was also thought that a deteriorating uterine environment played a major role in female reproductive senescence; for example, it was hypothesized that sclerotic changes in the uterine blood supply, loss of cell elasticity, and increasing amounts of collagen in the endometrium and myometrium contribute to failure of implantation and an increasing risk of pregnancy loss with maternal age (Woessner, 1963; Naeye, 1983; Gosden, 1985; Gostwamy et al., 1988). However, a recent study of women age 40–44 suffering from premature ovarian failure has shown that the older uterus is able to maintain pregnancy throughout gestation following artificial stimulation of the endometrium and transplantation of embryos from young oocyte donors (Sauer et al., 1990). Similarly, Navot et al. (1991) have applied the techniques of donor-oocyte treatment to infertile women age 40+ who were still experiencing menstrual cycles but whose previous attempts at in vitro fertilization with self-oocytes had failed; women in this study achieved a 56 percent pregnancy rate using donated oocytes from younger women, but only 3 percent with their own oocytes.

6.   At higher parities, selectivity bias resulting from heterogeneous fecundability obscures the relationship. This bias occurs because women who have had at least five children are a select subsample of all women.

7.   Elsewhere in this book fecundability is defined as a probability, not a hazard. Since there is no commonly accepted term for the hazard of conception, I have temporarily appropriated "fecundability" as a convenient label; I can only hope that my appropriation causes no confusion. At the low values of fecund-

ability typical of humans, the hazard and probability of conception are almost identical in value.

8. In keeping with the assumptions of the model, we eliminate women who report no living children at the time of their first use of any modern method of contraception, including surgical sterilization of either partner; fortunately, only about 2.7 percent of women are removed by this restriction, and little if any selectivity bias is likely to result from it. We also exclude premarital conceptions, specifically durations of fewer than eight months from marriage to first birth. These make up only 3.5 percent of first conceptions.

9. Note that we have not run models with λ_c, the common hazard of a conception, set at zero. Since at least one conception was observed for each age group, those particular reduced models could always be rejected a priori.

10. Elsewhere we have suggested that variances in effective fecundability that grow larger with age might be expected because of increasingly divergent risks of fetal loss among older women (Wood and Weinstein, 1990).

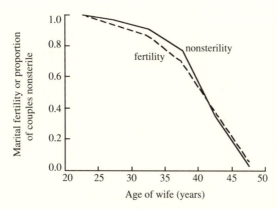

Figure 10.1. The relationship of adjusted natural marital fertility rates to proportions of couples not yet permanently sterile. Fertility rates are averages from 10 populations, adjusted so that the rate at ages 20–24 is one; proportions not yet sterile are the zero-order age-specific parity progression ratios for 14 German villages, 1750–1899. (Redrawn from Wilson et al., 1988)

Figure 10.2. Proportion of couples sterile by wife's age at marriage, according to Vincent's estimator. Based on family reconstitution studies of historical demographic data from several European and European-derived populations: 14 German villages, marriages of 1750–1899 (*circles*); 16 rural English parishes, mid-sixteenth to early nineteenth centuries (*triangles*); Ireland, 1911 census (*diamonds*); English Quakers (*inverted triangles*); rural Quebec women born before 1876 (*right-pointing triangles*); and Scotland, 1911 census (*squares*). (Redrawn from Menken and Larsen, 1986)

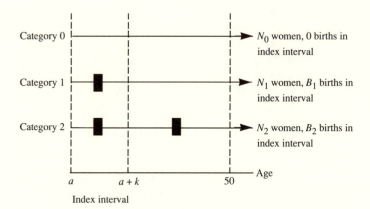

Figure 10.3. The three types of reproductive experience (beginning at age *a*) distinguished in Henry's (1961) method of
estimating age-specific prevalences of sterility. Heavy bar
denotes *at least one* live birth per woman in the age interval
in which the bar occurs.

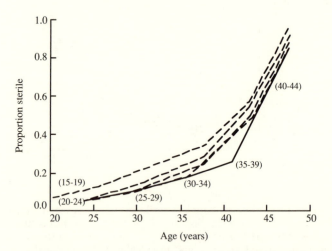

Figure 10.4. Estimated proportion of couples sterile by
wife's age (*x*-axis) and age at marriage (in parentheses), 16 rural English parishes from the mid-sixteenth
to the early nineteenth centuries. Wife's age has been
calibrated by microsimulation. *Solid line,* estimated
proportions with primary sterility; *broken lines,* estimated proportions with both primary and secondary
sterility. (Redrawn from Trussell and Wilson, 1985)

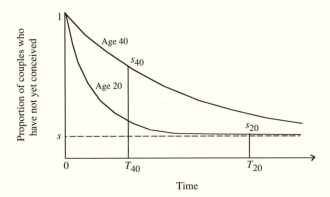

Figure 10.5. The effect of differential truncation acting upon two different distributions of waiting times to first fertile conception.
Women who marry at age 40 are assumed to have lower effective
fecundability and thus longer waiting times than those who marry
at age 20. It is also assumed for the sake of argument that the fraction of women who are permanently sterile at marriage is the
same for both ages at marriage and equal to *s* (*broken line*). Truncation is at wife's age 50, a point that occurs earlier in marriage for
women who marry at 40 (T_{40}) than for those who marry at age 20
(T_{20}). The net effect of earlier truncation and longer waiting times
is to bias upward the estimate of the proportion of couples with
primary sterility at wife's age 40 (s_{40}).

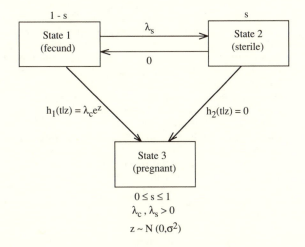

Figure 10.6. A multistate model of the first conception within marriage (see text for details). (Redrawn from Wood et al., 1994a)

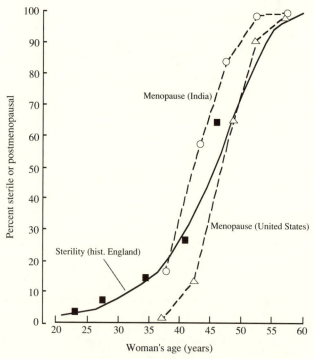

Figure 10.7. Fit of the Pittenger model to estimates of
primary sterility in 16 rural English parishes (mid-
sixteenth to early nineteeth centuries). ■, age-specific
prevalences of primary sterility in England calibrated
by microsimulation (equivalent to the solid line in
Figure 10.4); *solid curve*, the Pittenger hazards model
of primary sterility fit to the English data by nonlinear least squares. The *broken lines* show the cumulative incidence of menopause in two twentieth-cen-
tury populations, one with comparatively early
menopause (India) and the other with late
menopause (United States). (Indian data from Dandekar, 1959; U.S. data from MacMahon and Worcester, 1960; English data from Trussell and Wilson,
1985)

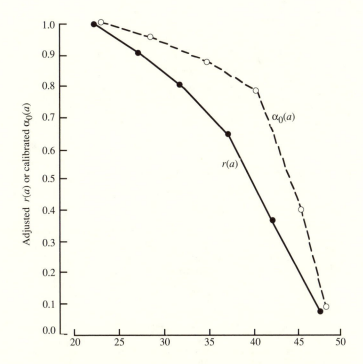

Figure 10.8. A second look at the relationship between adjusted natural marital fertility rates, $r(a)$, and propor- tions of couples not yet permanently sterile, $\alpha_0(a)$. In con- trast to Figure 10.1, data on both marital fertility and pro- portions sterile come from the same population (16 rural English parishes from the mid-sixteenth to the early nine- teenth centuries). In addition, the maternal ages to which the estimated proportions sterile apply have been cali- brated by microsimulation. Finally, the values of $\alpha_0(a)$ have been adjusted so that the value at age 20–24 is one; this adjustment is consistent with the adjustment in $r(a)$ values since an $r(20)$ of one is a fertility rate that neces- sarily excludes sterile women. (Fertility estimates from Wilson, 1984; sterility estimates from Trussell and Wil- son, 1985)

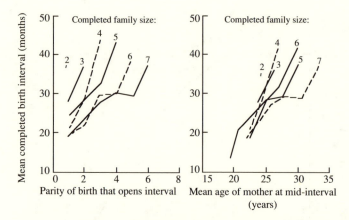

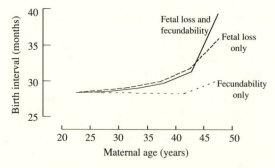

Figure 10.9. Mean completed birth intervals in Mexican-Americans (Laredo, Texas, 1890–1939) by parity (*left*) and mother's age (*right*). Only completed maternity histories (those in which both the woman and her husband survive to wife's age 50) are shown. Data are broken down by mother's completed family size (CFS); otherwise the relationship with parity is obscured because the sample becomes increasingly selected for short intervals at higher parities. Sample sizes are 115 (CFS = 2), 95 (CFS = 3), 60 (CFS = 4), 46 (CFS = 5), 34 (CFS = 6) and 18 (CFS = 7). Because of small samples, completed families of 8+ are not shown. (Unpublished data generously provided by A.V. Buchanan and K.M. Weiss)

Figure 10.10. Predicted changes in completed birth intervals by maternal age when both total fecundability and fetal loss vary with age (*solid line*), only fetal loss varies with age (*broken line*), and only total fecundability varies with age (*interrupted line*). In all three cases, coital frequency and the duration of lactational anovulation are held constant. (Redrawn from Wood, 1989)

11

Marriage and the Male Contribution

Until recently, fertility research has focused almost exclusively on the role of the female partner in reproduction. During the past decade, however, physiologists have become increasingly interested in male factors in reproduction, as can be seen from any recent issue of *Biology of Reproduction,* the *Journal of Clinical Endocrinology and Metabolism,* or the *Journal of Reproduction and Fertility.* Thus far, this renaissance of interest in the male on the part of basic reproductive scientists has had little impact on population-oriented research. The principal reasons for this lack of attention to males on the part of demographers are that conventional demographic techniques are poor at separating male and female factors affecting fertility (Goldman and Montgomery, 1989) and that two-sex demographic problems are analytically difficult (Pollak, 1990). Demographers have, however, devoted considerable attention to characteristics of the *couple,* the male and female partners considered jointly. Perhaps the most important of these characteristics for fertility research is the way in which couples are formed—that is, the way individuals pair off in stable sexual unions, which in most societies means marriage. This chapter takes up two related subjects. First, we discuss the one last proximate determinant of natural fertility that we have not yet considered in detail, namely, the age at marriage. Second, we ask whether any important component of fertility variation can be ascribed to factors operating solely in the male partner.

475

MARRIAGE AND FERTILITY

Demographers use the term *nuptiality* to refer to the frequency, characteristics, and dissolution of marriages studied at the population level (Pressat, 1985). From the point of view of fertility analysis, nuptiality is important because of the close linkage that exists between marriage and exposure to the risk of conception in most societies. This linkage can sometimes be investigated using biological techniques. In our research among the Gainj, for example, we conducted approximately 400 pregnancy tests using hCG assays and about 40 genetic exclusions based on 27 polymorphic blood group, serum protein, and red-cell enzyme systems; these tests revealed no cases of premarital pregnancy and only one case of extramarital conception (Wood, 1987b, 1992). Such tight regulation of conception by marriage is, however, far from universal. Among the Yanomama Indians of Venezuela, nonpaternity approaches 10 percent of all surviving offspring (Neel and Weiss, 1975). Similarly, family reconstitution studies suggest that preindustrial Europe was characterized by high if variable rates of premarital conception (Table 11.1). In sixteenth- and seventeenth-century England, to choose one example, between one-sixth and one-third of all first births were conceived before marriage (Hair, 1966, 1970).[1] Under the conditions of high fertility that prevailed in preindustrial Europe, however, such premarital conceptions represented only a small fraction of all births (last column in Table 11.1), and there is little evidence that extramarital conceptions were common once a marriage had been instituted. Perhaps the most extreme case on

Table 11.1. Premarital Conceptions as a Percentage of All First Births, and as a
Percentage of Births of All Orders, Various Historical Family Reconstitution
Studies (number of parishes reconstituted given in parentheses)

Country	Period	Premarital conceptions, % of all first births		Premarital conceptions, % of all births	
England	pre-1750	19.7	(14)	2.6	(24)
	1740–1790	37.3	(3)	4.3	(24)
	1780–1820	34.5	(2)	5.9	(24)
France	pre-1750	6.2	(9)	2.9	(8)
	1740–1790	10.1	(9)	4.1	(12)
	1780–1820	13.7	(6)	4.7	(6)
Germany	pre-1750	13.4	(12)	2.5	(11)
	1740–1790	18.5	(15)	3.9	(15)
	1780–1820	23.8	(8)	11.9	(8)

From Flinn (1981)

record of a societywide failure of marriage to influence the probability of conception is the Herero of Botswana, where Pennington and Harpending (1991) have found no significant effect of marital status on the length of birth intervals of any order. It is premature to generalize any of these findings to other preindustrial societies. However, although the Gainj and Herero may represent polar extremes, the available evidence suggests that most natural-fertility populations probably experience low to moderate rates of nonmarital conception. Marriage, in other words, appears to be a powerful but not perfect regulator of heterosexual behavior in most traditional societies.

To the extent that marriage does regulate sexual relations, the age at which a woman marries for the first time marks the beginning of her exposure to the risk of conception. Consequently, demographers have long been interested in the age dynamics of *marital formation*. For comparative purposes, it is useful to have a single number that expresses the average age at which women marry. If we have reliable prospective data on a cohort of women, all of them monitored over their entire life span, we can easily compute the mean of their ages at marriage. But although such data can sometimes be drawn from historical family reconstitution studies, they are almost never available for contemporary natural-fertility populations. It is desirable, therefore, to have a measure of the mean age at marriage that can be computed from cross-sectional observations. One widely used summary statistic is the *singulate mean age at marriage* (SMAM) developed by Hajnal (1953). This statistic can always be calculated from current-status data on the proportion of women married at each age, such as might be collected in a single census or survey; however, interpreting the SMAM as a true mean requires that we assume that the age pattern of marriage is constant and that the age-specific mortality and migration rates of married and unmarried women are identical.

If we could observe a true cohort, the difference between the proportions of women single (i.e., never married) at two different ages would directly reflect the number of marriages that took place between those ages (ignoring mortality and migration). It would also be possible to calculate the person-years lived by women before their first marriage; the average number of person-years lived in the single state among those women who eventually do marry is the true mean age at first marriage. When calculating the SMAM, the cross-sectional proportions of women single at each age are used to construct a *hypothetical* cohort, assuming in effect that the experience of women has not changed over recent decades. If $S(a)$ is the proportion of women who are single at age a, and ω is the end of the reproductive span (variously taken as 40, 45, or 50 years), then

$$\text{SMAM} = \frac{\sum_{a=0}^{\omega} S(a) - [S(\omega) \cdot \omega]}{1 - S(\omega)} \tag{11.1}$$

Summation of the $S(a)$ values provides an estimate of the total number of person-years lived in the single state before age ω, and $S(\omega) \cdot \omega$ estimates the person-years lived by those who are *still* single at age ω. Subtraction of $S(\omega) \cdot \omega$ and division by $1 - S(\omega)$ remove the women who never marry from the calculation. In practice, the values of $S(a)$ are often computed for five-year age intervals rather than single years of age, in which case their sum needs to be multiplied by five; in addition, summation is usually begun at the earliest age at which marriage occurs (e.g., 10–14 or 15–19 years).

Estimates of the true mean age at marriage from cohort studies and the singulate mean age at marriage from cross-sectional studies are given in Table 11.2 for several regions and periods. In much of Asia, the

Table 11.2.   Estimated Mean Age of Women at First Marriage, Selected Countries

Region or country	Period	Type of estimate	Mean age at marriage (years)
NEAR EAST AND NORTH AFRICA			
Libya	1964	Singulate	16.8
Morocco	1960	Singulate	17.5
Iran	1956	Singulate	19.1
Egypt	1960	Singulate	19.2
Algeria	1948	Singulate	19.8
Iraq	1957	Singulate	20.1
ASIA			
Pakistan	1961	Singulate	16.5
India	1961	Singulate	16.8
Malaysia	1957	Singulate	19.3
Taiwan	1956	Singulate	21.0
Thailand	1960	Singulate	21.9
Ceylon	1963	Singulate	22.0
Philippines	1960	Singulate	22.1
EUROPE			
France	1575–1749	Cohort	24.6
Belgium	1608–1749	Cohort	25.0
England	1550–1749	Cohort	25.0
Germany	1631–1749	Cohort	26.4
Scandinavia	1581–1740	Cohort	26.7

From Dixon (1971), Flinn (1981)

Near East, and North Africa, women marry on average before (and often well before) age 22, suggesting that exposure to the risk of conception usually begins during or soon after the end of adolescent subfecundity (see Chapter 9). Western Europe, in contrast, stands out as a region of comparatively late marriage, with the mean age of women at marriage generally exceeding 24 years. The uniqueness of the traditional European marriage pattern was first emphasized in a famous paper by Hajnal (1965), who pointed not only to mean ages at marriage that are typically late relative to puberty, but also to the large fraction of European women (up to 15–20 percent) who never marry.[2] These two variables, the mean age at marriage and the proportion never marrying, might generally be expected to vary in tandem across populations, since both reflect the overall rate at which marriages are formed at each age of life; and indeed, there does seem to be a fairly strong cross-cultural association between the two (Dixon, 1971). As shown in Figure 11.1, western Europe is clearly extreme along both dimensions.

It is not entirely clear why the traditional European pattern differed so markedly from that of other regions. The classic explanation was first offered by Homans (1941), who attributed late marriage in medieval England to a combination of impartible inheritance, under which the eldest surviving son inherits the entire parcel of land cultivated by his parents, and neolocal residence, wherein the newly married couple is expected to form its own, economically independent household. Under such a system, the younger siblings of the designated heir are likely to have restricted access to domestic resources and hence limited marriage prospects, resulting in a large fraction of individuals who marry late or never marry at all. However, more recent research suggests that noninheriting sons in medieval England were less restricted in their marriage opportunities than Homans supposed (Britton, 1976; Razi, 1980), and that the geographical association between impartible inheritance and late marriage in preindustrial England was far from perfect (Smith, 1984). In addition, inheritance of agricultural land in the medieval past would scarcely seem to explain the *persistence* of the European marriage pattern into the urban, industrialized present. Unfortunately, no other single explanation of the European marriage pattern seems to work any better.[3]

As useful as the mean age at marriage is as a summary statistic, a better understanding of the dynamics of marital formation can be gained by examining the entire age curve of marriage. Figure 11.2 shows the marriage curves for several populations, expressed both as the proportion of women ever married at each age (top panel) and as the distribution of ages of women at first marriage (bottom panel). Clearly, there are marked differences among populations. Taiwan, for example, shows a pattern of early and near-universal marriage: women begin marrying before age 15, and virtually all women are married by age 30; the highest frequencies of first marriage in Taiwan are tightly clustered between ages 15 and 25 (Figure 11.2, bottom). From a worldwide perspective, this pattern of early and universal marriage, or something very like it, appears to be the norm in most preindustrial societies. However, as might be expected, western European countries like Germany, the Netherlands, and Sweden all display patterns of late marriage, with large fractions of women never marrying before the end of the reproductive span (Figure 11.2). In such countries, the proportion of women ever married is zero until age 15 or so, rising only slowly after that age, and fully 10–15 percent of women are still unmarried at age 50; in these cases, the first marriage frequencies are spread out over a much wider age range than in Taiwan.

Despite these obvious differences, Coale (1971) has emphasized the common features shared by all these curves. In the case of the proportion of women ever married at each age, Coale points out that three factors account for essentially all the differences among populations: (1) the earliest age at which marriage occurs, (2) the rate at which the marriage curve rises after that age, and (3) the ultimate fraction of women marrying by the end of the reproductive span. When appropriate adjustments for these three factors are made, an extraordinary thing happens: all the observed age-patterns of marriage collapse to a single curve (Figure 11.3). This result is truly remarkable and is by no means an inevitable outcome of the adjustments; it suggests a profound commonality cutting across all the cultural differences separating human societies.

In an attempt to capture this common, underlying age structure of marriage, Coale and McNeil (1972) have developed a closed-form model for the dynamics of marital formation. Under their model, the probability density function for a woman's first marriage at age t is

$$f(t) = [\beta/\Gamma(\alpha/\beta)] \exp\{-\alpha(t - \theta) - \exp[-\beta(t - \theta)]\} \tag{11.2}$$

where α, β, and θ are parameters to be estimated, and $\Gamma(\cdot)$ is the gamma function (see Chapter 7). The parameter θ can be interpreted as the earliest age at which marriage can occur, while α and β together determine both the steepness of the cumulative distribution of marriage by age and the fraction of women who remain unmarried. Maximum-likelihood estimation procedures for this model have been developed by Rodriguez and Trussell (1980). Frequent application has shown that this model fits most (but not all) age patterns of marriage from many parts of the world and many different time periods (Coale, 1977; Rodriguez and Trussell, 1980; Chowdhury, 1983; Wood et al., 1985a). An example is shown in Figure 11.4.

Coale and McNeil (1972) went on to show that equation 11.2 is approximately equal to the convolution of a normal and a negative exponential density function. Coale (1977) interprets this result as suggesting that marriage universally involves two stages: entry into a state of "marriageability," the age of which is approximately normally distributed, and an exponentially distributed waiting time from entry into that state until actual marriage (Figure 11.5). If the exponential model for this second component is correct, it implies that women are homogeneous in the rate at which they exit the "marriageable" state—that is, become married—which is unexpected to say the least. It would be of enormous interest if social anthropologists were to test the cross-cultural validity of this interpretation, as well as determine what factors affect the age at which women become marriageable and the rate at which they marry once they have done so. Informal ethnographic investigations among the Gainj suggest that the marriageability of women may be closely tied to pubertal development—more precisely, to a stage of breast maturation that typically follows menarche by about a year (Wood et al., 1985a). Thus, in at least this one instance, the dynamics of marital formation may be linked in important ways to the underlying pace of somatic development. Whether the same holds true for other populations is as yet unknown, but it would be fascinating to find out.

Despite the common functional form of the cumulative marriage curve, it is still clear that the *actual* age pattern of marriage varies widely across societies, as is shown in Figure 11.2. Even though marriage is often an imperfect regulator of exposure to the risk of conception, this cross-cultural variation in marriage is known to translate directly into variation in the distribution of women's ages at first birth and hence in completed family sizes (Ruzicka, 1976; Bumpass and Mburugu, 1977; Bloom, 1982; Pebley et al., 1982; Udry and Cliquet, 1982; Trussell and Bloom, 1983; Mason and Entwisle, 1985). The question of how much interpopulation variation in fertility is induced by the observed levels of variation in marriage will be taken up in the next chapter.

Before leaving the subject, it is worth noting other aspects of nuptiality that potentially affect fertility. These include *marital dissolution* through divorce or widowhood (Menken et al., 1981a; Koo and Janowitz, 1983) and *spousal separation*, or the occasional geographic separation of husband and wife as may occur as a result of labor migration, hunting expeditions, or transhumant pastoralism (Millman and Potter, 1984). Although both these factors can in theory have significant depressive effects on natural fertility, they have received far less attention in empirical studies than has marital formation. In addition, we have already seen that *marital duration* appears to influence the frequency of sexual intercourse within marriage in every society where it has been investigated (Chapter 7). This effect of marital duration combines with the age pattern of marriage (which determines the distribution of marital durations observed at any given time) to set the actual age pattern of coital frequencies that obtains within any particular population (Wood and Weinstein, 1988).

A final aspect of nuptiality that has been of great interest to fertility researchers concerns the fertility-reducing effects of multiple marriages. It has long been conjectured that polygyny (marriage of a man to more than one woman) may reduce female fertility, principally by lowering the coital rate experienced by each wife in a sort of "dilution" effect (Muhsam, 1956; Dorjahn, 1959). This effect is likely to be most pronounced in regions such as sub-Saharan Africa where the husband often maintains separate residences for some of his wives—so-called town wives and village wives (Garenne and van de Walle, 1985); in such cases, each wife may experience extended periods of spousal separation depending upon where her husband is residing at any given time. Unfortunately, empirical evidence relating to this hypothesis (Olusanya, 1971; Smith and Kunz, 1976; Sembajwe, 1979; Chojnacka, 1980; Ahmed, 1986; Bean and Mineau, 1986; Aborampah, 1987; Roth and Kurup, 1988) has proven to be equivocal and difficult to interpret owing to a variety of analytical problems. In most analyses, there is little attempt to control for potential confounding factors, such as differences in the age at which women enter polygynous versus monogamous unions, the fact that most polygynous unions have an earlier history of monogamy, and the variable periods of exposure experienced by first, second, and higher-order wives (Smith and Kunz, 1976; Bean and Mineau, 1986). In addition, men often cite infertility as a reason for divorce or for taking another wife (Goldman and Montgomery, 1989). In sub-Saharan Africa, first wives who have been divorced because of their infertility may subsequently enter other households as junior wives (Pebley and Mbugua, 1989). As a result, women in polygynous unions may have lower lifetime fertility not because of any depression in coital frequency, but because of a higher likelihood that at least some women in such unions are sterile. Finally, polygynous men are often substantially older than monogamous men, and the husband's age tends to increase in direct proportion to the number of wives he has (Wood et al., 1985a). If male age exerts an influence on fecundity or coital frequency (see below), that influence may be difficult to separate analytically from the direct effects of polygyny per se (Garenne and van de Walle, 1985; Bean and Mineau, 1986). The best analyses to date suggest that polygyny may indeed have depressive effects on fertility, albeit weak and variable ones (Bean and Mineau, 1986; Garenne and van de Walle, 1985; Goldman and Montgomery, 1989; Pebley and Mbugua, 1989), but even those analyses have not provided an unambiguous explanation for why such effects exist.

THE MALE CONTRIBUTION

In keeping with long-standing demographic tradition, earlier chapters in this book have emphasized reproductive processes taking place in the female. Although Chapter 5 did discuss spermatogenesis and the sperm's role in fertilization, and Chapter 7 reviewed patterns of coital behavior, the inattentive reader from Mars might still conclude that humans reproduce asexually. Such is not the case. Ignoring the different sizes of the sex chromosomes, as well as the newly discovered phenomena of genetic imprinting and uniparental disomy,[4] the genetic contributions of the male and female parent to the production of a new offspring are essentially identical. Why, then, do most studies of fertility variation focus upon females rather than males? The reason has been spelled out in previous chapters: most of the rate-limiting steps in human reproduction take place in the female. Those steps include, most obviously, pregnancy and lactational infecundability. In addition, since the production of eggs differs from that of sperm in being episodic, the female cycle limits the rate of conception in a way that the more continuous gametogenesis of the male does not. Finally, menarche and menopause place more or less unambiguous, absolute outer boundaries on the period of potential reproduction; there is no such clear-cut, readily accessible marker of puberty in males, nor is there any age so advanced that all men who reach it are unable to reproduce. As noted in Chapter 3, differences in rate-limiting steps usually have a greater impact on overall fertility variation than do differences in other aspects of the reproductive process. Thus, in attempting to explain variation in fertility it has made sense to concentrate, at least initially, on factors affecting women. Still, in view of the joint participation of the sexes in reproduction, we need to ask whether fertility variation can be understood entirely without considering male factors.

Surprisingly little research has been undertaken to answer this question. A handful of historical demographers have investigated the effects of male age on levels of natural fertility. Using Irish census data before 1911, for example, Anderson (1975) found some evidence of a small decline in completed family size by husband's age, controlling for wife's age and the duration of marriage. Unfortunately, the published tabulations did not permit any detailed analysis of why such a decline might occur. Similarly, Mineau and Trussell (1982) decomposed marital fertility rates among the nineteenth-century Mormons by both wife's age and husband's age while controlling for marital duration (Figure 11.6). Although there does appear to be some depressive effect of husband's age on marital fertility, it is small compared with the effect of wife's age. While studies such as these confirm the primacy of female factors in determining patterns of natural fertility, they also suggest that there may be *something* going on in males to induce fertility variation.

But what is that "something"? The perspective adopted in this book provides a way at least to begin answering that question: identification of the specific proximate determinants that potentially reflect characteristics of the male partner may be a useful first step toward delimiting the scope of the problem. It should be emphasized that I am referring here strictly to *direct* effects. A father who commits infanticide may thereby disrupt breastfeeding and shorten the subsequent birth interval experienced by his wife, thus increasing her completed fertility. Such an action, however, would not be a proximate cause of increased fertility, but rather a second-order effect operating via lactational infecundability. Which, if any, proximate determinants of natural fertility can directly reflect male factors?

Table 11.3 lists the relevant determinants. Males can obviously influence the frequency of sexual intercourse and insemination. The male also has a direct effect upon the duration of the fertile period since the fertile period is partly determined by the fertile life span of the sperm cell (see Chapter 4). Several other characteristics of sperm, such as their concentration in the ejaculate, their motility, and their ability to undergo the acrosomal reaction, could influence the probability of conception resulting from a single insemination during the fertile period. And because of the male's genetic contribution to the zygote, male genes may influence the risk of fetal loss. Moreover, males can potentially affect the length of gestations ending in fetal loss; for example, if such losses are associated with a genetic defect inherited from the father, then the tim-

Table 11.3. Proximate Determinants of Natural Fertility that Potentially Can Be *Directly* Affected by Attributes of the Male

---

Frequency of insemination
Duration of the fertile period
Probability of conception from a single insemination in the fertile period
Probability of fetal loss
Length of gestation ending in fetal loss
Age at onset of permanent sterility

---

ing of fetal death will presumably reflect the stage of embryonic or fetal development when the defective gene is first expressed. Finally, as mentioned in Chapter 10, sterility is a characteristic of the couple and may reflect changes in the male partner as well as, or rather than, the female. These, then, are the factors that need to be investigated in order to assess the male contribution to variation in natural fertility.

Before turning to these factors, it is important to note that almost all previous attempts to isolate the effects of male variables have been restricted to age effects. For example, various authors have tested for the effects of paternal age on the risk of fetal loss or on the waiting time to conception. This fact is perhaps unsurprising since almost all demographic and epidemiological surveys record, at a minimum, the ages of subjects and their spouses; no other variable is included as consistently or as often in survey questionnaires, except perhaps for marital status. Because of the focus on age in demographic analysis, we will begin our discussion of the male contribution with what is known about the biology of male reproductive aging. Age effects do not, however, exhaust the potential contribution of male factors to variation in natural fertility. Moreover, age effects may be uniquely difficult to unravel because of the high correlation observed in most societies between the male and female partners' ages, and between both ages and marital duration. The preponderance of negative and ambiguous results in the studies reviewed below may well reflect these difficulties.

The Biology of Male Reproductive Aging[5]

In contrast to research on female reproductive senescence, work on age-related physiological changes in the male has yielded surprisingly few consistent and unambiguous findings. Contradictory results have emerged from studies of testosterone secretion, Leydig cell maintenance, hypothalamo-pituitary function, spermatogenesis, and the quality of sperm. These uncertainties reflect two important biological facts: first, the effects of reproductive aging are much more subtle in men than in women, and second, the male reproductive system is extremely variable in its responses to aging.

Although it has taken some effort to establish, the weight of evidence now points fairly consistently to reduced secretion of testosterone in older men (Tenover et al., 1988; Vermeulen et al., 1989). Ambiguous results in earlier studies probably reflected several factors. Some studies may have suffered from biased subject selection, especially since the health of older subjects appears to be of critical importance; there may also have been inadequate controls for circadian variation in testosterone secretion and for the testosterone-stimulating effects of sexual arousal. Recently, Nahoul and Roger (1990) have suggested that contradictory results may partly reflect the fact that some researchers measure total testosterone (protein-bound and free) and others only free testosterone; studies of free testosterone are more consistent in showing age-related declines (Davidson et al., 1983; Vermeulen et al., 1989). Typical results, illustrating both the age trend and the extreme variability at each age, are shown in Figure 11.7.

Although declining testosterone production in older men is probably attributable to some form of impaired Leydig cell function, studies comparing Leydig cell *numbers* in younger and older men have been inconclusive: some researchers have found an age-dependent reduction in the number of Leydig cells (Nieschlag and Michel, 1986), others have not (Kothari and Gupta, 1974). The fact that a variety of pathological processes can significantly affect the number of Leydig cells (Harbitz, 1973) has undoubtedly complicated the resolution of this issue. From studies of elderly men with prostate cancer, Takahashi et al. (1983) have suggested that declining testosterone levels are primarily attributable, not to a change in Leydig cell numbers, but rather to decreased production within the Leydig cells of pregnenolone, a precursor of testosterone.

In addition to declining testicular function as reflected by testosterone levels, a few studies suggest some deterioration in hypothalamo-pituitary function with age (Vermeulen et al., 1989). As detailed in Chapter 5, testosterone exerts a negative feedback on the release of GnRH by the male hypothalamus, analogous to negative feedback in the hypothalamo-pituitary-ovarian axis of women. As testosterone declines and negative feedback is relaxed, older men exhibit a rise in LH and FSH levels similar to the menopausal pattern of elevated gonadotropin secretion in women (Nahoul and Roger, 1990). Despite these elevated gonadotropin levels, however, some studies have found evidence of suppressed GnRH release by the hypothalamus and increased hypothalamo-pituitary sensitivity to steroid feedback (Urban et al., 1988; Kaufman et al., 1990). For example, administration of exogenous GnRH results in gonadotropin secretion that is significantly delayed in aging men in comparison with their younger counterparts (Winters and Troen, 1982). And exogenously administered dihydrotestosterone has a significantly greater effect in lowering gonadotropin levels in elderly men than in younger men (Vermeulen et al., 1989). Unfortunately, these findings leave the age-related increase in LH and FSH completely unexplained.

Based on autopsy material, Neaves et al. (1984) have suggested that spermatogenesis is reduced in older men, with a significant negative correlation between FSH levels and sperm production. Spermatogenesis may thus be linked to the increase in gonadotropin levels as well as the decline in testosterone with age, as might be expected. However, Nieschlag et al. (1982) found that spermatogenesis was similar in younger and older men in their study, whereas sperm motility and seminal fructose was significantly lower in older subjects. These latter differences may be explained by a declining frequency of ejaculation with age rather than as physiological symptoms of senescence. On the other hand, Schwartz et al. (1983) found a marked deterioration in both sperm morphology and motility in older men even after controlling for the time since previous sexual intercourse.

In sum, the emerging picture suggests that male reproductive senescence is subtle, variable, and probably of less functional significance than senescence in females. Declining testosterone production appears to be the most consistent feature of male reproductive senescence, although its causes remain uncertain. Declining hypothalamo-pituitary function may also occur, although it is far from established. The age-related reduction in testosterone in turn seems to lead to reduced spermatogenesis or impaired sperm motility and possibly a reduction in male libido (Tsitouras et al., 1982). Despite these changes, the numerous documented cases of successful male reproduction at extremely advanced ages (e.g., Seymour et al., 1935) reinforce the impression of a high degree of variability among individual men.

Male Effects on Fecundability

How do these age-related changes in male physiology affect fecundability and the waiting time to conception? To separate the effects of male and female age on fecundability, it is necessary to use data from regions where there is substantial variability in husband's age for each age of wife. In practice, this usually means drawing data from societies in which polygyny is common. However, since polygyny may have its own depressive effects on fecundability, it is important to implement appropriate statistical controls for those effects whenever such data are used. By far the best analysis of this sort has been done by Goldman and Montgomery (1989), who applied a parametric proportional hazards model to World Fertility Survey data from Africa and the Near East. Significant effects of the wife's age were found in all five countries examined, while significant marital duration effects were found in four (Table 11.4). In contrast, although increases in the age of the husband lowered fecundability in all five countries, those effects were significant at the 0.05 level or better in only two cases, Kenya and Syria. Moreover, the magnitude of the effect of husband's age was consistently and substantially less than that of wife's age and the duration of marriage (Figure

Table 11.4. Proportional Hazards Analysis of the Effects of Wife's Age, Husband's Age, and Marital Duration on the Hazard of Conception, Selected Countries or Regions[a]

Predictor variable	Relative Risk of a Conception				
	Ivory Coast	Ghana	Kenya	Northern Sudan	Syria
WIFE'S AGE					
under 35 years	1.000	1.000	1.000	1.000	1.000
over 35 years	0.656	0.724	0.867	0.684	0.769
HUSBAND'S AGE					
under 45 years	1.000	1.000	1.000	1.000	1.000
45–55 years	0.910	0.943	0.881	0.855	0.691
over 55 years	0.820	0.925	0.879	0.886	0.771
MARITAL DURATION					
0 years	1.000	1.000	1.000	1.000	1.000
5 years	1.050	0.973	0.971	0.948	1.070
10 years	1.030	0.883	0.867	0.857	1.030
15 years	0.951	0.748	0.711	0.737	0.901
20 years	0.821	0.592	0.536	0.604	0.715
25 years	0.664	0.437	0.371	0.472	0.514
30 years	0.503	0.301	0.236	0.351	0.335

[a] Adjusted for breastfeeding and type of union (polygynous versus nonpolygynous). Significant effects of wife's age were found in Ivory Coast and Kenya ($P < 0.05$ in both cases) and in Ghana, northern Sudan, and Syria ($P < 0.005$ in all three cases). Significant effects of husband's age were found only in Kenya and Syria ($P < 0.05$ in both countries). Marital duration or its square was found to have significant effects in Ghana and northern Sudan ($P < 0.05$) and in Kenya and Syria ($P < 0.005$).
From Goldman and Montgomery (1989)

11.8). The available evidence, limited though it is, thus suggests that the female partner's age and marital duration are far more important than the male's age in determining age patterns of fecundability, although male age may have some effect. As careful as Goldman and Montgomery's analysis is, however, it tells us nothing about the mechanisms through which any of these effects are operating. It is therefore necessary to turn to other studies in order to assess the individual fecundability factors listed in Table 11.3.

Frequency of Insemination

Studies of the male contribution to coital rates have been confined almost exclusively to analyses of the effects of the male partner's age.

And here too it is difficult to disentangle the collinearity of the male and female partners' ages and the duration of marriage. To date, only four studies have examined both spouses' ages and marital duration simultaneously in multivariate analyses, and all four have been confined to U.S. data. Trussell and Westoff (1980) report a small but significant negative effect of the husband's age but do not indicate whether the magnitude of this effect is greater or less than that of the wife's age or the duration of marriage. Udry (1980) found that marital duration was a much better predictor of coital rates than the age of either spouse, a result confirmed by two studies of panel data from the 1970 and 1975 National Fertility Surveys (Jasso, 1985; Kahn and Udry, 1986). Although none of these studies indicates a large effect of the male partner's age on the frequency of intercourse, that conclusion cannot be generalized until the analyses have been duplicated using data from other countries (for example, data from the worldwide Demographic and Health Survey).

Duration of the Fertile Period

As explained in Chapter 4, the fertile life span of sperm is one of the principal determinants of the length of the fertile period. If that life span is short, then the fertile period can begin only a short time before ovulation itself occurs; however, if the fertile life span of sperm is long, an insemination many days before ovulation may still result in conception. In contrast, the length of the portion of the fertile period *following* ovulation is determined principally by the fertile life span of the egg. As suggested in Chapter 4, the single best analysis of these relationships has been done by Royston (1982), who showed that the mean life span of sperm is about 1.5 days while that of the egg is about 0.7 days (see Chapter 4, Table 4.8). These results suggest that although the life span of the sperm is important, it is less of a limiting factor than that of the egg. It is notable, however, that Royston's analysis was based on a model that assumed that the survival probabilities of gametes from each sex are homogeneous. The analysis did allow estimation of a female age effect, but there was no parallel adjustment for male age. Thus, by its nature this analysis was completely uninformative about any contribution of male factors to *variation* in the fertile period. Despite sensitivity analyses suggesting that variability in the length the fertile period is potentially an important source of heterogeneous fecundability (Wood and Weinstein, 1988; Weinstein et al., 1990), no other research has yet been undertaken to isolate the male contribution to variation in this proximate determinant.

*Probability of Conception from a Single Insemination
in the Fertile Period*

In the clinical literature, many different characteristics of sperm and seminal fluid have been shown to vary among men, and many of these are believed by clinicians to affect the chances of conception (Glass, 1991). These characteristics include the volume of the ejaculate and the pH of the seminal fluid, as well as the concentration of sperm in the ejaculate. Some men may exhibit either a low sperm count (*oligospermia*) or a complete absence of sperm (*azoospermia*) because of primary defects in the hypothalamo-pituitary axis or because a varicose vein in the scrotum (a *varicocele*) has raised the temperature of the testes. In addition, sperm may be present in normal numbers but vary in either their total motility (ability to move at all) or progressive motility (ability to sustain vigorous forward motion). A variety of different head and tail defects have been identified, including short or coiled tails, multiple tails, and microcephalic heads. There are male-derived protein complexes on the surface of the sperm that may act as antigens, inducing an immune response from the female partner leading to premature destruction of the sperm cell within the female tract. There may also be defects in the proteins of the acrosome that reduce the sperm's ability to tunnel through the zona pellucida and engage in fertilization. Some of these variables—for example, ejaculate volume, pH, sperm concentration, motility, head and tail defects—are fairly easy to assess in most clinical settings, and the detection of any apparent abnormality is likely to result in a diagnosis of male infertility.

To what extent do these sperm characteristics actually contribute to fertility variation in the population at large? The answer to this question depends upon (1) the effect of abnormalities in these sperm traits on the probability of conception and (2) the frequency of such abnormalities in the general population. The first consideration has been addressed in only a few large-scale studies. In a series of classic papers (MacLeod and Gold, 1951a, 1951b, 1951c, 1952, 1953a, 1953b; MacLeod and Wang, 1979), MacLeod and his colleagues found that the percentage of motile sperm differed between fertile and infertile men and, in fertile men, appeared to have a large effect on the waiting time to conception. In contrast, variation in sperm morphology appeared to have no such effect, although collinearity with motility made it difficult to isolate the effects of morphology in any convincing way. Unfortunately, the analyses by MacLeod and his colleagues suffered three major shortcomings. First, all their data came from clinical studies, making it difficult to generalize their results to the larger population of males. Second, male subjects were dichotomized into normal/abnormal fractions based upon

their sperm characteristics, ignoring the more or less continuous variation that exists in these traits. And third, their analyses of the waiting
time to conception used only uncensored observations, thus introducing
an unknown degree of censoring bias.

Some of these problems have been circumvented in a recent study by
Polansky and Lamb (1988). Those authors examined the results of an
initial semen analysis in 1,089 couples who reported to the Stanford
University Infertility Clinic, not including couples in which the male
partner exhibited complete azoospermia. The couples were further subdivided according to whether or not the female was "normal", i.e., had
normal results of an endometrial biopsy and a laparoscopy or hysterosalpingography. Further studies were done on a subsample of 421 couples whose fertility status 36 months after the examination was known
from follow-up interviews; 313 of those couples had become pregnant
within 36 months (in none of these cases was in vitro fertilization or artificial insemination with donor sperm performed). Figure 11.9 compares
the distributions of various sperm traits in couples who conceived within the follow-up period (the upper distribution in each panel) and those
who did not (the lower distribution). For each trait, substantial variability exists among patients, but in no case does the distribution differ significantly between fertile and infertile men. Dichotomizing the sperm
characteristics according to the cutoff points listed in Table 11.5, Polansky and Lamb then compared the waiting times to conception for couples falling on each side of the cutoff (Figure 11.10). Again, no significant differences were found. To prevent the loss of information that
results from dichotomizing the predictor variables and discarding censored observations, Polansky and Lamb also performed a multivariate
proportional hazards analysis of the waiting time to conception using

Table 11.5. Cutoff Points Used to Define Subgroups of Men,
with Associated Sample Sizes

Semen characteristic	Cutoff point	Number in higher-value group	Number in lower-value group
Concentration, millions/ml	20	975	114
Volume, ml	1	1,035	54
Progressive motility, %	30	803	286
Total motility, %	50	663	426
Normal morphology, %	50	928	161
Head defects, %	25	336	753
Tail defects, %	25	126	963

From Polansky and Lamb (1988)

continuous predictor variables. Once again, no significant effect of any sperm characteristic on the time to conception was detected, even when attention was restricted to couples with an apparently normal female partner (Table 11.6). In sum, it appears from these analyses that *none* of the sperm or seminal fluid traits routinely measured in infertility clinics is of any use whatsoever in predicting ultimate conception. In view of these startling results, it is something of a relief to learn that men with azoospermia were excluded from the sample—for otherwise we might have to conclude that humans really *are* asexual!

The wider implications of these analyses are uncertain, principally because they are still based on clinical samples. As Figure 11.9 clearly demonstrates, there is enormous heterogeneity in sperm characteristics among men who are examined at infertility clinics. Unfortunately, since these traits are *only* measured in such men, no one knows what the corresponding distributions look like in the general population. Is the sample examined in this study extremely selective with respect to these traits, or does it overlap broadly with the male population at large? Until this question is answered, we will not know what to make of Polansky and Lamb's results. In addition, many other potential sperm defects cannot yet be measured routinely in the clinic (e.g., presence of immunoactive antigens or molecular defects in acrosomal proteins), and the effects of such defects on the probability of conception are unknown. Nonetheless, the studies that have been done to date do not provide strong evidence that existing levels of variation in sperm characteristics have any important effect on fecundability.

Table 11.6.   Proportional Hazards Analysis of the Effects of Various Sperm Factors on the Waiting Time to Conception, All Couples and Couples with a Presumptive Normal Female Partner

	All couples (N = 1,089)		Normal females (N = 210)	
Predictor variable	Estimated exp (β)	P value	Estimated exp (β)	P value
Volume	0.989	0.76	0.945	0.52
pH	0.870	0.51	1.812	0.38
Concentration	1.000	0.85	1.001	0.45
Progressive motility	1.003	0.32	1.009	0.24
Total motility	1.003	0.35	1.012	0.13
Normal morphology	1.000	0.94	1.002	0.84
Head defects	1.002	0.62	1.000	0.97
Tail defects	0.995	0.46	0.995	0.76

From Polansky and Lamb (1988)

Fetal Loss

There is some evidence suggesting that the paternal contribution to
fetal loss increases with age (Woolf, 1965; Selvin and Garfinkel, 1976;
Hook, 1986), similar to the maternal pattern described in Chapter 6. Perhaps the best statistical analysis to date of paternal effects was done by
Selvin and Garfinkel (1976) using data on approximately 19,000 losses in
almost 1.5 million pregnancies registered in New York state between
1959 and 1967 (Table 11.7). After controlling for maternal age and pregnancy order in a multivariate logistic regression analysis, they found a
significant positive effect of paternal age on the risk of fetal loss (Table
11.8). There were also significant main effects of maternal age and pregnancy order, and together these three predictor variables explained
approximately 41 percent of the variation in the data.[6] Although maternal age was the single most important predictor of loss in this analysis,
the effect of paternal age was surprisingly similar.

   In interpreting these results, it is important to recall a potential confounding problem first discussed in Chapter 6. The apparent effect of
maternal age on the risk of fetal loss may be, in part, an artifact of heterogeneity in risk among women coupled with selective fertility arising
from deliberate family planning. That is, women in controlled-fertility
settings who are still attempting to reproduce at more advanced ages
may be selected for high rates of fetal loss, thus inducing a spurious
association between maternal age and pregnancy wastage. An exactly
parallel process could be taking place in men if they are themselves heterogeneous for the risk of fetal loss. Had Selvin and Garfinkel examined
interactions between paternal and maternal age on the one hand and
pregnancy order on the other, they may have been able to control for
this artifact to some degree (compare the analyses of maternal age and
pregnancy order by Santow and Bracher, 1989).

   It might be thought that this issue could be clarified by restricting
attention to pregnancy losses of known etiology, especially those associated with chromosomal defects whose parental origin can be determined. But in fact, no clear picture of paternal effects has yet emerged
from such studies. Hatch (1983) found no association between paternal
age and trisomic abortions after controlling for maternal age; this negative result may be consistent with the finding that most live-born cases
of trisomy 21 (Down's syndrome) are maternal in origin (Stene et al.,
1981; Yates and Ferguson-Smith, 1983; Antonarakis et al., 1991). Hatch
(1983) did find some evidence for a paternal-age effect on the incidence
of monosomy X abortions after controlling for maternal age, but this
result has not been duplicated in other studies (Lauritsen, 1976; Warburton et al., 1980; Hassold et al., 1980). Results for triploid abortions

Table 11.7. Recorded Numbers of Pregnancies and Fetal Deaths and Estimated Crude Fetal Death Rates by Paternal Age, Maternal Age, and Pregnancy Order, New York State (excluding New York City), 1959–1967[a]

Predictor variable	Fetal deaths	Pregnancies	Rate/1,000 pregnancies
PATERNAL AGE (YEARS)			
< 20	358	28,064	12.8
20–24	3,005	310,228	9.7
25–29	4,388	433,962	10.1
30–34	4,275	348,652	12.3
35–39	3,576	214,767	16.7
40–44	2,112	95,094	22.3
45–49	875	30,250	28.9
50–54	266	8,473	31.4
≥ 55	108	3,356	32.2
Total	18,963	1,472,846	12.9
MATERNAL AGE (YEARS)			
< 15	2	133	15.0
15–19	1,289	117,185	11.0
20–24	4,448	475,579	9.4
25–29	4,687	417,190	11.2
30–34	4,033	275,247	14.7
35–39	3,146	146,590	21.5
40–44	1,256	38,934	32.3
≥ 45	102	1,988	51.3
Total	18,963	1,472,846	12.9
PREGNANCY ORDER			
1	4,450	368,790	12.1
2	3,268	367,045	8.9
3	3,496	300,392	11.6
4	2,792	196,460	14.2
5	1,934	111,384	17.4
6+	3,023	128,775	23.5
Total	18,963	1,472,846	12.9

[a] Legitimate births to white mothers only. Pregnancies ending in live births with congenital defects were excluded from the pregnancies at risk.
From Selvin and Garfinkel (1976)

have been even less consistent: both positive (Hatch, 1983) and negative (Hassold et al., 1980) effects of paternal age have been detected—and in one study of 288 abortuses, no effect whatsoever of paternal age was found (Lauritsen, 1976).

If unambiguous paternal age effects on fetal loss are eventually established, it is still possible that these effects are artifacts of declining coital frequency, as hypothesized by Guerrero and Rojas (1975). This hypothesized effect reflects the presumed "aging" or "overripening" of the

Table 11.8. Results of Multivariate Logistic Regression Analysis of the Data in Table 11.7

Predictor variable	Estimated coefficient (± s.e.)	Relative risk of fetal loss[a]			
		Age 20	Age 30	Age 40	Age 50
Paternal age	0.027 ± 0.001‡	1.00	1.30	1.70	2.21
Maternal age	0.032 ± 0.001‡	1.00	1.37	1.87	2.54
Pregnancy order	0.022 ± 0.004‡	—	—	—	—
$R^2 = 0.407$					

[a] Relative risks were computed using the probability of loss at age 20 as a baseline, assuming pregnancy order = 1 and partner's age = 20.
‡ $P < 0.001$
From Selvin and Garfinkel (1976)

gametes within the female tract—i.e., the idea that sperm (or eggs for that matter) residing in the female tract for long periods before conception may accumulate various forms of damage, perhaps including chromosomal defects that predispose to fetal loss. The period of "residence," in turn, is a reflection of coital frequency: the more often insemination occurs, the less time is spent on average by sperm in the female tract before conception. Therefore, higher risks of fetal loss could result from low coital rates. Although Potter and Millman (1986) have explored the possible implications of gametic aging for fecundability and early fetal loss, little empirical work has been done to support this hypothesis.

Finally, it is at least a theoretical possibility that male factors can influence the length of gestations resulting in fetal loss, thereby causing variation in the time added to the birth interval by such loss. For example, if a defective allele for a gene expressed very early in embryogenesis were inherited from the father, a loss would presumably occur much earlier than if the paternal gene were not switched on until late in fetal life. (Possible examples are the genes for the delta chain of embryonic hemoglobin and the gamma chain of fetal hemoglobin.) To my knowledge, this possible avenue for a male effect on natural fertility has received little if any attention.

Onset of Permanent Sterility

As in the female, causes of sterility in the male must be subdivided into pathological factors and those associated with the normal aging process. Several sexually transmitted diseases, including gonorrhea, chlamydia, and genital mycoplasmas, can cause inflammation of the urethra, accessory glands, vas deferens, and epididymis, leading to fibrous constrictions or obstructions of the ejaculatory ducts similar to the tubal occlusions observed in women (McFalls and McFalls, 1984). Although these complications have been most thoroughly studied in cases of gonorrhea, the basic inflammatory response is essentially the same in all these infections. Sterility resulting from obstructive azoospermia occurs most often when there is involvement of the epididymis, which takes place in approximately 15–20 percent of all untreated cases of gonorrhea (Kraus, 1972). Involvement of the accessory glands (e.g., the prostate and seminal vesicles) in sterility and subfecundity is less clear, although it has been suggested that inflammation of these glands may interfere with secretion of acid phosphatase, citric acid, fructose, and prostaglandins into the seminal fluid, compounds that are essential for sperm metabolism (Eliasson, 1976). Inflammation of the prostate and seminal vesicles appears to occur in fewer than 2 percent of untreated gonococcal infections (Pelouze, 1939).

On a worldwide scale, it is unclear to what extent sexually transmitted diseases in males contribute to variation in natural fertility. In Chapter 10, it was argued that pathological sterility was unlikely to be an important cause of female sterility, except in a few parts of the world where the prevalences of STDs are unusually high, for example, central Africa. The same argument would hold for male sterility as well. Meheus et al. (1980) estimate that the annual incidence of urethritis in African men is about 3,750/100,000, of which 80 percent is caused by gonorrhea. This incidence rate is perhaps the highest attained in any part of the developing world. Recall that only a few percent of cases of urethritis lead to involvement of the epididymis or accessory glands and hence to infertility; therefore, even allowing for significant underreporting of urethritis in African men, it is unlikely that pathological sterility in males is an important cause of worldwide variation in natural fertility.

Sterility associated with normal aging is potentially of wider significance. Complete and irreversible loss of reproductive capacity, such as occurs at menopause, is not inevitable in men no matter how long they survive. But there does seem to be an age-related decline in male libido, apparently associated with decreased production of testosterone (Tsitouras et al., 1982), and this change is reflected in an age-specific hazard of impotence that increases among older men (Figure 11.11). Unlike menopause, however, impotence is often reversible. In addition, the hazard of impotence lags some thirty years behind that of menopause (Figure 11.11), suggesting that impotence is unlikely to have much depressive effect on fertility except in rare instances in which elderly men are married to women who are several decades younger.

The Contrast between Male and Female Reproductive Aging

The studies summarized in this chapter are remarkably consistent in showing a greater effect of female age on fertility than male age. In large part, this fact reflects a dramatic difference in the pace of reproductive senescence between the sexes (Figure 11.12). Reproductive function in healthy women reaches its peak between ages 25 and 35, and it remains fairly constant during that decade. Declining oocyte quality and hormone production are evident from approximately age 35 onwards, with a marked threshold at 40. The mean age at menopause, when female reproductive function irrevocably ceases, occurs on average at about age 50 in western women, but fecundability is very low for several years preceding menopause. In contrast, reproductive aging in men is a gradual process that continues until death. In healthy men, senescent changes in reproductive function appear to have a negligible impact on potency and fertility even at quite advanced ages. In addition, reproductive senescence in males, but not females, occurs at a rate that is broadly similar to other senescent changes affecting the risk of death (Wood et al., 1994b). Female reproductive senescence thus appears to be markedly accelerated relative to other forms of senescence. The key feature explaining these differences lies in the differential processes of gametogenesis and the ultimate depletion of primary follicles and oocytes in women. Because of follicular depletion, reproductive function ceases decades earlier in women than in men—indeed, at no age do men show a comparable total loss of reproductive capacity. For this reason, female reproductive senescence is likely to be far more important in limiting fertility at the population level than are senescent changes occurring in males.

Despite the obvious difference between the sexes in the pace of gametogenesis, important similarities remain between men and women in the processes of reproductive aging. Both sexes exhibit an age-related decline in steroid production that is linked to degeneration of gonadal function, and both show a corresponding increase in gonadotropin levels. In addition, there is evidence that the acceleration in the risk of fetal loss with age in the two sexes may be similar. If only the effects of follicular depletion in females could be removed, it seems reasonable to suppose that reproductive function would decline with age at roughly similar rates in males and females—a hypothesis whose only shortcoming is that it is completely untestable.

*   *   *

The studies reviewed in this chapter suggest two general conclusions about the male contribution to variation in natural fertility. First, not much is known about it. Second, what little *is* known suggests that it is probably much smaller than the female contribution. It is unclear whether this second conclusion reflects reality or ignorance. And it would be unfortunate if the second conclusion reinforced the first. It seems premature, given our current lack of knowledge, to dismiss any important male contribution simply because our limited studies have not yet detected it. Further research on male factors is essential. Nonetheless, the best way to summarize the evidence currently available is as follows. (1) There appears to be some negative effect of male age on fecundability, but the magnitude of the effect is generally much less than that of female age or marital duration; the precise mechanisms through which male age influences fecundability are uncertain but probably include declining coital rates at a minimum. (2) The positive effect of male age on the risk of fetal loss may be almost as great as that of female age, but both apparent age effects may in part be artifacts of either heterogeneity and selective fertility or declining coital rates. (3) Male sterility is much less important for fertility variation than female sterility, except perhaps in regions where sexually transmitted diseases are unusually common. (4) Compared with aging in the female partner, male reproductive senescence places few constraints on fertility. At least as a first approximation, then, the conventional emphasis on female factors in reproduction appears to be warranted.

NOTES

1. It is intriguing that the fraction of first births conceived before marriage rose rather consistently throughout western Europe in the decades just before the modern decline in fertility (Table 11.1). This finding may point to an increase in unwanted births during this period, or a restriction in resources that made it more difficult for couples to establish new households (Flinn, 1981:81–83).

2. Although Hajnal referred to these two characteristics jointly as the "European marriage pattern," he noted that they are not observed in eastern Europe, broadly speaking, east of a line from Trieste to St. Petersburg. In addition, although Hajnal emphasized the importance of the European pattern in the preindustrial past, the same pattern can still be found throughout much of western Europe (Dixon, 1971).

3. Numerous authors (e.g., Flinn, 1981; Hajnal, 1982; Lesthaeghe, 1986) have suggested that fluctuations in the age pattern of marriage constituted one of the chief mechanisms whereby fertility levels in preindustrial Europe were adjusted to available resources. Insofar as this mechanism presupposes a tight linkage between marriage and impartible inheritance, this conjecture may have been rendered less plausible by recent research (Smith, 1988).

4. *Genetic imprinting* refers to a situation in which an autosomal allele has different phenotypic effects in the offspring depending upon which parent it

was inherited from (Reik, 1989; Hall, 1990). The classic example in humans is a small deletion in region q11q13 of chromosome 15, which results in Prader-Willi syndrome if inherited from the father and Angelman syndrome if inherited from the mother, two diseases with quite distinct symptoms (Knoll et al., 1989; Butler, 1990). *Uniparental disomy* is a condition in which an individual carries the usual two copies of a particular chromosome but both are inherited from a single parent. In humans, certain rare forms of cystic fibrosis and hemophilia are thought to reflect uniparental disomy (Thompson et al., 1991). On current evidence, both of these types of deviation from normal Mendelian inheritance appear to be rare.

5. I am grateful to Gillian Bentley for her significant contributions to this section, which summarizes part of a paper we wrote together (Wood et al., 1994b).

6. As explained in Chapter 6, the apparent effect of pregnancy order on the risk of loss probably reflects heterogeneity of risks among couples, since two couples of about the same age may differ in pregnancy order precisely because one of them has experienced more losses.

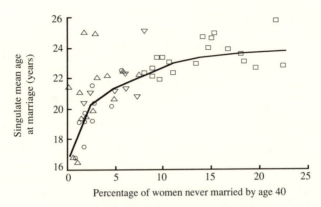

Figure 11.1. The relationship between the singulate mean age of women at marriage and the percentage of women who never marry by age 40 in various countries, based upon census data collected around 1960. ○, the Near East; △, Asia; ▽, eastern Europe; □, western Europe. *Solid line,* lowess smoothing of the points across all four regions. (Data from Dixon, 1971)

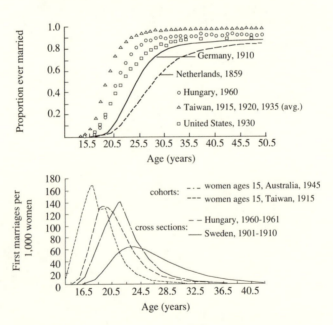

Figure 11.2. Age patterns of marital formation, various countries. *(Top)* proportion of women ever married at each age. *(Bottom)* distribution of women's ages at first marriage. (Redrawn from Coale, 1971)

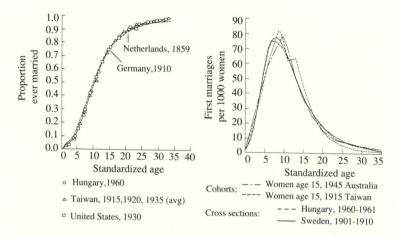

o Hungary, 1960

▵ Taiwan, 1915, 1920, 1935 (avg)

□ United States, 1930

Cohorts: — ·— Women age 15, 1945 Australia
          ---- Women age 15, 1915 Taiwan

Cross sections: - - - Hungary, 1960-1961
                —— Sweden, 1901-1910

Figure 11.3. Age patterns of marital formation adjusted for differences in origin and scale. The curves are the same as in Figure
11.2 but redrawn with a common starting point, a vertical scale
adjusted by a multiplicative constant so the proportion ever married by age 50 equals one, and a horizontal scale adjusted so the
overall rate of increase in the curve of proportions married is
about the same in all populations. (For a detailed technical explanation of the adjustments, see Coale, 1971:214)

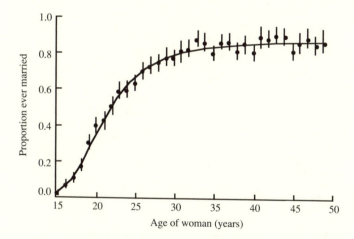

Figure 11.4. The fit of the Coale-McNeil model to observations on the proportion of women ever married, World
Fertility Survey data collected in Colombia, 1976. *Points*
are observed proportions (± two standard errors); *smooth
curve* is the fitted model. (Redrawn from Rodriguez and
Trussell, 1980)

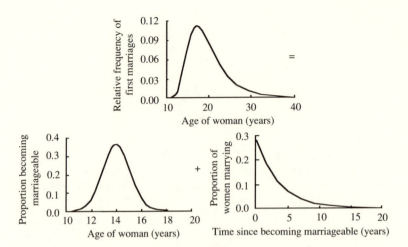

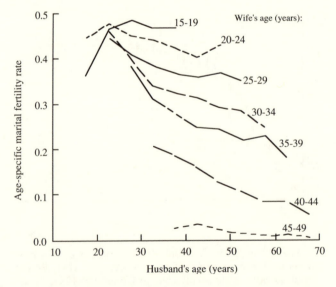

Figure 11.5. Decomposition of the Coale-McNeil model. The probability density function for first marriages by age (*top*) is approximately equal to the sum of two independent components: the density for entry of women into the marriageable state (*lower left*) and the density of waiting times from entry into marriageability until actual marriage (*lower right*).

Figure 11.6. Decomposition of age-specific marital fertility rates (adjusted for duration of marriage) by husband's and wife's ages, Utah Mormons, 1860–1879 birth cohort. (Redrawn from Mineau and Trussell, 1982)

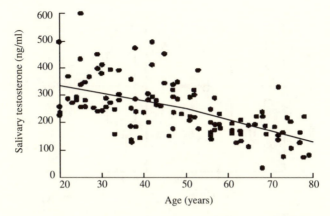

Figure 11.7. Salivary testosterone concentrations in 116 healthy U.S. men ages 20–79. Solid line is a quadratic regression fit to the observations ($R^2 = 0.339$, $P < 0.001$). Salivary testosterone is by its nature free, i.e., not protein-bound. (Data generously provided by Benjamin Campbell)

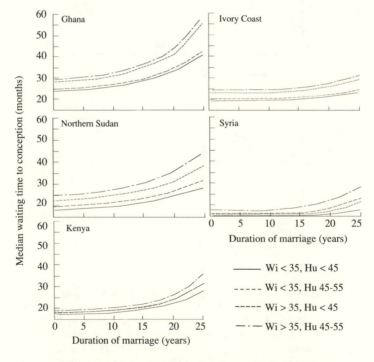

Figure 11.8. Estimated median waiting time to conception by marital duration, wife's age, and husband's age, selected countries. (Redrawn from Goldman and Montgomery, 1989)

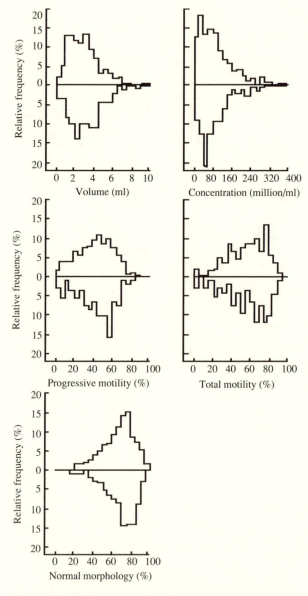

Figure 11.9. Frequency distributions of various sperm
and seminal-fluid variables in 313 fertile couples
(above the line in each panel) and 108 infertile
couples (below the line), where "fertile" is defined
as having achieved an unassisted pregnancy within 36 months of the semen analysis. (Redrawn
from Polansky and Lamb, 1988)

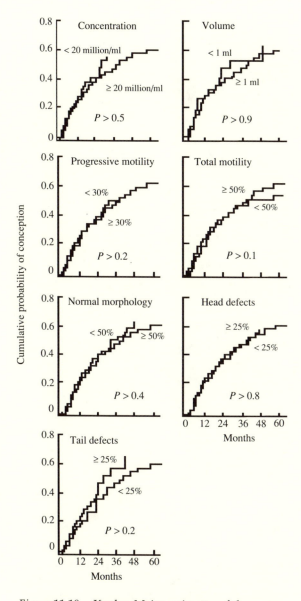

Figure 11.10. Kaplan-Meier estimates of the cumulative probability of conception for subgroups of males with various sperm characteristics. Also shown are attained significance levels for the difference between subgroups (*P*) according to a Wilcoxon test. (Redrawn from Polansky and Lamb, 1988)

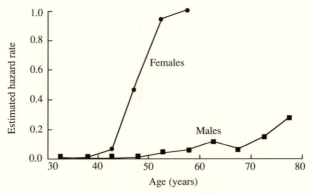

Figure 11.11. Estimated hazard of cessation of reproductive function in U.S. women and men. The marker for cessation of reproductive function in women
is taken to be menopause; in men, impotence. Sample sizes are 324 for women, 3,164 for men. (Data on
menopause from Treloar, 1974; data on impotence
from Kinsey et al., 1948)

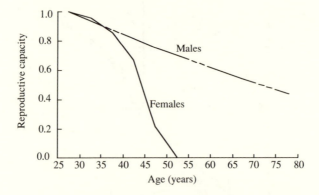

Figure 11.12. Reproductive senescence in human
males and females, relative to ages 25–29 years.
Reproductive capacity in males is taken as proportional to predicted circulating testosterone concentration (regression line in Figure 11.7 standardized to equal one at age 27.5). Female reproductive
capacity is proportional to the value of fecundability expected when coital frequency is held constant while the physiological determinants of fecundability are allowed to vary by the female
partner's age. (The model from which expected
fecundability was derived is described in Wood
and Weinstein, 1988; the predicted values are from
Wood, 1989)

PART III

Beyond the Proximate Determinants

12

Extensions and Elaborations

We have now discussed each proximate determinant of natural fertility in turn, reviewing what is known about the biological and behavioral mechanisms underlying it, and linking it as nearly as possible to the relevant birth-interval components. The reader should now have a strong impression of how each determinant works on its own to cause fertility variation. This piecemeal approach, though somewhat arbitrary, was the most reasonable way to organize the mass of material that needed to be presented. Nevertheless, although taking things one piece at a time may be a sensible way to begin any analysis, ultimately it is a fiction, an analytical and didactic convenience. In reality the proximate determinants do not work in splendid isolation but interact with each other in complex ways. Nor is the reproductive system as a whole any more isolated than its parts: it has an ecological and cultural context, as well as an evolutionary history. Granted, no matter how interesting these interactions, complexities, and contexts may be, it would have been foolish to have launched into them at the outset without first gaining some appreciation of the basic mechanisms involved. But having reviewed those mechanisms, it is time to adopt a more comprehensive perspective.

Several important questions could not be addressed using the piecemeal approach. Most obviously, how do the proximate determinants compare with each other in their effects on fertility? Which are most important in causing fertility variation? Are there any determinants whose effects are so trivial that they can be ignored in future research? A second set of questions concerns the way in which the proximate determinants interact with each other. For example, are fecundability and the risk of fetal loss, which reflect different but related aspects of ovarian function, correlated in important respects? In women who have just resumed ovulating after childbirth, is the probability of conception affected at all by the length of the preceding period of lactational anovulation? Finally, there are innumerable questions about the effects of more remote influences on fertility, effects mediated by the proximate determinants themselves. Although we cannot hope to answer all those questions, we should at least try to single out the ones that currently seem most interesting.

HOW DOES THE REPRODUCTIVE PROCESS WORK?

Before trying to address these issues, it will be useful to review the basic logic of the reproductive process. A fundamental message of this book is that reproduction must be understood as a time-dependent process: the level of fertility is determined by a series of major events distributed across the female reproductive life course, and by the time intervals separating those events. The proximate determinants of fertility are important precisely because they determine the distributions of those time intervals. Models of birth-interval components therefore provide the most realistic way to mimic the dynamics of the reproductive process.

To give a concrete example, lactational infecundability reflects a complex system of physiological pathways linking the suckling stimulus to ovarian quiescence; at the same time, a host of ecological, social, and economic forces influences the frequency with which the suckling signal is actually delivered to this system. With respect to their effects on fertility, however, all these complexities boil down to a question of timing: when during the postpartum period does full fecundity (ovulation and normal luteal function) resume? Variation in the physiological and ecological factors influencing this system, as well as purely random variation, gives rise to a *distribution* of times to the resumption of fecundity, and this distribution is in principle the source of all fertility variation associated with breastfeeding. If we can model the distribution, we will have captured the full impact of breastfeeding on fertility, even if we do not yet understand each and every causal factor underlying the distribution. Of course, if our model is truly an *etiologic* model—one that embodies what we think we know about the causal mechanisms involved—it may be especially enlightening to fit and test it against data. However, even an *empirical* model—one that fits well but is uninformative about causation—may be useful for measuring the effect of breastfeeding on fertility.

Figure 12.1 summarizes this perspective one last time, keyed to the proximate determinants and associated birth-interval models in Table 12.1. The models in that table have been drawn from all the previous chapters. For example, the choice of a lognormal model for menarche is justified by the data of Treloar (1974) examined in Chapter 9, and the Coale-McNeil model for first marriages was discussed in Chapter 11. As noted in Chapter 7, the waiting time to conception follows a compound geometric distribution when fecundability is heterogeneous, and demographers often use the beta distribution for $g(p)$, the density function describing the couple-to-couple variation in fecundability. The time added to the birth interval by fetal loss is perhaps the single most complicated piece of the puzzle; however, the new (and as yet unsolved) equation of Wood et al. (1992), presented in Chapter 6, can at least be used in computer simulations. The length of gestation, by comparison, is simple: as shown in Chapter 5, it is approximately normally distributed. The duration of lactational infecundability has been modeled in various ways, as summarized in Chapter 8. Table 12.1 combines the mixture model of Ford and Kim (1987) with the etiologic model of Ginsberg (1973); the assumed linkage between those models is that the two fractions of nursing mothers in the Ford and Kim specification differ from each other in the timing of supplementation and weaning, as parameterized in Ginsberg's model. Finally, following the argument of Chapter 10 menopause and the onset of permanent sterility have been taken to be equivalent, as they appear to be in most natural-fertility settings, and the model of Pittenger (1973) for permanent sterility is used for both in Table 12.1.

Together, Figure 12.1 and Table 12.1 represent the heart of the current book. Everything else is, in effect, explication and justification of the material in this figure and table. That is not to say that any of the models summarized in Table 12.1 is in any sense final. Alternative models are not only possible but will become essential as we learn more about the dynamics of reproduction. Nonetheless, I believe that the *logic* embodied in this table will survive any changes in model specification; that is, I think that dynamic modeling of birth-interval components is the best way to proceed, no matter what the precise models end up looking like. One major attraction of this approach, as argued in Chapter 3, is that hazards analysis provides a powerful and general way to estimate and test models of this sort based upon well-validated maximum-likelihood principles. While it would be narrow-minded and self-defeating to say that no other statistical approach to fertility analysis can

Table 12.1. Dynamic Models for the Proximate Determinants of Natural Fertility

Proximate determinant[a]	Probability density function	Source
1. Age at menarche	$f(t)$ approximately lognormal	Treloar, 1974
2. Age at marriage	$f(t) = [\beta/\Gamma(\alpha/\beta)]\exp\{-\alpha(t-\theta) - \exp[-\beta(t-\theta)]\}$, $\alpha,\beta > 0$ $\Gamma(\cdot) = $ gamma function	Coale and McNeil, 1972
3. Waiting time to conception[b]	$f(t) = \int_0^1 p(1-p)^{t-1}g(p)dp$, $0 \geq p \geq 1$ $g(p) = p^{\alpha-1}(1-p)^{\beta-1}/\int_0^1 y^{\alpha-1}(1-y)^{\beta-1}dy$, $\alpha,\beta > 0$	Sheps, 1964 Henry, 1964
4. Time added by fetal loss[c]	$f(t) = \int_0^1 \left\{\sum_{k=0}^{\infty} [\langle\zeta * \phi(t)\rangle_k * \zeta(t)q^k(1-q)]\right\}\psi(q)dq$ $\langle\zeta*\phi(t)\rangle_k = k$-fold convolution of ζ and ϕ	Wood et al., 1992
5. Length of gestation	$f(t)$ approximately normal	Hammes and Treloar, 1970
6. Duration of lactational infecundability[d]	$f(t) = qf_1(t) + (1-q)f_2(t)$, $0 \leq q \leq 1$ $f_i(t) = \lambda_i(t)e^{-\tau(t)}\tau(t)^{k-1}/(k-1)!$, $k \geq 1$, $i = 1,2$ $\tau(t) = \int_0^t \lambda(y)dy$ $\lambda(t) = \lambda_1$ up to supplementation, λ_2 up to weaning, λ_3 thereafter	Ford and Kim, 1987 Ginsberg, 1973
7. Age at menopause/onset of permanent sterility[e]	$f(t) = h(t_0)\phi^{t-t_0}\exp\left[\dfrac{h(t_0)(1-\phi^{t-t_0})}{\ln\phi}\right]$, $t \geq t_0$, $h(t_0)$, $\phi > 0$	Pittenger, 1973

[a] Numbers are keyed to Figure 12.1.
[b] Fecund waiting time to conception, determined by the value of fecundability (p) and the various fecundability factors.
[c] Combines both the probability of loss per conception (q) and the nonsusceptible period associated with each loss.
[d] Combines the Ford-Kim mixture model with mixing parameter q and the Ginsberg modified gamma model.
[e] Menopause and the onset of permanent sterility are assumed to be equivalent.

yield fruitful results, the central role of hazards analysis in current research cannot be denied.

But this is still the piecemeal approach. How can it be extended so that we can investigate the *joint* effects of the proximate determinants? The most useful and realistic framework for integrating the proximate determinants is the individual woman's own reproductive life course: different determinants and their associated birth-interval components act together precisely because they operate sequentially within the same woman over the course of her life (Figure 12.1). It would be helpful, therefore, to have a method for linking together the sequence of birth-interval components as they might actually be experienced by a particular woman. Restricting attention for the moment to a single interval between successive births, we need to string together several subintervals in the manner suggested by Figure 12.2. If we can assume that birth-interval components i and j (e.g., lactational infecundability and the subsequent waiting time to conception) are independent, then the probability density function of their sum can be found as the *convolution*

$$f_{i+j}(t) = \int_0^t f_i(t - y)f_j(y)dy \qquad (12.1)$$

Although this equation is often not solvable analytically, it can always be solved numerically on a computer. Applying equation 12.1 iteratively, we can build up entire birth intervals from their separate, sequential parts. Thus, once we have the sum of lactational infecundability and the waiting time to conception, we can tack on the time added by fetal loss; and once we have *that* sum, we can add the length of the next gestation (for an empirical application, see Wood et al., 1985a). Using precisely the same kind of convolution, we can add together a series of birth intervals, along with the events defining the beginning and end of the reproductive life span (menarche, marriage, and menopause), to yield an entire maternity history—or more precisely, a distribution of maternity histories as might be observed in an actual cohort of women.

An alternative approach is to link the birth-interval components together on the computer using *microsimulation* or *Monte Carlo* techniques. In this approach we create synthetic maternity histories for individual (and artificial) women according to the logic summarized in Figure 12.3. For each of n women we generate, in turn, an age at menarche, an age at marriage, a waiting time to first conception, etc., using Monte Carlo techniques.[1] Thus, we can build up an entire maternity history for each woman in the sample, a history that can be considered a specific random realization of the model on which the simulation is based. Once we have accumulated a large sample of such histories, we can use them to compute any of the demographic measures of fertility performance reviewed in Chapter 2, just as if we were analyzing data from a real population.

THE RELATIVE EFFECTS OF THE PROXIMATE DETERMINANTS

Such methods allow us to address a variety of questions. Perhaps the most important concerns the relative impact of the various proximate determinants on fertility variation. Which of the determinants are significant sources of interpopulation variation in natural fertility, and which are not? Do they all act equally, or do some have such small effects that we can safely ignore them in future research? And are some effects so outstandingly large that it behooves us to study them more thoroughly? One way to answer these questions is to perform a *sensitivity analysis* based upon the sorts of models outlined above. In this kind of analysis, we use our model incorporating all the proximate determinants to compute a quantitative measure of reproductive performance, for example, the total fertility rate. We then set all the proximate determinants at their "average" values for natural-fertility populations, at which point the model should yield the mean TFR observed in such populations—approximately equal to six according to Chapter 2. Finally, we allow a single proximate determinant to vary through its known or suspected range of interpopulation variation while holding all the other determinants constant at their average values, and then note the response in the TFR. If the range of the induced response in TFR is narrow, that particular proximate determinant is unlikely to be an important cause of interpopulation variation in natural fertility, and the converse can be concluded if the response is wide.

In a previous publication we presented a sensitivity analysis based upon microsimulation techniques (Campbell and Wood, 1988), and the reader is referred to that paper for technical details. (One important feature of the analysis was that we were interested in "normal" rather than pathological sterility, and therefore did not consider populations with primary sterility rates of $[1 - \alpha_0] \geq 0.15$.) Figure 12.4 shows the results of our analysis, along with a schematic representation of the observed pattern of variation in natural-fertility TFRs. As can be seen, variation in the duration of lactational infecundability is the single most important cause of interpopulation differences in natural fertility, with variation in the age pattern of marriage running a close second. In fact, either of those factors could by itself account for almost all the observed variation in natural fertility. To this extent, our sensitivity analysis agrees closely with one based on a different model investigated by Bongaarts and Potter (1983). However, our results differ from theirs in suggesting that variation in the fecund waiting time to conception (or equivalently, fecundability) may also be an important cause of fertility variation. In this respect our results are consistent with those from yet another sensitivity analysis carried out by Trussell (1979). Unfortunately, because of the difficulties in measuring fecundability discussed in Chapter 7, it is unclear whether the "observed" range of variation used in our analysis is real or merely represents estimation error. Nonetheless, our analysis indicates that the *potential* impact of fecundability is large, a point to which I return below.

As for the remaining proximate determinants, total fertility is moderately responsive to variation in the age at menarche, which may sometimes be important in explaining lower TFR values, and to variation in the prevalence of pathological sterility. It is important to reiterate, however, that we excluded the extreme prevalences of pathological sterility known to characterize such regions as central Africa. Had we included such extreme prevalences, we could have driven the range of responses in TFR downward well below three.

Total fertility is comparatively insensitive to variation in the age at menopause. There are two explanations for this result. First, interpopulation variation in the age at menopause is fairly modest (Chapter 9). Second, the ages at which menopause normally occurs in all populations are times of rapidly declining effective fecundability (Chapter 7). Consequently, the likelihood of successful reproduction after age 40 or so is low in all natural-fertility populations, regardless of when menopause itself occurs.

Finally, total fertility appears to be almost completely unresponsive to interpopulation variation in the risk of fetal loss and the length of gestation. In the case of gestation, this insensitivity reflects the fact that interpopulation variation in the duration of pregnancy resulting in live birth is quite restricted (see Chapter 5). In the case of fetal loss, however, the result may only reflect the fact that this particular proximate determinant has in the past been extremely difficult to measure reliably; hence, we know next to nothing about how it varies among human populations.

Returning for the moment to fecundability, and accepting for the moment that the impact of fecundability on variation in natural fertility is large, it is of interest to investigate the potential contributions of the various fecundability factors to interpopulation variation in fecundability itself. In Chapter 7 I described a sensitivity analysis of those factors based upon the Wood-Weinstein model of fecundability (Wood and Weinstein, 1988). Table 12.2 summarizes and extends the results of that

Table 12.2. Sensitivity Analysis of Fecundability Factors

Fecundability factor	Sensitivity	Interpopulation variability	Total effect
Proportion of cycles ovulatory	+++	+	++
Length of fertile period	++	?	?
Probability of conception per insemination in fertile period	++	?	?
Ovarian cycle length	+	+	+
Frequency of insemination	+	±	±

From Campbell and Wood (1988), Wood and Weinstein (1988)

analysis. The overall response of fecundability to variation in a particular fecundability factor depends upon two things: (1) *sensitivity* in the strict sense, i.e., the responsiveness of fecundability to small, fixed fluctuations in that factor, and (2) *variability*, or the actual amount of interpopulation variation observed in that factor. (The same is true of the response of total fertility to variation in the other proximate determinants, but for simplicity these two separate effects were combined in figure 12.4.) The total response of fecundability to a given factor will be large only if sensitivity and variability are *both* large.

According to our model, fecundability is quite sensitive to the proportion of all ovarian cycles that are ovulatory, and only slightly less sensitive to the duration of the fertile period within each ovulatory cycle and to the probability of conception resulting from one act of unprotected intercourse within the fertile period. Fecundability is only moderately sensitive to the overall length of the cycle, and surprisingly insensitive to the frequency of insemination.[2] Unfortunately, interpopulation variation in all these fecundability factors has been, at best, only poorly documented. As discussed in Chapter 4, recent research suggests that ovarian function, including cycle length and the frequency of anovulatory cycles, may vary rather more among human populations than was previously thought (Ellison et al., 1986; P. Johnson et al., 1987, 1990). However, the degree of variation in both the duration of the fertile period and the probability of conception per insemination in the fertile period is completely unknown; indeed, we still find it difficult to measure these two quantities even within our own population (Royston, 1991).

To make these sensitivity analyses less abstract, I have gathered together statistics on four specific natural-fertility populations spanning the observed range of variation in total fertility rates (Table 12.3). These four groups are by no means randomly selected from all possible cases of natural fertility; rather they were chosen strictly because detailed

Table 12.3.   Intergroup Contrasts in the Determinants of Natural Fertility[a]

| Population | TFR | Median Age (years) of Women at | | | | | Primary sterility rate | Median interbirth interval (mos.) | Median duration of lactational infecundability (mos.) |
		Menarche	Marriage	First birth	Last birth	Menopause			
Gainj	4.3	18.4	21.2	25.7	(40.0)	47.7	0.011	36.5	20.4
!Kung	4.6	17.1	17.4	20.9	37.0	?	0.135	35.1	(18.0)
Matlab	6.1	15.9	17.3	?	38.8	(44+)	?	33.3	17.3
Hutterites	9.8	(12–13)	22.0	(23.5)	39.0	?	0.024	19.6	(6.0)

[a] Figures in parentheses are estimated indirectly and are of dubious validity.

SOURCES Gainj: Wood et al. (1985a, 1985b); P. Johnson et al. (1987, 1990); Wood (1992). !Kung: Howell (1979); Konner and Worthman (1980); Harpending and Wandsnider (1982); Stern et al. (1986). Matlab: Chen et al. (1974); Chowdhury et al. (1977); Huffman et al. (1978a); Chowdhury (1983); Karim et al. (1985). Hutterites: Eaton and Mayer (1953); Tietze (1957); Sheps (1965) information is available on their reproductive patterns. The populations in question are, in increasing order of TFR: (1) the *Gainj*, a group of tribal swidden horticulturalists living in the highlands of Papua New Guinea with whom our research group has been working for the past 15 years (P. Johnson et al., 1987, 1990; Wood et al., 1985a, 1985b; Wood, 1992); (2) the *!Kung San*, nomadic hunter-gatherers of southern Africa's Kalahari Desert whose demography is known primarily through the work of Howell (1976, 1979), Lee (1979), Konner and Worthman (1980), and others (Lee and DeVore, 1976; Harpending and Wandsnider, 1982; Stern et al., 1986); (3) *Matlab*, a rural administrative unit in Bangladesh densely occupied by peasant rice cultivators and studied by several investigators associated with the International Centre for Diarrhœal Disease Research in Dhaka (for example, Chen et al., 1974; Chowdhury et al., 1977; Huffman et al., 1978a, 1978b, 1987; Chowdhury, 1983; Karim et al., 1985; Pebley et al., 1985; Becker et al., 1986; Foster et al., 1986; John et al., 1987; Riley et al., 1989); and (4) the *Hutterites*, a North American Anabaptist sect (Hostetler, 1974) whose fertility has been studied by Eaton and Mayer (1953), Tietze (1957), Sheps (1965), Robinson (1986), Heckman and Walker (1987, 1990), and Larsen and Vaupel (1993), among others.

Table 12.3 can be read either row by row to see how various proximate determinants combine within a single population to produce a particular level of achieved fertility, or column by column to see how a single proximate determinant varies among populations. For example, the low level of fertility observed among the Gainj results largely from late initiation of reproduction combined with exceptionally long interbirth intervals. Reproduction begins comparatively late in Gainj women because of late menarche and an associated late age at marriage, and because of an unusually long lag between marriage and first birth, which we tentatively ascribe to prolonged adolescent subfecundity and irregular sexual relations between newlyweds (Wood et al., 1985a; Wood, 1992); long birth intervals in the Gainj have been shown to be caused primarily by prolonged lactational infecundability (Wood et al., 1985b). Among the !Kung, in contrast, a similarly low level of fertility is achieved through a combination of long interbirth intervals, presumably reflecting the prolonged and frequent breastfeeding known to typify this group (Konner and Worthman, 1980), and a high primary sterility rate, apparently attributable to widespread gonorrhea (Howell, 1979:185–187). The high prevalence of pathological sterility among the !Kung is also reflected in the comparatively early age of women at the birth of their last child. The people of Matlab have a total fertility rate close to the mean for all natural-fertility populations, achieved through early marriage and a median birth interval that is slightly shorter than that of the Gainj and the !Kung.

Finally, it is clear that the Hutterites attain their extraordinarily high fertility mainly through shortened birth intervals associated with early weaning. Huntington and Hostetler (1966) report that most Hutterite children are at least partially weaned by six months of age, and that few are not fully weaned by the end of their first year. This pattern of early weaning stands in marked contrast to the Gainj, among whom half of all surviving children are still nursing at 39 months of age and at least some still nurse at 60 months (Wood et al., 1985b). Consistent with the early weaning of Hutterite children, indirect analyses suggest that the median duration of lactational infecundability among the Hutterites is unlikely to exceed six months (Sheps, 1965). Nonetheless, this brief period of infecundability is still some four to five months longer than expected in the complete absence of breastfeeding. Thus, breastfeeding does act at least to a limited degree as a brake on reproduction among the Hutterites. In addition, the Hutterites achieve their high fertility *despite* their comparatively late ages at marriage and first birth; late marriage, moreover, occurs despite an average age at onset of menarche that is probably similar to that of the general U.S. population. Had marriage been earlier and had breastfeeding been even less prolonged, there is no obvious reason the Hutterites could not have attained an even higher level of fertility, dubious though that achievement might have been.

When Table 12.3 is read column by column, the importance of birth spacing in general and breastfeeding in particular becomes evident. Differences in interbirth intervals account for most of the variation in TFRs in this table, while differences in lactational infecundability account for almost all the variation in interbirth intervals. Other proximate determinants may be important limiting factors for reproduction within particular populations (e.g., late menarche among the Gainj or sterility among the !Kung), but they tend to vary in complex, inconsistent ways across populations. And this observation raises an important caveat about our sensitivity analyses: such analyses may indicate the relative contribution of each proximate determinant across *all* natural-fertility populations but still tell us little about how important each determinant is as a limiting factor within a *particular* population. Moreover, the reproductive patterns observed at the population level may reveal nothing about interesting and important differences among individuals *within* populations. Ultimately, all these dimensions of variability will need to be taken into account.

Despite these limitations, our sensitivity analyses suggest a clear agenda for future research. Perhaps most important, we need to learn more about how and why breastfeeding patterns vary across populations and environments. As noted in Chapter 8, we now know that considerable variation exists and that it has important implications for reproductive output, but we have a very incomplete understanding of why breastfeeding patterns actually vary as much as they do. In addition, though anthropologists have long been interested in many different aspects of marriage in preindustrial society, including who may or may not marry and the political and economic ties that are cemented by marriage, remarkably little attention has been paid by anthropologists to the *age pattern* of marriage. Clearly there is great variation in the age at marriage across different cultural settings, variation with profound implications for fertility differentials—but *why* does such variation exist? With respect to fecundability and fetal loss, the most immediate goal is simply to determine how much interpopulation variation actually exists in humans; several important methodological innovations will be necessary to settle this question. If nontrivial variation does turn out to exist, it will become of prime importance to understand its sources.

HOW DO THE PROXIMATE DETERMINANTS INTERACT?

One respect in which the sensitivity analyses just described may be misleading is that the model on which they are based assumes that the proximate determinants vary independently of each other. In reality, the determinants interact in important ways—that is, the effects of any one determinant are contingent upon the actions of another. While far less is known about these interactions than about the main effects of the proximate determinants considered individually, certain important facts are now well established. For example, the ages of women at menarche and marriage are known to be positively correlated in a wide variety of cultural contexts (Udry and Cliquet, 1982). Since sexual maturation may be a near-universal criterion of marriageability, this association is perhaps unsurprising. In contrast, the ages at menarche and menopause appear to be genuinely independent at the individual level (Treloar, 1974), which may seem strange until it is realized that their underlying causes are entirely different (see Chapter 9). Because of adolescent and senescent subfecundity, the ages at both menarche and menopause have important interactions with various aspects of ovarian function including, at a minimum, cycle length, luteal sufficiency, and the likelihood of ovulation (Chapters 4, 9, and 10). And interactions of age at menarche with the risk of fetal loss now seem likely (Sandler et al., 1984), paralleling earlier findings on menopause (Chapter 6). With respect to fetal loss itself, it is very unlikely that the probability of loss is independent of the length of gestation at which loss occurs: zygotes carrying a major chromosomal aberration such as a monosomy, for example, are probably at risk not only for intrauterine death, but also for *early* death. And there is good evidence that lactational infecundability is involved in several complex interactions. For example, demographic and physiological research has shown that the duration of the fecund waiting time to conception following the postpartum resumption of ovulation is affected by the duration and intensity of suckling, as well as by the duration of postpartum anovulation itself (Jain et al., 1979; Brown et al., 1985).

Sexual behavior interacts in important ways with several other fecundability factors. We saw in Chapter 7 that the relationship between coital frequency and fecundability is nonlinear, and that the degree of curvature in that relationship depends upon the length of the fertile period and the probability of conception from one insemination in the fertile period. This example raises an extremely important point about the interactions among the proximate determinants: because of those interactions, it is impossible to separate fully the behavioral from the biological determinants of reproduction. Their effects are inextricably intertwined.

This list of potential interactions is far from exhaustive. To be incorporated realistically into our models of the reproductive process, each of these interactions will require something more complex than equation 12.1. Thus, not only do these interactions deserve further empirical study, they will entail some difficult theoretical work as well.

MORE REMOTE INFLUENCES

The primary objectives of this book have been to identify the proximate determinants of natural fertility, summarize what is known about them, and assess their contributions to variation in human fertility. Measuring those contributions, however, provides only a proximal explanation of fertility variation, not a final one. The focus on the proximate determinants can only be justified as a first step toward assessing the effects of more remote influences, which are often of greatest ultimate interest. Indeed, the proximate-determinants approach can be thought of as an accounting frame within which the precise mechanisms whereby remote influences act upon the reproductive process can be specified. Unfortunately, it would not be possible to provide a comprehensive treatment of the more remote influences without doubling the length of this book. In the remainder of this chapter, therefore, I wish simply to indicate some of the most active and interesting areas of current research on the remote influences, with an emphasis on biological variables that unapologetically reflects my own background and biases. My aim is not to provide a detailed discussion of any of these factors, but merely to highlight what I consider the most important and vexing issues.

Nutrition and Fertility

One area of lively interest and debate concerns the impact of maternal nutritional status on reproduction. Does poor nutritional status result in suppression of ovarian function and hence in reduction or obliteration of fertility? Is nutritional status per se the biologically relevant factor, or do negative changes in energy balance impair reproductive capacity regardless of the baseline level of nutrition? In the twenty years since Rose Frisch and her colleagues first posed these questions in their modern form (Frisch and Revelle, 1970; Frisch and McArthur, 1974; Frisch, 1975, 1978) most research has focused on three specific aspects of reproductive function. First, several investigators have tried to determine whether undernutrition delays reproductive maturation, as reflected in the age of women at menarche (Bhalla and Shrivatava, 1974; Foster et al., 1986) or the onset of luteal function (Boros et al., 1986). Second, field studies have been conducted in several parts of the world to estimate whether undernutrition prolongs lactational anovulation and hence intensifies the normal contraceptive effects of breastfeeding (Delgado et al., 1982; Huffman et al., 1987; John et al., 1987). And third, studies have been done in both the field and laboratory to determine if undernutrition adversely affects ovarian function in mature, cycling women, causing an elevated risk of anovulatory cycles, oligomenorrhea, or luteal insufficiency (Graham et al., 1979; Bates et al., 1982; Pirke et al., 1985; Ellison et al., 1986; Lager and Ellison, 1990; Cumming, 1993). As a result of this research, a number of physiologists have come to believe that the reproductive effects of maternal undernutrition are probably large (Ellison et al., 1986; Reid and van Vugt, 1987; Schweiger et al., 1989). Most demographers, in sharp contrast, discount these effects as unimportant (Bongaarts, 1980; Menken et al., 1981b; Gray, 1983; Hobcraft, 1985; Pebley et al., 1985; John et al., 1987) except perhaps in extreme conditions such as famines (Stein and Susser, 1978). Why do these experts disagree—and disagree so vehemently? There are several reasons, reflecting both conceptual and methodological problems.

At the most fundamental level, physiologists and demographers differ widely in what they consider important. In general, demographers dismiss any effect that does not have a clear, measurable impact at the population level, an impact that is large compared with other factors known to affect fertility. For example, several demographic researchers have found statistically significant effects of maternal nutritional status on the duration of lactational amenorrhea (Bongaarts and Delgado, 1979; Huffman et al., 1978a, 1978b, 1987), but these effects have been discounted on the grounds that the time added to the birth interval by undernutrition is small compared with the overall duration of amenorrhea (e.g., a few weeks added to amenorrhea that is already close to two years long). Physiologists, on the other hand, would find such results of considerable interest for what they may reveal about the biological mechanisms linking nutrition and ovarian function. And physiologists of an evolutionary bent would point out that this effect, though small in demographic terms, still represents a decrease in individual fitness that may need explaining. None of these perspectives is inherently incorrect, but each applies to a separate theoretical domain. In other words, "Does maternal nutrition affect fertility in any important way?" is not a single question but rather multiple questions with different possible answers. Physiologists may be perfectly correct to say "Yes" even while demographers are justified in responding "No."

In addition, individuals on both sides of the debate have strong motives for supporting their own hypotheses. For example, an important subset of physiologists studying the nutrition-fertility linkage are the self-styled *reproductive ecologists,* who stress the adaptive nature of the nutritional suppression of reproductive function (Ellison, 1990; Leslie and Fry, 1989; Bailey et al., 1992). In their view, nutritional suppression represents a signal that the current environment is suboptimal for successful reproduction and that it would be prudent to wait until the availability of resources improves. These researchers believe that evolutionary theory provides powerful a priori reasons for believing that nutritional suppression *must* occur. The concerns motivating demographers, in contrast, tend to be more practical in nature. Unlike most physiologists, demographers are often involved in international research, and much of their work has a direct bearing on public policy. These researchers admit privately to being worried about one possible policy-related implication of the nutritional suppression hypothesis— namely, that programs designed to improve maternal nutritional status might be discouraged for fear that they would lead to an increase in population growth. Because of these strong prior motivations, an unfortunate atmosphere has developed in which each participant believes that his or her favorite hypothesis must be correct. This atmosphere has not been conducive to critical thinking about what is, after all, an extremely complex set of issues.

With respect to methodology, the studies done to date by both physiologists and demographers have almost all had important limitations.[3]
One of these is shared by both sides: most studies rely upon what many
nutritionists now regard as problematic measures of nutritional status,
primarily anthropometric measures such as height, body mass, and skinfolds (Habicht et al., 1979; Willett, 1990). These measures can be judged
according to the standard criteria of reliability, validity, sensitivity, and
specificity. Reliability or replicability is a major problem for many
anthropometric measures. For example, stature may vary by several cen-
timeters simply because the subject is not standing perfectly straight,
and measurement of body mass may be influenced by how recently the
scale has been calibrated or how level the ground surface is under the
scale (Willett, 1990). Skinfold measures are notoriously unreliable;
numerous studies have indicated that there is often enormous inter-
observer error, and that intra-observer error in skinfolds is not necessarily negligible (Martorell et al., 1975; Mueller and Martorell, 1988).
These errors occur because it is often difficult to identify the appropriate site on the body for measurement and to avoid including underlying muscle in the skinfold, especially in poorly nourished subjects. Since
reliable measurement requires constant pressure to be applied to the
skinfold by the caliper, errors between instruments can be large, and
pressure can also vary when the same instrument is used unless it is calibrated frequently. New methods for assessing body fat in the field, such
as bioelectric impedance, hold great promise for overcoming these problems, but they still need to be tested for reliability.

Validity is also a serious problem with anthropometric measurements.
Given that anthropometrics are far removed from the pertinent sites of
biological action, what do they actually measure? Anthropometrics may
lag behind the relevant changes in dietary intake by weeks, months, or
even years. Measures such as stature tend to reflect chronic conditions
whereas body mass index (weight/stature^2) and skinfolds are more likely to be affected by acute changes (Willett, 1990). The type of measurement most relevant to reproductive function depends upon the hypothesized mechanisms, which are rarely specified with any precision in
empirical studies. Perhaps most important, anthropometric measures
tend to reflect gross protein-calorie balance, thus missing any possible
effects of micronutrients such as iron or iodine on reproduction. Numerous hints (but few conclusive studies) in the clinical literature suggest
that micronutrient intake may be important for reproduction in females.
For example, chronic iodine deficiency may be associated with an elevated risk of pregnancy loss (Greenman et al., 1962), and hypovita-
minosis E appears to *mitigate* the effects of falciparum malaria on pregnancy loss (Eaton et al., 1976). Despite these intriguing hints, the

possible effects of micronutrients have largely been ignored in population-related research.

Any diagnostic criteria, including those for detecting undernutrition, must be sensitive and specific. That is, they must detect as many subjects suffering the condition as possible while not misdiagnosing individuals who do not have the condition. Many demographic studies of the relationship between nutrition and fertility use a threshold value of weight, body mass index, or summed skinfolds to distinguish well- and poorly nourished women. Often the cutoff point is taken uncritically from previous research, including research on different study populations. But how sensitive and specific is this approach? If the cutoff is set high it is likely to be sensitive since most undernourished subjects will fall below the cutoff, but specificity will be poor since many well-nourished individuals will also be below the threshold—and the converse is true if the cutoff is low. Quite aside from the fact that sharp threshold effects of nutrition on reproduction have yet to be demonstrated physiologically, there is often a high degree of arbitrariness in the choice of threshold. Although it might seem that the use of the same threshold in all studies is a good way to ensure comparability, the actual biological threshold (if it exists) may vary across populations or subgroups within populations because of genetic differences, differences in the distribution of body fat, interactions with local pathogens, and so forth. Other approaches to the diagnosis of undernutrition may also have problems with sensitivity and specificity, but the use of preset threshold values seems especially problematic.

Most of the physiological studies of the nutrition-fertility linkage that have been conducted thus far have involved extremely small and highly selective samples (e.g., anorectic women, marathon runners, ballet dancers, or competition rowers), making it difficult to generalize their results to any larger population. Physiologists have also given insufficient attention to other aspects of study design, ignoring, for example, potential confounding variables such as ethnicity or psychological stress. In addition, almost all the physiological studies have used inappropriate models and techniques in the statistical analysis of their data. Indeed, it is difficult to find any physiological study that does not suffer from crippling statistical shortcomings.

Most important, physiological studies have suffered from a lack of attention to actual fertility outcomes. That is, although these studies may investigate the effects of nutritional status on some aspect of ovarian function (e.g., luteal adequacy or frequency of ovulation), they do not generally follow through to measure the impact on childbearing itself. While this failure may seem puzzling, it is in fact almost inevitable given that most physiologists work with subjects from controlled-fertil-

ity settings. Thus, any quantitative linkage between maternal nutritional status and fertility is largely obliterated by contraception and induced
abortion. Recently, a few researchers have adapted the physiological
approach to the field investigation of natural-fertility populations,
where more dramatic effects might be expected on both demographic
and nutritional grounds. Ironically, however, the one project that pio-
neered this reorientation and that has done the most thorough work on
this issue, the Harvard Ituri Forest Project in Zaire, happened to select
study populations with prevalences of pathological sterility that are
high even by central African standards: the Efe and Lese of the Ituri Forest have primarily sterility rates of almost 40 percent and total fertility
rates of only about 2.4, apparently reflecting widespread infection by
gonorrhea and chlamydia (Ellison, 1986). Here, too, it is well-nigh impossible to investigate the effects of ovarian suppression on fertility,
simply because there is not much fertility to study.[4] Since studies of
other natural-fertility populations have only been initiated quite recently, it is too early for them to have produced any definitive results.

Although previous demographic research has had the merit of working directly with fertility itself, or at least with the proximate determinants of fertility, it has suffered from other methodological shortcomings (see the review by Popkin et al., 1993). Most earlier studies, for
example, used cross-sectional data to test what are inherently dynamic
relationships, thus missing the importance of time-varying covariates
such as changes in dietary intake, as well as simultaneity problems such
as the feedback effects of reproduction on the nutritional status of the
mother. Only recently have demographers applied dynamic hazards
models to longitudinal observations (John et al., 1987; Popkin et al.,
1993; Jones and Palloni, 1993). Although sample selectivity has been less
of a problem in this research than in the physiological studies, much
demographic research has still paid insufficient attention to important
confounders, such as infant feeding patterns, mother's dietary intake, or
mother's activity level. Finally, in only a few cases have demographers
attempted to adjust for such modeling complexities as unobserved heterogeneity and the endogeneity of nutritional covariates (John et al.,
1987; Popkin et al., 1993).[5]

A final reason that physiologists and demographers have reached dia-
metrically opposing conclusions about the influence of maternal nutrition on reproduction is conceptual. Physiological and demographic
studies have been based upon partial (and only partially overlapping)
models of the relevant biological processes. In each case the preferred
model is usually quite simple, envisioning a single set of intervening
variables linking nutritional status and fertility (Figure 12.5). In demographic models, the intervening variables are a select set of proximate

determinants, such as fecundability or postpartum infecundability; because of the close relationships of the proximate determinants to fertility itself, and because of the availability of well-validated mathematical models specifying those relationships, once linkage 1 or 2 in Figure 12.5 (top panel) has been estimated, then linkage 3 or 4 follows immediately. This simple scheme, however, ignores the multiplicity of biological factors that may intervene between nutritional status and the proximate determinants; failure to include those factors leaves the demographic approach vulnerable to multiple confounding influences. For physiologists, the central intervening variable is usually some aspect of ovarian function (Figure 12.5, bottom panel), typically assessed through peripheral hormone levels. And at least some physiologists differ from almost all demographers in including the simultaneous effects of activity level on ovarian function. However, while many physiological studies have attempted to estimate linkage 1 or 2 in Figure 12.5 (bottom), little attention has been paid to linkage 3 for the reasons outlined above. Unfortunately, existing theory does not allow this linkage to be modeled with precision; as a consequence, it is difficult to translate the physiological findings about ovarian function into precise quantitative predictions about fertility itself.

Reality, of course, is far more complex than either of these models suggests, and Figure 12.6 is an attempt to summarize a few of the complexities. In the real world, nutritional status as measured by anthropometrics or clinical signs of undernutrition operates simultaneously with dietary intake and activity level. These three factors are highly correlated, both because of direct biological linkages and because of the prior effects of other, usually unmeasured factors ("the rest of the world" in Figure 12.6). Any statistical model must allow for these types of correlations. Nutritional status, dietary intake, and activity level may have multiple effects on different aspects of ovarian function, some of which may be independent but many of which are almost certain to interact; it may be extraordinarily difficult to disentangle those effects statistically. Ovarian function, in turn, acts upon several different proximate determinants simultaneously, including fecundability and lactational infecundability but also, perhaps, fetal loss. Although these proximate determinants translate more or less directly into effects on fertility, the precise translation may be confounded by unobserved heterogeneity unless appropriate models are used. Finally, a variety of fertility outcomes feed back upon a woman's activity level, dietary intake, and nutritional status. For example, an additional child may require an increase in a woman's subsistence labor, pregnancy may involve food taboos, or prolonged lactation may partially deplete a woman's fat reserves (Adair et al., 1983; Miller and Huss-Ashmore, 1989). These feed-

back relationships may result in simultaneity effects that make it difficult to distinguish cause from consequence, especially when using cross-
sectional data.

And in fact, Figure 12.6 is still only a partial model of the complexities. Figure 12.7 shows one recent attempt to elaborate the multiple relationships between dietary intake and ovarian function (Tracer, 1991).
There is some experimental evidence for each of the various components
of this model, although none has been established unequivocally. The
point, however, is not whether this model is correct, but rather that reality is likely to be *at least* this complicated. In view of these complexities,
and in view of the fact that each research group has of necessity
addressed only a tiny subset of the likely linkages, it is unsurprising that
no very clear picture has yet emerged.

In sum, a strong linkage between maternal nutritional status and fertility would be of enormous importance, both practical and theoretical,
if it could be established convincingly. The definitive refutation of such
a linkage would be of equal significance. The studies done to date, however, have been so deficient—and this includes both demographic and
physiological studies—that no firm conclusions are as yet possible save
that severe malnutrition probably does have an effect. As far as more
subtle forms of undernutrition are concerned, the question remains
open.

As a final point, I would note that the reproductive ecologists are
working under an additional burden, for they have to show not only
that nutrition has an important effect on fertility, but also that the effect
is of adaptive significance. This, it seems to me, will be extraordinarily
difficult to establish, even if it is generally true. It is easiest to make a
case that some system or process is adaptive when it has a clear positive effect on either fertility or mortality, the two components of genetic fitness. Nutritional suppression, however, involves a *negative* effect on
fertility, one that to be adaptive must be countered by an even larger
positive effect at some later time. Until this later benefit is demonstrated convincingly, it will be uncertain whether nutritional suppression (if
it occurs) represents an adaptation or a *failure* of adaption.

In this connection, there is an important cautionary tale to be drawn
from the study of growth and development. It has long been known that
extreme undernutrition during childhood is associated statistically with
short adult stature (Willett, 1990). During the 1960s and 1970s, several
researchers suggested that this association might be adaptive, since
short individuals may survive better than tall ones under conditions of
resource limitation because they require a smaller dietary intake (Malcolm, 1970). For a while, this hypothesis had considerable appeal based
largely upon a priori evolutionary reasoning. It was not until several

well-designed prospective studies showed unequivocally that nutritional stunting is associated with *increased* mortality at subsequent ages (Kielmann and McCord, 1978; Chen et al., 1980; Heywood, 1982) that this idea was finally laid to rest. Clearly, nutritional stunting is pathological, not adaptive. Is the same perhaps true of the nutritional suppression of ovarian function?

This question, which lies at the very heart of reproductive ecology, may be profoundly difficult to answer. Any adaptive effects of nutritional suppression must, by their nature, be long-term: according to the ecological hypothesis, a woman forgoes an opportunity to reproduce now in order to increase the likelihood of successful reproduction at some point in the future. Therefore, these relationships must be studied using long-term prospective data. Since any positive effects of nutritional suppression are very unlikely to appear wherever birth control is widespread, they must also be studied in natural-fertility settings. But the few remaining parts of the world where something like natural fertility still exists are the very ones where, for logistical reasons, it is most difficult to accumulate high-quality prospective data. And yet, until such data are collected, the theoretical foundations of reproductive ecology will remain untested.

Seasonal Factors

One dimension of fertility variation that appears to be of very general importance in human populations is seasonality: virtually everywhere the necessary data are available, monthly birth rates appear to vary in fairly regular patterns over the course of the year (Cowgill, 1966; Scaglion, 1978; Huss-Ashmore, 1988; Becker, 1981; Lam and Miron, 1991). In the few cases in which seasonal variation has not been detected, insufficient statistical power and inadequate observation periods may be to blame. Seasonal variation in births is found in both natural- and controlled-fertility settings and in both rural and urban environments. As two researchers have recently put it, "The fact that births in April are 31 percent greater than births in December in contemporary urban Sweden, for example, is a dramatic result that begs for an explanation" (Lam and Miron, 1991:76).

Accordingly, demographers, anthropologists, and physiologists have become very interested in providing an explanation of human birth seasonality. Their task is made more difficult by the fact that no single pattern of seasonality exists; instead, there seem to be several different regional patterns, some with a single peak and trough each year, others with multiple peaks and troughs. In the course of recent research, two major questions have emerged. First, is there a reasonably small number of seasonally varying factors that can account for most or all of the observed patterns of human birth seasonality, or are there so many different factors involved that each locale is essentially sui generis in its particular set of causative agents? Second, is seasonality generally *adaptive*, relating births to fluctuating resources in advantageous ways, or is it an epiphenomenon, an accidental by-product of other factors that happen to vary seasonally? The first question has turned out to be much easier to answer than the second.

Early research on birth seasonality held out the promise that simple explanations might work. For example, the discovery of unusually marked seasonality at high latitudes (Condon, 1982; Ehrenkranz, 1983) suggested to some researchers that photoperiod might be important in humans, as it is known to be in other mammalian species.[6] Similarly, the fact that the seasonal pattern of births in white settlers in New Zealand and South Africa is approximately the mirror image of that observed in Great Britain (Parkes, 1976) implied the involvement of simple environmental factors such as photoperiod or ambient temperature, whose yearly variation in the two hemispheres is offset by six months. In retrospect, these simple, global explanations remained plausible only so long as data were limited. As information became available on other populations throughout the world, the illusion of simple global patterns evaporated.

Before describing the patterns that are actually observed, it is necessary to mention some important analytical problems that may confound the detection and characterization of birth seasonality. First, many empirical studies report monthly changes in the *number* of births rather than in birth rates. This can be misleading unless corrections are made for possible seasonal changes in the number of women exposed to the risk of childbearing. For example, Wilmsen (1978) has reported marked seasonal fluctuations in the number of births occurring to !Kung women at one particular bush camp in northern Botswana. His results are difficult to interpret, however, because we are not told how many women were living at the camp each month, a special concern in a nomadic population such as this. In general, it is better to report monthly birth *rates*, adjusted appropriately for the base population at risk.

Second, data from a single year of observation are inadequate for characterizing seasonality. Seasonal patterns are by their nature recurrent, and multiple years of data are needed to show recurrence convincingly. In a single year, unique events such as wars, famines, or economic downturns can mask the normal pattern of seasonality or even induce *apparent* seasonality where none exists. At the same time, data from multiple years of observation should not be pooled uncritically. This is especially important when population sizes or fertility rates are changing rapidly, for unique events in years when the base population is large or fertility high will tend to swamp the pattern in other years.

A final problem has to do with the statistical tests chosen to detect birth seasonality. Most procedures used to test for seasonality actually test for departures from a uniform distribution of births across the year, not for regular cyclic variation. Consequently, these tests do not distinguish two very different situations (Figure 12.8): one in which a single portion of the year (e.g., a single month) has a significant excess or deficit of births while the rest of the year is more uniform, the other in which births vary regularly from month to month as might be described by a sine wave or some other periodic function. Clearly, these two situations are likely to have very different implications for our inferences about underlying causation. To detect regular, cyclic variation of the second kind, special statistical procedures such as Fourier analysis are required (Becker, 1981; Leslie and Fry, 1989).

Recently, a major study has been undertaken to estimate worldwide patterns of human birth seasonality (Lam and Miron, 1987, 1991). This study has drawn together as many sets of reliable observations as possible, from all regions of the world and many different periods, and has applied a consistent set of statistical analyses to them. The analyses are based on the following regression model:

$$\hat{B}_t = \hat{S}_t \hat{T}_t \hat{X}_t \tag{12.2}$$

$$\hat{S}_t = \exp\left(\sum_{m=1}^{12} \hat{\alpha}_m d_{mt}\right) \tag{12.3}$$

$$\hat{T}_t = \exp(\hat{\beta}_1 t + \hat{\beta}_2 t^2 + \hat{\beta}_3 t^3 + \hat{\beta}_4 t^4) \tag{12.4}$$

where B_t is the number of births per day in month t, S_t is the seasonal component of B_t, T_t is the "trend" component of B_t (e.g., associated with changing population size or declining fertility rates), and X_t is the "random error" (nonseasonal, nontrend) component of B_t. (The " ^ " indicates an estimated value.) In equation 12.3, d_{mt} is a seasonal dummy variable for each month; for example, d_{1t} is equal to one if an observed birth is in January and zero otherwise. The estimated value of α_m therefore gives the magnitude and direction of the seasonal factor in B_t for month m. If $\hat{\alpha}_m = 0$ for a particular month, then the number of births in that month is equal to the untrended mean monthly number over the period of observation as a whole. Note that while equation 12.2 is written in terms of the absolute number of births each day rather than the birth rate, estimation of T_t adjusts for changes in base population and fertility level, at least up to a fourth-degree polynomial. Note, too, that no restrictions are placed on how the values of α_m may vary; thus, no assumption is made that they follow a smooth pattern, such as a sine curve. However, since α_m is estimated month by month, graphical display of it will reveal any pattern that may exist.

Application of this model to high-quality demographic data has produced several interesting results (Lam and Miron, 1987, 1991). Significant seasonality (as indicated by at least one $\hat{\alpha}_m$ value differing significantly from zero at the 0.01 level) has been detected in every single case. Generally speaking, the seasonal component explains well over half, and often as much as 80 percent, of the nontrend variation in B_t. Peak-to-trough amplitudes are almost always greater than 10 percent and are often as high as 30 percent, even in urban, controlled-fertility settings. Thus, these analyses confirm that seasonality is a ubiquitous, nontrivial component of human fertility variation.

When more detailed patterns are examined, however, the results become less clear-cut. There is substantial geographical variation in seasonality, even at the same latitude and within the same country (Figure 12.9). The presence of a particular pattern is not determined in any obvious way by the hemisphere in which it occurs, nor does seasonality consistently follow ambient temperature or photoperiod (Figure 12.10). Wherever the data allow disaggregation, there are surprisingly few differences in seasonal patterns across urban/rural sectors or legitimate/illegitimate births within the same country (Figure 12.11). (Ethnicity and income produce conflicting results but generally also have small effects.) The overall impression is that no simple global patterns exist such as might be expected from climate or photoperiod. Each region seems to have its own idiosyncratic pattern. And the absence of consistent urban/rural differences suggests that these patterns are not, as a general rule, closely tied to seasonal variation in work patterns or other economic factors.

Perhaps the most striking results of these analyses have to do with time trends in seasonality. When historical demographic data are available, several regions display a remarkable persistence in the pattern of seasonality over prolonged periods (Lam and Miron, 1991). In the case of England, for example, the locations of the peaks and troughs persist over at least four centuries, although the amplitude of the seasonality appears to decline fairly consistently with time (Figure 12.12).[7] Such persistence is deeply mysterious, especially when coupled with the absence of an important urban/rural difference. What could a London shopkeeper in 1983 possibly have in common with a Devonshire ploughman in 1638 other than sheer nationality? Why should both display a primary birth peak in March and a secondary peak in September (Figure 12.12)? Such persistence seems to mock any attempt to explain seasonality in simple ecological terms.

Based upon their analyses, Lam and Miron (1991) suggest that it is probably impossible either to rule out or to support most hypotheses about the causes of seasonality simply by examining the overall patterns of seasonal variation in births. Instead, they advocate what is in effect a proximate-determinants approach, using data on variables other than birth itself and linking those variables to birth with appropriate mathematical models (for a good example, see Lam et al., 1994). Other authors have pointed to a wide variety of possible determinants of birth seasonality, including behavioral factors affecting coital frequency, such as marriage and spousal separation (Udry and Morris, 1967; Menken, 1979; Huss-Ashmore, 1988; Underwood, 1991).[8] Possible physiological influences have also been cited, including factors predisposing to fetal loss (Warren et al., 1986), variation in the age at menarche (Rodgers and Buster, 1994), temperature-dependent changes in plasma prolactin (Mills and Robertshaw, 1981), and variation in sperm count (Tjoa et al., 1982; Levine, 1994). Reproductive ecologists have pointed out that maternal nutritional status often varies with season in preindustrial societies and have attempted to link such variation with birth seasonality via changes in luteal function (Bailey et al., 1992). Other ecologically minded anthropologists have tried to show a relationship between resource availability and birth seasonality by associating conceptions with variation in rainfall or temperature (Condon, 1982; Leslie and Fry, 1989; Bailey et al., 1992).

It is important to emphasize that all these "explanations" use data aggregated at the population level. To pick one example, the Turkana study of Leslie and Fry (1989) relates the total number of births in any month to the total precipitation occurring at some earlier time; it does not attempt to relate how much rainfall each individual Turkana woman experiences to her own chances of childbearing. Similarly, in the Ituri Forest study of Bailey et al. (1992), populationwide seasonal changes in luteal function are said to parallel populationwide variation in inferred conception rates, but no attempt is made to associate the luteal function of each individual woman with her actual waiting time to conception. Unfortunately, studies that draw inferences about individual-level risks from population-level data are vulnerable to what epidemiologists call the *ecological fallacy*. This is a common error of inference that reflects a failure to distinguish among different levels of analysis (Last, 1988:40). Correlations based upon group characteristics ("ecological correlations," as epidemiologists call them) need not be reproduced, and may well be reversed, when the same variables are examined at the individual level (for a mathematical proof, see Robinson, 1950). The classic case of an ecological fallacy involves the relationship between pulmonary tuberculosis and overcrowding (Cowles and Chapman, 1935). At the community level (specifically, at the level of the U.S. census tract), a strong positive association between these two variables has been observed, leading to an inference that overcrowding contributes to the risk of tuberculosis. This relationship is, however, spurious, reflecting the confounding effects of socioeconomic status, which varies widely among census tracts. *Within* tracts, it turns out that individuals with tuberculosis are actually more likely to be living in less-crowded circumstances than their uninfected neighbors (Cowles and Chapman, 1935).

The ecological fallacy is a special danger for studies of temporal variation, for any two variables that change over the same time period may be correlated whether they are related or not. In one of the earliest applications of time-series analysis, the British statistician G. Udny Yule (1906) found an extremely close association between the number of births in England and Wales over the final decades of the nineteenth century and the estimated number of storks in those same countries. Presumably, the availability of storks has no direct causal effect on the number of babies born—not unless the previous eleven chapters of this book have been seriously mistaken. More likely (I hope), both changes were somehow related to trends in the industrializing and urbanizing European economy, trends simultaneously affecting the number of nesting sites for storks and the desirability or feasibility of contraception.

Yule's cautionary analysis should be kept firmly in mind by any investigator attempting to relate two variables that change with the seasons. Simply stated, *everything* varies seasonally, from rainfall and photoperiod to prices and employment levels to mobility patterns and marriage rates to preferred colors of clothing and the number of locally nesting nuthatches. Some of these variables may have direct causal links to birth seasonality, but many do not. However, all seasonally varying measures will be correlated, positively or negatively, with birth seasonality at the population level simply because they *are* seasonally varying, and the correlations may become quite high if our statistical model allows for time-lags between variables. But to infer causality from such correlations is to risk committing the ecological fallacy. Thus, the fact that the number of births occurring each month among the Turkana is positively correlated with the amount of precipitation 3.6 months before the inferred date of conception (Leslie and Fry, 1989) by itself demonstrates no necessary causal link between rainfall and conception. Similarly, the fact that the nutritional status of Lese women in the Ituri Forest varies in parallel with the amount of progesterone secreted during the luteal phase (Bailey et al., 1992) would only be indicative of a causal connection if other factors that vary seasonally could be discounted.

There are two general strategies for avoiding the ecological fallacy. The first is to link two seasonally varying measures by a well-validated mathematical model that makes biological sense (Lam et al., 1994). If the observed numerical relationship corresponds closely to that predicted by the model, then the evidence for a genuine causal connection is strengthened. A simple example can be drawn from household data on the Old Order Amish in rural Pennsylvania (LeFor et al., 1993). Amish fertility is only mildly seasonal, and measurable seasonality is largely confined to the first birth within each marriage; there is, however, a marked degree of seasonality in the distribution of marriages, representing an extreme example of an old European pattern (Hostetler, 1980; Bradley, 1970). Is the seasonality of marriages sufficient to explain the seasonality of first births? An inspection of the data in Figure 12.13 suggests that there may be a connection: the peak in births is roughly eight to ten months after the peak in marriages. But is this merely an ecological association without any causal basis? Or does the precise numerical pattern of births conform to what is expected from the seasonality of marriages? As it happens, we have a mathematical model to link the two: the model of the distribution of waiting times from marriage to first fertile conception discussed in Chapter 7. Because some fraction of Amish couples may be sterile at the time of marriage, we use the version of the model that adjusts for sterility (see Chapter 10). As we show elsewhere (Wood et al., 1994a), the estimated mean effective fecundability of fecund Amish couples is 0.246, the estimated variance in effective fecundability is 0.025, and the estimated fraction of couples sterile at the outset of marriage is 0.026. Substituting these values into the model-predicted probability density function for times to first birth, and combining that PDF with the observed distribution of marriages, we obtain the broken line in Figure 12.13. According to a conservative test for goodness-of-fit, the correspondence between the predicted and observed numbers of births each month is excellent (log-likelihood ratio test statistic = 9.98, df = 9, $0.25 < P < 0.50$). This is about as convincing a connection between seasonally varying measures as could ever be established using population-level data.

A second and more powerful strategy for avoiding the ecology fallacy is to limit the analysis to individual-level data. Perhaps the single best investigation of seasonality to adopt this strategy has been carried out in Matlab, Bangladesh. Based upon sophisticated statistical models, especially hazards models, it appears that birth seasonality in Matlab may be attributable partly to seasonal changes in breastfeeding patterns, which are associated in turn with variation in the availability of weaning foods and in women's work patterns (Huffman et al., 1980; Becker et al., 1986). There may be additional, though less important, effects of seasonal spousal separation associated with occupations like fishing that take men away from their homes for certain parts of the year (Chen et al., 1974).[9] The Matlab project is striking in its uniqueness: no study in any other part of the world has done such a thorough job of investigating birth seasonality using large-scale survey techniques, appropriate statistical models, and individual-level data. The Matlab study thus provides a paradigm for how this work ought to be done.

The larger question about birth seasonality is whether it represents an adaptive response to fluctuating resources or is merely an accidental by-product of other factors that happen to vary seasonally. Unfortunately, this is a difficult question to answer: all the obstacles facing the demonstration that nutritional suppression of ovarian function is adaptive apply with equal force to seasonality. Still, a few observations are possible. Lam and Miron's study shows convincingly that particular populations or regions tend to have their own idiosyncratic patterns of seasonality (Lam and Miron, 1991). Perhaps those idiosyncrasies reflect the vagaries of local resource fluctuation, but that supposition is far from established. The absence of a clear urban-rural difference in seasonality as well as the persistence of seasonality after the industrial revolution (Lam and Miron, 1991) would seem to argue strongly that resource availability is *not* an important general determinant. In many cases, as in the holiday effect and the seasonality of Amish marriages, local cultural factors seem to predominate. Perhaps these factors are indirectly linked to resources; the Amish, for example, say that marriages are clustered in the late fall because at other times of the year women are too involved in farm work to have the time needed to prepare food for weddings (Hostetler, 1980). But this linkage to the agricultural cycle is one of convenience rather than necessity, as indicated by the fact that other agriculture communities schedule weddings differently. On balance, there is little reason to suspect that seasonality is adaptive in most cases, although we cannot be certain until we know more about the mechanisms responsible for it. Human birth seasonality may yet turn out to be less interesting from an ecological perspective than many researchers have supposed.

Fertility-Mortality Interactions

Scrimshaw (1978) has suggested that the ultimate regulator of fertility levels in preindustrial societies is not reproductive biology, much less reproductive decision-making, but rather death—or at least death of a certain kind. According to this view, which I believe has considerable merit, the level of fertility is determined in important ways by early childhood mortality, which in turn is influenced by fertility. Two important interactions between fertility and mortality are relevant here: When a nursing child dies, lactation is terminated and the mother resumes ovulating and becomes pregnant sooner than she otherwise would have, a phenomenon known as "reproductive compensation"; thus, high levels of early childhood death can result in increased fertility. Conversely, when a child's nursing is disrupted by the birth of the next child and the nutritional and immunological benefits of human milk are withdrawn, it may be placed at an elevated risk of death (what might be called "nursing competition"). Depending upon their strength, these interactions or "trade-offs" can be important determinants of reproductive success and may be largely responsible for the evolution of the distinctive pattern of human reproduction.

Several attempts have recently been made to measure these interactions empirically (Hobcraft et al., 1983; Trussell and Pebley, 1984; Palloni and Tienda, 1986; Jones and Palloni, 1994). However, a sticky simultaneity problem often makes it difficult to separate reproductive compensation from nursing competition: on the one hand, childhood mortality affects birth spacing; on the other, birth spacing affects childhood mortality. As shown in Figure 12.14, the only way to untie this knot is to obtain precise information on the timing of death relative to the resumption of ovulation, weaning, and conception. This requirement almost necessarily demands collection of prospective observations. The analyses that have thus far been able to exploit such observations suggest that both reproductive compensation and nursing competition are likely to be important in many, perhaps most, natural-fertility settings, although the precise magnitude of each effect may vary substantially across environments. I suggest that elucidating the role of these processes in determining fertility levels should be a major goal of future research.

The Genetics of Reproduction

One area of fertility research that has received far too little attention is the genetic analysis of reproductive traits. Despite demographers' longstanding interest in fertility variation, few attempts have been made to estimate the contribution of genetic differences to variation both within and among human populations. For a few important determinants of fertility, such as menarche and fetal loss, genetic influences are already well-established (Kline et al., 1989; Meyer et al., 1991). There is no reason to think that genes do not play some role in the other determinants as well, except perhaps the age at marriage. (Even there, genetic effects on the age at menarche may play some indirect part.) In addition to helping explain fertility variation, studies of the genetics of reproduction could be enlightening for at least two other reasons. First, studies of quantitative genetics can be informative about patterns of natural selection that have been operating on the traits involved, a point that should be of special interest for reproductive ecologists. Second, identification and characterization of specific genes affecting reproduction—e.g., genes for peptide hormones, for the enzymes involved in steroidogenesis, or for hormone receptors—can play a central role in elucidating the basic physiology of reproduction. Compared with the work that has been done on the genetics of chronic disease and disability, however, research on traits of reproductive significance has lagged far behind.

As it turns out, there is an important difficulty in studying the genetics of reproduction, one that is by no means insurmountable but that may partly explain the dearth of previous research on this topic. It has to do with linking quantitative-genetics models to the *timing* of reproductive events, the variable of greatest interest to demographers. Geneticists have long been interested in phenotypic traits that vary among individuals in their timing or age at onset, such as the neurological disorder Huntington's chorea. Reproductive traits that fall under the same heading would include much of the subject matter of this book, for example, the time to first conception, the time to first postpartum ovulation, or the age at menopause. How do we estimate familial effects when the timing or age at onset of a reproductive event is the phenotype of interest? The naive solution is to treat the time variable or the individual's age of onset as the measurable phenotype and then correlate relatives' times or ages of onset to estimate genetic effects. For the most part, this approach has produced negative results, probably reflecting two general problems. The first of these problems is censoring: in some pairs of relatives, one or both members may not have experienced onset of the phenotype of interest by the time of the study or may drop out of the study before experiencing it. More fundamentally, this approach appears to be lacking in statistical power: stochastic variability in the timing of vital events appears to be great enough to swamp any correlation among relatives. This latter issue has been explored in detail by Vaupel (1988) using a proportional hazards model specified as

$$h_i(t) = z_i \cdot h_c(t) \tag{12.5}$$

where $h_c(t)$ is the "common" component of the hazard shared by everyone and z_i is the "liability" component specific to the ith individual. (In genetic terms, $h_c(t)$ may reflect genes that are fixed in the population, i.e., shared by everyone, while z_i may reflect polymorphic genes.) Under this model, all individual-level variation in the hazard arises from variation in the z-values; consequently, any correlation between relatives must itself be attributable to a correlation in z. Now, the correlation between relatives 1 and 2, say, in the actual age at onset or time to occurrence of the event of interest is

$$\rho(T_1, T_2) = \frac{\text{cov}(T_1, T_2)}{\sqrt{\text{var}(T_1) \cdot \text{var}(T_2)}} \tag{12.6}$$

The fundamental statistical problem here is that, although the covariance term in this expression is entirely attributable to the correlation in z-values, the two variance terms include variability both in z and inherent in $h_C(t)$. Unfortunately, for most realistic specifications of $h_C(t)$ the variance in T is extremely large and overwhelms any signal associated with z (Vaupel, 1988).

The problem, then, is how to estimate correlations in z, which cannot be measured directly, instead of in T, which can. Recently, Vaupel (1990a, 1990b) and Thomas et al. (1990) have provided methods appropriate for twin studies, based on the idea that monozygotic twins share identical positions on the liability distribution. A more general approach, applicable to relatives of any degree of relatedness, has been developed by Meyer and Eaves (1988) and Meyer (1989). The logic of this approach is as follows: Suppose we are working with a hazards model that takes the proportional-hazards form of equation 12.5. Following standard practice in statistical demography (Clayton, 1978; Vaupel et al., 1979; Oakes, 1982, Clayton and Cuzick, 1985; Gage, 1989; Vaupel, 1990a, 1990b), we might assume that z is distributed among individuals according to a gamma density function. Now, if we actually knew the value of z_i for the ith individual in our sample, then by analogy with equation 3.12 in Chapter 3 the likelihood of the ith observation would be

$$l_i(\theta \mid z_i) = f(t_i \mid z_i, \theta)^{d_i} S(t_i \mid z_i, \theta)^{1-d_i} \tag{12.7}$$

where θ is the set of parameters contained in $h_C(t)$, $f(t)$ and $S(t)$ are the density and survival functions associated with $h_C(t)$, and d_i is the censoring index. By its nature, however, z_i is unobservable. Still, observations on relatives may provide information about it. The total likelihood for data on n pairs of relatives is

$$L = \prod_{i=1}^{n} \int_{-\infty}^{\infty} \int_{-\infty}^{\infty} \phi(z_{i1}, z_{i2}; \rho) l_{i1}(\theta \mid z_{i1}) l_{i2}(\theta \mid z_{i2}) dz_{i1} dz_{i2} \tag{12.8}$$

where $\phi(z_{i1}, z_{i2}; \rho)$ is the bivariate gamma density function with mean one and correlation ρ (see Johnson and Kotz, 1973:217–218). Although this likelihood function may look formidable, it can be maximized numerically on the computer using standard algorithms.

With data on informative pairs of relatives, we can estimate ρ as a general measure of *familial aggregation* to see if z tends to "run" in families. Further, if certain simplifying assumptions are made, ρ can be partitioned into genetic and environmental sources of shared variation as shown by Meyer (1989). In one of few quantitative-genetic studies ever done on a human trait of reproductive interest, Meyer et al. (1991) applied this method to data on the age at menarche in 1,889 Australian twins (1,178 monozygotic and 711 dizygotic). They found not only that significant heterogeneity exists among women in the age at menarche, but also that this heterogeneity contains a strong familial component ($\hat{p}$-values > 0). Further partitioning of the familial aggregation suggests that the strict-sense heritability of menarche is about 0.17 and that strong dominance effects exist, of the sort that might be expected of a trait under sustained directional selection. As we have argued elsewhere (Wood et al., 1992), it is not clear that the specification of $h_c(t)$ used in their analysis is at all appropriate for the biological mechanisms underlying menarche. However, this lack of biological realism for the particular case of menarche does not detract from the more general usefulness of this approach. In particular, I believe that the likelihood function in equation 12.8 or something very like it should provide a powerful and general way to estimate familial aggregation for hazards models of reproductive processes. It can readily be modified for other genetic models, for example, a three-point distribution of liability consistent with a major-gene effect. In general, this approach promises to be a fruitful way to unite fertility analysis and genetics.

Statistical estimation of familial aggregation does not, of course, exhaust the interesting genetic questions to be asked about reproduction. Once we have established that such aggregation exists, the goal then becomes to identify and characterize specific genes affecting reproduction, to ascertain their chromosome location within the human genome, and to learn how their protein products function physiologically. Although these questions have only recently been formulated for reproductive processes, much of the necessary molecular and statistical methodology for doing this sort of work already exists (Ott, 1991; Khoury et al., 1993; Weiss, 1993), and indeed the genes for several proteins involved in reproduction, including LH and FSH, steroid receptors, and enzymes involved in steroidogenesis, have already been cloned, sequenced, and mapped (Pinsky and Kaufmann, 1987; Gwatkin, 1993). I predict that over the next decade, genetic investigations of human reproduction will take their proper place among the established approaches to fertility research.

SOME FINAL THOUGHTS

As we learn more about the beautifully tuned neuroendocrine system that controls reproduction, it becomes increasingly important to understand how this system operates in time. A central argument of this book has been that the temporal dynamics of the reproductive system must ultimately be studied in natural-fertility settings, where the pace of childbearing is least affected by contraception and induced abortion. The collection and analysis of field data on natural fertility will require an unprecedented collaborative effort, one that crosses traditional academic boundaries to draw together demographers, anthropologists, reproductive biologists, epidemiologists, geneticists, and biostatisticians. And such work must be done soon, because modern methods of fertility control are spreading every day to more and more parts of the world. This trend is unquestionably a desirable one since it increases the ability of people around the world to make intelligent decisions about a fundamental aspect of their lives; indeed, it could be argued that researchers in human fertility have a special obligation to encourage the spread of rational family planning. Nonetheless, broadening the access to contraceptives will eventually lead to the disappearance of natural fertility, except perhaps in a tiny number of very special and unrepresentative cases. The current cohort of researchers may well be the last one in human history to have an opportunity to study a wide range of relatively unacculturated natural-fertility populations. Their work should therefore be accorded the highest possible priority within fertility research.

This book has tried to focus attention not only on what is known about the proximate determinants of natural fertility, but also what is not yet known. While the basic facts about the proximate determinants—their natural history, as it were—are now fairly well established, much remains to be learned about how those determinants act in concert to govern fertility variation. And there is far more left to be learned about the influences of more remote factors that ultimately explain such variation. The proximate determinants approach, in conjunction with the kinds of dynamic birth-spacing models highlighted in this book, provides a powerful set of tools for linking these more remote influences to variation in human fertility. Although analysis of the remote influences will undoubtedly introduce complexities far beyond the scope of the present book, the ingenuity displayed thus far by researchers on human fertility gives hope that those complexities will ultimately be solved.

NOTES

1. In the Monte Carlo approach, a specific age or time at a given reproductive event is produced for an individual woman using a pseudorandom number generator in combination with the *inverse distribution function* associated with the

PDF for that event (for technical details, see Christensen, 1984:263–264). Although many of the PDFs in Table 12.1 do not have explicit, closed-form inverse distributions, various methods exist for approximating them (Bratley et al., 1983). These methods allow us, for example, to enter the lognormal distribution of ages at menarche at random to select a specific age at menarche for assignment to the ith simulated maternity history. We then proceed similarly for every other event making up the ith reproductive career.

2. Preliminary analyses suggest that fecundability may be more sensitive to the way intercourse is distributed across the cycle than to the average frequency of intercourse per random cycle day.

3. This discussion owes a great deal to the excellent reviews by John (1993) and Popkin et al. (1993).

4. This is a special instance of the problem noted in Chapter 4, that ovarian function is difficult to study in natural-fertility populations because most women who are exposed to the risk of conception are either pregnant or anovulatory because of breastfeeding. Widespread sterility among the Efe and Lese is precisely what makes it possible to study ovarian function in a large sample of women; unfortunately, it also makes it impossible to link changes in ovarian function to fertility. This problem is a profound and perhaps unavoidable one for this whole area of research.

5. *Endogeneity* refers to a situation in which some portion of the error structure in a regression model includes disturbances in the predictor variables because the predictors are related to each other by some implicit system of simultaneous equations or by the effects of other variables not included in the analysis (Spanos, 1986:608); failure to adjust for endogeneity can introduce substantial biases into the estimated regression coefficients. The potential biases arising from a failure to control for *unobserved heterogeneity* were discussed in Chapter 3.

6. *Photoperiod* refers to the duration of daylight during the diurnal light-dark cycle. Photoperiod varies predictably across the year, with an amplitude that increases with latitude: there is very little seasonal variation in photoperiod near the equator, much more near the poles.

7. This pattern of persistence is not uncommon, although in some cases the amplitude may remain unchanged and in at least one case (Sweden) it actually appears to increase in more recent periods.

8. A popular variant involves a "holiday effect," wherein increased leisure time associated with a seasonal holiday such as Christmas is followed nine months later by a birth peak (Rosenberg, 1966; Wrigley and Schofield, 1981). However, examination of the observed patterns (Figures 12.9–12.12) suggests that such holiday-related peaks are unlikely to account for more than a small fraction of the overall seasonal variation in births.

9. Breastfeeding and spousal separation do not entirely explain the observed birth seasonality in Matlab (McElrath, 1993). Researchers at Matlab have recently begun to investigate possible seasonal variation in luteal function (S. Becker, J. Menken, personal communication) and fetal loss (D. Holman, personal communication), but it is too early to tell what the contributions of those factors may be.

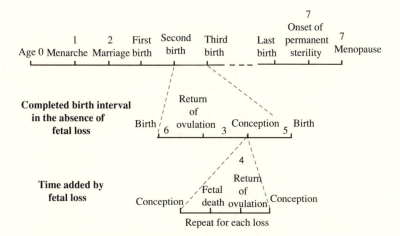

Figure 12.1. Components of the female reproductive life course, keyed to the proximate determinants in Table 12.1. (Adapted from Bongaarts and Potter, 1983:4)

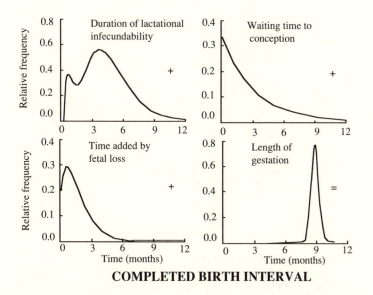

COMPLETED BIRTH INTERVAL

Figure 12.2. Adding up the components of the completed birth interval. Each completed interval between successive live births must include a time to first postpartum ovulation (the period of lactational infecundability), a fecund waiting time to the next conception, and a complete gestation. In addition, some time may be added to the interval by the occurrence of one or more fetal losses. The distributions shown here are all taken from Table 12.1. (Redrawn from Wood et al., 1992)

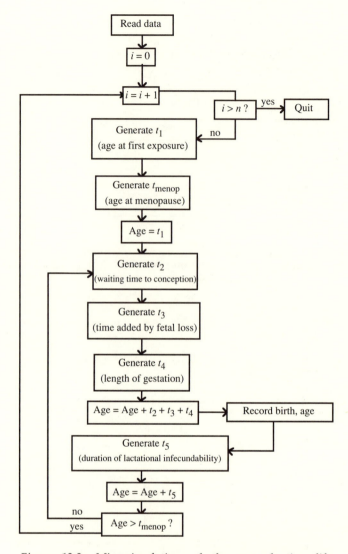

Figure 12.3. Microsimulation of the reproductive life course. The index *i* represents a particular woman from an artificial cohort of size *n*. Each woman is taken sequentially through the major components of reproductive life, and a time for completion of each component is chosen for her from an input distribution using Monte Carlo techniques. A running record of her age at each live birth is compiled, and the total collection of records can be analyzed as if they were real-world data on fertility.

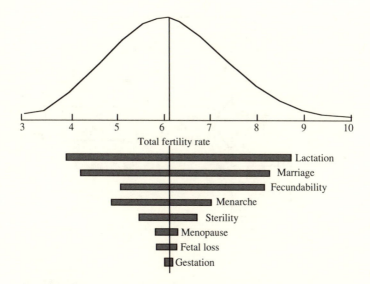

Figure 12.4. Sensitivity analysis of the proximate determinants of natural fertility. Curve (*top*) is a schematic representation of the distribution of total fertility rates among natural-fertility populations (see Figure 2.1). *Shaded bars,* range of response in TFR induced by varying each proximate determinant in turn while holding the others constant. (Redrawn from Campbell and Wood, 1988)

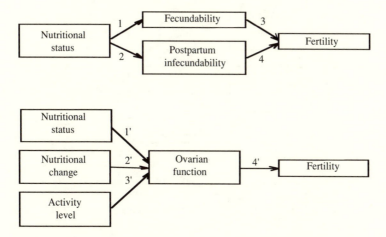

Figure 12.5. Two competing views of the relationship between maternal nutritional status and fertility: (*top*) the view generally adopted in demographic studies; (*bottom*) that used in most physiological research. (Redrawn from Wood, 1994)

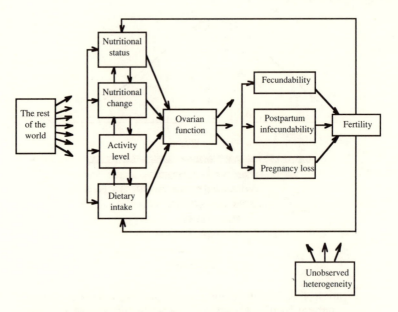

Figure 12.6. Additional complexities in the relationship between maternal nutritional status and fertility. (Redrawn from Wood, 1994)

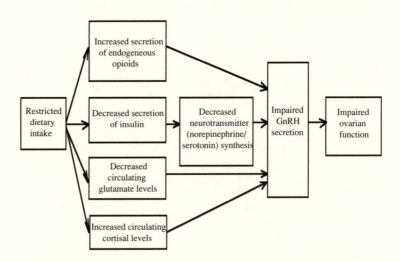

Figure 12.7. An attempt to specify the physiological mechanisms linking restricted dietary intake and ovarian function. (Redrawn from Tracer, 1991:118)

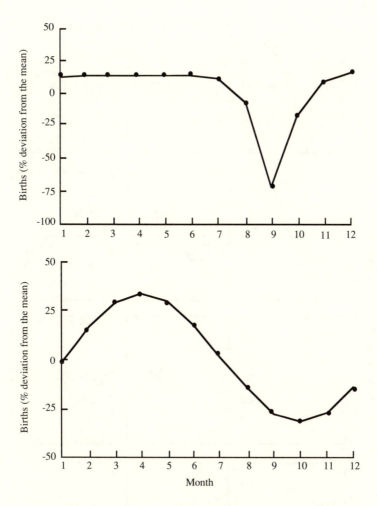

Figure 12.8. Two distinct patterns of birth seasonality: (*top*) a "sporadic" pattern in which one or a few months deviate sharply from the rest of the year, which is more nearly uniform; (*bottom*) a "cyclic" pattern in which the number of births changes more gradually from month to month so that successive months are highly autocorrelated. Births are expressed as percentage deviations from the mean monthly number.

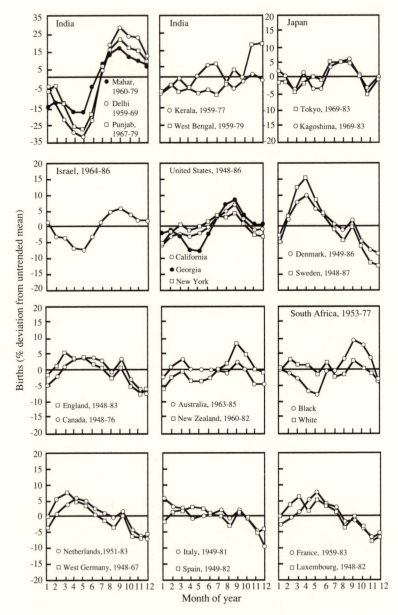

Figure 12.9. Typical patterns of birth seasonality from selected countries and periods. Births are expressed as the percentage deviation
from the mean monthly number after detrending for nonseasonal
changes in the number of births. (Redrawn from Lam and Miron,
1991)

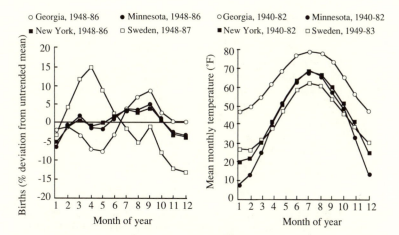

Figure 12.10. The lack of correspondence between birth seasonality (*left*) and seasonal changes in temperature (*right*), selected regions. (Redrawn from Lam and Miron, 1991)

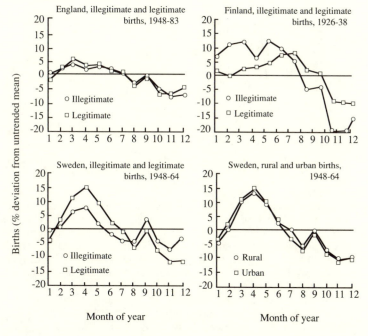

Figure 12.11. Intrapopulation variation in birth seasonality. (Redrawn from Lam and Miron, 1991)

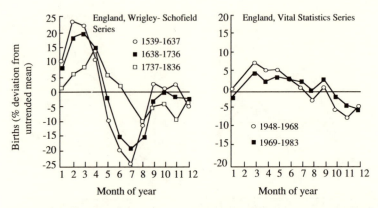

Figure 12.12. Changes in birth seasonality through time, England 1537-1983. Note that the locations of the peaks and troughs remain virtually unchanged, but that the amplitude of the seasonality decreases consistently with time. (Redrawn from Lam and Miron, 1991)

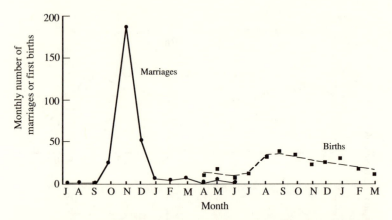

Figure 12.13. Seasonality of first marriages and first births among the Old Order Amish of rural Pennsylvania. Marriages and births are from the same period but are offset by nine months to highlight the relationship between the two. *Broken line,* predicted seasonality of first births given (1) the observed seasonality of marriages, (2) the predicted waiting time to first fertile conception, and (3) the estimated proportion of couples who are sterile at the time of marriage. The predicted and observed distributions of births do not differ significantly (likelihood ratio test statistic = 9.89, d.f. = 9, 0.25 < P < 0.5). (Redrawn from LeFor et al., 1993)

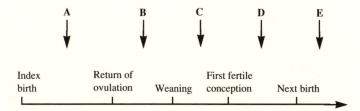

Figure 12.14. Timing considerations in the study of the interaction between birth spacing and early childhood mortality. A single completed birth interval is shown. Letters A-E refer to different possible times at which the index child (the child whose birth opens the interval) can die relative to the birth of the next child. Each of these times has different implications. Death during lactational anovulation (time A) will generally lead to an early resumption of ovulation and thus a shortened birth interval. Death at B may have a similar effect because of the residual contraceptive effects of breastfeeding, but this effect is expected to be smaller and more difficult to detect. If the index child is still nursing during the next pregnancy, a death at D could be partly attributable to that pregnancy since rising steroid levels in the mother can interfere with lactation. Death of the index child following the next birth (time E) may directly reflect nursing competition. Thus, deaths at A and B potentially increase fertility while conception and birth of the next child potentially increase the risk of death at times D and E. Only deaths during segment C are expected to be neutral with respect to the fertility-mortality interaction. In modeling and analyzing these effects, it is necessary to treat the different time segments separately insofar as possible.

Appendix: An Introduction to Quantitative Endocrinology

Kenneth L. Campbell *and* James W. Wood

Quantitative endocrinology is concerned with the measurement of variables associated with chemical communication among cells, including the concentration of hormones in biological fluids.[1] Over the past decade, measurements of hormone concentrations have come to play a central role in studies of childbearing patterns in human populations, thus moving quantitative endocrinology squarely into the traditional scientific domain of demography. A few of the research areas in which fundamental demographic insights have emerged from the analysis of endocrine data include the postpartum return of fecundity (Gray et al., 1993), the fecund waiting time to conception (Ellison et al., 1993; Weinstein and Stark, 1994), the fertility-reducing effects of fetal loss (Wilcox et al., 1988; Leslie et al., 1993), and the demographic impact of reproductive aging (Wood et al, 1993b; Campbell, 1994). (Other examples can be found throughout this book.) It is thus becoming increasingly important to understand something about how endocrinological measurements are made, even if one is interested purely in the demographic aspects of human reproduction.

Coming mainly from social science backgrounds, demographers sometimes take it for granted that endocrine measures are "harder"— easier to interpret and less subject to problems of validity and reliability—than most social measures. This is not necessarily so. The endocrinological measurement process is often complex and indirect, and issues of validity and replicability apply here as in any branch of science. In addition, since new endocrinological methods and materials are constantly being developed, validation and quality control are never-ending problems. This appendix attempts to explain endocrinological measurement in a way that will be intelligible to the nonendocrinologist, and to highlight some of the more important sources of error and bias. Since this material has provided the basis for many books and articles, the treatment here is necessarily somewhat superficial. Several recent reviews (Tijssen, 1985; Chan, 1987; Chard, 1987) can be consulted for more detail.

THEORETICAL FOUNDATIONS OF QUANTITATIVE ENDOCRINOLOGY

To understand endocrinological measurement, it is necessary to know something about the basic biology of hormones, their receptors, and the target-tissue response to them. Unlike some other biological molecules, which can be characterized fully by their molecular structure, hormones have classically been defined in functional terms, i.e., by their ability to carry a "message" from the cells that produce them and to induce a *biological response* in some defined set of target cells. The biological response can take many forms, but it always involves some change in the biochemistry, metabolism, gene expression, secretory activity, size, shape, or contractility of the target cell. As discussed below, the target-tissue response not only is the *raison d'être* of the hormone, but also provides a way to detect the hormone's presence.

Several classes of molecules can function as hormones: lipids, including steroids (which are derived from cholesterol) and prostaglandins and their derivatives (which are metabolites of the long-chain fatty acid arachidonic acid); polypeptides, which are oligopeptide chains or (if large enough) proteins; thyroid hormones and certain neurotransmitters such as the catecholamines, most of which are modified amino acids; and several locally generated gases such as nitrous oxide. In the areas of reproductive endocrinology most directly related to the present book, we are principally concerned with protein hormones such as luteinizing hormone (LH), follicle-stimulating hormone (FSH), and human chorionic gonadotropin (hCG), and with such steroids as estradiol, progesterone, and testosterone. Proteins and steroids differ in several features of their biochemistry. Steroids are small (low molecular weight), structurally simple, and hydrophobic (insoluble in water, but soluble in fat). Proteins are much larger, structurally more complex, and hydrophilic. Because of these differences, different protocols must often be used when steroids and proteins are being measured, as the following discussion will detail.

Hormones normally circulate in minute quantities, totaling only about 10^{-5} to 10^{-14} moles in the entire bloodstream (a *mole* is 6.023×10^{23} molecules). As a consequence, the target cell needs to have special molecular machinery to detect the weak hormonal signal and "amplify" it into a biological response. The first component of that machinery is a *protein receptor molecule*, located either on the surface of the target cell or in its interior. The function of the receptor is to identify and bind the hormone and, upon binding, to interact with other cellular components known as *signal transducers* and *effectors;* these components then translate the extracellular hormonal signal into the intracellular molecular language needed to induce a biological response.

The receptors for protein hormones such as LH, FSH, and hCG are located at the surface of the target cell. One functional domain of the receptor molecule protrudes into the extracellular fluid; it is that portion that actually binds the circulating hormone. A second domain transverses the lipid bilayer that constitutes the cell's outer membrane, and a third domain sticks into the intracellular medium and interacts with other molecules involved in signal transduction. The end result of signal transduction is usually the activation of enzymes, e.g., adenylate cyclase, phospholipase C, or, ultimately, one of the protein kinases, thereby altering the metabolism of the cell and perhaps affecting gene expression or cell division.

In contrast, receptors for steroid hormones such as estradiol, progesterone, or testosterone lie strictly within the cell, occurring principally in the nucleus. Upon binding, the steroid-receptor complex attaches to specific DNA sequences within the genome and modifies gene expression, probably by promoting or suppressing the activity of RNA polymerase and other molecules involved in transcription.

The difference in location of protein versus steroid receptors reflects the biochemical properties of the hormones they bind. Because steroids are small, uncharged lipids, they pass easily through the cell and nuclear membranes; proteins, in contrast, are large and often electrically charged and do not normally diffuse into the interior of the cell. Thus, the receptors for steroid hormones can lie within the nucleus, whereas those for protein hormones must have binding sites that are exposed to the extracellular environment.

Target cells in different tissues—or at different times within the same tissue—may differ in the suite of receptor, signal-transduction, and effector molecules they present to circulating hormones. Moreover, the response of a given target cell can be modulated by the presence of *other* hormones, in addition to the primary hormone inducing the response. As a consequence, a single hormone can induce radically different changes in different target cells. Estradiol, for example, can simultaneously suppress protein secretion by certain cells within the anterior pituitary while increasing it in the endometrium of the uterus. And the complexities of signal transduction allow a single hormone to induce multiple changes within a single target cell, a fact that should be borne in mind throughout the following discussion of the target-cell response.

A final feature of hormonal signalling systems that should be mentioned is the existence in the bloodstream of *carrier proteins*, which bind hormone molecules in reversible fashion. These carrier proteins, some of which bind particular hormones while others bind a range of different hormones, are involved in moving the hormone from its tissue of origin to its target tissue. Since the carrier proteins shield the hormone from degradation while it circulates in the bloodstream, binding to such proteins can greatly retard a hormone's metabolic clearance. Most water-soluble molecules, including protein hormones, circulate in the plasma in a free state, unassociated with carrier proteins. Hydrophobic molecules such as steroids, in contrast, are transported in the bloodstream by carrier proteins—either specific, high-affinity proteins, such as sex-hormone-binding globulin, or proteins such as albumin and prealbumin, which are much less specific in the hormones they bind.

The critical fact about the role of carrier proteins in hormonal signaling is that the hormone bound to such proteins is not bioactive. A hormone can interact with a target-cell receptor and induce a biological response only if it is *free*, i.e., not bound to a carrier protein. Although carrier proteins provide a pool of circulating hormones to draw upon, the hormone must dissociate from the carrier to become active. This point is important in evaluating assay systems, for some assays measure total circulating hormone (bound and free) whereas others measure free alone; only the latter directly reflect the concentration of bioactive hormone, which is what we usually want to know about. Measurements of total hormone are, however, frequently encountered. Not only are such measurements usually easier to perform, but the level of total hormone in the circulation is normally well-correlated with that of free hormone.

Dynamics of Hormone-Receptor Binding

The receptor molecule must bind the hormone with *high affinity* to make use of the low-concentration hormonal signal. High affinity is achieved through *noncovalent binding*, involving electrostatic and hydrophobic interactions between the receptor and hormone. Since noncovalent interactions are extremely sensitive to the distance between molecules, declining in strength exponentially as distance increases, the geometries of the hormone and receptor must "fit" each other snugly before binding can occur. This requirement imparts a considerable degree of *specificity* to the hormone-receptor bond: there are strict limits on the extent to which the biochemical structure of either molecule can be altered without compromising the ability to bind. Because of this specificity, a cell cannot respond to a hormone unless it expresses the particular receptors that bind that hormone. Indeed, the presence of a specific receptor is essentially what identifies a cell as a target for the corresponding hormone.

Noncovalent bonds are reversible. This reversibility can be diagramed as

$$H + R \underset{\kappa_r}{\overset{\kappa_f}{\rightleftharpoons}} HR$$

where H is the free (unbound) hormone, R is the free receptor, and HR is the bound hormone-receptor complex. The constants κ_f and κ_r describe, respectively, the rates at which the forward (association) and reverse (dissociation) reactions take place. If hormone binding is completely specific (i.e., only one class of hormone binds to the receptor), and if there is no competition or cooperation among receptors (i.e., binding at one receptor does not alter the likelihood of binding at another), then the rate of change in the concentration of hormone-receptor complex is governed by the Law of Mass Action:

$$d[HR]/dt = \kappa_f[H][R] - \kappa_r[HR] \tag{A.1}$$

where [·] denotes the molar concentration of the quantity inside the brackets. According to this equation, [HR] will change continuously, unless it is at an equilibrium point, making it too variable much of the time to be useful in endocrinological measurement. The quantitative endocrinologist relies upon the fact that, given enough time, this system will indeed reach a steady-state, dynamic equilibrium at which [HR] is no longer changing. To find the concentration of bound hormone at equilibrium, we set equation A.1 to zero and simplify:

$$[HR] = (\kappa_f/\kappa_r)[H][R] = [H][R]/K_d \tag{A.2}$$

In this expression, $K_d = \kappa_r/\kappa_f$ is known as the *equilibrium dissociation constant*. (Its inverse, written K_a, is the *equilibrium affinity constant*.) Empirical estimates of K_d for endocrine systems generally fall in the range 10^{-12} to 10^{-9}, small values compared to most other chemical binding phenomena. Thus, hormone-receptor complexes break apart readily only when hormone levels are very low, when biochemical processes alter the affinity of the hormone and receptor for each another, or when chemical agents such as solvents or detergents prevent the normal chemical interactions required for binding. But the fact that K_d is greater than zero is important to the endocrinologist who wishes to measure the hormone, for it allows a hormone tagged with some detectable "label" or signal to compete with the native, unlabeled hormone for binding to receptors during those brief periods when hormone and receptor are dissociated (see below).

Now, the total number of receptors is given by $[R_T] = [R] + [HR]$, the sum of the free and bound receptors. If we rearrange this as $[R] = [R_T] - [HR]$ and substitute in equation A.2, we have

$$[HR] = \frac{[H]([R_T] - [HR])}{K_d} = \frac{[R_T][H]}{K_d + [H]} \tag{A.3}$$

This expression is usually encountered in the form

$$B = \frac{B_{max}F}{K_d + F} \tag{A.4}$$

where B and F are the molar concentrations of bound and free hormones, respectively, and B_{max} is the maximum value of B obtained when all the receptor sites are occupied, that is, B_{max} is the total number of receptors. This equation describes a hyperbolic curve, as shown in Figure A.1. An important feature of this curve is that most of the "action" occurs at low hormone concentrations: the amount of binding is most sensitive to small changes in F when F is very low. This fact means that not only *can* the system make use of low concentrations of circulating hormones, it would actually respond with less sensitivity if hormone concentrations were high. If we set F equal to K_d, equation A.4 reduces to $B = \frac{1}{2}B_{max}$. Thus, K_d can be interpreted as the concentration of hormone required, at equilibrium, to occupy half the available receptor sites. The fact that B_{max} is finite means that the set of receptors is at least theoretically *saturable*, that is, it is possible for all receptor sites on a cell to be occupied by hormone. For most physiological systems, however, the concentration of hormone is less than K_d, so saturation is rarely approached. In such cases, it sometimes simplifies the mathematics of hormone-receptor dynamics to treat the system as unsaturable ($B_{max} \to \infty$).

Equation A.4 is one of the most important formulas in all of biochemistry. For some purposes, it is useful to substitute K_a^{-1} for K_d and rearrange the equation as

$$B/F = K_a B_{max} - K_a B \tag{A.5}$$

which is a negative linear function of B. Thus, an empirical plot of B/F against B values, known as a *Scatchard plot*, should be linear if the assumptions of steady-state equilibrium, complete specificity, no competition, no cooperation, and homogeneity of hormone and receptor preparations are correct. If a sufficiently wide range of B/F measurements are taken, nonlinear Scatchard plots can reveal the existence of disequilibrium, nonspecific binding, unexpected mixtures of related hormones or receptors, and complex interactions among receptor molecules (Figure A.2). Thus, Scatchard plots are among the basic tools used by quantitative endocrinologists to evaluate the reference standard preparations and reagents used in endocrinological measurement (see below). For good discussions of the interpretive problems associated with nonlinear Scatchard plots, as well as how to resolve them, see Clark et al. (1992) and Brown and Rothery (1993:349–356).

Dynamics of the Target-Cell Response

The formation and dissociation of receptor-hormone complexes have important implications for the rapidity, magnitude, and duration of the target-cell response. In turn, the characteristics of the biological response determine our ability to use such responses to measure hormones.

The biological response by the target cell depends upon the concentration of free hormone in its immediate vicinity, the number of receptors for the hormone carried by the cell, and the affinity of the hormone-receptor interaction.[2] Since most hormone-receptor systems are usually well below saturation, almost any increase in [H] results in an increase in the number of receptors occupied, permitting the target tissue to make a graded response even to very small changes in hormone concentration. This graded response, expressed as a *dose-response curve*, is typically sigmoidal when plotted against dose (hormone concentration) on a linear or log scale (Figure A.3).

It is important to understand that a given dose-response curve reflects one specific target-cell response. Target cells can respond in many different ways to changes in the concentration of one particular hormone. For example, as levels of estradiol rise during the follicular phase of the ovarian cycle, cells in the endometrium of the uterus respond by growing, altering their secretory activity, and dividing at a higher rate. Each of these responses follows its own dose-response curve, which may differ from the curves for other responses to the same hormone by the same cell.

For some protein hormones, the target-cell response is directly proportional to the number of occupied receptor sites. In these cases, the curves describing the biological response and the proportion of receptors occupied by hormone can be superimposed over the full range of hormone concentrations, and the maximum response is elicited at saturation, when 100 percent of the receptors are occupied (Figure A.4). As long as K_d does not change, the maximum biological response in such saturable systems is directly proportional to the total number of receptors.

For most hormones, however, the maximum response is observed at hormone concentrations far too low to occupy all the available receptors. For instance, in the granulosa cells of the antral follicle, maximum estradiol secretion occurs when only about one or two percent of the receptors are bound to gonadotropin. In this example, 98–99 percent of receptors are *spare receptors*. This term is not meant to imply that 98–99 percent of the receptors are never used, but rather that every receptor is unoccupied about 98–99 percent of the time. The degree of "spareness" of receptors for a given hormone varies from one biological response to another, which is one reason that different responses have differing dose-response curves (Mendelson, 1992).

Figure A.5 shows the hormone-binding and biological response curves for one particular response in a hypothetical target cell (an example provided by Mendelson, 1992). The cell normally carries 20,000 receptors for the hormone of interest. Since the maximum response is elicited at hormone concentrations sufficient to occupy only 5,000 receptors per cell, 75 percent of the receptors are spare receptors. When the number of receptors is reduced by anything less than 75 percent without a change in K_d, the maximum biological response in unchanged. However, any reduction in the number of receptors below 5,000 causes a proportionate reduction in the biological response. This example illustrates the principle that the greater the number of spare receptors for a given response, the more sensitive the target cell is to the hormone— that is, the lower the hormone concentration needed to achieve a 50 percent maximum response. In addition, spare receptors tend to prolong the biological response of a cell to short bursts of circulating hormone: the hormone-receptor complexes in excess of those required for the maximum response will maintain the response for longer periods as the concentration of hormone declines (Figure A.5, bottom panel).

These features of the biological response curve are important in the normal regulation of target-cell function, for they allow varied effects to be achieved simply by modulating the number of available receptor sites. They also help explain the variety of pathological responses that can result from defects in the hormone/receptor/signal-transduction pathway (Figure A.6). Individuals with hormone-resistance syndromes may, for example, exhibit a dose-response curve that indicates decreased *sensitivity* to the hormone, i.e., a shift of the curve to the right with no change in maximal response. This pattern suggests a defect in some step that is not rate limiting, perhaps a defective receptor in a cell with many spare receptors. In such cases of decreased, but not completely obliterated, sensitivity, an elevation in the hormone concentration can overcome the resistance and restore the normal biological response. In other cases, however, there may be evidence of decreased *responsiveness*, i.e., a decrease in the maximal response itself, that may or may not be accompanied by decreased sensitivity. Decreased responsiveness implies a defect at some rate-limiting step, often at some stage of signal transduction (Kahn et al., 1992). In such cases, increasing hormone concentrations induce a biological response only up to a point that falls well short of the normal maximal response. Finally, some extreme forms of hormone resistance are marked by a complete absence of biological response, no matter what the concentration of hormone. For example, testicular feminization (androgen-resistance syndrome) is a condition in which a person who is chromosomally male appears to be a phenotypically normal female. Most cases of this syndrome are attributable to one of several mutations in the gene for the testosterone receptor. These mutations completely abolish the receptor's affinity for androgen, rendering target tissues insensitive to masculinization (Pinsky and Kaufmann, 1987).

BIOASSAYS

We turn now to a review of the methods that have been devised for measuring the concentration and potency of hormones in biological specimens. The logical starting point for this discussion is the classic bioassay. Although other types of assay have become increasingly important in recent years, they have all built upon the empirical and theoretical foundations provided by the bioassay.

In a bioassay, the dose-response curve is an integral part of the measurement process. Either cultured target cells or cells in intact laboratory animals are exposed to a graded series of known hormone concentrations while a specific response is monitored. Once the associated dose-response curve has been determined, the same in vitro or in vivo system can be exposed to specimens containing an *unknown* hormone concentration, and the resulting response used to infer the hormone concentration (Figure A.3). A classic example of such a bioassay is the so-called rabbit test that was once used for the clinical diagnosis of pregnancy. In this test, either known amounts of hCG, or samples of urine from women thought to be pregnant, are injected into a female rabbit, and the resulting changes in its ovaries (e.g., the rupturing of follicles or the formation of corpora lutea) are observed.

For a bioassay to be reliable, we need access to a well-defined, predictable biological response system. Often the chosen target is a tissue or organ that degenerates in the absence of the hormone we wish to measure. For example, male secondary reproductive organs, such as the prostate and seminal vesicles, undergo marked involution following removal of the testes (Mainwaring, 1979). Implantation of testicular tissue can restore the normal histology of those organs. Plainly, the testis makes something the other tissues need—namely, testosterone. Similarly, after removal of the ovaries the uterus undergoes involution as estrogens are cleared from the bloodstream: in the female rat, the adult uterus looks infantile within a few days of oophorectomy. Involution of the gonads, thyroid, or adrenals is measurable soon after blocking the pituitary stalk, thus depriving the pituitary gland of hypothalamic releasing- hormones (Harris, 1955). In addition to gross changes in tissue size and histology, more subtle metabolic effects can be used to measure the presence or amount of a hormone. To cite some examples of reproductive interest, ovarian tissue becomes depleted of ascorbic acid following injection of purified LH (Parlow, 1961), the myometrial muscle of rat uterus previously exposed to estrogens displays strong contractions when purified oxytocin is added in vitro (Fuchs and Dawood, 1980), and Leydig cells from the testis respond to anterior pituitary extracts or purified LH by producing testosterone (Dufau et al., 1974). All these target-cell responses can be used to measure hormone concentration.

Bioassays are now done mostly in in vitro systems using sophisticated new methods for monitoring cell growth and metabolic changes. These newer forms of bioassay have much to offer by way of precision, sensitivity, and standardization. There is, however, a cost. As the measured end-points become more narrowly defined, the resulting picture of the target-cell response grows more myopic. The improvement in precision and reproducibility gained by using a completely defined, in vitro bioassay system is bought at the expense of a more comprehensive view of the spectrum of biological responses to the hormone of interest.

Sources of Variability in Bioassay

The relationship between hormone concentration and biopotency (ability to induce a biological response) is fairly straightforward in low-molecular-weight compounds, such as steroids. In such a compound, alteration of a single atom or the angle of a single chemical bond can change the molecule's biochemical properties so completely that it becomes a different hormone. For example, testosterone and 5α-dihydrotestosterone differ by only two atoms of hydrogen, yet their melting points, solubility, and biological potency differ so thoroughly that they act as distinct forms of androgen and are given different names.

For protein hormones, several factors complicate the relationship between concentration and biopotency. When a batch of purified hormone is being prepared for use in constructing a dose-response curve, some molecules in it may be irreversibly altered by the chemical procedures involved in purification; this subpopulation of molecules may no longer be able to bind to receptors. Different preparations of protein hormone can differ in potency, even when isolated by similar methods from similar source materials. Variable potency can also occur in batches of protein hormone stored for differing periods of time, even when maintained under optimal conditions.

Although many problems result from the artificial manipulation of protein hormones during isolation, purification, and storage, variable potency can also be biological in origin. Many protein hormones are actually glycoproteins—polypeptides to which sugar side-chains have been attached covalently during post-translational processing within the cell of origin. Glycoprotein formation and degradation is a dynamic process. A given population of hormone molecules, for example FSH within a single gonadotroph cell, will exhibit a spectrum of forms that differ in *glycosylation*, i.e., the number and size of their sugar side-chains. The same is true of protein hormones in plasma and urine or in purified reference standard preparations (see below), and it is true even of protein hormones produced by genetically engineered cell lines if the cells involved are capable of glycosylation. Although the extent of variation can be restricted during purification and isolation, it can rarely be eliminated entirely. Moreover, if the hormone is introduced into a living biological system, much of the heterogeneity may reappear while the hormone circulates.

Other potential causes of variable hormone response can confound bioassays, including the genetic strain of the animals used, the nutrients fed to those animals or the pathogens to which they were exposed, and the culture medium in in vitro systems. Whatever their origin, even slight variations can affect the response observed in a bioassay system, no matter how pure the hormone preparation being tested. Thus, there will always be errors when estimating unknown hormone concentrations, even under well-controlled conditions. The quantitative endocrinologist must, of course, try to minimize such errors but must also be prepared to evaluate their magnitude since they can never be eliminated altogether.

REFERENCE STANDARD PREPARATIONS

Since the dose-response curve is an essential part of any bioassay, the bioassay approach requires preparations of known hormone concentration and biopotency so the curve can be estimated with as little error as possible. As detailed below, other forms of assay require similar preparations so other kinds of standard curves can be set up. These hormone preparations are known as *reference standard preparations*, and they need to be calibrated using *reference methods*. The main purpose for establishing, validating, and disseminating these methods and standards is to maintain comparability of assay results both among assays and across laboratories. Hormone assays should always use standard preparations that have been calibrated against such reference standards and methods. Indeed, there are now laws requiring clinical laboratories to use such materials. As new assays are developed, they must be validated against earlier assays based upon reference standards and methods. It is desirable, therefore, to make the reference methods less expensive and more accessible to laboratories that are routinely developing new forms of endocrine measurement. This goal is especially important for laboratories involved in population-related research, since such labs must often devise new methods for handling specimens collected under unusual field conditions (Campbell, 1994).

The choice of method for validating a standard reference preparation depends upon the chemical nature of the hormone involved. Small molecules, such as steroids, can be purified by crystallization and then measured using instruments of exquisite precision and reliability, such as mass or nuclear magnetic resonance spectrometers or infrared spectrophotometers. Some smaller polypeptides, such as GnRH (gonadotropin-releasing hormone) or oxytocin, can also be measured by spectrophotometric methods or by quantitative chemical methods, if they are first isolated by some separation procedure such as high-performance liquid or gas chromatography. Since these physical and chemical methods are not subject to the kinds of biological variability that can affect bioassays, as well as assays using antibodies (see below), their responses to purified hormone preparations are more accurate and reproducible. The major drawback of the strictly instrumental reference methods is that they usually require samples that have been purified or at least are much purer than what is found in biological fluids such as serum or urine.

For protein hormones, standard reference methods based upon physical or chemical measurement are not yet routinely used. Part of the reason is a lack of consensus about which molecular forms to purify. As mentioned above, protein hormones can vary in structure in several ways, including their degree of glycosylation. Many of these structural variants exist in the intact organism and may play differing physiological roles. There are, in other words, many different options (and little consensus) in choosing a protein hormone for purification and use as a reference standard. In the absence of agreed-upon physical or chemical methods, quantitative endocrinologists have had to decide which bioassay or immunological method is to be adopted as the standard against which all other methods should be validated. Special conferences and workshops have been held to reach consensus on which bioassay methods to adopt for specific protein hormones, and on how to produce the most uniform standard preparations possible given the available biochemical methods. At those same conferences and workshops, decisions have also been made about what mass of the reference standard preparation constitutes one "unit dose" or international unit (IU) of hormone.

History has thus played a large part in the development and acceptance of reference preparations. The lab that first publishes information on a highly purified, uniform, and biologically potent hormone preparation that can be stored in a reasonably stable form often provides other researchers with batches of the hormone for use as a provisional reference preparation. Aliquots of the preparation are then tested in other labs for consistent biopotency in similar assay systems. If a preparation proves reliable (albeit within rather wide limits, sometimes on the order of 100–200 percent inter-laboratory error!) it may be adopted as the final reference standard. It will then serve as the ultimate calibration standard for all subsequent preparations and all assays that claim to be specific for the same hormone.

Some of the earliest reference preparations, developed between 1950 and 1970, have since been found to contain not only multiple biopotent forms of the hormone of interest but other contaminants as well, including inactive forms of the hormone (Birken et al., 1991) and other, often structurally related hormones. Since some of the contaminants are antagonists of the hormone of interest (Pekönen et al., 1988), they can interfere with the response of a bioassay system to the reference preparation. These contaminants bias the assay results by suppressing the ratio of biopotent hormone mass to the total mass of the hormone present in the reference preparation.

The original reference preparations were produced in limited quantities. Aliquots were made available to qualified researchers and clinicians worldwide through the World Health Organization's Expert Committee on Biological Standardization, with the aid of the Department of Biological Standards at the National Institute for Medical Research (United Kingdom) and the National Pituitary Program of the National Institutes of Health (United States). Not surprisingly, most of the original reference preparations have long since run out, and a second generation has been needed—indeed, in the case of the anterior pituitary hormones, several generations have now passed. Although pains have been taken to maintain potency levels across successive generations, there have undoubtedly been some changes along the way, reflecting such factors as the level of inbreeding of the animals used as sources of the response systems, the development of newer purification methods, and the adoption of in vitro reference bioassays (Wang, 1988). Final agreement on reference assay systems and the development of genetically engineered reference preparations would greatly help maintain quality.

RADIOIMMUNOASSAYS

After the bioassay, the next most important tool in the history of quantitative endocrinology was the *radioimmunoassay* (RIA), developed in the late 1950s and early 1960s by Rosalyn Yalow and Solomon Berson (Yalow, 1978). RIAs were first applied to hormones of reproductive interest in the mid-1960s (Midgley, 1966; Midgley and Jaffe, 1966). Along with the other immunoassays descended from them (see below), RIAs have played a key role in the rapid advance of reproductive endocrinology over the past thirty years. It is no exaggeration to say that this book could not have been written had the RIA never been invented.

Where the bioassay requires the hormone being measured to bind to its normal target-cell receptor, the immunoassay calls for binding of the hormone to antibodies raised against it. More precisely, the RIA involves competition between *radio-labeled hormone* and *unlabeled hormone* for binding to the same antibody. Radio-labeling involves the insertion, by chemical methods, of radioactive isotopes into the hormone of interest; the resulting labeled hormone, known as a *tracer*, emits a radioactive signal (either a beta particle or a gamma ray) that can be detected by a scintillation counter. The unlabeled hormone in an RIA is present either in a reference standard preparation or in the biological specimen to be measured. An antibody is usually produced by injecting an intact animal, such as a rabbit, with purified hormone and then "harvesting" the polyclonal antibodies against the hormone generated by the animal's immune system.[3]

If the labeled hormone is denoted by H^*, the competition between the labeled and unlabeled hormones for the antibody can be diagrammed as

$$H^* + Ab \underset{\kappa_r}{\overset{\kappa_f}{\rightleftharpoons}} H^*Ab$$

versus

$$H + Ab \underset{\kappa_r}{\overset{\kappa_f}{\rightleftharpoons}} HAb$$

where *Ab* is the antibody raised against the hormone *H*. The fact that the rate constants κ_f and κ_r are assumed to be the same for both competing reactions implies that the affinity of the antibody for the labeled and unlabeled hormone is the same—something that often needs to be verified for a new RIA. While $\kappa_r \ll \kappa_f$ for most antigen-antibody reactions, the reaction is still reversible ($\kappa_r > 0$). At equilibrium, the amount of hormone bound by the antibody is determined by the combined rates of association and dissociation, entirely analogous to the binding of the hormone by its target-cell receptor.

When developing an RIA, we first need to establish a *standard RIA curve*, which will later be used as a scaling device for estimating unknown hormone concentrations. We estimate the standard curve by examining a series of assay tubes corresponding to different points along the curve (Figure A.7). Predetermined, constant amounts of antiserum (or purified antibody) and tracer are first added to the tubes. Antibody is added in limiting quantities, and tracer in amounts just large enough to saturate the antibody in the absence of any unlabeled hormone. We then add, to the same series of tubes, increasing amounts of unlabeled hormone drawn from a standard preparation, each tube receiving its own particular, known quantity. These quantities begin with a "zero control" containing no unlabeled hormone and end with a large (>300-fold) excess of unlabeled over labeled hormone. Following mixture and an appropriate incubation period, during which the competing antibody-hormone interactions come to equilibrium, the antibody-bound hormone is separated from unbound or free hormone. (There are several ways to achieve this separation; for example, if the hormone is a steroid, dextran-coated charcoal can be used to adsorb the free hormone.) Since both [*Ab*] and [*H**] are fixed, the amount of antibody-bound tracer is strictly a function of the concentration of unlabeled hormone in the assay tube: *the higher the concentration of unlabeled hormone, the greater the competition for antibody, and the lower the quantity of tracer bound to the fixed amount of antibody.* After the radioactivity in each tube is measured with a scintillation counter, the radioactive counts for the various tubes are plotted against the amount of standard present, usually as a semi-log plot. The resulting standard curve (lower panel of Figure A.7) can then be used for quantification of *unknown* samples, just as the dose-response curve is used in bioassay.[4]

Although Figure A.7 shows the unadjusted level of bound tracer along the *y*-axis, it is more common to plot the ratio of *B*, the amount of labeled tracer bound in the presence of the standard, to B_0, the amount bound in the complete absence of the standard. Scaling the curve by B_0 corrects for changes in the absolute level of radioactivity in the tracer preparation, which necessarily declines over time as a result of radioactive decay. Since the signal emitted by the tracer grows weaker as the tracer preparation ages, assays performed at different times using the same tracer preparation will not be strictly comparable. Dividing by B_0 adjusts for this effect since the total signal in the zero control tube is a reflection of the remaining radioactivity of the tracer. This adjustment may, however, introduce problems of interpretation because of correlated errors in B and B_0 reflecting the limits of measurement precision at varying count levels.

Scaling B by B_0 produces the kind of standard RIA curve shown in Figure A.8. The curve is sigmoid, starting at one and approaching a lower asymptote as the amount of standard increases. The comparative flatness of the curve at both low and high doses of standard makes the assay less sensitive at the extremes. The most reliable measures of hormone concentration are obtained near the middle of the curve, where the dose-response relationship is approximately linear.

By substituting an antibody for the normal hormone receptor and the competitive-binding curve for the biological response curve, immunoassays obviate the need to identify a precisely measurable target-cell response to the hormone of interest. But they impose other conditions. For example, a sufficiently pure standard hormone preparation must exist. In other words, when the preparation is injected into an animal from a different species, the resulting immune response must yield a fairly homogeneous set of serum-borne antibodies that can be harvested and used as reagents in a well-defined assay system. (Often, however, the polyclonal origin of naturally produced antibodies limits their homogeneity.) In addition, purified hormone must be radio-labeled so it can be used as a competitive tracer whose binding to antibody is observable (unlike unlabeled hormone, which remains invisible). Finally, a standard curve must be established, which requires appropriate statistical models.

Antibody Preparation

One might expect that the structural heterogeneity of protein hormones would be an insurmountable barrier to the production of antibodies that are sufficiently homogeneous for use in an RIA. Nonetheless, the rabbits used in the earliest RIAs managed to generate polyclonal antibodies that bound to *epitopes* (antibody-recognition sites) that were either unique to the major hormonal form present or shared among the heterogeneous forms of the hormone. The researchers screened all the antisera produced and chose for use the ones that bound with the highest apparent affinity and generated smooth standard curves. For practical purposes, such antisera could be considered homogeneous. Some of the antisera contained antibody of such high concentration and affinity that they could be diluted massively for wide dissemination. One of the first antibodies against LH, for example, was used at dilutions of over 1:800,000, so the few hundred milliliters of serum harvested from one rabbit were sufficient to supply much of the endocrine community with antibody for more than a decade (Niswender et al., 1968).

But no matter how thinly diluted, the antiserum produced by an individual animal must eventually run out. Some antibody preparations that provided the "gold standard" in earlier analyses are no longer available. An example from Chapter 6 of this book is the highly specific antiserum to hCG used in the original studies of fetal loss by Allen Wilcox and his colleagues (Wilcox et al., 1988). Unfortunately, there are limits to what can be done to guarantee that a new host animal will produce antibodies as concentrated, sensitive, and specific as the original antibody. All that can be done is to inoculate multiple hosts and compare the antibodies they produce—comparisons that may themselves take years. The most promising way to circumvent this problem is development of *monoclonal antibodies* produced in vitro, either by β-lymphocytes from mammalian spleen tissue immortalized by fusion with myeloma cells (Goding, 1986) or, more recently, by genetically engineered bacteria cells transfected with DNA from immunologically competent eukaryote leukocytes (Lerner et al., 1992). Unlike the polyclonal antibodies secreted by intact animals, supplies of monoclonal antibodies from cultured cells are, for all practical purposes, inexhaustible: subpopulations of the immortalized cell lines can be frozen, thawed, and reestablished at will. Moreover, the antibodies produced by a single clone of cells are identical, barring mutation; thus, monoclonal antibodies of practically perfect purity, directed against specific epitopes, can be generated.

In the case of steroids, the same molecular forms are found in virtually all vertebrates (and often in many invertebrates and even plants). They cannot, therefore, be antigenic on their own account. In these instances, it is necessary to link the hormone to a larger protein that can induce an immune response. Usually this is done by reacting the hormone with a molecular linking arm that is then attached covalently to a large protein, such as limpet hemocyanin or bovine thyroglobulin. After removal of nonreacted hormone, the complexes are injected into host animals and the resulting antisera are harvested.

The fact that many compounds are vital to all mammalian species may limit the ability to raise antibodies against them. Antibodies against neurotransmitters, for example, are rare precisely because neurotransmitters play a central role in the maintenance of life; high-affinity antibodies against them would rapidly kill the host animal. By the same reasoning, the relative ease with which high-affinity antibodies can be raised against most of the steroid hormones involved in reproduction suggests either that those hormones are not essential for life or that the host organism can somehow compensate for massive reductions in the concentration of free, bioactive hormone in the circulation. Both explanations probably contain some truth since immature, castrated, and postreproductive individuals are all able to live quite contentedly in the absence of certain steroids.

Tracer Preparation

In addition to an antibody, RIAs require a radio-labeled form of the hormone to use as a tracer. For steroids, the tracer is usually produced by synthesizing the hormone in vitro and substituting a radioactive isotope such as tritium (^3H), carbon-14 (^{14}C), or sulphur-35 (^{35}S) for one of the atoms normally present in the molecule. Except for very subtle differences, the radio-labeled steroid behaves identically in the assay system to its unlabeled counterpart, and antibodies generally cannot distinguish them, which is what we want.

Labeled and unlabeled protein hormones are usually assumed to be equally indistinguishable, but this assumption is almost certainly wrong in many cases. Proteins are typically labeled by adding radioactive atoms or chemical groups to the purified hormone. If ^3H, ^{14}C, or ^{35}S is used as a label, it is normally attached in the form of a small reactive group, such as acetate or mercaptoethane, which modifies an amino acid or sugar side-chain (Fox, 1976). In radioiodide labeling using ^{125}I, one or more bulky atoms of iodine are added to accessible tyrosine residues, often located on the outer surface of the polypeptide chain. In either case, the result is a "blemish" on the surface of the tracer molecule that can compromise the affinity of the antibody raised against the unlabeled hormone. Since antibody affinity can be affected by features as small as a single atom of oxygen (Haber and Novotny, 1985), it is likely that there will often be some difference in the ability of an antibody to bind to labeled versus unlabeled protein hormone. This issue can be addressed during the validation of a particular RIA system by testing whether the assay response curves of the reference standard and the tracer are parallel, indicating similar interactions with the antibody (Feldkamp and Smith, 1987). If parallelism is verified, then differences in the affinity of the antibody for labeled and unlabeled hormone are likely to be small and unimportant.

The Standard RIA Curve and Its Error Structure

Assuming that a suitable antiserum and tracer are available, the next step in an immunoassay is to establish a standard curve similar to that shown in Figure A.8. The curve itself is not "observed" in any real sense. Rather, discrete points that presumably fall along it are estimated from the series of assay tubes, each point corresponding to one particular concentration of standard. The curve itself is provided by a statistical model fit to those points. The proper choice of a model has important implications for the error distribution associated with the curve and, thus, for the reliability of the estimated hormone concentrations obtained from the assay. Several models for the curvilinear immunoassay response curve have been suggested (for a review, see Maciel, 1985). The one presented here is based on the widely used method of Rodbard et al. (1968).

We saw in equation A.4 that

$$B = \frac{B_{max}F}{K_d + F} \tag{A.6}$$

This expression for the concentration of bound hormone at equilibrium (B) was originally derived for the binding of a hormone to its natural receptor, but it applies equally well to the binding of hormone to antibody. In a radioimmunoassay we can simplify the expression by making F so large relative to the amount of antibody that we are effectively operating at the upper asymptote of the curve, so $B \simeq B_{max}$. Using the subscripts L and U to denote labeled and unlabeled hormone, respectively, we have

$$B_L + B_U = B_{max} \tag{A.7}$$

Next, we assume that

$$B_L/(B_L + B_U) = Q_L/(Q_L + Q_U) \tag{A.8}$$

where Q represents the total concentration of bound plus free hormones, either labeled or unlabeled as indicated by the subscript. Equation A.8 says simply that the fraction of the bound hormone that is labeled is the same as the fraction of the *total* hormone that is labeled. Now B_L is observed: it is what we measure by the radioactive count of the bound fraction. And Q_L is known because we add a predetermined amount of labeled tracer to the assay tube. If we set B_U to zero in equation A.7, B_{max} turns out to be equal to B_0, the amount of labeled tracer in the zero control tube. Combining equations A.6–A.8, we thus have

$$B_L = \frac{B_0 Q_L}{Q_U + Q_L} = \frac{B_0(Q_L/Q_U)}{1 + (Q_L/Q_U)} \tag{A.9}$$

As noted earlier, we ordinarily standardize B_L by dividing it by B_0. By convention, we also omit the subscript from B_L on the understanding that B refers strictly to bound *tracer*. Noting that $Q = \exp(\ln Q)$, we can recast equation A.9 as

$$\frac{B}{B_0} = \frac{e^{\ln Q_L - \ln Q_U}}{1 + e^{\ln Q_L - \ln Q_U}} \tag{A.10}$$

This equation provides the basic functional relationship between B/B_0 and Q_U. To allow for other sources of variability that can affect the assay results (e.g., nonspecific binding or incomplete separation of bound and free hormone), Rodbard et al. (1968) generalize equation A.10 as

$$\frac{B}{B_0} = \beta_1 + \beta_2 \cdot \frac{e^{\ln Q_L - \beta_3 \cdot \ln Q_U}}{1 + e^{\ln Q_L - \beta_3 \cdot \ln Q_U}} \tag{A.11}$$

Thus, we have a conventional logit-logistic regression equation with three parameters, β_1, β_2, and β_3, to be estimated. With the data obtained from the reference hormone preparation, in which Q_L (which is just a constant) and Q_U are both known, this model can be fit by nonlinear least squares or maximum likelihood using almost any statistical package, including the software routinely provided with RIA instrumentation.[5] The estimated model can then be used to derive an "inverse prediction" of Q_U in an unknown test preparation, along with its standard error (Finney, 1971). A nice feature of this model is that it neatly links the dynamic theory of hormone-binding systems to straightforward and easily implemented regression methods. Rodbard (1974) presents further generalizations of equation A.11 that may fit a wider range of assay results, and Normolle (1993) reviews alternative model specifications and estimation procedures.

However it is estimated, the standard curve has several features that are important in interpreting the results estimated from it (Figures A.7 and A.8). In combination with B_0, the lower asymptote of the curve (B_∞) defines the full range of binding responses. B_0 is estimated from the zero control tube, i.e., the assay tube with no unlabeled standard. Since RIAs use limiting amounts of antibody and excess amounts of tracer, the tracer should saturate the available antibody binding sites when no unlabeled hormone is present. Thus, the value of B_0 largely reflects the concentration and affinity of the available antibody. The value of B_∞, in contrast, reflects the presence of materials that bind nonspecifically to antibody (or even tracer that binds nonspecifically to the side of the assay tube) and cannot be displaced by hormone; indeed, B_∞ is often denoted NSB, for "nonspecific binding." In general, the challenge is to make B_0 and B_∞ as far apart as possible to increase the range of informative assay responses.

The error structure associated with a particular standard curve depends upon the absolute level of radioactivity measured over the full range of the curve, the reliability of the counting instrument, and the efficiency of the process used to separate the free and bound hormone. The absolute number of radioactive decay events counted at various points along the standard curve has a major influence on the error structure. Radioactive decay is a random process: the number of counts in a given assay tube is approximately normally distributed, with a variance that tends to decline as the mean increases. Thus, the greater the amount of tracer per tube, the *better* the precision of any estimate of that amount. At least in theory, then, the precision of estimation at the B_0 end of the curve should always be better than at the B_∞ end if counting is done over the same time interval for both segments of the curve, as is usually the case. Partly offsetting this factor is the fact that variation in the separation and recovery of antibody-bound hormone results in larger absolute errors at the B_0 end of the curve, because the number of radioactive decay events per mass of total hormone is greatest in the tubes that contain the least nonradioactive hormone. The balance between these errors usually results in a parabolic error profile similar to the one shown in Figure A.9. The "explosion" of errors at the upper and lower segments of the curve means that estimates falling in those regions are at best semi-quantitative. Because of this distribution of errors, exact RIA results are usually reported only for the portion of the curve that falls between about 10–15 percent and 80–90 percent of B_0. Values outside those limits are described merely as falling "below minimum" or "above maximum," and the offending samples are reassayed at new dilutions.

Another way to assess the error structure of the standard curve is provided by the *precision profile* (Figure A.10). Here, variation in the responses for known standards, spanning the diagnostic range of interest, is plotted against the hormone concentration. The usual measure of variation is the coefficient of variation (CV), or the standard deviation of the responses expressed as a percentage of the mean response. The concentrations at which the CV is close to its minimum define the useful analytical range of the assay. The CV can be computed either from multiple internal standards within a single assay (the *intra-assay CV*) or from separate, independent assays of the same reference hormone preparation (the *inter-assay CV*). Since the inter-assay CV is almost always larger than the intra-assay CV, many journals have stopped publishing both values, preferring the inter-assay CV as a more conservative estimate of assay precision.

Other Quality-Control Issues

The *assay detectability* or *limit of detection* is the minimum concentration of hormone that can be measured reliably in a given immunoassay.

Defining $\hat{\mu}_0$ and $\hat{\sigma}_0^2$ as the estimated mean and sampling variance, respectively, of B_0, we can operationalize the limit of detection as the point on the x-axis of the standard RIA curve corresponding to $\hat{\mu}_0 - 2\hat{\sigma}_0$ or, more conservatively, $\hat{\mu}_0 - 3\hat{\sigma}_0$. In other words, the limit of detection is basically a question of how reliably we can estimate the zero control parameter, B_0.

Sensitivity is the ability of the assay to discriminate small differences in hormone concentration. It too is affected by precision, but it also depends upon the slope of the assay response curve across the useful range of the assay. In general, the steeper the response curve, the more sensitive the assay since small differences on the dose scale translate into large differences on the B/B_0 scale. In theory, a step function would have infinite sensitivity at the step, but its useful range would be nil. Conversely, an extremely shallow curve would have a broad range of responses but its sensitivity would be poor. Endocrinologists seek a middle ground, aiming for a useful range covering two to three orders of magnitude of dose while keeping the assay slope high enough to allow discrimination of hormone concentrations within 5 percent of standard deviation for a given assay, or 10–15 percent of standard deviation across assays.

The most important underlying chemical factors that determine the sensitivity and limit of detection of an RIA are the affinity of the antibody being used and the molar ratio of tracer to antibody (McArthur and Colton, 1970). If the affinity of the antibody is high, it will bind the hormone at very low concentrations and provide a measurable signal even in the presence of small amounts of tracer; in addition, the minimum amount of unlabeled hormone needed to produce detectable competition for the antibody will also be small. If the mass of tracer is only marginally in excess of the amount of antibody present, competition will quickly lead to a diminution in detectable signal. The minimal useful mass of tracer is determined by the amount of potential signal it carries in each molecule. Since the rate of radioactive decay is about 450 times greater for ^{3}H than for ^{14}C, steroid assays normally use [3]H. If an even lower limit of detection is needed, ^{125}I can be used since its decay rate is about 70 times that of [3]H. To approach the detection limits attainable for steroids, assays for protein hormones must use [125]I.

Accuracy or *validity* is the ability of an assay to estimate the actual concentration of hormone in a known standard. It is quite possible for an assay to measure a standard concentration reliably but to provide estimates that disagree consistently with those generated by other methods or laboratories. An assay, in other words, can be reliably inaccurate. Since accuracy determines to a large degree whether an assay system can be trusted to provide useful results, it is the primary criterion of most quality-control assessments.

For steroids, specimens can be analyzed by both immunoassay and a physical or chemical reference technique (see above). If the immunoassay results agree with the reference analyses across the full range of concentrations relevant to the endocrinologist, the assay can be considered accurate. For protein hormones, accuracy is usually determined in one of two ways. First, the RIA results can be compared with the potency of hormone in a reference preparation as originally evaluated in a well-established bioassay, a value agreed upon by the laboratories that originally characterized the reference preparation. Alternatively, a group of independent investigators using the same RIA protocol can reach a consensus about the amount of hormone found in a set of shared standard preparations or quality-control samples.

The concept of *bias* is related to accuracy. Bias is any *systematic* or *nonrandom* deviation from known or consensus concentrations of hormones in standard preparations. One important source of bias in RIAs and other immunoassays is *cross-reactivity*, or the binding of an antibody to molecules that are functionally distinct from the hormone of interest. Many protein hormones share some of their polypeptide subunits or segments, e.g., LH and hCG or prolactin and growth hormone. If the epitope recognized by an antibody is located at one of the shared sites, the antibody may be unable to distinguish reliably between the related hormones. For example, many early immunoassays for hCG used antibodies that bound nearly as well to LH, causing a high rate of false positive diagnoses of pregnancy in studies of early fetal loss (see Chapter 6). It was not until the development of more specific antibodies, which recognized epitopes found *only* on hCG, that this problem was corrected. Table A.1 lists some of the cross-reactivities estimated for 17β-estradiol using commercial antibodies, and for LH using antibodies distributed by the NIH National Pituitary Program. In the case of estradiol, the cross reactivity of estriol and estrone could complicate interpretation of the assay results, unless chromatographic separation of the three estrogens was done first. For LH, cross-reactivity will often be less of a problem unless pregnancy, medication, or some pathology grossly elevates the concentration of hCG or TSH in the circulation.

Parallelism has already been mentioned: the dilution curve of a test preparation must be parallel to the standard curve before we can be confident that the test and reference preparations are similar in their capacity to bind antibody and displace tracer. Immunoassays are best run on replicated serial dilutions of each unknown sample, in addition to the reference preparations used to generate the standard curve. Serial dilu-

Table A.1.   Variation in cross-reactivity using readily available RIA reagents

Hormone	Cross-reactive molecule	Percent cross-reactivity[a]
17β-Estradiol[b]	17β-estradiol	100.00
	estriol	2.46
	7α-estradiol	1.32
	estrone	1.32
	estrone sulfate	0.21
	ethanyl estradiol[c]	0.11
	diethystilbesterol[c]	<0.01
	cholesterol	<0.01
Luteinizing hormone[d]	luteinizing hormone	100.00
	chorionic gonadotropin	100.00
	thyroid-stimulating hormone	10.00
	follicle-stimulating hormone	<1.00
	growth hormone	<1.00
	prolactin	<1.00

[a] Assessed as (hormone mass at 50 percent B/B_0) ÷ (cross-reacting mass at 50 percent B/B_0).
[b] Measured with RSL rabbit anti-estradiol-17β-3-CME:BSA serum 1580 (Radioassay Systems Laboratory, Carson, California).
[c] Synthetic estrogens.
[d] Measured with RSL rabbit anti-hLH serum 1830 (Radioassay Systems Laboratory, Carson, California).

tions of the unknown samples allow comparison of their response slope to that obtained for the standards. If the slopes are not parallel across a significant portion of the response curve (with due allowance for statistical error), the assay is unlikely to be valid.

The slope of the assay response in a sample of unknown concentration can differ from that of the standard curve for a number of reasons. The antibody may be responding to a cross-reactive molecule, either by itself or mixed in with the hormone of interest. The sample may contain carrier proteins that compete with the antibody for the binding of the hormones, both labeled and unlabeled. Whatever the cause, nonparallelism usually takes the form of a slope that is lower in the unknown response curve than in the standard curve (Figure A.11). This difference in slope may cause underestimation of hormone concentration; it may, for example, take more unlabeled hormone to displace the tracer from a combination of antibody and carrier proteins than from antibody alone (line A in Figure A.11). Or it may lead to more complex forms of bias if B_0 is also altered by the binding of tracer to carrier proteins (line B), or if the binding of tracer to antibody is inhibited in the presence of cross-reactive molecules (lines C and D).

Biphasic Responses and the Low-dose Hook

Conceptually, the competitive immunoassay response curve (Figure A.8) has much in common with the bioassay dose-response curve (Figure A.3). There are, however, some important differences. In bioassays, responses are often *biphasic:* the response curve can actually reverse direction as the hormone dose changes. An example of central importance to the present book is the biphasic response of LH to increasing levels of estradiol during the follicular phase of the ovarian cycle. Early in the cycle, rising estradiol depresses LH levels, but just before ovulation estradiol induces a rapid elevation in LH—the midcycle LH surge (see Chapter 4). The causes of such biphasic responses vary. They may reflect the existence of competing control systems within the same tissue or organism; autocrine or paracrine factors that modulate the primary effects of hormones; or changes through time in the number of receptors on the target cell. When biphasic responses occur in a bioassay, the relationship between a given response and the hormone concentration needed to achieve it becomes ambiguous.

In competitive immunoassays, a phenomenon somewhat analogous to the biphasic bioassay response is the *low-dose hook.* Sometimes the binding of tracer to antibody in the zero-dose tube is actually *lower* than that observed in the tubes with the lowest nonzero doses of unlabeled hormone, the reverse of what is normally expected. (The response curve usually becomes better behaved at higher doses of unlabeled hormone.) The low-dose hook most commonly occurs when a tracer with a high *specific activity* (ratio of radio-signal to mass) is used with a highly diluted antibody. Since the Law of Mass Action governs the ultimate binding equilibrium, the combination of low tracer mass and low antibody concentration may result in an equilibrium state in which not all antibody binding sites are saturated with tracer. If unlabeled hormone is then added in small increments, a measurable increase in the amount of antibody-hormone complex can occur before true competition sets in. The ambiguity associated with this phenomenon is similar to that in the biphasic bioassay response curve—two distinct doses can yield the same response. There are two ways to remedy the low-dose hook: either the dilutions of antibody and tracer can be readjusted, or test preparations can be measured at multiple dose levels to clarify where they actually fall on the standard curve.

OTHER COMPETITIVE IMMUNOASSAYS

Over the past decade, radioisotopes have become very expensive, concerns about the environmental and health effects of low-level radiation have risen sharply, and state and federal guidelines for the disposal of radioactive waste have become increasingly stringent. In response, quantitative endocrinologists have sought alternative methods that do not require radioactive tracers. Many of the resulting methods are still competitive immunoassays, based upon the same underlying logic as the RIA but using nonradioactive labels. A few of the most promising of these methods will be mentioned here, with appropriate references where more detail can be found.

Among the earliest substitutes for radioactive tracers were fluorescent labels, particularly fluorescein and rhodamine (Nairn, 1969). Instead of using a scintillation counter to measure radioactive decay, fluorescent immunoassays use a photometer to detect photons emitted by fluorescent labels (the labels must first be excited by light of slightly shorter wavelengths than those being monitored). An advantage of this form of labeling is that the fluorescent tracer is more stable than radioactive material and, when properly stored, can provide consistent assay performance for many months or even years. The yield of photons can approach one per molecule of fluorescent label, and each molecule can theoretically cycle repeatedly between the ground and excited (photon emitting) state over microsecond intervals of time. Thus, there seem to be few theoretical limits to the sensitivity of fluorescent immunoassays. (Practical limits may arise because of materials in the samples that reduce the efficiency of photon emission or restrict the number of signal-producing cycles that a label molecule can actually undergo.) A recent advance known as *time-resolved fluorescence* (Lövegren et al., 1985) allows for even greater sensitivity and tracer stability.

In enzyme immunoassays (EIAs), the "label" is the measurable activity of an enzyme attached to the tracer molecule. After the binding equilibrium for hormone, antibody, and tracer has been reached, and bound and free hormone molecules have been separated, the normal substrate for the enzyme is added to the assay tube, and bound tracer is measured (usually spectrophotometrically) by the amount of substrate converted to detectable end-product. Because enzymes are extremely efficient at catalyzing the conversion of substrate, the enzyme label serves, in effect, as an amplification system that generates many thousands of signals for each molecule of bound tracer. Enzymes commonly used in enzyme immunoassays include horseradish peroxidase, β-galactosidase, alkaline phosphatase, β-glucuronidase, urease, and glucose oxidase (Tijssen, 1985; Goding, 1986). All these enzymes can be isolated in pure and active form and can be purchased inexpensively from commercial sources. They are, moreover, extremely stable under long-term storage, with few of the decay problems associated with radioactive labels. Since the intensity of signal is primarily a function of the time allowed for the enzyme to act upon its substrate, enzyme labeling provides excellent limits of detection in many assay systems.

One other form of immunoassay, the luminescent immunoassay, is becoming increasingly popular (Wilchek and Bayer, 1990). In this type of competitive assay the tracer is labeled with a molecule that, upon oxidation, generates an electronically excited intermediate molecule, which then relaxes to ground state and emits a photon of detectible (often visible) light. Photodetectors transform the light signal into electric currents, which can be used to assess the peak height or integrated area of the signal for the few seconds that it lasts. This system enables detection of very low levels of labeled tracer, so long as background interference is kept to a minimum. Recent results with luminescent immunoassays suggest useful response ranges that are 20–200 times greater than for other competitive immunoassays (DeLuca, 1988).

The recent proliferation of nonradiometric competitive immunoassays, including many types not mentioned here, will almost certainly yield further improvements in sensitivity, range, and precision. In comparison with classic RIAs, many of these assays are less expensive, more portable, and more robust to changes in assay conditions. Such features make these methods very useful for work at remote field sites or developing-country institutions. However, the proliferation of assay types does require the researcher to document carefully which system is being used, how well it performs, and how its results compare with those from reference methods or more commonly used assays when applied to reference standard preparations.

NONCOMPETITIVE ASSAYS

The other major format for the immunological quantification of hormone concentrations is the noncompetitive immunometric assay (Baker et al., 1985). Examples include the immunoradiometric assay (IRMA), the enzyme-linked immunosorbent assay (ELISA), the fluoroimmunometric assay (FIMA), and the luminescent immunometric assay (LIMA). In all these assays, a sample or standard preparation is first incubated with an antibody directed at one particular epitope on the hormone of interest. The antibody must be present in amounts too large to be saturated, so it binds essentially all the available hormone. This "capture" antibody is usually immobilized on an insoluble matrix, allowing the unbound antibody to be washed away with a buffer solution. A second antibody directed at a *different* epitope is then incubated with the captive hormone, again in amounts too great for saturation. The second antibody is labeled, typically with one of the radioactive, fluorescent, or enzyme-linked labels already discussed, and thus serves as tracer. Following the second incubation, the amount of tracer bound to the original hormone-antibody complex is measured, providing a direct assessment of the hormone concentration. Note that this approach does not involve competition between labeled and unlabeled hormone for binding to an antibody; it uses a labeled antibody itself as tracer.

One limitation of noncompetitive immunometric assays is that their use is restricted to large molecules such as proteins, because only large molecules possess multiple epitopes. It is essential that the two epitopes recognized in the assay do not overlap or interfere with each other's binding to the antibody, a requirement that is most easily met if the epitopes are widely separated on the molecule.

In compensation, the use of two antibodies increases the specificity of these assays over that of competitive immunoassays. Since the hormone must undergo two separate immunological screens, any lack of specificity in one of the antibodies is largely corrected by the other. Although this increase in specificity is usually desirable, it can sometimes result in underestimation of the bioactive hormone present in a sample. This bias can occur when the pair of chosen antibodies is "hyperspecific" and selects against hormonal forms that may still retain biological potency, even though they differ in minor ways from the preparation used to raise the antibodies. This problem can be avoided by using a mixture of capture antibodies raised to the multiple molecular forms that are bioactive in the system under investigation. This may involve using either a polyclonal antiserum or a mixture of monoclonal antibodies.

In noncompetitive assays, the hormone is the limiting reagent. All other reagents, including the two antibodies, are present in excess. This feature has several implications. First, since binding reactions are forced to completion by the large quantity of antibody, the assays often reach equilibrium more quickly than competitive immunoassays and are therefore faster to run. In addition, more hormone molecules are bound at equilibrium, and more signal-generating tracer molecules remain at the detection stage, than would usually be the case in immunoassays. And because the first step in the immunometric assay involves the capture of hormone by matrix-bound antibody, these assays can concentrate hormone from much larger fluid volumes than are normally used in competitive assays, allowing more dilute samples to be assayed. As a result of all these factors, noncompetitive assays often have significantly lower limits of detection than their competitive counterparts. These low limits of detection have allowed studies of physiological states that are inaccessible to RIA. For example, the earliest possible detection of pregnancy using hCG is currently being pushed back from 9–10 days to about 5–7 days postovulation (John O'Connor, personal communication,

1994). Other advantages of noncompetitive assays are discussed by Campbell (1988) and by Strasburger et al. (1988).

The dose-response curve for noncompetitive assays (Figure A.12) differs in several important respects from that of competitive assays, most obviously in the fact that, like the bioassay response curve, it rises with most increases in hormone concentrations. Because the response increases monotonically with hormone concentration across the useful analytical range, the absolute response error is correlated with the response itself, a correlation that needs to be corrected when a statistical model is fit to the response data. On the other hand, the ease with which non-bound tracer can be washed away, and the definition of the remaining signal by one limiting reagent (the hormone in noncompetitive assays) rather than two tightly balanced reagents (antibody and tracer in competitive assays), result in intra- and inter-assay CVs that are often much smaller than in competitive assays.

If the two incubations are done at the same time, as often happens, a high-dose hook may be observed at higher hormone concentrations in noncompetitive assays (Figure A.12). This occurs because any hormone in excess of the amount needed to saturate the capture antibody may compete with captured hormone for binding to the *second*, labeled antibody. This complication can be avoided by use of a two-incubation protocol or by estimation at two or more dilutions of the sample in the same assay.

Most noncompetitive assays have extraordinarily wide useful analytical ranges, defined as falling between the limit of detection and the high-dose hook (points *A* and *B* in Figure A.12). The lower limit of the useful range for these assays is generally set at the mean zero dose response plus two or three standard deviations of the zero dose response. The upper limit is the mean dose response at the highest concentration of standard, or at the peak of the high-dose hook, minus 2–3 s.d. of that response. One example of a wide analytical range is a recent FIMA assay for hCG, in which concentrations from 1 to 10,000 IU/L can be measured in a single assay (Lövegren et al., 1985). This system allows all but the highest levels of hormone produced over the full course of pregnancy to be monitored without any dilution of the sample.

CONCLUSION

Bioassays, competitive immunoassays, and noncompetitive immunometric assays are the methods now most commonly used for measuring hormone concentrations. Although new approaches are already being developed and will doubtless offer numerous advantages, the existing methods will continue to be important in providing the standards against which newer techniques are to be judged. In this appendix we have attempted to review the basic logic of the existing assays, paying particular attention to the measurement problems that can sometimes confound the interpretation of endocrine data. Such problems of validity and reliability should never be ignored, for the sophisticated use of hormonal data requires a full understanding of the data's potential limitations. At the same time, however, these problems should be kept in perspective: they are essentially the same kinds of difficulties that affect every form of scientific measurement. Quantitative endocrinologists are keenly aware of the potential measurement problems and are constantly developing new protocols to limit their scope. The profound and enduring impact on the study of human reproduction of the methods they have already developed is itself the best measure of just how well these methods normally work.

NOTES

1. The biological fluids most commonly examined in reproductive endocrinology are plasma or serum, urine, and saliva. Plasma is blood minus its cellular fraction (red and white cells, which normally make up 40–60 percent of the total blood volume). Serum is plasma from which fibrinogen has been eliminated by allowing the blood to clot before removal of the cells. Urinary assays often measure particular *metabolites,* or metabolically altered forms, of hormones; for example, many of the steroids present in urine have been altered in the liver by conjugation with glucuronic or sulfuric acid. Salivary assays are possible only for small, lipidlike hormones such as steroids, since these hydrophobic molecules diffuse readily through the cell membranes of the salivary-gland epithelium and thereby pass from serum into saliva.

2. Of course, the response also depends upon the nature of the signal-transduction system and many other aspects of the cell's interior biochemistry, most of which are not controlled when measurements of the response are made. The models of the biological response presented here assume, in effect, that these factors are invariant in the face of changing hormone concentrations and receptor numbers.

3. These antibodies are polyclonal because several different cell lineages or "clones" in the animal's immune system contribute to them, as would be true of the normal immune response to any foreign antigen.

4. Note, however, that the standard curve for a competitive immunoassay has a negative slope, unlike the bioassay response curve.

5. Since Q_L and Q_U are generally known with error, it may be better to fit the model using weighted least squares, with the weights proportional to the inverse of the sampling variances of the estimated Q-values.

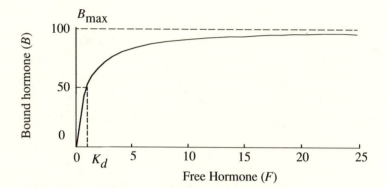

Figure A.1. The hyperbolic relationship between the concentration of bound (B) and free (F) hormone in a hormone-receptor system with high specificity and no interaction among receptors. As F increases, B approaches an upper asymptote equal to B_{max}, the total number of receptors. Fifty percent of the receptor sites are occupied when F equals K_d, the equilibrium dissociation constant.

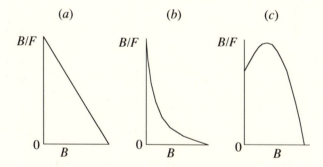

Figure A.2. Scatchard plots (B/F versus B) for (a) hormone-receptor systems with only one class of hormone and no interaction among receptors, (b) systems exhibiting heterogeneous forms of hormone, non-specific binding, or competitive interactions among receptors, and (c) systems exhibiting cooperative interactions among receptors. (Redrawn from Brown and Rothery, 1993:351)

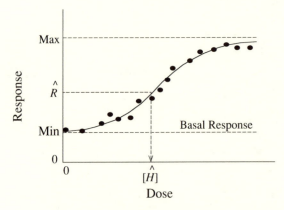

Figure A.3. A dose-response curve, showing the typical sigmoid shape. Once the curve is known, an unknown hormone concentration can be inferred from the magnitude of the response it induces (*broken line*). This type of measurement is known as a bioassay.

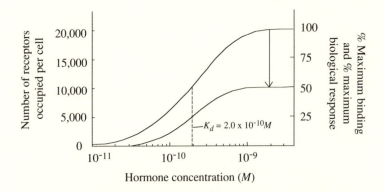

Figure A.4. Hormone binding and biological response curves when no spare receptors are present. Note that the two curves are superimposable over the entire range of hormone concentrations. A 50 percent reduction in receptor number, with no change in K_d, will result in a 50 percent reduction in maximum biological response. (Redrawn from Mendelson, 1922:30)

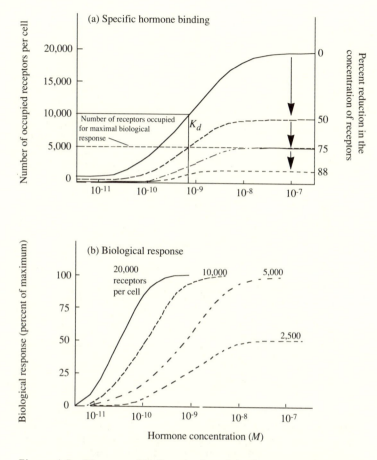

Figure A.5. Hormone binding (*top*) and biological response (*bottom*) curves when spare receptors are present. In this example, when the target cell contains its full compliment of 20,000 receptors, a maximum biological response is elicited at hormone concentrations required to occupy as few as 25 percent of the receptors. When the number of receptors is reduced to 10,000 or 5,000 without a change in K_d, a maximum response can still be attained, although progressively higher concentrations of hormone are needed. When the number of receptors drops below 25 percent (e.g., to 12 percent in this example), the maximum response is reduced proportionately. (Redrawn from Mendelson, 1992:31)

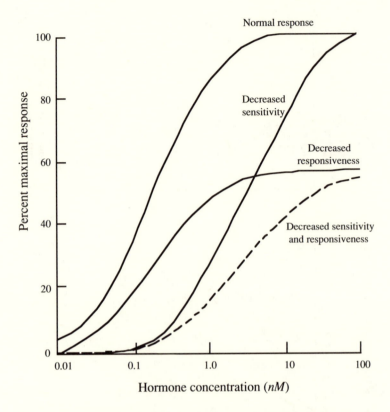

Figure A.6. Impaired target-cell responses in hormone resistance syndromes. In cases of decreased sensitivity, the dose-response curve is shifted to the right, but the maximum response can still be reached if hormone concentrations are high enough. Decreased responsiveness, in contrast, involves a reduction in the maximum biological response. Impairments in both sensitivity and responsiveness can occur in the same individual (*broken line*). (Redrawn from Kahn et al., 1992:102)

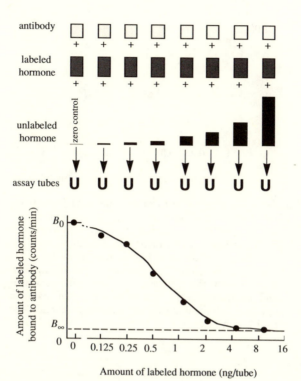

antibody

labeled
hormone

unlabeled
hormone

zero control

assay tubes

Amount of labeled hormone bound to antibody (counts/min)

B_0

B_∞

0

0   0.125 0.25 0.5   1   2   4   8   16

Amount of labeled hormone (ng/tube)

Figure A.7. The basic design of the competitive radioimmunoassay. Antibody, labeled hormone (tracer), and unlabeled hormone (standard) are added to each of a series of assay tubes. The amount of antibody is the same in all tubes, as is that of tracer. Antibody is present in limiting quantity in each tube, and the amount of tracer is in excess of that needed to saturate the antibody when no unlabeled hormone is present. Unlabeled hormone is added to the tubes in a series of known, increasing amounts. The tube receiving no unlabeled hormone is referred to as the *zero control*. After incubation and removal of hormone (tracer and standard) that is not bound to antibody, the amount of bound tracer in each tube is measured with a scintillation counter. Since tracer and standard compete for antibody binding sites, competitive suppression of the binding of tracer to antibody is a direct reflection of the amount of standard. Thus, the greater the amount of standard in each tube, the less the bound tracer there is to be measured. The results are plotted (*lower panel*), and a statistical model is fit to them to produce the *standard RIA curve*. In this figure, the amounts of labeled and unlabeled hormone are not to scale. For example, the amount of unlabeled hormone required to reduce the curve to 50 percent of B_0 should be equal to the amount of tracer in each tube.

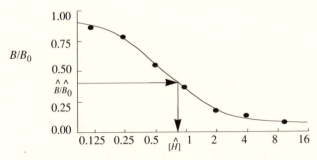

Amount of unlableled hormone (ng/tube)

Figure A.8. The usual form of the standard RIA curve. The raw count of labeled tracer (B) has been scaled by the count in the zero control tube (B_0) to correct for any differences from one assay run to another. Once the standard curve has been established, it can be used to estimate the hormone concentration in an unknown test sample, $[H]$, from its observed level of standardized bound tracer, B/B_0, using a statistical prcedure known as "inverse prediction" or "backfitting" (*arrows*).

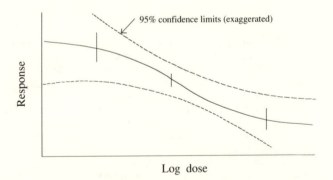

Log dose

Figure A.9. The distribution of random errors around the standard curve for a competitive immunoassay. Note that the confidence band expands rapidly near the ends of the curve as theoretical and technical limits on the precision of both sample handling and signal counting dominate the measurement process.

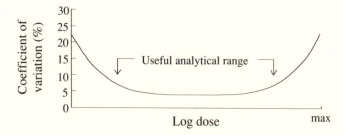

Figure A.10. The precision profile for a competitive immunoassay. The coefficient of variation (*CV*) for responses to given doses of hormone is plotted against the concentration of the hormone being measured. The inter-assay *CV* shown here is estimated from data accumulated across multiple replications of the same assay. Because of the sharp increase in errors at both low and high hormone concentrations (corresponding to the upper and lower portions, respectively, of the assay standard curve), the useful analytical range is limited to intermediate concentrations. Unknowns with concentrations outside that range must be concentrated or diluted to fall within the useful range.

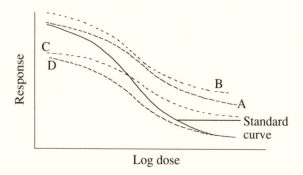

Figure A.11. The problem of non-parallelism. In competitive immunoassays, several factors can cause serial dilutions of unknown samples to follow curves (*broken lines*) that are not parallel to the standard curve (*solid line*). Those factors can include cross-reactive molecules, carrier proteins that compete with the antibody for binding to the hormone, differences in salt, pH, lipid, or protein content in the sample and assay standard, and lytic agents that destroy antibody, tracer, or unlabeled hormone in the sample but not in the standard.

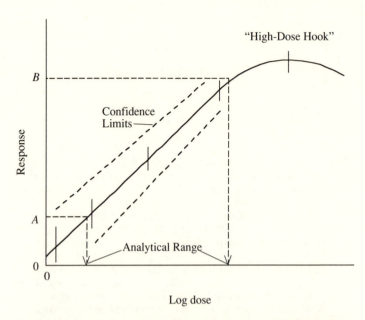

Figure A.12. A typical standard curve for a noncompetitive immunometric assay. Across a broad range of hormone concentrations, the response is approximately linear and precision is fairly uniform. The high-dose hook reflects the biphasic response often encountered in assays with a single incubation of sample, tracer, and capture antibody. The useful analytical range is defined by points *A* and *B*, where *A* = mean zero dose response + 2 *SD* of zero dose response, and *B* = mean response at the apex of the high dose hook − 2 *SD* of that response. *A* and *B* are usually estimated empirically during establishment of the standard curve.

References

Aborampah, O.-M. 1987. Plural Marriages and Fertility Differentials: A Study of the Yoruba of Western Nigeria. *Human Organization* 46:29–38.

Adair, L. S., E. Pollitt, and W. H. Mueller. 1983. Maternal Anthropometric Changes during Pregnancy and Lactation in a Rural Taiwanese Population. *Human Biology* 55:771–787.

Adams, D. B., A. R. Gold, and A. D. Burt. 1978. Rise in Female-initiated Sexual Activity at Ovulation and Its Suppression by Oral Contraceptives. *New England Journal of Medicine* 299:1145–1150.

Ahmed, J. 1986. Polygyny and Fertility Differentials among the Yoruba of Western Nigeria. *Journal of Biosocial Science* 18:63–73.

Allison, P. D. 1984. *Event History Analysis: Regression for Longitudinal Event Data.* Sage University Paper Series on Quantitative Applications in the Social Sciences, No. 07-046. London: Sage Publications.

Anderson, A. N., and V. Schioler. 1982. Influence of Breastfeeding Pattern on Pituitary-ovarian Axis of Women in an Industrialized Community. *American Journal of Obstetrics and Gynecology* 143:673–677.

Anderson, B. A. 1975. Male Age and Fertility: Results from Ireland Prior to 1911. *Population Index* 41:561–566.

Anderton, D. L., N. O. Tsuya, L. L. Bean, and G. P. Mineau. 1987. Intergenerational Transmission of Relative Fertility and Life Course Patterns. *Demography* 24:467–480.

Ann, T. B., R. Othman, W. P. Butz, and J. DaVanzo. 1983. Age at Menarche in Peninsular Malaysia: Time Trends, Ethnic Differentials and Association with Ages at Marriage and First Birth. *Malaysian Journal of Reproductive Health* 1:91–108.

Antonarakis, S. E., and the Down Syndrome Collaborative Group. 1991. Parental Origin of the Extra Chromosome in Trisomy 21 as Indicated by Analysis of DNA Polymorphisms. *New England Journal of Medicine* 324:872–876.

Apter, D., and R. Vihko. 1983. Early Menarche, a Risk Factor for Breast Cancer, Indicates Early Onset of Ovulatory Cycles. Journal of Clinical Endocrinology and Metabolism 57:82–86.

Apter, D., L. Viinikka, and R. Vihko. 1978. Hormonal Patterns of Adolescent Menstrual Cycles. *Journal of Clinical Endocrinology and Metabolism* 47:944–954.

Armstrong, C. E., P. H. Ehrlich, S. Birken, J. P. Schlatterer, E. Siris, W. C. Hembree, and R. E. Canfield. 1984. Use of a Highly Sensitive and Specific Immunoradiometric Assay for Detection of Human Chorionic Gonadotropin in Urine of Normal, Non-pregnant and Pregnant Individuals. *Journal of Clinical Endocrinology and Metabolism* 59:867–874.

Arnold, F. 1991. An Assessment of Data Quality in the Demographic and Health Surveys. Pp. 785–806 in *Proceedings of the Demographic and Health Surveys World Conference, Washington, D.C., 1991*, Volume 2. Columbia, Md.: Institute for Resource Development/Macro International.

Auletta, F. J., D. L. Kamps, S. Pories, J. Bisset, and M. Gibson. 1984. An Intracorpus Luteum Site for the Luteolytic Action of Prostaglandin $F_{2\alpha}$ in the Rhesus Monkey. *Prostaglandins* 27:285–298.

Austin, C. R., and M. W. H. Bishop. 1957. Fertilization in Mammals. *Biological Reviews* 32:296–348.

Backer, J. 1965. *Ekteskap, fodsler og vandringer i Norge 1856–1960*. Oslo: Statistik Sestralbyra.

Bailey, J., and J. Marshall. 1970. The Relationship of the Post-ovulatory Phase of the Menstrual Cycle to Total Cycle Length. *Journal of Biosocial Science* 2: 123–132.

Bailey, R. C., M. R. Jenike, P. T. Ellison, G. R. Bentley, A. M. Harrigan, and N. R. Peacock. 1992. The Ecology of Birth Seasonality among Agriculturalists in Central Africa. *Journal of Biosocial Science* 24:393–412.

Baird, D. D., C. R. Weinberg, A. J. Wilcox, D. R. McConnaughey, P. I. Musey, and D. C. Collins. 1991. Hormonal Profiles of Natural Conception Cycles Ending in Early, Unrecognized Pregnancy Loss. *Journal of Clinical Endocrinology and Metabolism* 72:793–800.

Baird, D. T. 1983a. Factors Regulating the Growth of the Preovulatory Follicle in the Sheep and Human. *Journal of Reproduction and Fertility* 69:343–352.

———. 1983b. Prediction of Ovulation: Biophysical, Physiological and Biochemical Coordinates. Pp. 1–17 in S. L. Jeffcoate (ed.), *Ovulation: Methods for Its Prediction and Detection*. New York: John Wiley.

———. 1984. The Ovary. Pp. 91–114 in C. R. Austin and R. V. Short (eds.), *Reproduction in Mammals, Volume 3: Hormonal Control of Reproduction*, 2nd edition. Cambridge: Cambridge Univ. Press.

———. 1987. A Model for Follicular Selection and Ovulation: Lessons from Superovulation. *Journal of Steroid Biochemistry* 27:15–23.

Baird, D. T., A. S. McNeilly, R. S. Sawers, and R. M. Sharpe. 1979. Failure of Estrogen-induced Discharge of Luteinizing Hormone in Lactating Women. *Journal of Clinical Endocrinology and Metabolism* 49:500–506.

Baker, T. G. 1971. Radiosensitivity of Mammalian Oocytes with Particular Reference to the Human Female. *American Journal of Obstetrics and Gynecology* 110:746–761.

———. 1972. Oogenesis and Ovulation. Pp. 14–45 in C. R. Austin and R. V. Short (eds.), *Reproduction in Mammals, Volume 1: Germ Cells and Fertilization*. Cambridge: Cambridge Univ. Press.

————. 1982. Oogenesis and Ovulation. Pp. 17–45 in C. R. Austin and R. V. Short (eds.), *Reproduction in Mammals, Volume 1: Germ Cells and Fertilization,* 2nd edition. Cambridge: Cambridge Univ. Press.

Baker, T. S., S. R. Abbott, S. G. Daniel, and J. F. Wright. 1985. Immunoradiometric Assays. Pp. 59–76 in W. P. Collins (ed.), *Alternative Immunoassays.* New York: John Wiley.

Balakrishnan, T. R. 1979. Probability of Conception, Conception Delay, and Estimates of Fecundability in Rural and Semiurban Areas of Certain Latin American Countries. *Social Biology* 26:226–231.

Baldwin, W. H., and C. W. Nord. 1984. *Delayed Childbearing in the U.S.: Facts and Fictions.* Washington: Population Reference Bureau.

Balinsky, B. I. 1975. *Introduction to Embryology,* 4th edition. Philadelphia: W. B. Saunders.

Barad, D. H. 1991. Epidemiology of Infertility. *Infertility and Reproductive Medicine Clinics of North America* 2:255–266.

Barlow, W. E., and R. L. Prentice. 1988. Residuals for Relative Risk Regression. *Biometrika* 75:65–74.

Barrett, J. C. 1969. A Monte-Carlo Simulation of Human Reproduction. *Genus* 25:1–22.

————. 1970. An Analysis of Coital Patterns. *Journal of Biosocial Science* 2:351–357.

————. 1971. Fecundability and Coital Frequency. *Population Studies* 25:309–313.

————. 1972. Use of a Fertility Simulation Model to Refine Measurement Techniques. *Demography* 8:481–490.

————. 1985. The Fertile Period, Frequency of Intercourse and Risk of Conception. *IRCS Medical Science* 13:513.

————. 1986. The Estimation of Natural Sterility. *Genus* 42:23–31.

Barrett, J. C., and J. Marshall. 1969. The Risk of Conception on Different Days of the Menstrual Cycle. *Population Studies* 23:455–461.

Bartholemew, D. J. 1963. The Sampling Distribution of an Estimate Arising in Life Testing. *Technometrics* 5:361–374.

Bassetti, M., A. Spada, G. Pezzo, and G. Giannattasio. 1984. Bromocriptine Treatment Reduces the Cell Size in Human Macroprolactinomas: A Morphometric Study. *Journal of Clinical Endocrinology and Metabolism* 58:268–273.

Bates, G. W., S. R. Bates, and N. S. Whitworth. 1982. Reproductive Failure in Women who Practice Weight Control. *Fertility and Sterility* 37:373–378.

Beall, C. M. 1983. Ages at Menopause and Menarche in a High-altitude Himalayan Population. *Annals of Human Biology* 10:365–370.

Bean, L. L., and G. P. Mineau. 1986. The Polygyny-fertility Hypothesis: A Reevaluation. *Population Studies* 40:67–81.

Becker, S. 1981. Seasonal Patterns of Fertility Measures: Theory and Data. *Journal of the American Statistical Association* 76:249–259.

————. 1993. The Determinants of Adolescent Fertility with Special Reference to Biological Variables. Pp. 21–49 in R. H. Gray, H. Leridon, and A. Spira (eds.), *Biomedical and Demographic Determinants of Reproduction.* Oxford: Clarendon Press.

Becker, S., A. Chowdhury, and H. Leridon. 1986. Seasonal Patterns of Reproduction in Matlab, Bangladesh. *Population Studies* 40:457–472.

Bedford, J. M. 1970. Sperm Capacitation and Fertilization in Mammals. *Biology of Reproduction*, Supplement 2:128–158.

———. 1982. Fertilization. Pp. 128–163 in C. R. Austin and R. V. Short (eds.), *Reproduction in Mammals, Volume 1: Germ Cells and Fertilization*, 2nd edition. Cambridge: Cambridge Univ. Press.

Bedford, J.M., and G. W. Cooper. 1978. Membrane Fusion Events in the Fertilization of Vertebrate Eggs. Pp. 66–125 in G. Poste and G. L. Nicolson (eds.), *Cell Surface Reviews, Volume 5: Membrane Fusion*. Amsterdam: Elsevier/North-Holland.

Beer, A. E., and R. Billingham. 1974. The Embryo as Transplant. *Scientific American* 230(4):36–46.

Behrman, H., D. H. Grinwich, M. Hichens, and G. J. MacDonald. 1978. Effect of Hypophysectomy, Prolactin and Prostaglandin $F_{2\alpha}$ on Gonadotropin Binding *in vivo* and *in vitro* in the Corpus Luteum. *Endocrinology* 103:349–357.

Beling, C. G., S. L. Marcus, and S. M. Markham. 1970. Functional Activity of the Corpus Luteum Following Hysterectomy. *Journal of Clinical Endocrinology and Metabolism* 30:30–39.

Belsey, M. A. 1979. Biological Factors Other than Nutrition and Lactation which may Influence Natural Fertility: Additional Notes with Particular Reference to Sub-Saharan Africa. Pp. 253–272 in H. Leridon and J. Menken (eds.), *Natural Fertility*. Liège: Ordina Editions.

Bendel, J.-P., and C. Hua. 1978. An Estimate of the Natural Fecundability Ratio Curve. *Social Biology* 25:210–217.

Bentley, G. R. 1985. Hunter-gatherer Energetics and Fertility: A Reassessment of the !Kung San. *Human Ecology* 13:79–109.

Bentley, G. R., T. Goldberg, and G. Jasie_ska. 1993. The Fertility of Agricultural and Non-agricultural Traditional Societies. *Population Studies* 47:269–281.

Bentley, G. R., A. M. Harrigan, and P. T. Ellison. 1990. Ovarian Cycle Length and Days of Menstruation of Lese Horticulturalists. Presented at the annual meeting of the American Association of Physical Anthropologists, Miami.

Berman, M., K. Hanson, and I. L. Hellman. 1972. Effect of Breast-feeding on Postpartum Menstruation, Ovulation, and Pregnancy in Alaskan Eskimos. *American Journal of Obstetrics and Gynecology* 114:524–534.

Berquo, E. S., R. M. Marques, M. L. Milanesi, J. S. Martins, E. Pinho, and I. Simon. 1968. Levels and Variations in Fertility in Sao Paulo. *Milbank Memorial Fund Quarterly* 46:167–185.

Bhalla, M., and J. R. Shrivatava. 1974. A Prospective Study of the Age of Menarche in Kampur Girls. *Indian Pediatrics* 11:487–493.

Bierman, J. M., E. Siegel, F. E. French, and K. Simonian. 1965. Analysis of the Outcome of All Pregnancies in a Community. *American Journal of Obstetrics and Gynecology* 91:37–45.

Billewicz, W. Z. 1979. The Timing of Post-partum Menstruation and Breast Feeding: A Simple Formula. *Journal of Biosocial Science* 11:141–151.

Billewicz, W. Z., H. M. Fellowes, and A. M. Thomson. 1980. Post-menarcheal Menstrual Cycles in British (Newcastle-upon-Tyne) Girls. *Annals of Human Biology* 7:177–180.

Birdsell, J. B. 1968. Some Predictions for the Pleistocene Based on Equilibrium

Systems among Recent Hunter-gatherers. Pp. 229–240 in R. B. Lee and I. DeVore, *Man the Hunter*. Chicago: Aldine.

Birken, S., M. A. Gawinowicz, A. Kardana, and L. A. Cole. 1991. The Heterogeneity of Human Chorionic Gonadotropin (hCG). II. Characteristics and Origins of Nicks in hCG Reference Standards. *Endocrinology* 129:1551–1558.

Biswis, S. 1963. A Study of Amenorrhea after Childbirth and Its Relationship to Lactation Period. *Indian Journal of Public Health* 7:9–14.

Blaha, G. C. 1964. Effect of Age of the Donor and Recipient on the Development of Transferred Golden Hamster Ova. *Anatomical Records* 150:413–416.

Blake, J. 1985. The Fertility Transition: Continuity or Discontinuity with the Past? Pp. 393–414 in *Proceedings of the International Union for the Scientific Study of Population International Population Conference, Florence*. Liège: Ordina Editions.

Blazar, A. S., J. Harlin, A. A. Zaizi, and E. Diczfalusy. 1980. Differences in Hormonal Patterns during the First Postabortion Menstrual Cycle after Two Techniques of Termination of Pregnancy. *Fertility and Sterility* 33:493–500.

Bleil, J. D., and P. M. Wassarman. 1980. Mammalian Sperm and Egg Interaction: Identification of a Glycoprotein in Mouse-egg Zonae Pellucidae Possessing Receptor Activity for Sperm. *Cell* 20:873–882.

———. 1986. Autoradiographic Visualization of the Mouse Egg's Sperm Receptor Bound to Sperm. *Journal of Cell Biology* 102:1363–1371.

Block, E. 1952. Quantitative Morphological Investigations of the Follicular System in Women: Variations at Different Ages. *Acta Anatomica* 14:108–123.

Bloom, D. E. 1982. Age Patterns of Women at First Birth. *Genus* 38:101–128.

Bloom, W., and D. W. Fawcett. 1969. *A Textbook of Histology*. Philadelphia: W. B. Saunders.

Blossfeld, H.-P., A. Hamerle, and K. U. Mayer. 1989. *Event History Analysis: Statistical Theory and Application in the Social Sciences*. Hillsdale, N. J.: Lawrence Erlbaum Associates.

Boklage, C. E. 1990. Survival Probability of Human Conceptions from Fertilization to Term. *International Journal of Fertility* 35:75–94.

Bongaarts, J. 1975. A Method for the Estimation of Fecundability. *Demography* 12:645–660.

———. 1976. Intermediate Fertility Variables and Marital Fertility Rates. *Population Studies* 30:227–241.

———. 1978. A Framework for Analyzing the Proximate Determinants of Fertility. *Population and Development Review* 4:105–132.

———. 1980. Does Malnutrition Affect Fertility? A Summary of the Evidence. *Science* 208:564–569.

———. 1982a. Fertility after Age 30: A False Alarm. *Family Planning Perspectives* 14:75–78.

———. 1982b. The Fertility-inhibiting Effects of the Intermediate Fertility Variables. *Studies in Family Planning* 13:179–189.

———. 1983. The Proximate Determinants of Natural Marital Fertility. Pp. 103–138 in R. A. Bulatao and R. D. Lee (eds.), *Determinants of Fertility in Developing Countries, Volume 1: Supply and Demand for Children*. New York: Academic Press.

Bongaarts, J., and H. Delgado. 1979. Effects of Nutritional Status on Fertility in Rural Guatemala. Pp. 107–133 in H. Leridon and J. Menken (eds.), *Natural Fertility*. Liège: Ordina Editions.

Bongaarts, J., and R. G. Potter. 1983. *Fertility, Biology, and Behavior: An Analysis of the Proximate Determinants*. New York: Academic Press.

Bongaarts, J., O. Frank, and R. Lesthaeghe. 1984. The Proximate Determinants of Fertility in Sub-Saharan Africa. *Population and Development Review* 10:511–537.

Bonnar, J., M. Franklin, P. N. Nott, and A. S. McNeilly. 1975. Effects of Breast-feeding on Pituitary-ovarian Function after Childbirth. *British Medical Journal* 4:82–84.

Bonte, M., and J. van Balen. 1969. Prolonged Lactation and Family Spacing in Rwanda. *Journal of Biosocial Science* 1:97–100.

Boros, A., L. G. Lampe, A. Balogh, J. Csonknyay, and F. Ditroi. 1986. Ovarian Function Immediately after the Menarche. *International Journal of Gynaecology and Obstetrics* 24:239–242.

Boué, J., A. Boué, and P. Lazar. 1975. Retrospective and Prospective Epidemiological Studies of 1500 Karyotyped Spontaneous Human Abortions. *Teratology* 12:11–26.

Bracher, M., and G. Santow. 1981. Some Methodological Considerations in the Analysis of Current Status Data. *Population Studies* 35:425–437.

Bradley, L. 1970. An Enquiry into Seasonality of Baptisms, Marriages, and Burials. *Local Population Studies* 5:18–35.

Brass, W. 1958. The Distribution of Births in Human Populations. *Population Studies* 12:51–72.

———. 1975. *Methods for Estimating Fertility and Mortality from Limited and Defective Data*. Chapel Hill: Carolina Population Center, Univ. of North Carolina.

Bratley, P., B. L. Fox, and L. E. Schrage. 1983. *A Guide to Simulation*. New York: Springer-Verlag.

Brenner, R. M., and I. A. Maslar. 1988. The Primate Oviduct and Endometrium. Pp. 303–329 in E. Knobil, J. D. Neill, L. L. Ewing, G. S. Greenwald, C. L. Markert, and D. W. Pfaff (eds.), *The Physiology of Reproduction*, Volume 1. New York: Raven Press.

Britton, E. 1976. The Peasant Family in Fourteenth-Century England. *Peasant Studies* 5:2–7.

Broström, G. 1983. *Estimation in a Model for Marital Fertility*. Statistical Research Report, Institute of Mathematical Statistics, Univ. of Umea, Sweden.

———. 1985. Practical Aspects on the Estimation of the Parameters in Coale's Model for Marital Fertility. *Demography* 22:625–631.

Brown, D., and P. Rothery. 1993. *Models in Biology: Mathematics, Statistics, and Computing*. New York: John Wiley.

Brown, J. B., P. Harrison, and M. A. Smith. 1978. Oestrogen and Pregnanediol Excretion through Childhood, Menarche and First Ovulation. *Journal of Biosocial Science*, Supplement 5:43–62.

———. 1985. A Study of Returning Fertility after Childbirth and during Lactation by Measurement of Urinary Oestrogen and Pregnanediol Secretion and Cervical Mucus Production. *Journal of Biosocial Science*, Supplement 9:5–23.

Brown, P. E. 1966. The Age at Menarche. *British Journal of Preventive and Social Medicine* 20:9–14.

Bugos, P. E., and L. M. McCarthy. 1984. Ayoreo Infanticide: A Case Study. Pp. 503–520 in G. Hausfater and S. B. Hrdy (eds.), *Infanticide: Comparative and Evolutionary Perspectives.* New York: Aldine de Gruyter.

Bullough, V. L. 1981. Age at Menarche: A Misunderstanding. *Science* 213:365–366.

Bumpass, L. L., and E. K. Mburugu. 1977. Age at Marriage and Completed Family Size. *Social Biology* 24:31–36.

Butler, M. G. 1990. Prader-Willi Syndrome: Current Understanding of Cause and Diagnosis. *American Journal of Medical Genetics* 35:319–332.

Byskov, A. G. 1982. Primordial Germ Cells and Regulation of Meiosis. Pp. 1–16 in C. R. Austin and R. V. Short (eds.), *Reproduction in Mammals, Volume 1: Germ Cells and Fertilization,* 2nd edition. Cambridge: Cambridge Univ. Press.

Calder, W. A. 1984. *Size, Function, and Life History.* Cambridge, Mass.: Harvard Univ. Press.

Caldwell, J. C., and P. Caldwell. 1983. The Demographic Evidence for the Incidence and Cause of Abnormally Low Fertility in Tropical Africa. *World Health Statistics Quarterly* 36:2–34.

Caldwell, P., and J. C. Caldwell. 1981. The Function of Child-spacing in Traditional Societies and the Direction of Change. Pp. 73–92 in H. J. Page and R. Lesthaeghe (eds.), *Child-spacing in Tropical Africa.* New York: Academic Press.

Cameron, I. T., and D. T. Baird. 1988. The Return to Ovulation Following Early Abortion: A Comparison between Vacuum Aspiration and Prostaglandin. *Acta Endocrinologica* 118:161–167.

Campbell, A. A. 1983. *Manual of Fertility Analysis.* Edinburgh: Churchill Livingstone.

Campbell, K. L. 1985. Methods of Monitoring Ovarian Function and Predicting Ovulation. *Research Frontiers in Fertility Regulation* 3:1–16.

――――. 1988. Solid State Assays: Reagents and Film Technology for Dip-stick Assays. Pp. 237–287 in F. P. Haseltine and B. D. Albertson (eds.), *Non-radiometric Assays: Technology and Application in Polypeptide and Steroid Hormone Detection.* New York: Alan R. Liss.

――――. 1994. Blood, Urine, Saliva, and Dip-Sticks: Experiences in Africa, New Guinea, and Boston. *Annals of the New York Academy of Sciences* 709:312–332.

Campbell, K. L., and J. W. Wood. 1988. Fertility in Traditional Societies. Pp. 39–69 in P. Diggory, M. Potts, and S. Teper (eds.), *Natural Human Fertility: Social and Biological Determinants.* London: Macmillan.

Cantrelle, P., and B. Ferry. 1979. Approche de la Fécondité Naturelle dans les Populations Contemporaines. Pp. 315–370 in H. Leridon and J. Menken (eds.), *Natural Fertility.* Liège: Ordina Editions.

Carfagna, M., E. Figurelli, G. Matarese, and S. Matarese. 1972. Menarcheal Age of Schoolgirls in the District of Naples, Italy, in 1969–70. *Human Biology* 44:117–125.

Carlson, B. M. 1988. *Patten's Foundations of Embryology,* 5th edition. New York: McGraw-Hill.

Carr, B. R., R. K. Sakler, D. B. Rochell, M. A. Stalmach, R. C. McDonald, and E. R. Simpson. 1981. Plasma Lipoprotein in Regulation of Progesterone Biosynthesis by Human Corpus Luteum in Organ Culture. *Journal of Clinical Endocrinology and Metabolism* 58:875–881.

Carter, C. O. 1982. Contribution of Gene Mutations to Genetic Disease in

Humans. Pp. 1–8 in K. C. Born (ed.), *Progress in Mutation Research 3*. Amsterdam: Elsevier Biomedical Press.

Cartwright, A. 1976. *How Many Children?* London: Routledge and Kegan Paul.

Casagrande, J. T., M. C. Pike, and B. E. Henderson. 1982. Re: Menarcheal Age and Spontaneous Abortion: A Causal Connection? (letter) *American Journal of Epidemiology* 115:481–483.

Casper, R. F., and S. S. F. Yen. 1981. Menopausal Flushes: Effects of Pituitary Gonadotropin Desensitization by a Potent Luteinizing Hormone-releasing Factor Agonist. *Journal of Clinical Endocrinology and Metabolism* 53:1056–1058.

Casterline, J. B. 1989. Maternal Age, Gravidity, and Pregnancy Spacing Effects on Spontaneous Fetal Mortality. *Social Biology* 36:186–212.

Casterline, J. B., J. Cleland, and S. Singh. 1983. The Proximate Determinants of Fertility: Cross-national and Sub-national Variations. Presented at annual meeting of the Population Association of America, Pittsburgh.

Cates, W., T. M. M. Farley, and P. J. Rowe. 1985. Worldwide Patterns of Infertility: Is Africa Different? *The Lancet* 8455:596–598.

Cates, W., R. T. Rolfs, and S. O. Aral. 1993. The Pathophysiology and Epidemiology of Sexually Transmitted Diseases in Relation to Pelvic Inflammatory Disease and Infertility. Pp. 101–125 in R. H. Gray, H. Leridon, and A. Spira (eds.), *Biomedical and Demographic Determinants of Reproduction*. Oxford: Clarendon Press.

Chakraborty, R. 1989. Family Size Distributions and Correlations between the Number of Boys and Girls in Families of Variable Sizes. *Journal of Quantitative Anthropology* 1:261–277.

Chan, D. W. (ed.). 1987. *Immunoassay: A Practical Guide*. Orlando, Florida: Academic Press.

Charbonneau, H. 1970. *Tourouvre-au-Perche au XVIIème et XVIIIème Siècles: Etude de Démographie Historique*. Paris: Presses Universitaires de France.

————. 1979. Les Régimes de Fécondité Naturelle en Amérique du Nord: Bilan et Analyse des Observations. Pp. 443–491 in H. Leridon and J. Menken (eds.), *Natural Fertility*. Liège: Ordina Editions.

Chard, T. 1987. *An Introduction to Radioimmunoassays and Related Techniques*, 3rd edition. Amsterdam: Elsevier.

Chen, L. C., S. Ahmed, M. Gesche, and W. H. Mosley. 1974. A Prospective Study of Birth Interval Dynamics in Rural Bangladesh. *Population Studies* 28:277–297.

Chen, L. C., A. K. Chowdhury, and S. L. Huffmann. 1980. Anthropometric Assessment of Energy-protein Malnutrition and Subsequent Risk of Mortality among Preschool-aged Children. *American Journal of Clinical Nutrition* 33: 1836–1843.

Chiazze, L., F. T. Brayer, J. J. Macisco, M. P. Parker, and B. J. Duffy. 1968. The Length and Variability of the Human Menstrual Cycle. *Journal of the American Medical Association* 203:377–380.

Chojnacka, H. 1980. Polygyny and the Rate of Population Growth. *Population Studies* 34:91–107.

Chowdhury, A. K. M. A. 1983. Application of a Marriage Model in Rural Bangladesh. *Journal of Biosocial Science* 15:281–287.

Chowdhury, A. K. M. A., S. L. Huffman, and G. T. Curlin. 1977. Malnutrition, Menarche and Marriage in Rural Bangladesh. *Social Biology* 24:316–325.

Christensen, R. 1984. *Data Distributions: A Statistical Handbook.* Lincoln, Mass.: Entropy Limited.

Claman, H. N. 1993. *The Immunology of Human Pregnancy.* Totowa, N. J.: Humana Press.

Clark, J. H., W. T. Schrader, and B. W. O'Malley. 1992. Mechanisms of Action of Steroid Hormones. Pp. 35–90 in J. D. Wilson and D. W. Foster (eds.), *Williams Textbook of Endocrinology,* 8th edition. Philadelphia: W. B. Saunders.

Clayton, D. 1978. A Model for Association in Bivariate Life Tables and Its Application in Epidemiological Studies of Familial Tendency in Chronic Disease Incidence. *Biometrika* 65:141–151.

Clayton, D., and J. Cuzick. 1985. Multivariate Generalisations of the Proportional Hazards Model. *Journal of the Royal Statistical Society* 148:82–117.

Cleland, J. G., and V. C. Chidambaram. 1981. The Contribution of the World Fertility Survey to an Understanding of Fertility Determinants and Trends. Presented at the general meeting of the International Union for the Scientific Study of Population, Manila.

Clermont, Y., and C. P. Leblond. 1955. Spermiogenesis of Man, Monkey, and Other Animals as Shown by the Periodic Acid-Schiff Technique. *American Journal of Anatomy* 96:229–253.

Coale, A. J. 1971. Age Patterns of Marriage. *Population Studies* 25:193–214.

———. 1972. *The Growth and Structure of Human Populations: A Mathematical Investigation.* Princeton: Princeton Univ. Press.

———. 1973. The Demographic Transition Reconsidered. *Proceedings of the International Union for the Scientific Study of Population International Population Conference* 1:53–72.

———. 1977. The Development of New Models of Nuptiality and Fertility. *Population* 32:131–154.

———. 1986. The Decline in Fertility in Europe since the Eighteenth Century as a Chapter in Demographic History. Pp. 1–30 in A. J. Coale and S. Watkins (eds.), *The Decline of Fertility in Europe.* Princeton: Princeton Univ. Press.

Coale, A. J., and D. R. McNeil. 1972. The Distribution by Age of the Frequency of First Marriage in a Female Cohort. *Journal of the American Statistical Association* 67:743–749.

Coale, A. J., and J. Trussell. 1974. Model Fertility Schedules: Variations in the Age Structure of Childbearing in Human Populations. *Population Index* 40:185–258.

———. 1978. Technical Note: Finding the Two Parameters that Specify a Model Schedule of Marital Fertility. *Population Index* 44:203–213.

Coale, A. J., B. Anderson, and E. Harm. 1979. *Human Fertility in Russia Since the Nineteenth Century.* Princeton: Princeton Univ. Press.

Cohen, A. C., and B. J. Whitten. 1988. *Parameter Estimation in Reliability and Life Span Models.* New York: Dekker.

Cohen, M. N. 1989. *Health and the Rise of Civilization.* New Haven: Yale Univ. Press.

Condon, R. G. 1982. Inuit Natality Rhythms in the Central Canadian Arctic. *Journal of Biosocial Science* 14:167–177.

Corrie, J. E. T., W. M. Hunter, and J. S. Macpherson. 1981. A Strategy for Radioimmunoassay of Plasma Progesterone with Use of a Homologous-Site I125-Labeled Radioligand. *Clinical Chemistry* 27:594–599.

Corsini, C. A. 1979. Is the Fertility-reducing Effect of Lactation Really Substantial? Pp. 195–215 in H. Leridon and J. Menken (eds.), *Natural Fertility*. Liège: Ordina Editions.

Cowgill, U. M. 1966. The Season of Birth in Man. *Man* 1:232–241.

Cowles, A., and E. N. Chapman. 1935. Statistical Study of Climate in Relation to Pulmonary Tuberculosis. *Journal of the American Statistical Association* 30:517–536.

Cox, D. R. 1962. *Renewal Theory*. London: Methuen.

———. 1972. Regression Models and Life-tables. *Journal of the Royal Statistical Society* (series B) 34:187–220.

Cox, D. R., and D. Oakes. 1984. *Analysis of Survival Data*. London: Chapman and Hall.

Creasy, M. R., J. A. Crolla, and E. D. Alberman. 1976. A Cytogenetic Study of Human Spontaneous Abortions Using Banding Techniques. *Human Genetics* 31:177–196.

Cross, H. E., and V. A. McKusick. 1970. Amish Demography. *Social Biology* 17:83–101.

Crowley, W. F., M. Filicori, D. I. Spratt, and N. F. Santoro. 1985. The Physiology of Gonadotrophin-releasing Hormone (GnRH) Secretion in Men and Women. *Recent Progress in Hormone Research* 41:473–526.

Cumming, D. C. 1993. The Effects of Exercise and Nutrition on the Menstrual Cycle. Pp. 132–156 in R. H. Gray, H. Leridon, and A. Spira (eds.), *Biomedical and Demographic Determinants of Reproduction*. Oxford: Clarendon Press.

Curtis, A., and J. Huffman. 1950. *A Textbook of Gynecology*. Philadelphia: W. B. Saunders.

Dacou-Voutetakis, C., D. Klontza, P. Lagos, A. Tzonou, E. Katsarou, S. Antoniadis, G. Papazisis, G. Papadopoulos, and N. Matsaniotis. 1983. Age of Pubertal Stages including Menarche in Greek Girls. *Annals of Human Biology* 10:557–563.

Daly, M., and M. Wilson. 1983. *Sex, Evolution, and Behavior*, 2nd edition. Boston: Willard Grant Press.

Damon, A., S. T. Damon, R. B. Reed, and I. Valadian. 1969. Age at Menarche of Mothers and Daughters, with a Note on Accuracy of Recall. *Human Biology* 41:161–175.

Dandekar, K. 1959. Intervals between Confinements. *Eugenics Quarterly* 6:180–186

Dann, T. C., and D. F. Roberts. 1984. Menarcheal Age in University of Warwick Students. *Journal of Biosocial Science* 16:511–519.

DaVanzo, J., W. P. Butz, and J.-P. Habicht. 1983. How Biological and Behavioural Influences on Mortality in Malaysia Vary during the First Year of Life. *Population Studies* 37:381–402.

Davenport, W. 1965. Sexual Patterns and Their Regulation in a Society of the

Southwest Pacific. Pp. 164–207 in F. A. Beach (ed.), *Sex and Behavior*. New York: John Wiley.

David, P. A., T. A. Mroz, W. D. Sanderson, K. W. Wachter, and D. R. Weir. 1988. Cohort Parity Analysis: Statistical Estimates of the Extent of Fertility Control. *Demography* 25:163–188.

Davidson, J. M., J. J. Chen, L. Crapo, G. D. Gray, W. J. Greenleaf, and J. A. Catania. 1983. Hormonal Changes and Sexual Function in Aging Men. *Journal of Clinical Endocrinology and Metabolism* 57:71–77.

Davis, K., and J. Blake. 1956. Social Structure and Fertility: An Analytic Framework. *Economic Development and Cultural Change* 4:211–235.

DeCherney, A. H., and G. S. Berkowitz. 1982. Female Fecundity and Age. *New England Journal of Medicine* 306:424–426.

Delaine, G. n.d. *Observation des Variables Intermédiaires dans le Cadre de la Fécondité Naturelle: Etude de l'Allaitement*. Unpublished doctoral dissertation. Paris: Institute de Démographie, Université de Paris.

Delgado, H. L., A. Lechtig, E. Brineman, R. Martorell, C. Yarbrough, and R. E. Klein. 1978. Nutrition and Birth Interval Components: The Guatemalan Experience. Pp. 385–399 in W. H. Mosley (ed.), *Nutrition and Human Reproduction*. New York: Plenum Press.

Delgado, H. L., R. Martorell, and R. E. Klein. 1982. Nutrition, Lactation, and Birth Interval Components in Rural Guatemala. *American Journal of Clinical Nutrition* 35:1468–1476.

DeLuca, M. 1988. Bioluminescence Assays. Pp. 47–60 in B. Albertson and F. Haseltine (eds.), *Non-Radiometric Assays: Technology and Application in Polypeptide and Steroid Hormone Detection*. New York: Alan R. Liss.

Delvoye, P., and C. Robyn. 1980. Breast-feeding and Post Partum Amenorrhea in Central Africa. *Journal of Tropical Pediatrics* 26:184–189.

Delvoye, P., M. Demaegd, Uwayitu-Nyampeta, and C. Robyn. 1978. Serum Prolactin, Gonadotropins, and Estradiol in Menstruating and Amenorrheic Mothers during Two Years' Lactation. *American Journal of Obstetrics and Gynecology* 130:635–639.

Demeny, P. 1975. Letter to the Editor. *Scientific American* 232:6

Devereux, G. 1955. *A Study of Abortion in Primitive Societies*. New York: Julian Press.

Díaz, S., O. Peralta, G. Juez, A. M. Salvatierra, M. E. Casado, E. Duran, and H. B. Croxatto. 1982. Fertility Regulation in Nursing Women, I: The Probability of Conception in Full Nursing Women in an Urban Setting. *Journal of Biosocial Science* 14:329–341.

Díaz, S., M. Serón-Ferré, H. Cárdenas, V. Schiappacasse, A. Brandeis, and H. B. Croxatto. 1989. Circadian Variation of Basal Plasma Prolactin, Prolactin Response to Suckling, and Length of Amenorrhea in Nursing Women. *Journal of Clinical Endocrinology and Metabolism* 68:946–955.

Dixon, R. B. 1971. Explaining Cross-cultural Variations in Age at Marriage and Proportions Never Marrying. *Population Studies* 25:215–233.

Dobbins, J. G. 1980. Implication of a Time-dependent Model of Sexual Intercourse within the Menstrual Cycle. *Journal of Biosocial Science* 12:133–140.

Donnet, M. L., P. W. Howie, M. Marnie, W. Cooper, and M. Lewis. 1990. Return of Ovarian Function Following Spontaneous Abortion. *Clinical Endocrinology* 33:13–20.

Döring, G. K. 1963. Uber die Relative Haufigkeit des Anovulatorischen Cyclus im Leben der Frau. *Archiv Gynaekologie* 199:115–131.

———. 1969. The Incidence of Anovular Cycles in Women. *Journal of Reproduction and Fertility*, Supplement 6:77–81.

Dorjahn, V. 1959. Fertility, Polygyny and Their Interrelationships in Temne Society. *American Anthropologist* 60:838–860.

Doyle, L. L., D. L. Barclay, G. W. Duncan, and K. T. Kirton. 1971. Human Luteal Function Following Hysterectomy as Assessed by Plasma Progestin. *American Journal of Obstetrics and Gynecology* 110:92–97.

Drife, J. O., and K. Baynham. 1988. Breast Sensitivity during Lactation: A Predictor of Returning Fertility? Report to the Task Force on Natural Methods for the Regulation of Fertility.

Duchen, M. R., and A. S. McNeilly. 1980. Hyperprolactinaemia and Long-term Lactational Amenorrhoea. *Clinical Endocrinology* 12:621–627.

Dufau, M. L., C. Mendelson, and K. J. Catt. 1974. A Highly Sensitive In Vitro Bioassay for Luteinizing Hormone and Chorionic Gonadotropin: Testosterone Production by Dispersed Leydig Cells. *Journal of Clinical Endocrinology and Metabolism* 39:610–613.

Dumond, D. E. 1975. The Limitation of Human Population: a Natural History. *Science* 187:713–721.

Dym, M. 1977. The Male Reproductive System. Pp. 979–1038 in L. Weiss and R. O. Greep (eds.), *Histology*, 4th edition. New York: McGraw-Hill.

Early, J. D. 1985. Low Forager Fertility: Demographic Characteristic or Methodological Artifact? *Human Biology* 57:387–399.

Eaton, J., J. Eckman, E. Berger, and H. Jacob. 1976. Suppression of Malaria Infection by Oxidant-sensitive Host Erythrocytes. *Nature* 264:758–759.

Eaton, J. W., and A. J. Mayer. 1953. The Social Biology of Very High Fertility among the Hutterites: The Demography of a Unique Population. *Human Biology* 25:206–264.

Eaton, S. B., M. C. Pike, R. V. Short, J. Trussell, R. A. Hatcher, J. W. Wood, C. M. Worthman, N. Blurton-Jones, M. J. Konner, K. Hill, and R. Bailey. 1992. Women's Reproductive Cancers in Evolutionary Context. Presented at the annual meeting of the Population Association of America, Denver.

Edmonds, D. K., K. S. Lindsay, J. F. Miller, E. Williamson, and P. J. Woods. 1982. Early Embryonic Mortality in Women. *Fertility and Sterility* 38:447–453.

Edwards, R. G. 1980. *Conception in the Human Female*. New York: Academic Press.

Efron, B. 1982. *The Jackknife, the Bootstrap and Other Resampling Plans*. Philadelphia: Society for Industrial and Applied Mathematics.

Ehrenkranz, J. R. L. 1983. Seasonal Breeding in Humans: Birth Records of the Labrador Eskimo. *Fertility and Sterility* 40:485–489.

Eisenberg, J. F. 1981. *The Mammalian Radiations*. Chicago: Univ. of Chicago Press.

Elandt-Johnson, R. C., and N. L. Johnson. 1980. *Survival Models and Data Analysis*. New York: John Wiley.

Elias, M. F., J. Teas, J. Johnston, and C. Bora. 1986. Nursing Practices and Lactation Amenorrhea. *Journal of Biosocial Science* 18:1–10.

Eliasson, R. 1976. Accessory Glands and Seminal Plasma with Special Reference to Infertility as a Model for Studies on Induction of Sterility in the Male. *Journal of Reproduction and Fertility*, Supplement 24:163–174.

Ellison, P. T. 1982. Skeletal Growth, Fatness, and Menarcheal Age: A Comparison of Two Hypotheses. *Human Biology* 54:269–281.

———. 1986. Reproductive Physiology of Lese Women. Presented at the annual meeting of the American Association of Physical Anthropologists, Albuquerque.

———. 1988. Human Salivary Steroids: Methodological Considerations and Applications in Physical Anthropology. *Yearbook of Physical Anthropology* 31:115–142.

———. 1990. Human Ovarian Function and Reproductive Ecology: New Hypotheses. *American Anthropologist* 92:933–952.

Ellison, P. T., and C. Lager. 1986. Moderate Recreational Running is Associated with Lowered Salivary Progesterone Profiles in Women. *American Journal of Obstetrics and Gynecology* 154:1000–1004.

Ellison, P. T., C. Panter-Brick, S. F. Lipson, and M. T. O'Rourke. 1993. The Ecological Context of Human Ovarian Function. *Human Reproduction* 8:2248–2258.

Ellison, P. T., N. R. Peacock, and C. Lager. 1986. Salivary Progesterone and Luteal Function in Two Low-fertility Populations of Northeast Zaire. *Human Biology* 58:473–483.

———. 1989. Ecology and Ovarian Function among Lese Women of the Ituri Forest, Zaire. American Journal of Physical Anthropology 78:519–526.

Erhardt, C. L. 1963. Pregnancy Losses in New York City, 1960. *American Journal of Public Health* 53:1337–1352.

Evans, M. D. R. 1986. Fertility Patterns in Black and White Cohorts Born 1903–1956. *Population and Development Review* 12:267–293.

Eveleth, P. B., and J. M. Tanner. 1976. *Worldwide Variation in Human Growth.* Cambridge: Cambridge Univ. Press.

Ewens, W. J. 1979. *Mathematical Population Genetics.* Berlin: Springer-Verlag.

Farid-Coupal, N., M. L. Contreras, and H. M. Castellano. 1981. The Age at Menarche in Carabobo, Venezuela, with a Note on Secular Trend. *Annals of Human Biology* 8:283–288.

Farris, E. J. 1956. Human Ovulation and Fertility. Philadelphia: J. B. Lippincott.

Fawcett, D. W. 1975. The Mammalian Spermatozoan. *Developmental Biology* 44:394–436.

Fédération CECOS, D. Schwartz, and M.-J. Mayaux. 1982. Female Fecundity as a Function of Age. *New England Journal of Medicine* 306:404–406.

Federici, N., and L. Terrenato. 1982. Biological Determinants of Early Life Mortality. Pp. 331–361 in S. H. Preston (ed.), *Biological and Social Aspects of Mortality and the Length of Life.* Liège: Ordina Editions.

Feeney, G. 1983. Population Dynamics Based on Birth Intervals and Parity Progression. *Population Studies* 37:75–89.

Feinman, M. A. 1991. Genetic Factors and Infertility. *Infertility and Reproductive Medicine Clinics of North America* 2:277–285.

Feldkamp, C. S., and S. W. Smith. 1987. Practical Guide to Immunoassay Method Evaluation. Pp. 49–95 in D. W. Chan (ed.), *Immunoassay: A Practical Guide.* Orlando, Florida: Academic Press.

Fenster, L., B. Eskenazi, G. C. Windham, and S. H. Swann. 1991. Coffee Consumption during Pregnancy and Spontaneous Abortion. *Epidemiology* 2: 168–174.

Fildes, V. A. 1988. *Wet Nursing: A History from Antiquity to the Present.* Oxford: Basil Blackwell.

Findlay, A. L. R. 1974. The Role of Suckling in Lactation. Pp. 453–477 in J. B. Josimovich, M. Reynolds, and E. Cobo (eds.), *Lactogenic Hormones, Fetal Nutrition, and Lactation.* New York: John Wiley.

Finney, D. J. 1971. *Statistical Method in Biological Assay,* 2nd edition. London: Griffin.

Fisher, A. (ed.). 1988. *Sri Lanka Demographic and Health Survey.* Columbia, MD: Institute for Resource Development/Macro International.

Fisher, R. A. 1956. *Statistical Methods and Scientific Inference.* Edinburgh: Oliver and Boyd.

Flinn, M. W. 1981. *The European Demographic System, 1500–1820.* Baltimore: Johns Hopkins Univ. Press.

Florman, H. M., and P. M. Wassarman. 1985. O-linked Oligosaccharides in Mouse Egg ZP3 Account for Its Sperm Receptor Activity. *Cell* 41:313–324.

Ford, C. S. 1945. *A Comparative Study of Human Reproduction.* New Haven: Yale Univ. Press.

Ford, K., and Y. Kim. 1987. Distributions of Postpartum Amenorrhea: Some New Evidence. *Demography* 24:413–430.

Ford, K., S. L. Huffman, A. K. M. A. Chowdhury, S. Becker, H. Allen, and J. Menken. 1989. Birth-interval Dynamics in Rural Bangladesh and Maternal Weight. *Demography* 26:425-437.

Foster, A., J. A. Menken, A. K. M. A. Chowdhury, and J. Trussell. 1986. Female Reproductive Development: A Hazards Model Analysis. *Social Biology* 33:183–198.

Foster, D. L., S. M. Yen, and D. H. Osler. 1985. Internal and External Determinants of the Timing of Puberty in the Female. *Journal of Reproduction and Fertility* 75:327–344.

Fox, B. W. 1976. *Techniques of Sample Preparation for Liquid Scintillation Counting.* Amsterdam: North-Holland.

France, J. T. 1981. Overview of the Biological Aspects of the Fertile Period. *International Journal of Fertility* 26:143–152.

Franceschini, R., P. L. Venturini, A. Cataldi, T. Barreca, N. Ragni, and E. Rolandi. 1989. Plasma Beta-endorphin Concentrations during Suckling in Lactating Women. *British Journal of Obstetrics and Gynaecology* 96:711–713.

Frank, O. 1983. Infertility in Sub-Saharan Africa: Estimates and Implications. *Population and Development Review* 9:137–144.

Frank, O., P. G. Bianchi, and A. Campana. 1992. The End of Fertility: Age, Fecundity and Fecundability in Women. Presented at the annual meeting of the Population Association of America, Denver.

Freedman, R., P. K. Whelpton, and A. Campbell. 1959. *Family Planning, Sterility, and Population Growth.* New York: McGraw-Hill.

French, F. E., and J. M. Bierman. 1962. Probabilities of Fetal Mortality. *Public Health Report* 77:835–847.

Fricke, T. E. 1986. *Himalayan Households: Tamang Demography and Domestic Processes*. Ann Arbor: UMI Research Press.

Frisch, R. E. 1975. Demographic Implications of the Biological Determinants of Female Fecundity. *Social Biology* 22:17–22.

———. 1978. Population, Food Intake, and Fertility. *Science* 199:22–30.

Frisch, R. E., and J. W. McArthur. 1974. Menstrual Cycles: Fatness as a Determinant of Minimum Weight for Height Necessary for Their Maintenance and Onset. *Science* 185:949–951.

Frisch, R. E., and R. Revelle. 1970. Height and Weight at Menarche and a Hypothesis of Critical Body Weights and Adolescent Events. *Science* 169: 397–399.

Fuchs, A.-R., and M. Y. Dawood. 1980. Oxytocin Release and Uterine Activation during Parturition in Rabbits. *Endocrinology* 107:1117–1126.

Gage, T. B. 1989. Bio-mathematical Approaches to the Study of Human Variation in Mortality. *Yearbook of Physical Anthropology* 32:185–214.

Gaisie, S. K. 1984. *The Proximate Determinants of Fertility in Ghana*. WFS Scientific Report No. 53. Voorburg, Netherlands: International Statistical Institute.

Galdikas, B. M. F., and J. W. Wood. 1990. Birth Spacing Patterns in Humans and Apes. *American Journal of Physical Anthropology* 75:219–232.

Galton, F. 1889. *Natural Inheritance*. London: Macmillan.

Ganiage, J. 1960. *La Population Européene de Tunis au Milieu de XIXème Siècle*. Paris: Presses Universitaires de France.

———. 1963. *Trois Villages d'Ile-de-France au XVIIème Siècle*. Paris: Presses Universitaires de France.

Garenne, M., and E. van de Walle. 1985. Polygyny and Fertility among the Sereer of Senegal. Unpublished manuscript.

Garland, H. O. 1985. Maternal Adjustments to Pregnancy. Pp. 1–19 in R. M. Case (ed.), *Variations in Human Physiology*. Manchester: Manchester Univ. Press.

Gautier, E., and L. Henry. 1958. *La Population de Crulai, Paroisse Normande: Etude Historique*. Institut National d'Etudes Démographiques, Travaux et Documents, Cahier No. 33. Paris: Presses Universitaires de France.

Gehan, E. A. 1969. Estimating the Survival Function from the Life Table. *Journal of Chronic Diseases* 21:629–644.

Geissler, A. 1885. Ueber den Einfluss der Sauglingsterblichkeit und die Eheliche Fruchtbarkeit. *Zeitschrift des Karl Sachsischen Statistischen Bureau* 31:27–52.

Gelehrter, T. D., and F. S. Collins. 1990. *Principles of Medical Genetics*. Baltimore: Williams and Wilkins.

Gilbert, S. F. 1988. *Developmental Biology*, 2nd edition. Sunderland, Mass.: Sinauer Associates.

———. 1991. *Developmental Biology*, 3rd edition. Sunderland, Mass: Sinauer Associates.

Gini, C. 1924. Premières Recherches sur la Fécondabilité de la Femme. *Proceedings of the International Mathematical Congress* 2:889–892.

Ginsberg, R. B. 1973. The Effect of Lactation on the Length of the Postpartum Anovulatory Period: An Application of a Bivariate Stochastic Model. *Theoretical Population Biology* 4:276–299.

Girard, P. 1959. Aperçus de la Démographie de Sotteville-lès-Rouen vers la Fin du XVIIIème Siècle. *Population* 14:485–508.

Gladen, B. 1986. On the Role of "Habitual Aborters" in the Analysis of Spontaneous Abortion. *Statistics in Medicine* 5:557–564.

Glasier, A., A. S. McNeilly, and D. T. Baird. 1986. Induction of Ovarian Activity by Pulsatile Infusion of LHRH in Women with Lactational Amenorrhoea. *Clinical Endocrinology* 24:243–252.

Glasier, A., A. S. McNeilly, and P. W. Howie. 1983. Fertility after Childbirth: Changes in Serum Gonadotrophin Levels in Bottle and Breast Feeding Women. *Clinical Endocrinology* 24:243–252.

———. 1984. Pulsatile Secretion of LH in Relation to the Resumption of Ovarian Activity Post Partum. *Clinical Endocrinology* 19:493–501.

Glass, D. V., and E. Grebenik. 1954. *The Trend and Pattern of Fertility in Great Britain.* Papers of the Royal Commission on Population, Volume 6. London: Her Majesty's Stationery Office.

Glass, M. R., B. T. Rudd, S. S. Lynch, and W. R. Butt. 1981. Oestrogen: Gonadotrophin Feedback Mechanisms in the Puerperium. *Clinical Endocrinology* 14:257–267.

Glass, R. H. 1991. Infertility. Pp. 689–709 in S. S. C. Yen and R. B. Jaffe (eds.), *Reproductive Endocrinology.* Philadelphia: W. B. Saunders.

Glasser, J. H., and P. A. Lachenbruch. 1968. Observations on the Relationship between Frequency and Timing of Intercourse and the Probability of Conception. *Population Studies* 22:399–407.

Goding, J. W. 1986. *Monoclonal Antibodies: Principles and Practice,* 2nd edition. London: Academic Press.

Golbeck, A. L. 1981. A Probability Mixture Model of Completed Parity. *Demography* 18:645–658.

Goldman, N., and M. Montgomery. 1989. Fecundability and Husband's Age. *Social Biology* 36:146–166.

Goldman, N., C. F. Westoff, and L. E. Paul. 1985. Estimation of Fecundability from Survey Data. *Studies in Family Planning* 16:252–259.

———. 1987. Variations in Natural Fertility: The Effect of Lactation and Other Determinants. *Population Studies* 41:127–146.

Goluboff, L. G., and C. Ezrin. 1969. Effect of Pregnancy on the Somatotroph and the Prolactin Cell of the Human Adenohypophysis. *Journal of Clinical Endocrinology and Metabolism* 29:1533–1538.

Goodman, M. J., A. Estioko-Griffin, P. B. Griffin, and J. S. Grove. 1985. Menarche, Pregnancy, Birth Spacing and Menopause among the Agta Women Foragers of Cagayan Province, Luzon, the Philippines. *Annals of Human Biology* 12:169–177.

Gordon, K., M. B. Renfree, R. V. Short, and I. J. Clarke. 1987. Hypothalamo-pituitary Portal Blood Concentrations of Beta-endorphin during Suckling in the Ewe. *Journal of Reproduction and Fertility* 79:397–408.

Gosden, R. G. 1985. Maternal Age: A Major Factor Affecting the Prospects and Outcome of Pregnancy. *Annals of New York Academy of Sciences* 442:45–57.

Gostwamy, R. K., G. Williams, and P. C. Steptoe. 1988. Decreased Uterine Perfusion: A Cause of Infertility. *Human Reproduction* 3:955–959.

Gould, K. G., M. Flint, and C. E. Graham. 1981. Chimpanzee Reproductive Senescence: A Possible Model for Evolution of the Menopause. *Maturitas* 3:157–166.

Graham, C. E. 1981. Menstrual Cycle of the Great Apes. Pp. 1–43 in C. E. Graham (ed.), *Reproductive Biology of the Great Apes.* New York: Academic Press.

Graham, R. L., D. L. Grimes, and R. D. Campbell. 1979. Amenorrhea Secondary to Voluntary Weight Loss. *Southern Medical Journal* 72:1259–1264.

Grainger, C. R. 1980. The Age of Menarche in Schoolgirls in Mahe, Seychelles. *Transactions of the Royal Society of Tropical Medicine and Hygiene* 74:123–124.

Gray, R. H. 1976. The Menopause: Epidemiological and Demographic Considerations. Pp. 25–40 in R. J. Beard (ed.), The Menopause: *A Guide to Current Research and Practice.* Lancaster, United Kingdom: MTP Press.

————. 1979. Biological Factors Other than Nutrition and Lactation Which May Influence Natural Fertility: A Review. Pp. 217–251 in H. Leridon and J. Menken (eds.), *Natural Fertility.* Liège: Ordina Editions.

————. 1981. Birth Intervals, Postpartum Sexual Abstinence and Child Health. Pp. 93–110 in H. J. Page and R. Lesthaeghe (eds.), Child-spacing in Tropical Africa: Traditions and Change. New York: Academic Press.

————. 1983. The Impact of Health and Nutrition on Natural Fertility. Pp. 139–162 in R. Bulatao and R. Lee (eds.), *Determinants of Fertility in Developing Countries, Volume 1: Supply and Demand for Children.* New York: Academic Press.

Gray, R. H., R. Apelo, O. Campbell, S. Eslami, H. Zacur, and M. Labbock. 1993. The Return of Ovarian Function during Lactation: Results of Studies from the United States and the Philippines. Pp. 428–445 in R. H. Gray, H. Leridon, and A. Spira (eds.), *Biomedical and Demographic Determinants of Reproduction.* Oxford: Clarendon Press.

Gray, R. H., O. M. Campbell, R. Apelo, S. S. Eslami, H. Zacur, R. M. Ramos, J. C. Gehret, and M. H. Labbok. 1990. The Risk of Ovulation during Lactation. *The Lancet* 335:25–29.

Gray, R. H., O. M. Campbell, H. A. Zacur, M. H. Labbock, and S. L. MacRae. 1987. Postpartum Return of Ovarian Activity in Nonbreastfeeding Women Monitored by Urinary Assays. *Journal of Clinical Endocrinology and Metabolism* 64:645–650.

Grech, E., J. Everett, and F. Mukasa. 1973. Epidemiological Aspects of Acute Pelvic Inflammatory Disease in Uganda. *Tropical Doctor* 3:123–132.

Greenman, G. W., N. O. Gabrielson, J. Howard-Flanders, and M. A. Wessel. 1962. Thyroid Dysfunction in Pregnancy: Fetal Loss and Follow-up Evaluation of Surviving Infants. *New England Journal of Medicine* 267:1442–1444.

Greenwald, G. S., and P. F. Terranova. 1988. Follicular Selection and Its Control. Pp. 387–445 in E. Knobil, J. D. Neill, L. L. Ewind, G. S. Greenwald, C. L. Markert, and D. W. Pfaff (eds.), *The Physiology of Reproduction,* Volume 1. New York: Raven Press.

Griffin, J. E. 1992. Assessment of Endocrine Function. Pp. 61–74 in J. E. Griffin and S. R. Ojeda (eds.), *Textbook of Endocrine Physiology,* 2nd edition. New York: Oxford Univ. Press.

Griffin, J. E., and S. R. Ojeda (eds.). 1992. *Textbook of Endocrine Physiology,* 2nd edition. New York: Oxford Univ. Press.

Griffin, J. E., and J. D. Wilson. 1989. The Androgen Resistance Syndromes: 5α-Reductase Deficiency, Testicular Feminization, and Related Disorders. Pp.

1919–1944 in C. R. Scriver, A. L. Beaudet, W. S. Sly, and D. Valle (eds.), *The Metabolic Basis of Inherited Disease*, 6th edition. New York: McGraw-Hill.

Grimm, F. M., and D. Diderot. 1813. *Correspondance Litteraire, Philosophique et Critique*, Volume 5. Paris: Garnier Fréres.

Gross, B. A. 1981. The Hormonal and Ecological Correlates of Lactation Infertility. *International Journal of Fertility* 26:209–218.

Gross, B. A., and C. J. Eastman. 1985. Prolactin and the Return to Ovulation in Breast-feeding Women. *Journal of Biosocial Science*, Supplement 9:25–32.

Gross, B. A., C. J. Eastman, K. M. Bowen, and A. P. McElduff. 1979. Integrated Concentrations of Prolactin in Breast Feeding Mothers. *Australian and New Zealand Journal of Obstetrics and Gynecology* 19:150–153.

Grudzinskas, J. G., and A. M. Nysenbaum. 1985. Failure of Human Pregnancy after Implantation. *Annals of the New York Academy of Sciences* 442:38–44.

Guerrero, R., and C. A. Lanctot. 1970. Aging of Fertile Gametes and Spontaneous Abortion. *American Journal of Obstetrics and Gynecology* 107:263–267.

Guerrero, R., and O. I. Rojas. 1975. Spontaneous Abortion and Aging of Human Ova and Spermatozoa. *New England Journal of Medicine* 293:573–575.

Gwatkin, R. B. L. 1993. *Genes in Mammalian Reproduction*. New York: Wiley-Liss.

Haber, E., and J. Novotny. 1985. The Antibody Combining Site. Pp. 57–76 in T. A. Springer (ed.), *Hybridoma Technology in the Biosciences and Medicine*. New York: Plenum Press.

Habicht, J.-P., J. DaVanzo, W. P. Butz, and L. Meyers. 1985. The Contraceptive Role of Breastfeeding. *Population Studies* 39:213–232.

Habicht, J.-P., C. Yarbrough, and R. Martorell. 1979. Anthropometric Field Methods: Criteria for Selection. Pp. 365–387 in D. B. Jelliffe and E. F. P. Jelliffe (eds.), *Human Nutrition, A Comprehensive Treatise, Volume 2: Nutrition and Growth*. New York: Plenum Press.

Hafez, E. S. E. 1975. *Scanning Electron Microscopic Atlas of Mammalian Reproduction*. New York: Springer-Verlag.

Hair, P. E. H. 1966. Bridal Pregnancy in Earlier Centuries. *Population Studies* 20:233–243.

———. 1970. Bridal Pregnancy in Rural England Further Examined. *Population Studies* 24:59–70.

Hajnal, J. 1953. Age at Marriage and Proportions Marrying. *Population Studies* 7:111–136.

———. 1965. European Marriage Patterns in Perspective. Pp. 101–143 in D. V. Glass and D. E. C. Eversley (eds.), *Population in History: Essays in Historical Demography*. London: Edward Arnold.

———. 1982. Two Kinds of Preindustrial Household Formation System. *Population and Development Review* 8:449–494.

Hall, J. G. 1990. Genomic Imprinting: Review and Relevance to Human Disease. *American Journal of Human Genetics* 46:857–873.

Hamilton, W. J., J. D. Boyd, and H. W. Mossman. 1945. *Human Embryology*. Cambridge: Heffer.

Hammes, L. M., and A. E. Treloar. 1970. Gestational Interval from Vital Records. *American Journal of Public Health* 60:1496–1505.

Handwerker, W. P. 1983. The First Demographic Transition: An Analysis of Subsistence Choices and Reproductive Consequences. *American Anthropologist* 85:5–27.

Harbitz, T. B. 1973. Morphometric Studies of the Leydig Cells in Elderly Men with Special Reference to the Histology of the Prostate. *Acta Pathologica, Microbiologica et Immunologica Scandinavica* 81A:301–313.

Harcourt, A. H., D. Fossey, K. J. Stewart, and D. P. Watts. 1980. Reproduction in Wild Gorillas and Some Comparisons with Chimpanzees. *Journal of Reproduction and Fertility*, Supplement 28:59–70.

Harlap, S., P. H. Shiono, and S. Ramcharan. 1980. A Life Table of Spontaneous Abortions and the Effects of Age, Parity, and Other Variables. Pp. 145–158 in I. H. Porter and E. B. Hook (eds.), *Human Embryonic and Fetal Death*. New York: Academic Press.

Harpending, H. C., and L. Wandsnider. 1982. Population Structures of Ghanzi and Ngamiland !Kung. Pp. 29–50 in M. H. Crawford and J. H. Mielke (eds.), *Current Developments in Anthropological Genetics*, Volume 2: Ecology and Population Structure. New York: Plenum Press.

Harper, M. J. K. 1982. Sperm and Egg Transport. Pp. 102–127 in C. R. Austin and R. V. Short (eds.), *Reproduction in Mammals, Volume 1: Germ Cells and Fertilization*, 2nd edition. Cambridge: Cambridge Univ. Press.

Harris, G. W. 1955. The Function of the Pituitary Stalk. *Bulletin of the Johns Hopkins Hospital* 97:358–375.

Harris, M., and E. B. Ross. 1987. Death, Sex, and Fertility: *Population Regulation in Preindustrial and Developing Societies*. New York: Columbia Univ. Press.

Harvey, P. H., and T. H. Clutton-Brock. 1985. Life History Variation in Primates. *Evolution* 39:559–581.

Hassold, T., N. Chen, J. Funkhouser, T. Jooss, B. Manuel, J. Matsuura, A. Matsuyama, C. Wilson, J. A. Yamane, and P. A. Jacobs. 1980. A Cytogenetic Study of 1000 Spontaneous Abortions. *Annals of Human Genetics* 44:151–178.

Hatch, M. C. 1983. Paternal Risk Factors for Spontaneous Abortion. Unpublished doctoral dissertation. New York: Columbia Univ.

Hatcher, R. A., G. K. Stewart, F. Guest, R. Finkelstein, and C. Godwin. 1976. *Contraceptive Technology* 1976–1977, 8th edition. New York: Irvington Publishers.

Heap, R. B., and A. P. F. Flint. 1984. Pregnancy. Pp. 153–194 in C. R. Austin and R. V. Short (eds.), *Reproduction in Mammals, Volume 3: Hormonal Control of Pregnancy*, 2nd edition. Cambridge: Cambridge Univ. Press.

Heckman, J. J., and B. Singer. 1984. A Method for Minimizing the Impact of Distributional Assumptions in Econometric Models for Duration Data. *Econometrica* 52:271–320.

Heckman, J. J., and J. R. Walker. 1987. Using Goodness of Fit and Other Criteria to Choose among Competing Duration Models: A Case Study of Hutterite Data. Pp. 247–307 in C. C. Clogg (ed.), *Sociological Methodology* 17. New York: American Sociological Association.

———. 1990. Estimating Fecundability from Data on Waiting Times to First Conception. *Journal of the American Statistical Association* 85:283–294.

Hedricks, C. A. 1993. Hormones and Sexual Behavior: A Field Study from Zimbabwe. Presented at the New York Academy of Sciences Conference on Human Reproductive Ecology, Research Triangle Park, N. C.

Hedricks, C. A., L. J. Piccinino, J. R. Udry, and T. H. K. Chimbira. 1987. Peak Coital Rate Coincides with Onset of Luteinizing Hormone Surge. *Fertility and Sterility* 48:234–238.

Hellman, L., J. Pritchard, and R. Wynn (eds.). 1971. *Williams' Obstetrics,* 14th edition. New York: Appleton-Century-Crofts.

Helm, P., and S. Helm. 1984. Decrease in Menarcheal Age from 1966 to 1983 in Denmark. *Acta Obstetrica et Gynaecologica Scandinavica* 63:633–635.

Hendershot, G. E., W. D. Mosher, and W. F. Pratt. 1982. Infertility and Age: An Unresolved Issue. *Family Planning Perspectives* 14:287–289.

Henderson, M., and J. Kay. 1967. Differences in Duration of Pregnancy: Negro and White Women of Low Socioeconomic Class. *Archives of Environmental Health* 14:904–911.

Hennart, P., Y. Hofvander, H. Vis, and C. Robyn. 1985. Comparative Study of Nursing Mothers in Africa (Zaîre) and in Europe (Sweden): Breastfeeding Behaviour, Nutritional Status, Lactational Hyperprolactinaemia and Status of the Menstrual Cycle. *Clinical Endocrinology* 22:179–187.

Henripin, J. 1954a. La Fécondité des Ménages Canadiens au Début du XVIIIème Siècle. *Population* 9:61–84.

———. 1954b. *La Population Canadienne au Début du XVIIIème Siècle.* Paris: Presses Universitaires de France.

Henry, L. 1953a. *Fécondité des Mariages: Nouvelle Methode de Mesure.* Institut National d'Etudes Démographiques, Travaux et Documents No. 16. Paris: Presses Universitaires de France.

———. 1953b. Fondements théoriques des mesures de la fécondité naturelle. *Revue de l'Institut International de Statistique* 21:135–151.

———. 1956. *Anciennes Familles Genevoises: Etudes Démographique XVIème-XXème Siècles.* Paris: Institut National d'Etudes Démographiques-Presses Universitaires de France.

———. 1958. Intervals between Confinements in the Absence of Birth Control. *Eugenics Quarterly* 5:200–211.

———. 1961. Some Data on Natural Fertility. *Eugenics Quarterly* 8:81–91.

———. 1964. Mortalité Intra-utérine et Fécondabilité. *Population* 19:899–940.

———. 1965. French Statistical Work in Natural Fertility. Pp. 333–350 in M. C. Sheps and J. C. Ridley (eds.), *Public Health and Population Change: Current Research Issues.* Pittsburgh: Univ. of Pittsburgh Press.

———. 1976. *Population: Analysis and Models.* London: Edward Arnold.

———. 1979. Concepts actuels et résultats empiriques sur la fécondité naturelle. Pp. 15–28 in H. Leridon and J. Menken (eds.), *Natural Fertility.* Liège: Ordina Editions.

Hertig, A. T., J. Rock, E. C. Adams, and M. C. Menken. 1959. Thirty-four Fertilized Human Ova, Good, Bad and Indifferent, Recovered from 210 Women of Known Fertility. *Pediatrics* 23:202–211.

Hewlett, B. S. 1987. Sexual Selection and Paternal Investment among Aka Pygmies. Pp. 263–276 in L. L. Betzig, M. Borgerhoff Mulder, and P. W. Turke (eds.), *Human Reproductive Behaviour: A Darwinian Perspective.* Cambridge: Cambridge Univ. Press.

Heywood, P. 1982. The Functional Significance of Malnutrition: Growth and

Prospective Risk of Death in the Highlands of Papua New Guinea. *Journal of Food and Nutrition* 39:13–19.

Hilgers, T. W., and A. J. Bailey. 1980. Natural Family Planning, II: Basal Body Temperature and Estimated Time of Ovulation. *Obstetrics and Gynecology* 55:333–339.

Hill, J. P. 1931. A Young Human Embryo (Embryo Dobbin) with Head-process and Prochordal Plate. *Philosophical Transactions of the Royal Society of London* (series B) 219:443–486.

Hill, K., and H. Kaplan. 1987. Tradeoffs and Female Reproductive Strategies among the Ache: Part 1. Pp. 277–290 in L. L. Betzig, M. Borgerhoff Mulder, and P. W. Turke (eds.), *Human Reproductive Behaviour: A Darwinian Perspective*. Cambridge: Cambridge Univ. Press.

Himes, N. E. 1963. *Medical History of Contraception*. New York: Gamut Press.

Hobcraft, J. 1985. Comments from a Demographer. Pp. 129–137 in J. Dobbing (ed.), *Maternal Nutrition and Lactational Infertility*. New York: Raven Press.

Hobcraft, J., and R. J. A. Little. 1984. Fertility Exposure Analysis: A New Method for Assessing the Contribution of Proximate Determinants to Fertility Differentials. *Population Studies* 38:21–45.

Hobcraft, J., J. W. McDonald, and S. Rutstein. 1983. Child-spacing Effects on Infant and Early Child Mortality. *Population Index* 49:585–618.

Hodgen, G. D., A. L. Goodman, A. O'Connor, and D. K. Johnson. 1977. Menopause in Rhesus Monkeys: Model for Study of Disorders in the Human Climacteric. *American Journal of Obstetrics and Gynecology* 127:581–584.

Hogarth, P. J. 1978. *Biology of Reproduction*. Glasgow: Blackie.

———. 1982. *Immunological Aspects of Mammalian Reproduction*. Glasgow: Blackie.

Holman, D. J., B. Shell, J. W. Wood, K. L. Campbell, and P. L. Johnson. 1992. Physiological Correlates of Long Ovarian Cycles in Women of Highland New Guinea. Unpublished ms.

Holmberg, I. 1970. *Fecundity, Fertility, and Family Planning: Applications of Demographic Micromodels*, Volume 1. Univ. of Gothenburg, Demographic Institute, Report 10. Gothenburg, Sweden: Almqvist and Wiksell.

Homans, G. C. 1941. *English Villagers of the Thirteenth Century*. Cambridge, Mass.: Harvard Univ. Press.

Hook, E. B. 1986. Paternal Age and Effects on Chromosomal and Specific Locus Mutations and on Other Genetic Outcomes in Offspring. Pp. 117–145 in L. Mastroianni and C. A. Paulsen (eds.), *Aging, Reproduction, and the Climacteric*. New York: Plenum Press.

Hook, E. B., and I. H. Porter. 1980. Terminological Conventions, Methodological Considerations, Temporal Trends, Specific Genes, Environmental Hazards, and Some Other Factors Pertaining to Embryonic and Fetal Deaths. Pp. 1–18 in I. H. Porter and E. B. Hook (eds.), *Human Embryonic and Fetal Death*. New York: Academic Press.

Hornsby, P. P., and A. J. Wilcox. 1989. Validity of Questionnaire Information on Frequency of Coitus. *American Journal of Epidemiology* 130:94–99.

Hoshi, H., and M. Kouchi. 1981. Secular Trend of the Age at Menarche of Japanese Girls with Special Regard to the Secular Acceleration of the Age at Peak Height Velocity. *Human Biology* 53:593–598.

Hostetler, J. A. 1974. *Hutterite Society.* Baltimore: Johns Hopkins Univ. Press.

———. 1980. *Amish Society,* 3rd edition. Baltimore: Johns Hopkins Univ. Press.

Howell, N. 1976. The Population of the Dobe Area !Kung. Pp. 137–151 in R. B. Lee, and I. Devore (eds.), *Kalahari Hunter-gatherers: Studies of the !Kung San and Their Neighbors.* Cambridge, Mass.: Harvard Univ. Press.

———. 1979. *Demography of the Dobe !Kung.* New York: Academic Press.

Howie, P. W., and A. S. McNeilly. 1982. Effect of Breastfeeding Patterns on Human Birth Intervals. *Journal of Reproduction and Fertility* 65:545–557.

Howie, P. W., A. S. McNeilly, M. J. Houston, A. Cook, and H. Boyle. 1982. Fertility after Childbirth: Infant Feeding Patterns, Basal Prolactin Levels and Postpartum Ovulation. *Clinical Endocrinology* 17:315–322.

Huffmann, S. L., A. K. M. A. Chowdhury, J. Chakraborty, and W. H. Mosley. 1978a. Nutrition and Post-partum Amenorrhoea in Rural Bangladesh. *Population Studies* 32:251–260.

Huffmann, S. L., A. K. M. A. Chowdhury, and W. H. Mosley. 1978b. Postpartum Amenorrhea: How Is It Affected by Maternal Nutritional Status? *Science* 200:1155–1157.

Huffmann, S. L., A. K. M. A. Chowdhury, J. Chakraborty, and N. K. Simpson. 1980. Breast-feeding Patterns in Rural Bangladesh. *American Journal of Clinical Nutrition* 33:144–154.

Huffmann, S. L., K. Ford, H. A. Allen, and P. Streble. 1987. Nutrition and Fertility in Bangladesh: Breastfeeding and Post Partum Amenorrhea. *Population Studies* 41:447–462.

Hunt, E. E., and N. W. Newcomer. 1984. The Timing and Variability of Menarche, Cumulative Fertility and Menopause: A Symmetrical and Parsimonious Bioassay Model. *Human Biology* 56:47–62.

Huntington, G. E., and J. A. Hostetler. 1966. A Note on Nursing Practices in an American Isolate with a High Birth Rate. *Population Studies* 19:321–324.

Hurley, L. S. 1980. *Developmental Nutrition.* Englewood Cliffs, N. J.: Prentice-Hall.

Huss-Ashmore, R. 1988. Seasonal Patterns of Birth and Conception in Rural Highland Lesotho. *Human Biology* 60:493–506.

Hyrenius, H. 1958. Fertility and Reproduction in a Swedish Population without Family Limitation. *Population Studies* 12:121–230.

Hytten, F. E., and I. Leitch. 1964. *The Physiology of Human Pregnancy.* Oxford: Blackwell Scientific.

Hytten, R., and G. Chamberlain. 1980. *Clinical Physiology in Obstetrics.* Oxford: Blackwell Scientific.

Jacobs, P. E. 1982. Human Population Cytogenetics: The First Twenty-five Years. *American Journal of Human Genetics* 34:689–698.

Jain, A. K. 1969a. Fecundability and Its Relation to Age in a Sample of Taiwanese Women. *Population Studies* 23:69–85.

———. 1969b. Pregnancy Outcome and the Time Required for Next Conception. *Population Studies* 23:421–433.

Jain, A. K., and T. H. Sun. 1972. Inter-relationship between Socio-demographic Factors, Lactation and Postpartum Amenorrhea. *Demography* (India) 1:1–15.

Jain, A. K., A. I. Hermalin, and T. H. Sun. 1979. Lactation and Natural Fertility.

Pp. 149–194 in H. Leridon and J. Menken (eds.), *Natural Fertility*. Liège: Ordina Editions.

Jain, A. K., T. C. Hsu, R. Freeman, and M. C. Chang. 1970. Demographic Aspects of Lactation and Postpartum Amenorrhea. *Demography* 7:255–271.

Jain, S. P. 1967. Post-partum Amenorrhea in Indian Women. Pp. 378–388 in *Contributed Papers, International Union for the Scientific Study of Population Sydney Conference*. Liège: Ordina Editions.

James, W. H. 1971. The Distribution of Coitus within the Human Intermenstruum. *Journal of Biosocial Science* 3:159–171.

———. 1979. The Causes of the Decline in Fecundability with Age. *Social Biology* 26:330–334.

———. 1980. Implication of a Time-dependent Model of Sexual Intercourse within the Menstrual Cycle: A Comment. *Journal of Biosocial Science* 12:495–496.

———. 1981. Distributions of Coital Rates and of Fecundability. *Social Biology* 28:334–341.

———. 1983. Decline in Coital Rate with Spouses' Ages and Duration of Marriage. *Journal of Biosocial Science* 15:83–87.

Janowitz, B., J. H. Lewis, A. Parnell, F. Henawi, M. N. Younis, and G. A. Serour. 1981. Breast-feeding and Child Survival in Egypt. *Journal of Biosocial Science* 13:287–297.

Jasso, G. 1985. Marital Coital Frequency and the Passage of Time: Estimating the Separate Effects of Spouses' Ages and Marital Duration, Birth and Marriage Cohorts, and Period Influences. *American Sociological Review* 50:224–241.

Jeffcoate, S. L. 1983. Use of Rapid Hormone Assays in the Prediction of Ovulation. Pp. 67–82 in S. L. Jeffcoate (ed.), *Ovulation: Methods for Its Prediction and Detection*. New York: John Wiley.

Johansson, E. D. B., V. Larsson-Cohn, and C. Gemzell. 1972. Monophasic Basal Body Temperature in Ovulatory Menstrual Cycles. *American Journal of Obstetrics and Gynecology* 113:933–937.

John, A. M. 1993. Statistical Evidence of Links between Maternal Nutrition and Post-partum Infertility. Pp. 372–382 in R. H. Gray, H. Leridon, and A. Spira (eds.), *Biomedical and Demographic Determinants of Reproduction*. Oxford: Clarendon Press.

John, A. M., J. A. Menken, and A. K. M. A. Chowdhury. 1987. The Effects of Breastfeeding and Nutrition on Fecundability in Rural Bangladesh: A Hazards-model Analysis. *Population Studies* 41:433–446.

Johnson, G., D. Roberts, R. Brown, E. Cox, Z. Evershed, P. Goutam, P. Hassan, R. Robinson, A. Sahdev, K. Swan, and C. Sykes. 1987. Infertility or Childless by Choice? A Multipractice Survey of Women Aged 35 and 50. *British Medical Journal* 294:804–806.

Johnson, M. H., and B. J. Everitt. 1988. *Essential Reproduction*, 3rd edition. Oxford: Blackwell Scientific.

Johnson, N. L., and S. Kotz. 1969. *Discrete Distributions*. New York: John Wiley.

———. 1973. *Distributions in Statistics: Continuous Multivariate Distributions*. New York: John Wiley.

Johnson, P. L., J. W. Wood, and M. Weinstein. 1990. Female Fecundity in High-land Papua New Guinea. *Social Biology* 37:26–43.

Johnson, P. L., J. W. Wood, K. L. Campbell, and I. A. Maslar. 1987. Long Ovarian Cycles in Women of Highland New Guinea. *Human Biology* 59:837–845.

Johnston, F. E., A. R. Roche, L. M. Schell, and H. B. Wettenhal. 1975. Critical Weight at Menarche: Critique of a Hypothesis. *American Journal of Diseases of Childhood* 129:19–23.

Jones, E. C., and P. L. Krohn. 1961. The Effect of Hypophysectomy on Age Changes in the Ovaries of Mice. *Journal of Endocrinology* 21:497–509.

Jones, R. E. 1988a. A Biobehavioral Model for Breastfeeding Effects on Return to Menses Postpartum in Javanese Women. *Social Biology* 35:307–323.

———. 1988b. A Hazards Model Analysis of Breastfeeding Variables and Maternal Age on Return to Menses Postpartum in Rural Indonesian Women. *Human Biology* 60:853–871.

———. 1989. Breastfeeding and Postpartum Amenorrhea in Indonesia. *Journal of Biosocial Science* 21:83–100.

Jones, R. E., and A. Palloni. 1994. Investigating the Determinants of Postpartum Amenorrhea Using a Multistate Hazards Model Approach (abstract). *Annals of the New York Academy of Sciences* 709:227–230.

Kahn, C. R., R. J. Smith, and W. W. Chin. 1992. Mechanisms of Action of Hormones that Act at the Cell Surface. Pp. 91–134 in J. D. Wilson and D. W. Foster (eds.), *Williams Textbook of Endocrinology*, 8th edition. Philadelphia: W. B. Saunders.

Kahn, J. R., and J. R. Udry. 1986. Marital Coital Frequency: Unnoticed Outliers and Unspecified Interactions Lead to Erroneous Conclusions. *American Sociological Review* 51:734–742.

Kalbfleisch, J. D., and R. L. Prentice. 1980. *The Statistical Analysis of Failure Time Data*. New York: John Wiley.

Kalule-Sabiti, I. 1984. Bongaarts' Proximate Determinants of Fertility Applied to Group Data from the Kenya Fertility Survey 1977/78. *Journal of Biosocial Science* 16:205–218.

Kaplan, E. L., and P. Meier. 1958. Nonparametric Estimation from Incomplete Observations. *Journal of the American Statistical Association* 53:457–481.

Kaplan, H. S. 1986. Sexual Relationships in Middle Age: Comparative Physiologic Changes in Women and Men. *Journal of Clinical Practice in Sexuality* 2:21–28.

Karim, A., A. K. M. A. Chowdhury, and M. Kabir. 1985. Nutritional Status and Age at Secondary Sterility in Rural Bangladesh. *Journal of Biosocial Science* 17:497–502.

Karsch, F. J. 1984. The Hypothalamus and Anterior Pituitary Gland. Pp. 1–20 in C. R. Austin and R. V. Short (eds.), *Reproduction in Mammals, Volume 3: Hormonal Control of Reproduction*, 2nd edition. Cambridge: Cambridge Univ. Press.

Kaufman, J. M., J. P. Delsypere, M. Giri, and A. Vermeulen. 1990. Neuroendocrine Regulation of Pulsatile Luteinizing Hormone Secretion in Elderly Men. *Journal of Steroid Biochemistry and Molecular Biology* 37:421–430.

Kelch, R. P., J. C. Marshall, S. Sauder, N. J. Hopwood, and N. E. Reame. 1983. Gonadotropin Regulation during Human Puberty. Pp. 229–256 in R. L. Norman (ed.), *Neuroendocrine Aspects of Reproduction*. New York: Academic Press.

Kendall, M., and A. Stuart. 1979. *The Advanced Theory of Statistics, Volume 2: Inference and Relationship*, 2nd edition. New York: Macmillan.

Keyfitz, N. 1985. *Applied Mathematical Demography*, 2nd edition. New York: Springer-Verlag.

Khalifa, M. A. 1986. Determinants of Natural Fertility in Sudan. *Journal of Biosocial Science* 18:325–336.

Khoury, M. J., T. H. Beatty, and B. H. Cohen. 1993. *Fundamentals of Genetic Epidemiology*. New York: Oxford Univ. Press.

Kielmann, A., and C. McCord. 1978. Weight for Age as an Index of Risk of Death in Children. *The Lancet* i:1247–1250.

King, J. 1975. Protein Metabolism during Pregnancy. *Clinical Perinatology* 2:243–252.

King, J. C., D. H. Calloway, and S. Margen. 1973. Nitrogen Retention, Total Body 40K and Weight Gain in Teenage Pregnant Girls. *Journal of Nutrition* 103: 772–785.

King, J. L. 1926. Menstrual Records and Vaginal Smears in a Selected Group of Normal Women. *Contributions to Embryology* 18:81–94.

Kinsey, A. C., W. B. Pomeroy, and C. E. Martin. 1948. *Sexual Behavior in the Human Male*. Philadelphia: W. B. Saunders.

Kinsey, A. C., W. B. Pomeroy, C. E. Martin, and P. H. Gebhard. 1953. *Sexual Behavior in the Human Female*. Philadelphia: W. B. Saunders.

Kline, J. 1986. Maternal Occupation: Effects on Spontaneous Abortions and Malformations. In Z. A. Stein and M. C. Hatch (eds.), *Reproductive Problems in the Workplace*. Philadelphia: Hanley and Belfos.

Kline, J., and Z. Stein. 1987. Epidemiology of Chromosomal Anomalies in Spontaneous Abortion: Prevalence, Manifestation and Determinants. Pp. 29–50 in M. J. Bennett and D. K. Edmonds (eds.), *Spontaneous and Recurrent Abortion*. Oxford: Blackwell Scientific.

Kline, J., Z. Stein, and M. Susser. 1989. *Conception to Birth: Epidemiology of Prenatal Development*. New York: Oxford Univ. Press.

Knobil, E., and J. Hotchkiss. 1988. The Menstrual Cycle and Its Neuroendocrine Control. Pp. 1971–1994 in E. Knobil, J. D. Neill, L. L. Ewing, G. S. Greenwald, C. L. Markert, and D. W. Pfaff (eds.), *The Physiology of Reproduction*, Volume 2. New York: Raven Press.

Knobil, E., J. D. Neill, L. L. Ewing, G. S. Greenwald, C. L. Markert, and D. W. Pfaff (eds.). 1988. *The Physiology of Reproduction*, Volume 2. New York: Raven Press.

Knodel, J. 1968. Infant Mortality and Fertility in Three Bavarian Villages: An Analysis of Family Histories from the Nineteenth Century. *Population Studies* 22:297–318.

———. 1978a. European Populations in the Past: Family-level Relations. Pp. 21–45 in S. H. Preston (ed.), *The Effects of Infant and Child Mortality on Fertility*. New York: Academic Press.

———. 1978b. Natural Fertility in Pre-industrial Germany. *Population Studies* 32:481–510

———. 1983. Natural Fertility: Age Patterns, Levels, and Trends. Pp. 61–102 in R. A. Bulatao and R. D. Lee (eds.), *Determinants of Fertility in Developing*

Countries, Volume 1: Supply and Demand for Children. New York: Academic Press.

———. 1986. Demographic Transitions in German Villages. Pp. 337–389 in A. J. Coale and S. C. Watkins (eds.), *The Decline of Fertility in Europe.* Princeton: Princeton Univ. Press.

———. 1988. *Demographic Behavior in the Past: A Study of Fourteen German Village Populations in the Eighteenth and Nineteenth Centuries.* Cambridge: Cambridge Univ. Press.

Knodel, J., and A. I. Hermalin. 1984. Effects of Birth Rank, Maternal Age, Birth Interval, and Sibship Size on Infant and Child Mortality: Evidence from 18th and 19th Century Reproductive Histories. *American Journal of Public Health* 74: 1098–1106.

Knodel, J., and E. van de Walle. 1986. Lessons from the Past: Policy Implications of Historical Fertility Studies. Pp. 390–419 in A. J. Coale and S. C. Watkins (eds.), *The Decline of Fertility in Europe.* Princeton: Princeton Univ. Press.

Knoll, J. H., R. D. Nichols, R. E. Mageniz, J. M. Graham, M. Lalande, and S. Latt. 1989. Angelman and Prader-Willi Syndromes Share a Common Chromosome 15 Deletion But Differ in Parental Origin of the Deletion. *American Journal of Medical Genetics* 32:285–290.

Kolodny, R. C., L. S. Jacobs, and W. H. Daughaday. 1972. Mammary Stimulation Causes Prolactin Secretion in Non-lactating Women. *Nature* 238:284–286.

Konner, M., and C. Worthman. 1980. Nursing Frequency, Gonadal Function, and Birth Spacing among !Kung Hunter-gatherers. *Science* 207:788–791.

Koo, H. P., and B. K. Janowitz. 1983. Interrelationships between Fertility and Marital Dissolution: Results of a Simultaneous Logit Model. *Demography* 20: 129–145.

Kothari, K. L., and A. S. Gupta. 1974. Effect of Aging on the Volume, Structure and Total Leydig Cell Content of the Human Testis. *International Journal of Fertility* 19:140–146.

Kraus S. J. 1972. Complications of Gonococcal Infection. *Medical Clinics of North America* 56:1115–1125.

Krohn, P. L. 1962. Review Lectures on Senescence, II: Heterochronic Transplantation in the Study of Aging. *Proceedings of the Royal Society of London* (series B) 157:128–147.

Lachenbruch, P. A. 1967. Frequency and Timing of Intercourse: Its Relation to the Probability of Conception. *Population Studies* 21:23–31.

Lager, C., and P. T. Ellison. 1987. Effects of Moderate Weight Loss on Ovulation Frequency and Luteal Function in Adult Women. *American Journal of Physical Anthropology* 72:221–222.

———. 1990. Effect of Moderate Weight Loss on Ovarian Function Assessed by Salivary Progesterone Measurements. *American Journal of Human Biology* 2: 303–312.

Lähteenmäki, P., and T. Luukkainen. 1978. Return of Ovarian Function after Abortion. *Clinical Endocrinology* 8:123–132.

Lam, D. A., and J. A. Miron. 1987. *The Seasonality of Births in Human Populations.* Research Report 87-114. Ann Arbor: Population Studies Center, Univ. of Michigan.

————. 1991. Seasonality of Births in Human Populations. *Social Biology* 38:51–78.

Lam, D. A., J. A. Miron, and A. P. Riley. 1994. Modeling Seasonality in Fecundability, Conceptions and Births. *Demography* 31:321–346.

Lancaster, T. 1990. *The Econometric Analysis of Transition Data*. Cambridge: Cambridge Univ. Press.

Langer, W. L. 1974. Infanticide: A Historical Survey. *History of Childhood Quarterly* 1:353–365.

Langman, J. 1981. *Medical Embryology*, 4th edition. Baltimore: Williams and Wilkins.

Larsen, U., and J. Menken. 1989. Measuring Sterility from Incomplete Birth Histories. *Demography* 26:185–201.

Larsen, U., and J. Vaupel. 1993. Hutterite Fecundability by Age and Parity: Strategies for Frailty Modeling of Event Histories. *Demography* 30:81–102.

Last, J. M. (ed.). 1988. *A Dictionary of Epidemiology*, 2nd edition. New York: Oxford Univ. Press.

Lauritsen, J. G. 1976. Aetiology of Spontaneous Abortion: A Cytogenetic and Epidemiological Study of 288 Abortuses and Their Parents. *Acta Obstetrica et Gyneacologica Scandinavica*, Supplement 52:1–29.

Lee, E. T. 1992. *Statistical Methods for Survival Data Analysis*, 2nd edition. New York: John Wiley.

Lee, R. B. 1979. *The !Kung San: Men, Women, and Work in a Foraging Society*. Cambridge: Cambridge Univ. Press.

Lee, R. B., and I. DeVore. 1976. *Kalahari Hunter-gatherers*. Cambridge, Mass.: Harvard Univ. Press.

LeFor, B. L., D. J. Holman, A. V. Buchanan, and J. W. Wood. 1993. Seasonality of Marriage and First Birth in the Old Order Amish. Presented at the annual meeting of the American Association of Physical Anthropologists, Toronto.

Lehtovirta, P., P. Arjomaa, T. Ranta, T. Laatikainen, E. Hirvonen, and M. Seppala. 1979. Prolactin Levels and Bromocriptine Treatment of Short Luteal Phase. *International Journal of Fertility* 24:57–60.

Leistøl, K. 1980. Menarcheal Age and Spontaneous Abortion: A Causal Connection? *American Journal of Epidemiology* 111:753–758.

Lenton, E. A., C. H. Gelsthorp, and R. Harper. 1988. Measurement of Progesterone in Saliva: Assessment of the Normal Fertile Range Using Spontaneous Conception Cycles. *Clinical Endocrinology* 28:637–646.

Lenton, E. A., B.-M. Landgren, and L. Sexton. 1984a. Normal Variation in the Length of the Luteal Phase of the Menstrual Cycle: Identification of the Short Luteal Phase. *British Journal of Obstetrics and Gynaecology* 91:685–689.

Lenton, E. A., B.-M. Landgren, L. Sexton, and R. Harper. 1984b. Normal Variation in the Length of the Follicular Phase of the Menstrual Cycle: Effect of Chronological Age. *British Journal of Obstetrics and Gynaecology* 91:681–684.

Leridon, H. 1967. Les Intervalles entre Naissancces: Nouvelles Données d'Observation. *Population* 22:821–840.

————. 1976. Facts and Artifacts in the Study of Intrauterine Mortality: A Reconsideration from Pregnancy Histories. *Population Studies* 30:319–335.

————. *Human Fertility: The Basic Components*. Chicago: Univ. of Chicago Press.

Lerner, R. A., A. S. Kang, J. D. Bain, D. R. Burton, and C. F. Barbas. 1992. Antibodies without Immunization. *Science* 258:1313–1314.

Leslie, P. W., and P. H. Fry. 1989. Extreme Seasonality of Births among Turkana Pastoralists. *American Journal of Physical Anthropology* 74:103–115.

Leslie, P. W., K. L. Campbell, and M. A. Little. 1993. Pregnancy Loss in Nomadic and Settled Women in Turkana, Kenya: A Prospective Study. *Human Biology* 65:237–254.

Lesthaeghe, R. 1986. On the Adaptation of Sub-Saharan Systems of Reproduction. Pp. 212–238 in D. Coleman and R. Schofield (eds.), *The State of Population Theory*. Oxford: Basil Blackwell.

Lesthaeghe, R., and H. J. Page. 1980. The Post-partum Non-susceptible Period: Development and Application of Model Schedules. *Population Studies* 34: 143–169.

Lesthaeghe, R., P. O. Ohadike, J. Kocher, and H. J. Page. 1981. Child-spacing and Fertility in Lagos. Pp. 147–179 in H. J. Page and R. Lesthaeghe (eds.), *Child-spacing in Tropical Africa: Traditions and Change*. New York: Academic Press.

Levine, R. J. 1991. Seasonal Variation in Human Semen Quality. Pp. 89–96 in A. W. Zorgniotti (ed.), *Temperature and Environmental Effects on the Testis*. New York: Plenum Press.

————. 1993. Male Factors Contributing to the Seasonality of Human Reproduction. *Annals of the New York Academy of Sciences* 709:29–45.

Levine, R. J., R. M. Mathew, C. B. Chenault, M. H. Brown, M. E. Hurtt, K. S. Bentley, K. L. Mohr, and P. W. Working. 1990. Differences in the Quality of Semen in Outdoor Workers during Summer and Winter. *New England Journal of Medicine* 323:12–16.

Lewis, P. R., J. B. Brown, M. B. Renfree, and R. V. Short. 1991. The Resumption of Ovulation and Menstruation in a Well-nourished Population of Women Breastfeeding for an Extended Period of Time. *Fertility and Sterility* 55:529–536.

Lincoln, D. W., H. M. Fraser, G. A. Lincoln, G. B. Martin, and A. S. McNeilly. 1985. Hypothalamic Pulse Generators. *Recent Progress in Hormone Research* 41: 369–411.

Lipson, S. F., and P. T. Ellison. 1992. Normative Study of Age Variation in Salivary Progesterone Profiles. *Journal of Biosocial Science* 24:233–244.

Liu, J. H., and H. Park. 1988. Gonadotropin and Prolactin Secretion Increases during Sleep during the Puerperium in Nonlactating Women. *Journal of Clinical Endocrinology and Metabolism* 66:839–841.

Livi-Bacci, M. 1986. Social-group Forerunners of Fertility Control in Europe. Pp. 182–200 in A. J. Coale and S. C. Watkins (eds.), *The Decline of Fertility in Europe*. Princeton: Princeton Univ. Press.

Livson, N., and D. McNeill. 1962. The Accuracy of Recalled Age at Menarche. *Human Biology* 34:218–221.

Longo, F. J. 1987. *Fertilization*. London: Chapman and Hall.

Lopez, L. C., E. M. Bayna, D. Litoff, N. L. Shaper, J. H. Shaper, and B. D. Shur. 1985. Receptor Function of Mouse Sperm Surface Galactosyltransferase during Fertilization. *Journal of Cellular Biology* 101:1501–1510.

Lövegren, T., I. Hemmilä, K. Pettersson, and P. Halonen. 1985. Time-resolved Fluorometry in Immunoassay. Pp. 203–217 in W. P. Collins (ed.), *Alternative Immunoassays*. New York: John Wiley.

MacArthur, R. H., and E. O. Wilson. 1967. *The Theory of Island Biogeography*. Princeton: Princeton Univ. Press.

Macfarlane, A., and M. Mugford. 1984. *Birth Counts: Statistics of Pregnancy and Childbirth.* London: Her Majesty's Stationery Office.

Maciel, R. J. 1985. Standard Curve Fitting in Immunodiagnostics: A Primer. *Journal of Clinical Endocrinology* 8:98–106.

Macleod, J., and R. Z. Gold. 1951a. The Male Factor in Fertility and Infertility, II: Spermatozoan Counts in 1000 Men of Known Fertility and in 1000 Cases of Infertile Marriage. *Journal of Urology* 66:436–449.

———. 1951b. The Male Factor in Fertility and Infertility, III: An Analysis of Motile Activity in the Spermatozoa of 1000 Fertile Men in Infertile Marriages. *Fertility and Sterility* 2:187–204.

———. 1951c. The Male Factor in Fertility and Infertility, IV: Sperm Morphology in Fertile and Infertile Marriages. *Fertility and Sterility* 2:394–414.

———. 1952. The Male Factor in Fertility and Infertility, V: Effect of Continence on Semen Quality. *Fertility and Sterility* 3:297–315.

———. 1953a. The Male Factor in Fertility and Infertility, VI: Semen Quality and Certain Other Factors in Relation to Ease of Conception. *Fertility and Sterility* 4:10–33.

———. 1953b. The Male Factor in Fertility and Infertility, VII: Semen Quality in Relation to Age and Sexual Activity. *Fertility and Sterility* 4:194–209.

Macleod, J., and Y. Wang. 1979. Male Fertility Potential in Terms of Semen Quality: A Review of the Past, a Study of the Present. *Fertility and Sterility* 31:103–116.

MacMahon, B., and J. Worcester. 1966. *Age at Menopause: United States, 1960–62.* Washington: National Center for Health Statistics, Public Health Service.

MacMahon, B., D. Trichopoulos, P. Cole, and J. Brown. 1982. Cigarette Smoking and Urinary Estrogens. *New England Journal of Medicine* 307:1062–1065.

Madhavan, S. 1965. Age at Menarche of South Indian Girls. *Indian Journal of Medical Research* 53:669–673.

Mainwaring, W. I. P. 1979. The Androgens. Pp. 117–156 in C. R. Austin and R. V. Short (eds.), *Reproduction in Mammals, Volume 7: Mechanisms of Hormone Action.* Cambridge: Cambridge University Press.

Majumdar, H., and M. C. Sheps. 1970. Estimators of a Type I Geometric Distribution from Observations on Conception Times. *Demography* 7:349–360.

Malaurie, J., L. Tabah, and J. Sutter. 1952. L'isolat equimau de Thule (Groenland). *Population* 7:675–712.

Malcolm, L. A. 1970. Growth and Development of the Bundi Child of the New Guinea Highlands. *Human Biology* 42:293–328.

Malik, S. L., and R. C. Hauspie. 1986. Age at Menarche among High Altitude Bods of Ladakh (India). *Human Biology* 58:541–548.

Marchbanks, P. A., H. B. Peterson, G. L. Rubin, P. A. Wingo, and the Cancer and Steroid Hormone Study Group. 1989. Research on Infertility: Definition Makes a Difference. *American Journal of Epidemiology* 130:259–267.

Mark, W. H., K. Signorelli, and E. Lacy. 1985. An Inserted Mutation in a Transgenic Mouse Line Results in Developmental Arrest at Day 5 of Gestation. *Cold Spring Harbor Symposia on Quantitative Biology* 50:453–463.

Marshall, W. A. 1970. Physical Growth and Development. Pp. 1–34 in *Brennemann's Practice of Pediatrics.* New York: Harper and Row.

Marshall, W. A., and J. M. Tanner. 1970. Variation in the Pattern of Pubertal Changes in Boys. *Archives of Diseases in Childhood* 45:13–23.

Martorell, R., J.-P. Habicht, C. Yarbrough, G. Guzmán, and R. E. Klein. 1975. The Identification and Evaluation of Measurement Variability in the Anthropometry of Preschool Children. *American Journal of Physical Anthropology* 43:347–352.

Masnick, G. S. 1979. The Demographic Impact of Breastfeeding: A Critical Review. *Human Biology* 51:109–125.

Mason, W. M., and B. Entwisle. 1985. Cross-national Variability in Age at First Birth: Theory and Evidence. Presented at the annual meeting of the Population Association of America, Boston.

Matsumoto, S., Y. Nogami, and S. Ohkuri. 1962. Statistical Studies on Menstruation. *Gunma Journal of Medical Science* 11:294–318.

Matsumoto, S., M. Ozawa, Y. Nagomi, and H. Ohashi. 1963. Menstrual Cycle in Puberty. *Gunma Journal of Medical Science* 12:119–143.

Mattioli, M., F. Conte, G. Graleati, and E. Seren. 1986. Effect of Naloxone on Plasma Concentrations of Prolactin and LH in Lactating Sows. *Journal of Reproduction and Fertility* 76:167–173.

McArthur, J. W., and T. Colton (eds.). 1970. *Statistics in Endocrinology, with a Supplement on Competitive Protein-Binding Assays.* Cambridge: MIT Press.

McCance, R. A., M. C. Luff, and E. E. Widdowson. 1937. Physical and Emotional Periodicity in Women. *Journal of Hygiene* 37:571–611.

McElrath, T. 1992. A Survey of Determinants of Extreme Birth Seasonality in Rural Bangladesh. Unpublished ms.

McFalls, J. A., and M. H. McFalls. 1984. *Disease and Fertility.* New York: Academic Press.

McIntosh, J. E. A., C. D. Matthews, J. M. Crocker, T. J. Broom, and L. W. Cox. 1980. Predicting the Luteinizing Hormone Surge: Relationship between the Duration of the Follicular and Luteal Phases and the Length of the Human Menstrual Cycle. *Fertility and Sterility* 34:125–130.

McKinlay, S. M., N. L. Bifano, and J. B. McKinlay. 1985. Smoking and Age at Menopause in Women. *Annals of Internal Medicine* 103:350–356.

McLaren, A. 1972. The Embryo. Pp. 1–42 in C. R. Austin and R. V. Short (eds.), *Reproduction in Mammals, Volume 2: Embryonic and Fetal Development.* Cambridge: Cambridge Univ. Press.

———. 1981. *Germ Cells and Soma: A New Look at an Old Problem.* New Haven: Yale Univ. Press.

———. 1982. The Embryo. Pp. 1–25 in C. R. Austin and R. V. Short (eds.), *Reproduction in Mammals, Volume 2: Embryonic and Fetal Development,* 2nd edition. Cambridge: Cambridge Univ. Press.

McLaren, A. 1990. *A History of Contraception: From Antiquity to the Present Day.* Oxford: Basil Blackwell.

McNatty, K. P. 1977. Prolactin and Ovarian Steroidogenesis (abstract). *Acta Endocrinologica,* Supplement 212:521.

———. 1979. Relationship between Plasma Prolactin and the Endocrine Microenvironment of the Developing Human Antral Follicle. *Fertility and Sterility* 32:433–438.

McNatty, K. P., P. Neal, and T. G. Baker. 1976. Effect of Prolactin on the Production of Progesterone by Mouse Ovaries *in vitro. Journal of Reproduction and Fertility* 47:155–156.

McNatty, K. P., R. S. Sawers, and A. S. McNeilly. 1974. A Possible Role for Prolactin in Control of Steroid Secretion by the Human Graafian Follicle. *Nature* 250:653–655.

McNeilly, A. S. 1993. Breastfeeding and Fertility. Pp. 391–412 in R. H. Gray, H. Leridon, and A. Spira (eds.), *Biomedical and Demographic Determinants of Reproduction*. Oxford: Clarendon Press.

McNeilly, A. S., A. F. Glasier, and P. W. Howie. 1985. Endocrine Control of Lactational Infertility. Pp. 1–16 in J. Dobbing (ed.), *Maternal Nutrition and Lactational Infertility*. New York: Raven Press.

McNeilly, A. S., A. F. Glasier, P. W. Howie, M. J. Houston, A. Cook, and H. Boyle. 1983. Fertility after Childbirth: Pregnancy Associated with Breast Feeding. *Clinical Endocrinology* 19:167–174.

McNeilly, A. S., A. Glasier, J. Jonassen, and P. W. Howie. 1982. Evidence for Direct Inhibition of Ovarian Function by Prolactin. *Journal of Reproduction and Fertility* 65:559–569.

Medlar, E. 1955. *The Behavior of Pulmonary Tuberculosis Lesions*. New York: Metropolitan Life.

Meheus, A., R. Ballard, M. Dlamini, J. P. Ursi, E. Van Dyck, and P. Piot. 1980. Epidemiology and Aetiology of Urethritis in Swaziland. *International Journal of Epidemiology* 9:239–245.

Mendelson, C. R. 1992. Mechanisms of Hormone Action. Pp. 28–60 in J. E. Griffin and S. R. Ojeda (eds.), *Textbook of Endocrine Physiology*, 2nd edition. New York: Oxford University Press.

Menken, J. 1979. Seasonal Migration and Seasonal Variation in Fecundability: Effects on Birth Rates and Birth Intervals. *Demography* 16:103–119.

Menken, J., and U. Larsen. 1986. Fertility Rates and Aging. Pp. 147–166, in L. Mastroianni and C. A. Paulsen (eds.), *Aging, Reproduction, and the Climacteric*. New York: Plenum Press.

Menken, J., J. Trussell, and U. Larsen. 1986. Age and Infertility. *Science* 233:1389–1394.

Menken, J., J. Trussell, D. Stempel, and O. Babakol. 1981. Proportional Hazards Life Table Models: An Illustrative Analysis of Socio-demographic Influences on Marriage Dissolution in the United States. *Demography* 18:181–200.

Menken, J., J. Trussell, and S. Watkins. 1981. The Nutrition-fertility Link: An Evaluation of the Evidence. *Journal of Interdisciplinary History* 11:425–441.

Messinis, I. E., and A. A. Templeton. 1991. Evidence that Gonadotropin Surge-attenuating Factor Exists in Man. *Journal of Reproduction and Fertility* 92:217–223.

Metcalf, M. G., R. A. Donald, and J. H. Livesey. 1981. Pituitary-ovarian Function in Normal Women during the Menopausal Transition. *Clinical Endocrinology* 14:245–255.

———. 1982. Pituitary-ovarian Function before, during and after the Menopausal Transition: A Longitudinal Study. *Clinical Endocrinology* 17:484–489.

Meyer, J. M. 1989. Modeling the Inheritance of Time to Onset. Unpublished Ph.D. Dissertation. Richmond: Medical College of Virginia, Virginia Commonwealth Univ.

Meyer, J. M., and L. J. Eaves. 1988. Estimating Genetic Parameters of Survival Distributions: A Multifactorial Model. *Genetic Epidemiology* 5:265–275.

Meyer, J. M., L. J. Eaves, A. C. Heath, and N. G. Martin. 1991. Estimating Genetic Influences on the Age-at-menarche: A Survival Analysis Approach. *American Journal of Medical Genetics* 39:148–154.

Meyerhoff, P.G., and Y. Masui. 1977. Ca and Mg Control of Cytostatic Factors from Rana pipiens Oocytes which Cause Metaphase and Cleavage Arrest. *Developmental Biology* 61:214–229.

Midgley, A. R. 1966. Radioimmunoassay: A Method for Human Chorionic Gonadotropin and Human Luteinizing Hormone. *Endocrinology* 79:10–18.

Midgley, A. R., and R. B. Jaffe. 1966. Human Luteinizing Hormone during the Menstrual Cycle: Determination by Radioimmunoassay. *Journal of Clinical Endocrinology and Metabolism* 26:1375–1381.

Miller, J. E., and R. Huss-Ashmore. 1989. Do Reproductive Patterns Affect Maternal Nutritional Status? An Analysis of Maternal Depletion in Lesotho. *American Journal of Human Biology* 1:409–419.

Miller, J. F., E. Williamson, J. Glue, Y. B. Gordon, J. G. Grudzinskas, and A. Sykes. 1980. Fetal Loss after Implantation. *The Lancet* ii:554–556.

Millman, S. R., and R. G. Potter. 1984. The Fertility Impact of Spousal Separation. *Studies in Family Planning* 15:121–126.

Mills, D. E., and D. Robertshaw. 1981. Response of Plasma Prolactin to Changes in Ambient Temperature and Humidity in Man. *Journal of Clinical Endocrinology and Metabolism* 52:279–283.

Mineau, G. P., and J. Trussell. 1982. A Specification of Marital Fertility by Parents' Age, Age at Marriage and Marital Duration. *Demography* 19:335–350.

Mineau, G. P., L. L. Bean, and M. Skolnick. 1979. Mormon Demographic History, II: The Family Life Cycle and Natural Fertility. *Population Studies* 33:429–446.

Molitch, M. E., and S. Reichlin. 1980. The Amenorrhea, Galactorrhea and Hyperprolactinemia Syndromes. *Advances in Internal Medicine* 26:37–65.

Monesi, V. 1972. Spermatogenesis and the Spermatozoa. Pp. 46–84 in C. R. Austin and R. V. Short (eds.), *Reproduction in Mammals, Volume 1: Germ Cells and Fertilization.* Cambridge: Cambridge Univ. Press.

Montagu, M. F. A. 1957. *The Reproductive Development of the Female.* New York: Julian Press.

Mornex, R., J. Orgiazzi, B. Hugues, J.-C. Gagnaire, and B. Claustrat. 1978. Normal Pregnancies after Treatment of Hyperprolactinemia with Bromoergocryptine, Despite Suspected Pituitary Tumors. *Journal of Clinical Endocrinology and Metabolism* 47:290–295.

Morris, N. M., and J. R. Udry. 1976. Incidence of Coitus during Menstruation. *Medical Aspects of Human Sexuality* 9(January):31, 95.

————. 1983. Menstruation and Marital Sex. *Journal of Biosocial Science* 15:173–181.

Morton, N. E., C. S. Chung, and M. P. Mi. 1967. *Genetics of Interracial Crosses in Hawaii.* New York: S. Karger.

Mosher, W. D. 1985. Reproductive Impairments in the United States, 1965–1982. *Demography* 22:415–430.

Mosher, W. D., and W. F. Pratt. 1982. *Reproductive Impairments among Married Couples: United States.* Hyattsville, Md.: National Center for Health Statistics, Public Health Service.

————. 1990a. *Contraceptive Use in the United States,* 1973–83. Hyattsville, Md.: National Center for Health Statistics, Public Health Service.

————. 1990b. *Fecundity and Infertility in the United States, 1965–88.* Hyattsville, Md.: National Center for Health Statistics, Public Health Service.

Mueller, W. H., and R. Martorell. 1988. Reliability and Accuracy of Measurement. Pp. 83–86 in T. G. Lohman, A. F. Roche, and R. Martorell (eds.), *Anthropometric Standardization Reference Manual.* Champaign, Ill.: Human Kinetics.

Muggeridge, M. 1965. The American Way of Sex. *Radio Times* 169:57.

Muhsam, H. V. 1956. Fertility of Polygamous Marriages. *Population Studies* 10: 3–16.

Muir, D. G., and M. A. Belsey. 1980. Pelvic Inflammatory Disease and Its Consequences in the Developing World. *American Journal of Obstetrics and Gynecology* 138:913–928.

Naeye, R. L. 1983. Maternal Age, Obstetric Complications, and the Outcome of Pregnancy. *Obstetrics and Gynecology* 61:210–216.

Nahoul, K., and M. Roger. 1990. Age-related Decline of Plasma Bioavailable Testosterone in Adult Men. *Journal of Steroid Biochemistry* 35:293–299.

Nairn, R. C. 1969. *Fluorescent Protein Tracing,* 3rd edition. Baltimore: Williams and Wilkins.

Nakano, M. 1989. Fractionation and Characterization of the Glycoproteins of Zona Pellucida. Pp. 75–98 in J. Dietl (ed.), *The Mammalian Egg Coat: Structure and Function.* Berlin: Springer-Verlag.

Napier, J. R., and P. H. Napier. 1967. *A Handbook of Living Primates.* New York: Academic Press.

National Academy of Sciences. 1980. *Recommended Dietary Allowances,* rev. edition. Washington: United States Government Printing Office.

Navot, D., P. A. Bergh, M. A. Williams, G. J. Garrisi, I. Guzman, B. Sandler, and L. Grunfeld. 1991. Poor Oocyte Quality Rather than Implantation Failure as a Cause of Age-related Decline in Female Fertility. *The Lancet* 337:1375–1377.

Naylor, A. F. 1974. Sequential Aspects of Spontaneous Abortion: Maternal Age, Parity, and Pregnancy Compensation Artifact. *Social Biology* 21:195–204.

Neaves, W. B., L. Johnson, J. C. Porter, C. R. Parker, and C. S. Petty. 1984. Leydig Cell Numbers, Daily Sperm Production, and Serum Gonadotropin Levels in Aging Men. *Journal of Clinical Endocrinology and Metabolism* 59:756–763.

Neel, J. V., and K. M. Weiss. 1975. The Genetic Structure of a Tribal Population, the Yanomama Indians. *American Journal of Physical Anthropology* 42:25–52.

Nelson, J. F., and L. S. Felicio. 1987. Reproductive Aging in the Female: An Etiological Perspective Updated. *Review of Biological Research in Aging* 3: 359–381.

Nelson, W. 1982. *Applied Life Data Analysis.* New York: John Wiley.

Newell, C. 1988. *Methods and Models in Demography.* New York: Guilford Press.

Nicosia, S. V., D. P. Wolf, and M. Inoue. 1977. Cortical Granule Distribution and Cell Surface Characteristics in Mouse Eggs. *Developmental Biology* 57:56–74.

Nieschlag, E., and E. Michel. 1986. Reproductive Functions in Grandfathers. Pp. 59–71 in L. Mastroianni and C. A. Paulsen (eds.), *Aging, Reproduction and the Climacteric.* New York: Plenum Press.

Nieschlag, E., V. Lommers, C. W. Freishem, K. Langer, and E. J. Wickings. 1982.

Reproductive Function in Young Fathers and Grandfathers. *Journal of Clinical Endocrinology and Metabolism* 55:676–681.

Niswender, G. D., and T. M. Nett. 1988. The Corpus Luteum and Its Control. Pp. 489–525 in E. Knobil, J. D. Neill, L. L. Ewing, G. S. Greenwald, C. L. Markert, and D. W. Pfaff (eds.), *The Physiology of Reproduction*, Volume 1. New York: Raven Press.

Niswender, G. D., A. R. Midgley, S. E. Monroe, and L. E. Reichert. 1968. Radioimmunoassay for Rat Luteinizing Hormone with Antiovine LH Serum and Ovine LH-^{131}I. *Proceedings of the Society for Experimental Biology and Medicine* 128:807–812.

Noel, G. L., H. K. Suh, and A. G. Frantz. 1974. Prolactin Release during Nursing and Breast Stimulation in Postpartum and Nonpostpartum Subjects. *Journal of Clinical Endocrinology and Metabolism* 38:413–423.

Norman, A. W., and G. Litwack. 1987. *Hormones.* New York: Academic Press.

Normolle, D. P. 1993. An Algorithm for Robust Non-Linear Analysis of Radioimmunoassays and Other Bioassays. *Statistics in Medicine* 12:2025–2042.

Norton, S. L. 1980. The Vital Question: Are Reconstructed Families Representative of the General Population? Pp. 11–22 in B. Dyke and W. T. Morrill (eds.), *Genealogical Demography.* New York: Academic Press.

Oakes, D. 1982. A Model for Association in Bivariate Survival Data. *Journal of the Royal Statistical Society* (series B) 44:414–422.

Ojo, O., A. Onifade, E. Akande, and R. Bannerman. 1971. The Pattern of Female Genital Tuberculosis in Ibadan. *Israeli Journal of Medical Science* 7:280–286.

Olusanya, P. O. 1971. The Problem of Multiple Causation in Population Analysis, with Particular Reference to the Polygamy-fertility Hypothesis. *Sociological Review* 19:165–178.

Osteria, T. S. 1973. Lactation and Postpartum Amenorrhea in a Rural Community. *Acta Medica Philippina* 9:144–151.

Ott, J. 1991. *Analysis of Human Genetic Linkage,* 2nd edition. Baltimore: Johns Hopkins Univ. Press.

Page, H. J. 1977. Patterns Underlying Fertility Schedules: A Decomposition by Both Age and Marriage Duration. *Population Studies* 31:85–106.

Palloni, A. 1984. Assessing the Effects of Intermediate Variables on Birth Interval-specific Measures of Fertility. *Population Index* 50:623–657.

Palloni, A., and S. Millman. 1986. Effects of Inter-birth Intervals and Breastfeeding on Infant and Early Childhood Mortality. *Population Studies* 40:215–236.

Palloni, A., and M. Tienda. 1986. The Effects of Breastfeeding and Pace of Childbearing on Mortality at Early Ages. *Demography* 23:31–52.

Palmore, J. A., and R. W. Gardner. 1983. *Measuring Mortality, Fertility and Natural Increase.* Honolulu: East-West Center.

Panter-Brick, C. 1991. Lactation, Birth Spacing and Maternal Work-loads among Two Castes in Rural Nepal. *Journal of Biosocial Science* 23:137–154.

———. 1992. Women's Working Behaviour and Maternal-child Health in Rural Nepal. Pp. 190–206 in N. Norgan (ed.), *Physical Activity and Health.* Cambridge: Cambridge Univ. Press.

Parkes, A. S. 1976. *Patterns of Sexuality and Reproduction.* Oxford: Oxford Univ. Press.

Parlow, A. F. 1961. Bioassay of Pituitary Luteinizing Hormone by Depletion of Ovarian Ascorbic Acid. Pp. 300–310 in A. Albert (ed.), *Human Pituitary Gonadotropins.* Springfield, Ilinois: Charles C. Thomas.

Patton, S., and R. G. Jensen. 1976. *Biomedical Aspects of Lactation.* Oxford: Pergamon Press.

Pawson, I. G. 1976. Growth and Development in High Altitude Populations: A Review of Ethiopian, Peruvian, and Nepalese Studies. *Proceedings of the Royal Society of London* (series B) 194:83–98.

Pearl, R. 1939. *The Natural History of Population.* New York: Oxford Univ. Press.

Pebley, A. R., and J. DaVanzo. 1988. Maternal Depletion and Child Survival in Guatemala and Malaysia. Presented at the annual meeting of the Population Association of America, New Orleans.

Pebley, A. R., and W. Mbugua. 1989. Polygyny and fertility in Sub-Saharan Africa. Pp. 338–364 in R. J. Lesthaeghe (ed.), *Reproduction and Social Organization in Sub-Saharan Africa.* Berkeley: Univ. of California Press.

Pebley, A. R., J. B. Casterline, and J. Trussell. 1982. Age at First Birth in 19 Countries. *International Family Planning Perspectives* 8:2–7.

Pebley, A. R., S. L. Huffman, A. K. M. A. Chowdhury, and P. W. Stupp. 1985. Intra-uterine Mortality and Maternal Nutritional Status in Rural Bangladesh. *Population Studies* 39:425–440.

Pekönen, F., H. Alfthan, U. H. Stenman, and O. Ylikorkäla. 1988. Human Chorionic Gonadotropin (hCG) and Thyroid Function in Early Human Pregnancy: Circadian Variation and Evidence for Intrinsic Thyrotropic Activity of hCG. *Journal of Clinical Endocrinology and Metabolism* 66:853–856.

Pelouze, P. 1939. *Gonorrhea in the Male and Female.* Philadelphia: W. B. Saunders.

Pennington, R., and H. Harpending. 1991. Infertility in Herero Pastoralists of Southern Africa. *American Journal of Human Biology* 3:135–153.

Pérez, A. 1981. Natural Family Planning: Postpartum Period. *International Journal of Fertility* 26:219–221.

Pérez, A., P. Vela, R. G. Potter, and G. Masnick. 1971. Timing and Sequence of Resuming Ovulation and Menstruation after Childbirth. *Population Studies* 25:491–503.

Perona, R. M., and P. M. Wassarman. 1986. Mouse Blastocysts Hatch in vitro by Using a Trypsin-like Proteinase Associated with Cells of Mural Trophectoderm. *Developmental Biology* 114:42–52.

Persky, H., H. I. Lief, D. Strauss, W. R. Miller, and C. P. O'Brien. 1978. Plasma Testosterone Level and Sexual Behavior of Couples. *Archives of Sexual Behavior* 7:157–173.

Pescovitz, O. H., F. Comite, F. Cassorla, A. J. Dwyer, M. A. Poth, M. A. Sterling, K. Hench, A. McNemar, M. Skerda, D. L. Loriaux, and G. B. Cutler, Jr. 1984. True Precocious Puberty Complicating Congenital Adrenal Hyperplasia: Treatment with Luteinizing Hormone-releasing Hormone Analog. *Journal of Clinical Endocrinology and Metabolism* 58:857–861.

Peters, H., and K. P. McNatty. 1980. *The Ovary.* Berkeley: Univ. of California Press.

Peters, R. H. 1983. *The Ecological Implications of Body Size.* Cambridge: Cambridge Univ. Press.

Pettersson, F. 1968. *Epidemiology of Early Pregnancy Wastage: Biological and Social Correlates of Abortion—An Investigation Based on Materials Collected within Uppsala County, Sweden.* Stockholm: Svenska Bokforlaget.

Phillips, D. M., and R. M. Shalgi. 1980. Surface Properties of the Zona Pellucida. *Journal of Experimental Zoology* 213:1–8.

Pike, M. C. 1988. Fertility and Its Effects on Health. Pp. 161–189 in P. Diggory, M. Potts, and S. Teper (eds.), *Natural Human Fertility: Social and Biological Determinants.* London: Macmillan.

Pinsky, L., and M. Kaufmann. 1987. Genetics of Steroid Receptors and Their Disorders. *Advances in Human Genetics* 16:299–472.

Pirke, K. M., U. Schweiger, W. Lemmel, J. C. Krieg, and M. Berger. 1985. The Influence of Dieting on the Menstrual Cycle of Healthy Young Women. *Journal of Clinical Endocrinology and Metabolism* 60:1174–1179.

Pirke, K. M., W. Wuttke, and U. Schweiger (eds.). 1989. *The Menstrual Cycle and Its Disorders: Influences of Nutrition, Exercise and Neurotransmitters.* Berlin: Springer-Verlag.

Pittenger, D. B. 1973. An Exponential Model of Female Sterility. *Demography* 10:113–121.

Plant, T. M. 1988. Puberty in Primates. Pp. 1763–1788 in E. Knobil, J. D. Neill, L. L. Ewing, G. S. Greenwald, C. L. Markert, and D. W. Pfaff (eds.), *The Physiology of Reproduction,* Volume 2. New York: Raven Press.

Plant, T. M., E. Schallenberger, D. L. Hess, J. T. McCormack, L. Dufy-Barbe, and E. Knobil. 1980. Influence of Suckling on Gonadotropin Secretion in the Female Rhesus Monkey. *Biology of Reproduction* 23:760–766.

Poindexter, A. A. N., M. B. Ritter, and P. K. Besch. 1983. The Recovery of Normal Plasma Progesterone Levels in the Post-partum Female. *Fertility and Sterility* 39:494–498.

Polansky, F. F., and E. J. Lamb. 1988. Do the Results of Semen Analysis Predict Future Fertility? A Survival Analysis Study. *Fertility and Sterility* 49:1059–1065.

Polgar, S. 1972. Population History and Population Policies from an Anthropological Perspective. *Current Anthropology* 13:203–277.

Pollak, R. A. 1990. Two-sex Population Models and Classical Stable Population Theory. Pp. 317–333 in J. Adams, D. A. Lam, A. I. Hermalin, and P. E. Smouse (eds.), *Convergent Issues in Genetics and Demography.* New York: Oxford Univ. Press.

Popkin, B. M., D. K. Guilkey, J. S. Akin, L. S. Adair, J. R. Udry, and W. Flieger. 1993. Nutrition, Lactation and Birth Spacing in Filipino Women. *Demography* 30:333–352.

Potter, R. G. 1961. Length of the Fertile Period. *Milbank Memorial Fund Quarterly* 39:132–162.

Potter, R. G., and F. E. Kobrin. 1981. Distributions of Amenorrhea and Anovulation. *Population Studies* 35:85–99.

Potter, R. G., and S. R. Millman. 1985. Fecundability and the Frequency of Marital Intercourse: A Critique of Nine Models. *Population Studies* 39:461–470.

———. 1986. Fecundability and the Frequency of Marital Intercourse: New Models Incorporating the Aging of Gametes. *Population Studies* 40:159–170.

Potter, R. G., and M. P. Parker. 1964. Predicting the Time Required to Conceive. *Population Studies* 18:99–116.

Potter, R. G., T. K. Burch, and S. Matsumoto. 1967. Long Cycles, Late Ovulation, and Calendar Rhythm. *International Journal of Fertility* 12:127–140.

Potter, R. G., M. L. New, J. B. Wyon, and J. E. Gordon. 1965a. Applications of Field Studies to Research on the Physiology of Human Reproduction: Lactation and Its Effects upon Birth Intervals in Eleven Punjab Villages, India. *Journal of Chronic Diseases* 18:1125–1140.

Potter, R. G., J. B. Wyon, M. New, and J. E. Gordon. 1965b. Fetal Wastage in Eleven Punjab Villages. *Human Biology* 37:262–273.

Potter, R. G., J. B. Wyon, M. Parker, and J. E. Gordon. 1965c. A Case Study of Birth Interval Dynamics. *Population Studies* 19:81–96.

Prakash, S., and G. Pathmanathan. 1984. Age at Menarche in Sri Lankan Tamil Girls in Jaffna. *Annals of Human Biology* 11:463–466.

Pratt, W. F., W. D. Mosher, C. A. Bachrach, and M. C. Horn. 1984. *Understanding U. S. Fertility: Findings from the National Survey of Family Growth, Cycle III.* Washington: Population Reference Bureau.

Press, W. H., B. P. Flannery, S. A. Teukolsky, and W. T. Vetterling. 1989. *Numerical Recipes in Pascal: The Art of Scientific Computing.* Cambridge: Cambridge Univ. Press.

Pressat, R. 1985. *The Dictionary of Demography* (ed. by C. Wilson). Oxford: Basil Blackwell.

Presser, H. B. 1974. Temporal Data Relating to the Human Menstrual Cycle. Pp. 145–168 in M. Ferin, F. Halbert, R. M. Richart, and R. L. Van de Wiele (eds.), *Biorhythms and Human Reproduction.* New York: John Wiley.

Propping, P., and J. Krüger. 1976. Uber die Haufigkeit von Zwillingsgeburten. *Deutsch Medizinische Wochenschrift* 101:506–512.

Rahman, O., and J. Menken. 1993. Age at Menopause and Fecundity Preceding Menopause. Pp. 65–84 in R. H. Gray, H. Leridon, and A. Spira (eds.), *Biomedical and Demographic Determinants of Reproduction.* Oxford: Clarendon Press.

Ralt, D., M. Goldenberg, P. Fetterolf, D. Thompson, J. Dor, S. Mashiach, D. L. Garbers, and M. Eisenbach. 1991. Sperm Attraction to a Follicular Factor(s) Correlates with Human Egg Fertilizability. *Proceedings of the National Academy of Sciences* 88:2840–2844.

Razi, Z. 1980. *Life, Marriage, and Death in a Medieval Parish: Economy, Society, and Demography in Halesowen, 1270–1400.* Cambridge: Cambridge Univ. Press.

Rebar, R. W. 1991. Practical Evaluation of Hormonal Status. Pp. 830–887 in S. S. C. Yen and R. B. Jaffe (ed.), *Reproductive Endocrinology.* Philadelphia: W. B. Saunders

Reid, R. L., and D. A. van Vugt. 1987. Weight-related Changes in Reproductive Function. *Fertility and Sterility* 48:905–913.

Reik, W. 1989. Genomic Imprinting and Genetic Disorders in Man. *Trends in Genetics* 5:331–336.

Renfree, M. B. 1982. Implantation and Placentation. Pp. 26–69 in C. R. Austin and R. V. Short (eds.), *Reproduction of Mammals, Volume 2: Embryonic and Fetal Development,* 2nd edition. Cambridge: Cambridge Univ. Press.

Resseguie, L. J. 1974. Pregnancy Wastage and Age of Mother among the Amish. *Human Biology* 46:633–639.

Riad-Fahmy, D. 1984. Salivary Progesterone for Investigating Ovarian Activity. *Frontiers of Oral Physiology* 5:110–123.

Riad-Fahmy, D., G. F. Read, R. F. Walker, S. M. Walker, and K. Griffiths. 1987. Determination of Ovarian Steroid Hormone Levels in Saliva: An Overview. *Journal of Reproductive Medicine* 32:254–272.

Richardson, S. J., V. Senikas, and J. F. Nelson. 1987. Follicular Depletion during the Menopausal Transition: Evidence for Accelerated Loss and Ultimate Exhaustion. *Journal of Clinical Endocrinology and Metabolism* 65:1231–1237.

Riley, A. P. 1990. Dynamic and Static Measures of Growth among Pre- and Postmenarcheal Females in Rural Bangladesh. *American Journal of Human Biology* 2:255–264.

Riley, A. P., S. L. Huffman, and A. K. M. A. Chowdhury. 1989. Age at Menarche and Postmenarcheal Growth in Rural Bangladeshi Females. *Annals of Human Biology* 16:347–359.

Riley, A. P., J. L. Samuelson, and S. L. Huffman. 1993. The Relationship of Age at Menarche and Fertility in Undernourished Adolescents. Pp. 50–64 in R. H. Gray, H. Leridon, and A. Spira (eds.), *Biomedical and Demographic Determinants of Reproduction*. Oxford: Clarendon Press.

Rivera, R., M. Barrera, K. I. Kennedy, P. P. Bhiwandiwala, and E. Ortiz. 1988. Breast-feeding and the Return to Ovulation in Durango, Mexico. *Fertility and Sterility* 49:780–787.

Robinson, J. E., and R. V. Short. 1977. Changes in Breast Sensitivity at Puberty, during the Menstrual Cycle, and at Parturition. *British Medical Journal* i:1188–1191.

Robinson, W. C. 1986. Another Look at the Hutterites and Natural Fertility. *Social Biology* 33:65–76.

Robinson, W. S. 1950. Ecological Correlations and the Behavior of Individuals. *American Sociological Review* 15:351–357.

Rodbard, D. 1974. Statistical Quality Control and Routine Data Processing for Radioimmunoassays and Immunoradiometric Assays. *Clinical Chemistry* 20:1255–1270.

Rodbard, D., P. L. Rayford, J. Cooper, and G. T. Ross. 1968. Statistical Quality Control of Radioimmunoassays. *Journal of Clinical Endocrinology and Metabolism* 28:412–436.

Rodgers, J. L., and M. Buster. 1993. Seasonality of Menarche among U.S. Females: Correlates and Linkages (abstract). *Annals of the New York Academy of Sciences* 709:196.

Rodriguez, G., and J. Cleland. 1988. Modelling Marital Fertility by Age and Duration: An Empirical Appraisal of the Page Model. *Population Studies* 42:241–257.

Rodriguez, G., and J. Trussell. 1980. *Maximum Likelihood Estimation of the Parameters of Coale's Model Nuptiality Schedule from Survey Data*. WFS Technical Bulletin 7. London: World Fertility Survey.

Róheim, G. 1933. Women and Their Life in Central Australia. *Journal of the Royal Anthropological Institute* 63:207–265.

Rolfe, B. E. 1982. Detection of Fetal Wastage. *Fertility and Sterility* 37:655–660.

Romaniuk, A. 1980. Increase in Natural Fertility during the Early Stages of Modernization: Evidence from an African Case Study, Zaire. *Population Studies* 34:293–310.

————. 1981. Increase in Natural Fertility during the Early Stages of Modernization: Canadian Indians Case Study. *Demography* 18:157–172.

Rosenberg, H. M. 1966. *Seasonal Variation of Births, United States 1933–63.* Vital Health Statistics, Series 21, No. 9. Hyattsville, Md.: National Center for Health Statistics, Public Health Service.

Rosetta, L. 1989. Breast Feeding and Post-partum Amenorrhea in Serere Women in Senegal. *Annals of Human Biology* 16:311–320.

Ross, J. L., J. Blangero, M. C. Goldstein, and S. Schuler. 1986. Proximate Determinants of Fertility in Kathmandu Valley, Nepal: An Anthropological Case Study. *Journal of Biosocial Science* 18:179–196.

Roth, E. A. 1985. A Note on the Demographic Concomitants of Sedentism. *American Anthropologist* 87:380–382.

Roth, E. A., and K. B. Kurup. 1988. Demography and Polygyny in a Southern Sudanese Agro-pastoralist Society. *Culture* 8:67–73.

Royston, J. P. 1982. Basal Body Temperature, Ovulation and the Risk of Conception, with Special Reference to the Lifetimes of Sperm and Egg. *Biometrics* 38:397–406.

————. 1991. Identifying the Fertile Phase of the Human Menstrual Cycle. *Statistics in Medicine* 10:221–240.

Ruzicka, L. T. 1976. Age at Marriage and Timing of the First Birth. *Population Studies* 30:527–538.

Ruzicka, L. T., and S. Bhatia. 1982. Coital Frequency and Sexual Abstinence in Rural Bangladesh. *Journal of Biosocial Science* 14:397–420.

Salber, E. J., M. Feinleib, and B. Macmahon. 1966. The Duration of Postpartum Amenorrhea. *American Journal of Epidemiology* 82:347–358.

Sandler, D. P., A. J. Wilcox, and L. F. Horney. 1984. Age at Menarche and Subsequent Reproductive Events. *American Journal of Epidemiology* 119:765–774.

Santow, G. 1987. Reassessing the Contraceptive Effect of Breastfeeding. *Population Studies* 41:147–160.

Santow, G., and M. Bracher. 1989. Do Gravity and Age Affect Pregnancy Outcome? *Social Biology* 36:9–22.

Sassin, J. F., A. G. Frantz, E. D. Weitzman, and S. Kapen. 1972. Human Prolactin: 24 Hour Pattern with Increased Release during Sleep. *Science* 177:1205–1207.

Sauer, M. V., R. J. Paulson, and R. A. Lobo. 1990. A Preliminary Report on Oocyte Donation Extending Reproductive Potential to Women over 40. *New England Journal of Medicine* 323:1157–1160.

Scaglion, R. 1978. Seasonal Births in a Western Abelam Village, Papua New Guinea. *Human Biology* 50:313–323.

Schallenberger, E., D. W. Richardson, and E. Knobil. 1981. Role of Prolactin in the Lactational Amenorrhea of the Rhesus Monkey (Macaca mulatta). *Biology of Reproduction* 25:370–374.

Schwartz, D., P. D. M. MacDonald, and V. Heuchel. 1980. Fecundability, Coital Frequency, and the Viability of the Ova. *Population Studies* 34:397–400.

Schwartz, D., M.-J. Mayaux, A. Spira, M.-L. Moscato, P. Jouannet, F. Czyglik, and G. David. 1983. Semen Characteristics as a Function of Age in 833 Fertile Men. *Fertility and Sterility* 39:530–535.

Schweiger, U., M. Schwingenschloegel, R. Laessle, M. Schweiger, H. Pfister, K. M. Pirke, and C. Hoehl. 1989. Diet-induced Menstrual Irregularities: Effects of Age and Weight Loss. *Fertility and Sterility* 48:746–751.

Scragg, R. F. R. 1973. Menopause and Reproductive Life Span in Rural New Guinea. Presented at the annual symposium of the Papua New Guinea Medical Society, Port Moresby.

Scrimshaw, S. C. M. 1978. Infant Mortality and Behavior in the Regulation of Family Size. *Population and Development Review* 4:383–403.

Sehgal, B. S., and S. R. Singh. 1966. *Breastfeeding, Amenorrhea, and Rates of Conception in Women.* Lucknow, India: Planning, Research, and Action Institute.

Selvin, S., and J. Garfinkel. 1976. Paternal Age, Maternal Age and Birth Order and the Risk of Fetal Loss. *Human Biology* 48:223–230.

Sembajwe, I. 1979. Effect of Age at First Marriage, Number of Wives, and Type of Marital Union on Fertility. *Journal of Biosocial Science* 11:346–351.

Seppälä, M., H.-A. Unnerus, E. Hirvonen, and T. Ranta. 1976. Bromocriptine Increases Plasma Estradiol-17β Concentration in Amenorrheic Patients with Normal Serum Prolactin. *Journal of Clinical Endocrinology and Metabolism* 43:474–477.

Setchell, B. P. 1982. Spermatogenesis and Spermatozoa. Pp. 63–101 in C. R. Austin and R. V. Short (eds.), *Reproduction in Mammals, Volume 1: Germ Cells and Fertilization,* 2nd edition. Cambridge: Cambridge Univ. Press.

Seymour, F. I., C. Duffy, and A. Koernev. 1935. A Case of Authenticated Fertility in a Man of 94. *Journal of the American Medical Association* 105:1423–1424.

Shapiro, S., and D. Bross. 1980. Risk Factors for Fetal Death in Studies of Vital Statistics Data: Inference and Limitations. Pp. 89–106 in I. H. Porter and E. B. Hook (eds.), *Human Embryonic and Fetal Death.* New York: Academic Press.

Shapiro, S., H. S. Levine, and M. Abramowicz. 1970. Factors Associated with Early and Late Fetal Loss. *Advances in Planned Parenthood* 6:45–63.

Sheps, M. C. 1964. On the Time Required for Conception. *Population Studies* 18:85–97.

———. 1965. An Analysis of Reproductive Patterns in an American Isolate. *Population Studies* 19:65–80.

Sheps, M. C., and J. Menken. 1972. Distribution of Birth Intervals According to the Sampling Frame. *Theoretical Population Biology* 3:1–26.

———. 1973. *Mathematical Models of Conception and Birth.* Chicago: Univ. of Chicago Press.

Short, R. V. 1976. The Evolution of Human Reproduction. *Proceedings of the Royal Society of London* (series B) 195:3–24.

———. 1984a. Breast Feeding. *Scientific American* 250:35–42.

———. 1984b. Oestrous and Menstrual Cycles. Pp. 115–152 in C. R. Austin and R. V. Short (eds.), *Reproduction in Mammals, Volume 3: Hormonal Control of Reproduction,* 2nd edition. Cambridge: Cambridge Univ. Press.

———. 1984c. Why Menstruate? Unpublished ms.

Shorter, E., J. Knodel, and E. van de Walle. 1971. The Decline of Non-marital Fertility in Europe, 1880–1940. *Population Studies* 25:375–393.

Shur, B. D., and N. G. Hall. 1982a. A Role for Mouse Sperm Surface Galactosyltransferase in Sperm Binding for the Egg Zona Pellucida. *Journal of Cellular Biology* 95:574–579.

———. 1982b. Sperm Surface Galactosyltransferase Activities during in vitro Capacitation. *Journal of Cellular Biology* 95:567–573.

Sibly, R. M., and P. Calow. 1986. *Physiological Ecology of Animals: An Evolutionary Approach*. Oxford: Blackwell Scientific.

Simpson, J. L. 1985. Genes and Chromosomes that Cause Female Infertility. *Fertility and Sterility* 44:725–739.

Simpson-Hebert, M., and S. L. Huffman. 1981. The Contraceptive Effect of Breastfeeding. *Studies in Family Planning* 12:125–133.

Simpson-Hebert, M., and L. P. Makil. 1985. Breast-feeding in Manila, Philippines: Preliminary Results from a Longitudinal Study. *Journal of Biosocial Science, Supplement* 9:137–146.

Singarimbun, M., and C. Manning. 1976. Breastfeeding, Amenorrhea, and Abstinence in a Javanese Village: A Case Study of Mojolama. *Studies in Family Planning* 7:175–179.

Singh, S., J. B. Casterline, and J. G. Cleland. 1985. The Proximate Determinants of Fertility: Sub-national Variations. *Population Studies* 39:113–135.

Skolnick, M., L. L. Bean, D. May, V. Arbon, K. De Nevers, and P. Cartwright. 1978. Mormon Demographic History, I: Nuptiality and Fertility of Once-married Couples. *Population Studies* 32:5–19.

Smart, Y. C., T. K. Roberts, R. L. Clancy, and A. W. Cripps. 1981. Early Pregnancy Factor: Its Role in Mammalian Reproduction. *Fertility and Sterility* 35:397–402.

Smith, J. E., and P. R. Kunz. 1976. Polygyny and Fertility in Nineteenth-Century America. *Population Studies* 30:465–480.

Smith, R. M. 1984. Families and Their Land in an Area of Partible Inheritance, Redgrave, Suffolk 1260–1320. Pp. 135–195 in R. M. Smith (ed.), *Land, Kinship and Life-cycle*. Cambridge: Cambridge Univ. Press.

———. 1988. Human Resources. Pp. 188–212 in G. Astill and A. Grant (eds.), *The Countryside of Medieval England*. Oxford: Basil Blackwell.

Smith, T. E. 1960. The Cocos-Keeling Islands: A Demographic Laboratory. *Population Studies* 14:91–130.

Snedecor, G. W. 1954. Biometry: Its Makers and Concepts. Pp. 3–10 in O. Kempthorne, T. A. Bancroft, J. W. Gowen, and J. L. Lush (eds.), *Statistics and Mathematics in Biology*. Ames: Iowa State Univ.

Spanos, A. 1986. *Statistical Foundations of Econometric Modelling*. Cambridge: Cambridge Univ. Press.

Speroff, L., R. H. Glass, and N. G. Kase. 1986. *Clinical Gynecologic Endocrinology and Infertility*, 4th edition. Baltimore: Williams and Wilkins.

Srinathsinghji, D. J. S., and L. Martini. 1984. Effects of Bromocriptine and Naloxone on Plasma Levels of Prolactin, LH, and FSH during Suckling in the Female Rat: Responses to Gonadotropin Releasing Hormone. *Journal of Endocrinology* 100:175–182.

Stanhope, R. 1989. The Endocrine Control of Puberty. Pp. 191–199 in J. M. Tanner and M. A. Preece (eds.), *The Physiology of Human Growth*. Society for the Study of Human Biology Symposium 29. Cambridge: Cambridge Univ. Press.

Stanhope, R., J. Adams, and C. G. Brook. 1988. Evolution of Polycystic Ovaries in a Girl with Delayed Menarche: A Case Report. *Journal of Reproductive Medicine* 33:482–484.

Statistique Générale de la France. 1907. *Statistique Générale du Mouvement de la Population 1749–1905*. Paris: Statistique Générale de la France.

Stearns, S. C. 1992. *The Evolution of Life Histories*. Oxford: Oxford Univ. Press.

Stein, Z., and M. Susser. 1978. Famine and Fertility. Pp. 123–145 in W. H. Mosley (ed.), *Nutrition and Human Reproduction*. New York: Plenum Press.

Steinberger, A., and D. N. Ward. 1988. Inhibin. Pp. 567–583 in E. Knobil, J. D. Neill, L. L. Ewing, G. S. Greenwald, C. L. Markert, and D. W. Pfaff (eds.), *The Physiology of Reproduction*, Volume 1. New York: Raven Press.

Stene, J., E. Stene, S. Stengel-Rutkowski, and J. D. Murken. 1981. Paternal Age and Down Syndrome: Data from Prenatal Diagnoses. *Human Genetics* 59: 119–124.

Stern, J. M., M. Konner, T. N. Herman, and S. Reichlin. 1986. Nursing Behaviour, Prolactin and Post-partum Amenorrhoea during Prolonged Lactation in American and !Kung Mothers. *Clinical Endocrinology* 25:247–258.

Stevenson, A. C., H. A. Warnock, M. Y. Dudgeon, and J. H. McClure. 1958. Observations on the Results of Pregnancies in Women Resident in Belfast. *Annals of Human Genetics* 23:382–420.

Stewart, K. J. 1988. Suckling and Lactational Anoestrous in Wild Gorillas (*Gorilla gorilla*). *Journal of Reproduction and Fertility* 83:627–634.

Stone, L. 1977. *The Family, Sex, and Marriage in England 1500–1800*. London: Weidenfeld and Nicholson.

Strasburger, C. J., Y. Amir-Zaltsman, and F. Kohen. 1988. The Avidin-Biotin Reaction as a Universal Amplification System in Immunoassays. Pp. 79–100 in B. Albertson and F. Haseltine (eds.), *Non-Radiometric Assays: Technology and Application in Polypeptide and Steroid Hormone Detection*. New York: Alan R. Liss.

Styne, D. M., and M. M. Grumbach. 1991. Disorders of Puberty in the Male and Female. Pp. 511–554 in S. S. C. Yen and R. B. Jaffe (eds.), *Reproductive Endocrinology*. Philadelphia: W. B. Saunders.

Suchindran, C. M., and P. A. Lachenbruch. 1975. Estimates of Fecundability from a Truncated Distribution of Conception Times. *Demography* 12:291–301.

Sundararaj, N., M. Chern, L. Gatewood, L. Hickman, and R. McHugh. 1978. Seasonal Behavior of Human Menstrual Cycles: A Biometric Investigation. *Human Biology* 50:15–31.

Sweet, R. L., M. Blankfort-Doyle, M. O. Robbie, and J. Schachter. 1985. The Occurrence of Chlamydial and Gonococcal Salpingitis during the Menstrual Cycle. *Journal of the American Medical Association* 255:2062–2064.

Takahashi, J., Y. Higashi, A. LaNasa, K. I. Yoshida, S. J. Winters, H. Oshima, and P. Troen. 1983. Studies of the Human Testis, XVIII: Simultaneous Measurement of Nine Intratesticular Steroids—Evidence for Reduced Mitochondrial Function in Testes of Elderly Men. *Journal of Clinical Endocrinology and Metabolism* 56:1178–1187.

Tanner, J. M. 1962. *Growth at Adolescence*, 2nd edition. Oxford: Blackwell Scientific.

————. 1981. Menarcheal Age (letter). *Science* 214:604.

————. 1990. *Fetus into Man: Physical Growth from Conception to Maturity*, rev. edition. Cambridge, Mass.: Harvard Univ. Press.

Taylor, W. F. 1970. The Probability of Fetal Death. Pp. 307–320 in F. C. Fraser, V. A. McKusick, and R. Robinson (eds.), *Congenital Malformations*. Amsterdam: Excerpta Medica.

Tenover, J. S., A. M. Matsumoto, D. K. Clifton, and W. J. Bremner. 1988. Age-related Alterations in the Circadian Rhythms of Pulsatile Luteinizing Hormone and Testosterone Secretion in Healthy Men. *Journal of Gerontology* 43:M163–M169.

Thapa, S., R. V. Short, and M. Potts. 1988. Breast Feeding, Birth Spacing and Their Effects on Child Survival. *Nature* 335:679–682.

Therneau, T. M., P. M. Grambsch, and T. R. Fleming. 1990. Martingale-based Residuals for Survival Models. *Biometrika* 77:147–160.

Thomas, D. C., B. Langholz, W. Mack, and B. Floderus. 1990. Bivariate Survival Models for Analysis of Genetic and Environmental Effects in Twins. *Genetic Epidemiology* 7:121–135.

Thompson, M. W., R. R. McInnes, and H. F. Willard. 1991. *Thompson and Thompson Genetics in Medicine*, 5th edition. Philadelphia: W. B. Saunders.

Thorner, M. O., and I. S. Login. 1981. Prolactin Secretion as an Index of Brain Dopaminergic Function. *Advances in Biochemical Psychopharmacology* 28:503–520.

Tietze, C. 1956. Statistical Contributions to the Study of Human Fertility. *Fertility and Sterility* 7:88–94.

————. 1957. Reproductive Span and Rate of Reproduction among Hutterite Women. *Fertility and Sterility* 8:89–97.

————. 1960. Probability of Pregnancy Resulting from a Single Unprotected Coitus. *Fertility and Sterility* 11:484–488.

Tijssen, P. 1985. *Practice and Theory of Enzyme Immunoassays*. Amsterdam: Elsevier Science.

Tjoa, W. S., M. H. Smolensky, B. P. Hsi, E. Steinberger, and K. D. Smith. 1982. Circannual Rhythm in Human Sperm Count Revealed by Serially Independent Sampling. *Fertility and Sterility* 38:454–459.

Tombes, R. M., and B. M. Shapiro. 1985. Metabolite Channeling: A Phosphocreatine Shuttle to Mediate High Energy Phosphate Transport between Sperm Mitochondrion and Tail. *Cell* 41:325–334.

Tracer, D. P. 1991. The Interaction of Nutrition and Fertility among Au Forager-horticulturalists of Papua New Guinea. Unpublished doctoral dissertation. Ann Arbor: Department of Anthropology, Univ. of Michigan.

Treloar, A. E. 1974. Menarche, Menopause, and Intervening Fecundability. *Human Biology* 46:89–107.

Treloar, A. E., R. E. Boynton, B. G. Behn, and B. W. Brown. 1967. Variation of the Human Menstrual Cycle through Reproductive Life. *International Journal of Fertility* 12:77–126.

Trussell, J. 1978. Menarche and Fatness: Re-examination of the Critical Body Composition Hypothesis. *Science* 200:1506–1509.

————. 1979. Natural Fertility: Measurement and Use in Fertility Models. Pp.

29–64 in H. Leridon and J. Menken (eds.), *Natural Fertility*. Liège: Ordina Editions.

Trussell, J., and D. E. Bloom. 1983. Estimating the Co-variates of Age at Marriage and First Birth. *Population Studies* 37:403–416.

Trussell, J., and A. R. Pebley. 1984. The Potential Impact of Changes in Fertility on Infant, Child, and Maternal Mortality. *Studies in Family Planning* 15: 267–280.

Trussell, J., and T. Richards. 1985. Correcting for Unmeasured Heterogeneity in Hazards Models Using the Heckman-Singer Procedure. Pp. 111–132 in N. B. Tuma (ed.), *Sociological Methodology 1985*. San Francisco: Jossey-Bass.

Trussell, J., and G. Rodriguez. 1990. Heterogeneity in Demographic Research. Pp. 111–132 in J. Adams, D. A. Lam, A. I. Hermalin, and P. E. Smouse (eds.), *Convergent Issues in Genetics and Demography*. Oxford: Oxford Univ. Press.

Trussell, J., and C. F. Westoff. 1980. Contraceptive Practice and Trends in Coital Frequency. *Family Planning Perspectives* 12:246–249.

Trussell, J., and C. Wilson. 1985. Sterility in a Population with Natural Fertility. *Population Studies* 39:269–286.

Trussell, J., R. A. Hatcher, W. Cates, F. H. Stewart, and K. Krost. 1990. A Guide to Interpreting Contraceptive Efficacy Studies. Obstetrics and Gynecology 76:558–567.

Tsitouras, P. D., C. E. Martin, and S. M. Harman. 1982. Relationship of Serum Testosterone to Sexual Activity in Healthy Elderly Men. *Journal of Gerontology* 37:288–293.

Tuan, C.-H. 1958. Reproductive Histories of Chinese Women in Rural Taiwan. *Population Studies* 12:40–50.

Tuchmann-Duplessis, H., G. David, and P. Haegel. 1971. *Illustrated Human Embryology*, Volume 1. New York: Springer-Verlag.

Tuma, N. B., and M. T. Hannan. 1984. *Social Dynamics: Models and Methods*. New York: Academic Press.

Turner, C. D. 1966. *General Endocrinology*, 5th edition. Philadelphia: W. B. Saunders.

Tutin, C. E. G. 1980. Reproductive Behaviour of Wild Chimpanzees in the Gombe National Park, Tanzania. *Journal of Reproduction and Fertility*, Supplement 28:43–57.

Tutin, C. E. G., and P. R. McGinnis. 1981. Chimpanzee Reproduction in the Wild. Pp. 239–264 in C. E. Graham (ed.), *Reproductive Biology of the Great Apes*. New York: Academic Press.

Udry, J. R. 1980. Changes in the Frequency of Marital Intercourse from Panel Data. *Archives of Sexual Behavior* 9:319–325.

Udry, J. R., and R. L. Cliquet. 1982. A Cross-cultural Examination of the Relationship between Ages at Menarche, Marriage, and First Birth. *Demography* 19:53–63.

Udry, J. R., and N. M. Morris. 1967. Seasonality of Coitus and Seasonality of Birth. *Demography* 4:673–679.

———. 1968. Distribution of Coitus in the Menstrual Cycle. *Nature* 220:593–596.

———. 1978. Relative Contribution of Male and Female Age to the Frequency of Marital Intercourse. *Social Biology* 25:128–134.

Udry, J. R., F. R. Deven, and S. J. Coleman. 1982. A Cross-national Comparison of the Relative Influence of Male and Female Age on the Frequency of Marital Intercourse. *Journal of Biosocial Science* 14:1–6.

Underwood, J. H. 1991. Seasonality of Vital Events in a Pacific Island Population. *Social Biology* 38:113–126.

United Nations (UN). 1954. *Demographic Yearbook, 1954.* New York: United Nations.

Urban, R. J., J. D. Veldhius, R. M. Blizzard, and M. L. Dufau. 1988. Attenuated Release of Biologically Active Luteinizing Hormone in Healthy Aging Men. *Journal of Clinical Investigation* 81:1020–1029.

Vaessen, M. 1984. *Childlessness and Infecundity.* World Fertility Survey Comparative Studies, No. 31. Voorburg, Netherlands: International Statistical Institute.

Valmary, P. 1965. *Familles Paysannes au XVIIIème Siècle en Bas-Quercy: Etude Démographique.* Paris: Institut National d'Etudes Démographiques-Presses Universitaires de France.

Vander, A. J., J. H. Sherman, and D. S. Luciano. 1980. *Human Physiology: The Mechanisms of Body Function,* 3rd edition. New York: McGraw-Hill.

———. 1990. *Human Physiology: The Mechanisms of Body Function,* 5th edition. New York: McGraw-Hill.

van de Walle, E., and F. van de Walle. 1993. Post-partum Sexual Abstinence in Tropical Africa. Pp. 446–460 in R. H. Gray, H. Leridon, and A. Spira (eds.), *Biomedical and Demographic Determinants of Reproduction.* Oxford: Clarendon Press.

van Keep, P. A., P. C. Brand, and P. Lehert. 1979. Factors Affecting the Age at Menopause. *Journal of Biosocial Science,* Supplement 6:37–55.

van Look, P. F. A., H. Lothian, W. M. Hunter, E. A. Michie, and D. T. Baird. 1977. Hypothalamic-pituitary-ovarian Function in Perimenopausal Women. *Clinical Endocrinology* 7:13–31.

van Noord-Zaadstra, M. Boukye, C. W. N. Looman, H. Alsbach, J. Dik, F. Habbema, E. R. de Velde, and J. Karbaat. 1991. Delayed Childbearing: Effect of Age on Fecundity and Outcome of Pregnancy. *British Medical Journal* 302:1361–1365.

van Steenbergen, W. M., J. A. Kusin, and M. M. van Rens. 1981. Lactation Performance of Akamba Mothers, Kenya: Breast Feeding Behavior, Breast Milk Yield, and Composition. *Journal of Tropical Pediatrics* 27:155–161.

Vaupel, J. W. 1988. Inherited Frailty and Longevity. *Demography* 25:277–287.

———. 1990a. Kindred Lifetimes: Frailty Models in Population Genetics. Pp. 155–170 in J. Adams, D. A. Lam, A. I. Hermalin, and P. E. Smouse (eds.), *Convergent Issues in Genetics and Demography.* Oxford: Oxford Univ. Press.

———. 1990b. Relatives' Risks: Frailty Models of Life History Data. *Theoretical Population Biology* 37:220–234.

Vaupel, J. W., and A. I. Yashin. 1985a. Heterogeneity's Ruses: Some Surprising Effects of Selection on Population Dynamics. *American Journal of Statistics* 39:176–185.

———. 1985b. The Deviant Dynamics of Death in Heterogeneous Populations. Pp. 179–211 in N. B. Tuma (ed.), *Sociological Methodology 1985.* San Francisco: Jossey-Bass.

Vaupel, J. W., K. G. Manton, and E. Stallard. 1979. The Impact of Heterogeneity in Individual Frailty on the Dynamics of Mortality. *Demography* 16:439–454.

Vermer, H. M., and R. Rolland. 1982. The Influence of Exogenous Oestradiol Benzoate on the Pituitary Responsiveness to LHRH during the Puerperium in Women. *Clinical Endocrinology* 16:251–258.

Vermeulen, A., A. Deslypere, and K. de Meirleir. 1989. A New Look to the Andropause: Altered Function of the Gonadotrophs. *Journal of Steroid Biochemistry* 32:163–165.

Vihko, R., and D. Apter. 1984. Endocrine Characteristics of Adolescent Menstrual Cycles: Impact of Early Menarche. *Journal of Steroid Biochemistry* 20: 231–236.

Vincent, P. 1950. La Stérilité Physiologique de Populations. Population 5:45–64.

———. 1956. Données Biométriques sur la Conception et la Grossesse. *Population* 11:1–29.

Vitzthum, V. J. 1989. Nursing Behavior and Its Relation to Duration of Post-partum Amenorrhea in an Andean Community. *Journal of Biosocial Science* 21: 145–160.

Vogel, F., and A. G. Motulsky. 1986. *Human Genetics: Problems and Approaches,* 2nd edition. Berlin: Springer-Verlag.

Vollman, R. F. 1953. Uber fertilitat und sterilitat der frauinnerhalb des menstruations cycles. *Archiv fur Gynakologie* 187:602–622.

Vollman, R. F. 1977. *The Menstrual Cycle.* Philadelphia: W. B. Saunders.

vom Saal, F. S., and C. E. Finch. 1988. Reproductive Senescence: Phenomena and Mechanisms in Mammals and Selected Vertebrates. Pp. 2351–2413 in E. Knobil, J. D. Neill, L. L. Ewing, G. S. Greenwald, C. L. Markert, and D. W. Pfaff (eds.), *The Physiology of Reproduction,* Volume 2. New York: Raven Press.

Wang, C. 1988. Bioassays of Follicle-Stimulating Hormone. *Endocrine Reviews* 9:374–377.

Warburton, D., and F. C. Fraser. 1964. Spontaneous Abortion Risks in Man: Data from Reproductive Histories Collected in a Medical Genetics Unit. *American Journal of Human Genetics* 16:1–25.

Warburton, D., Z. Stein, J. Kline, and M. Susser. 1980. Chromosome Abnormalities in Spontaneous Abortion: Data from the New York City Study. Pp. 261–287 in I. H. Porter and E. B. Hook (eds.), *Human Embryonic and Fetal Death.* New York: Academic Press.

Warren, C. W., M. L. Gwinn, and G. L. Rubin. 1986. Seasonal Variation in Conception and Various Pregnancy Outcomes. *Social Biology* 33:116–126.

Wassarman, P. M. 1987. The Biology and Chemistry of Fertilization. Science 235:553–560.

———. 1988. Fertilization in Mammals. *Scientific American* 259(6):78–84.

Wassarman, P. M., J. Bleil, C. Fimiani, H. Florman, J. Greve, R. Kinloch, C. Moller, S. Mortillo, R. Roller, G. Salzmann, and M. Vazquez. 1989. The Mouse Egg Receptor for Sperm: A Multifunctional Zona Pellucida Glycoprotein. Pp. 18–37 in J. Dietl (ed.), *The Mammalian Egg Coat: Structure and Function.* Berlin: Springer-Verlag.

Watkins, S. C. 1989. The Fertility Transition: Europe and the Third World Compared. Pp. 27–55 in J. M. Stycos (ed.), *Demography as an Interdiscipline.* New Brunswick, N. J.: Transaction Publishers.

Weatherall, D. J. 1991. *The New Genetics and Clinical Practice*, 3rd edition. Oxford: Oxford Univ. Press.

Wehmann, R. E., S. M. Harman, S. Birken, R. E. Canfield, and B. C. Nisula. 1981. Convenient Radioammunoassay that Measures Urinary Human Choriogonadotropin in the Presence of Urinary Human Lutropin. *Clinical Chemistry* 27:1997–2001.

Weinberg, C. R., and B. C. Gladen. 1986. The Beta-geometric Distribution Applied to Comparative Fecundability Studies. *Biometrics* 42:547–560.

Weinstein, M., and M. Stark. 1994. Behavioral and Biological Determinants of Fecundability. *Annals of the New York Academy of Sciences* 709:128–144.

Weinstein, M., J. W. Wood, and M.-C. Chang. 1993a. Age Patterns of Fecundability. Pp. 209–227 in R. H. Gray, H. Leridon, and A. Spira (eds.), *Biomedical and Demographic Determinants of Reproduction*. Oxford: Clarendon Press.

Weinstein, M., J. W. Wood, and D. D. Greenfield. 1993b. How Does Variation in Fetal Loss Affect the Distribution of Waiting Times to Conception? *Social Biology* 40:106–130.

Weinstein, M., J. W. Wood, M. A. Soto, and D. D. Greenfield. 1990. Components of Age-specific Fecundability. *Population Studies* 44:447–467.

Weir, W. C., and D. R. Weir. 1961. The Natural History of Infertility. *Fertility and Sterility* 12:443–451.

Weiss, K. M. 1990. The Biodemography of Variation in Human Frailty. *Demography* 27:185–206.

———. 1993. *Genetic Variation and Human Diseases: Principles and Evolutionary Approaches*. Cambridge: Cambridge Univ. Press.

Weitzman, E. D., R. M. Boyar, S. Kapen, and L. Hellman. 1975. The Relationship of Sleep and Sleep Stages to Neuroendocrine Secretion and Biological Rhythms in Man. *Recent Progress in Hormone Research* 31:399–446.

Wentz, A. C., and G. E. S. Jones. 1973. Transient Luteolytic Effect of Prostaglandin $F_{2\alpha}$ in the Human. *Obstetrics and Gynecology* 42:172–181.

West, C. P., and A. S. McNeilly. 1979. Hormone Profiles in Lactating and Nonlactating Women Immediately after Delivery and Their Relationship to Breast Engorgement. *British Journal of Obstetrics and Gynaecology* 86:501–506.

Westoff, C. F., and A. R. Pebley. 1981. Alternative Measures of Unmet Need for Family Planning in Developing Countries. *International Family Planning Perspectives* 7:126–135.

Westoff, C. F., R. G. Potter, P. C. Sagi, and E. G. Mishler. 1961. *Family Growth in Metropolitan America*. Princeton: Princeton Univ. Press.

Weström, L. 1975. Effect of Acute Pelvic Inflammatory Disease on Fertility. *American Journal of Obstetrics and Gynecology* 121:707–713.

Whitehead, R. G., A. G. M. Rowland, M. Hutton, A. M. Prentice, E. Müller, and A. Paul. 1978. Factors Influencing Lactation Performance in Rural Gambian Mothers. *The Lancet* ii:178–181.

Widdowson, E. M. 1976. Changes in the Body and Its Organs during Lactation: Nutritional Implications. Pp. 103–118 in *Breast-feeding and the Mother*. Ciba Foundation Symposium 45 (new series). Amsterdam: Elsevier North-Holland.

Wilchek, E. A., and E. A. Bayer (eds.). 1990. *Avidin-Biotin Technology*. Methods in Enzymology No. 184. San Diego: Academic Press.

Wilcox, A. J., and B. C. Gladen. 1982. Spontaneous Abortion: The Role of Heterogeneous Risk and Selective Fertility. *Early Human Development* 7:165–178.

Wilcox, A. J., A. E. Treloar, and D. P. Sandler. 1981. Spontaneous Abortion over Time: Comparing Occurrences in Two Cohorts of Women a Generation Apart. *American Journal of Epidemiology* 114:548–553.

Wilcox, A. J., C. R. Weinberg, and D. D. Baird. 1990. Risk Factors for Early Pregnancy Loss. *Epidemiology* 1:382–385.

Wilcox, A. J., C. R. Weinberg, D. D. Baird, and R. E. Canfield. 1993. Endocrine Detection of Conception and Early Foetal Loss. Pp. 316–328 in R. H. Gray, H. Leridon, and A. Spira (eds.), *Biomedical and Demographic Determinants of Reproduction*. Oxford: Clarendon Press.

Wilcox, A. J., C. R. Weinberg, J. F. O'Connor, D. D. Baird, J. P. Schlatterer, R. E. Canfield, E. G. Armstrong, and B. C. Nisula. 1988. Incidence of Early Loss of Pregnancy. *New England Journal of Medicine* 319:189–194.

Wilcox, A. J., C. R. Weinberg, R. E. Wehmann, E. G. Armstrong, R. E. Canfield, and B. C. Nisula. 1985. Measuring Early Pregnancy Loss: Laboratory and Field Methods. *Fertility and Sterility* 44:366–374.

Wild, C. J. 1983. Failure Time Models with Matched Data. *Biometrika* 70:633–641.

Wildt, L., G. Marshall, and E. Knobil. 1980. Experimental Induction of Puberty in the Infantile Female Rhesus Monkey. *Science* 207:1373–1375.

Wilkins, A. S. 1993. *Genetic Analysis of Animal Development*, 2nd edition. New York: Wiley-Liss.

Wilkinson, M., and R. Bhanot. 1982. A Puberty-related Attenuation of Opiate Peptide-induced Inhibition of LH Secretion. *Endocrinology* 110:1046–1048.

Willett, W. 1990. *Nutritional Epidemiology*. Oxford: Clarendon Press.

Williams, G. C. 1975. *Sex and Evolution*. Princeton: Princeton Univ. Press.

Wilmsen, E. N. 1978. Seasonal Effects of Dietary Intake on Kalahari San. *Proceedings of the Federation of American Societies for Experimental Biology* 37:65–72.

———. 1986. Biological Determinants of Fecundity and Fecundability: An Application of Bongaarts' Model to Forager Fertility. Pp. 59–82 in P. Handwerker (ed.), *Culture and Reproduction*. Boulder: Westview.

Wilson, C. 1984. Natural Fertility in Pre-industrial England, 1600–1799. *Population Studies* 38:225–240.

Wilson, C., J. Oeppen, and M. Pardoe. 1988. What is Natural Fertility? The Modeling of a Concept. *Population Index* 54:4–20.

Winters, S. J., and P. Troen. 1982. Episodic Luteinizing Hormone (LH) Secretion and the Response of LH and Follicle-stimulating Hormone to LH-releasing Hormone in Aged Men: Evidence for Coexistent Primary Testicular Insufficiency and an Impairment in Gonadotropin Secretion. *Journal of Clinical Endocrinology and Metabolism* 55:560–565.

Witschi, E. 1948. Migration of the Germ Cells of Human Embryos from the Yolksac to the Primitive Gonadal Fold. *Contributions to Embryology* 32:67–80.

Woessner, J. F. 1963. Age-related Changes of the Human Uterus and Its Connective Tissue Framework. *Journal of Gerontology* 18:220–226.

Wood, J. W. 1980. Mechanisms of Demographic Equilibrium in a Small Human Population, the Gainj of Papua New Guinea. Unpublished doctoral dissertation. Ann Arbor: Department of Anthropology, Univ. of Michigan.

————. 1987a. Problems of Applying Model Fertility and Mortality Schedules to Data from Papua New Guinea. Pp. 371–397 in T. McDevitt (ed.), *The Survey Under Difficult Conditions: Demographic Data Collection in Papua New Guinea.* New Haven: HRAF Press.

————. 1987b. The Genetic Demography of the Gainj of Papua New Guinea, 2: Determinants of Effective Population Size. *American Naturalist* 129:165–187.

————. 1989. Fecundity and Natural Fertility in Humans. *Oxford Reviews of Reproductive Biology* 11:61–109.

————. 1990. Fertility in Anthropological Populations. *Annual Review of Anthropology* 19:211–242.

————. 1992. Fertility and Reproductive Biology in Papua New Guinea. Pp. 93–118 in R. D. Attenborough and M. P. Alpers (eds.), *Human Biology in Papua New Guinea: The Small Cosmos.* Oxford: Clarendon Press.

————. 1994. Maternal Nutrition and Reproduction: Why Physiologists and Demographers Disagree about a Fundamental Relationship. *Annals of the New York Academy of Sciences* 709:101–116.

Wood, J. W., and M. Weinstein. 1988. A Model of Age-specific Fecundability. *Population Studies* 42:85–113.

————. 1990. Heterogeneity in Fecundability: The Effect of Fetal Loss. Pp. 171–188 in J. Adams, D. A. Lam, A. I. Hermalin, and P. E. Smouse (eds.), *Convergent Issues in Genetics and Demography.* New York: Oxford Univ. Press.

Wood, J. W., D. J. Holman, K. M. Weiss, A. V. Buchanan, and B. Lefor 1992. Hazards Models for Human Population Biology. *Yearbook of Physical Anthropology* 35:43–87.

Wood, J. W., P. L. Johnson, and K. L. Campbell. 1985a. Demographic and Endocrinological Aspects of Low Natural Fertility in Highland New Guinea. *Journal of Biosocial Science* 17:57–79.

Wood, J. W., D. Lai, P. L. Johnson, K. L. Campbell, and I. A. Maslar. 1985b. Lactation and Birth Spacing in Highland New Guinea. *Journal of Biosocial Science, Supplement* 9:159–173.

Wood, J. W., D. J. Holman, A. I. Yashin, R. J. Peterson, M. Weinstein, and M.-C. Chang. 1994a. A Multistate Model of Fecundability and Sterility. *Demography* (in press)

Wood, J. W., S. C. Weeks, G. R. Bentley, K. M. Weiss. 1994b. Human Population Biology and the Evolution of Aging. Pp. 19–75 in D. E. Crews and R. M. Garruto (eds.), *Biological Anthropology and Aging: Perspectives on Human Variation over the Life Span.* Oxford: Oxford Univ. Press.

Woolf, C. M. 1965. Stillbirths and Parental Age. *Obstetrics and Gynecology* 26:1–8.

World Health Organization (WHO). 1977. Recommended Definitions, Terminology and Format for Statistical Tables Related to the Perinatal Period and Use of a New Certificate for Cause of Perinatal Deaths. *Acta Obstetrica et Gynaecologica Scandinavica* 56:247–253.

————. 1978. *International Classification of Disease,* 9th revision. Geneva: World Health Organization.

————. Task Force on Methods for the Determination of the Fertile Period. 1980. Temporal Relationships Between Ovulation and Defined Changes in the Concentrations of Plasma Estradiol-17_, Luteinizing Hormone, Follicle-stimulating

Hormone and Progesterone, I: Probit Analysis. *American Journal of Obstetrics and Gynecology* 138:388–390.

———. 1986a. World Health Organization Multicenter Study on Menstrual and Ovulatory Patterns in Adolescent Girls, I: A Multicenter Cross-sectional Study of Menarche. *Journal of Adolescent Health Care* 7:229–235.

———. 1986b. World Health Organization Multicenter Study on Menstrual and Ovulatory Patterns in Adolescent Girls, II: Longitudinal Study of Menstrual Patterns in the Early Postmenarcheal Period, Duration of Bleeding Episodes and Menstrual Cycles. *Journal of Adolescent Health Care* 7:236–244.

Worthington-Roberts, B. S., J. Vermeersch, and S. R. Williams. 1985. *Nutrition in Pregnancy and Lactation,* 3rd edition. St. Louis: Times Mirror/Mosby College Publishing.

Worthman, C. M., J. F. Stallings, and D. Gubernick. 1991. Measurement of Hormones in Blood Spots: A Non-isotopic Assay for Prolactin (abstract). *American Journal of Physical Anthropology* 85:186–187.

Worthman, C. M., C. L. Jenkins, J. F. Stallings, and D. Lai. 1993. Attenuation of Nursing-related Ovarian Suppression and High Fertility in Well-nourished, Intensively Breast-feeding Amele Women of Lowland Papua New Guinea. *Journal of Biosocial Science* 25:425–443.

Wrigley, E. A., and R. Schofield. 1981. *The Population History of England 1541–1871: A Reconstruction.* Cambridge, Mass.: Harvard Univ. Press.

Wu, L. L. 1990. Simple Graphical Goodness-of-fit Tests for Hazard Rate Models. Pp. 184–199 in K. U. Mayer and N. B. Tuma (eds.), *Event History Analysis in Life Course Research.* Madison: Univ. of Wisconsin Press.

Wyon, J. B., and J. E. Gordon. 1971. *The Khanna Study: Population Problems in the Rural Punjab.* Cambridge, Mass.: Harvard Univ. Press.

Wyshak, G. 1983. Age at Menarche and Unsuccessful Pregnancy Outcome. *Annals of Human Biology* 10:69–73.

Wyshak, G., and R. E. Frisch. 1982. Evidence for a Secular Trend in Age of Menarche. *New England Journal of Medicine* 306:1033–1035.

Yalow, R. S. 1978. Radioimmunoassay: A Probe for the Fine Structure of Biologic Systems. *Science* 200:1236–1245.

Yamaguchi, K. 1991. *Event History Analysis.* London: Sage Publications.

Yanagimachi, R. 1981. Mechanisms of Fertilization in Mammals. Pp. 81–182 in L. Mastroianni and J. D. Biggers (eds.), *Fertilization and Embryonic Development in Vitro.* New York: Plenum Press.

Yates, J. R. W., and M. A. Ferguson-Smith. 1983. No Evidence for a Paternal Age Effect on the Frequency of Trisomy 21 at Amniocentesis in 13,300 Pregnancies: An Analysis of Data from a European Collaborative Study (abstract). *Journal of Medical Genetics* 20:457.

Yerushalmy, J., J. M. Bierman, D. H. Kemp, A. Connor, and F. E. French. 1956. Longitudinal Studies of Pregnancy on the Island of Kauai, Territory of Hawaii, I: Analysis of Previous Reproductive History. *American Journal of Obstetrics and Gynecology* 71:80–96.

Yu, Xie. 1991. What is Natural Fertility? The Remodeling of a Concept. *Population Index* 56:656–663.

Yuen, B. H., W. Cannon, L. Sy, J. Booth, and P. Burch. 1982. Regression of Pituitary Microadenoma during and following Bromocriptine Therapy: Persistent Defect in Prolactin Regulation before and throughout Pregnancy. *American Journal of Obstetrics and Gynecology* 142:634–639.

Yule, G. U. 1906. On the Changes in the Marriage- and Birth-rates in England and Wales during the Past Half Century. *Journal of the Royal Statistical Society* 69:88–106.

Zabin, L. S., E. S. Smith, M. B. Hirsch, and J. B. Hardy. 1986. Ages of Physical Maturation and First Intercourse in Black Teenage Males and Females. *Demography* 23:595–605.

Zablan, Z. C. 1985. Breast-feeding and Fertility among Philippine Women: Trends, Mechanisms, and Impact. *Journal of Biosocial Science,* Supplement 9:147–158.

Zacharias, L., W. Rand, and R. Wurtman. 1976. A Prospective Study of Sexual Development and Growth in American Girls. *Obstetrics and Gynecology Survey* 31:325–337.

Index